Perspectives in Molecular Toxinology

Perspectives in Molecular Toxinology

Edited by

A. Ménez
CEA/SACLAY

JOHN WILEY & SONS, LTD

Copyright © 2002 John Wiley & Sons, Ltd
    Baffins Lane, Chichester,
    West Sussex, PO19 1UD, England
    National   01243 779777
    International   (+44) 1243 779777

e-mail (for orders and customer service enquiries): cs-books@wiley.co.uk
Visit our Home Page on http://www.wiley.co.uk or http://www.wiley.com

Other Wiley Editorial Offices

John Wiley & Sons, Inc., 605 Third Avenue,
New York, NY 10158-0012, USA

Wiley-VCH Verlag GmbH
Pappelallee 3, D-69469 Weinheim, Germany

John Wiley & Sons Australia, Ltd
33 Park Road, Milton, Queensland 4064, Australia

John Wiley & Sons (Asia) Pte Ltd, 2 Clementi Loop #02-01,
Jin Xing Distripark, Singapore 129809

John Wiley & Sons (Canada) Ltd, 22 Worcester Road,
Rexdale, Ontario, M9W 1L1, Canada

Library of Congress Cataloging-in-Publication Data

Perspectives in molecular toxinology / edited by A. Menez.
        p. cm.
    Includes bibliographical references and index.
    ISBN 0-471-49503-4 (alk. paper)
        1. Toxins. 2. Venom. I. Ménez, A. (André)

    QP631 .P457 2002
    615′.373–dc21

                                                                    2001057389

British Library Cataloguing in Publication Data

A catalogue record for this book is available from the British Library

ISBN 0-471-49503-4

Typeset in 10/12pt Times by Kolam Information Services Pvt Ltd, Pondicherry, India.
Printed and bound in Great Britain by TJ International, Padstow, Cornwall.
This book is printed on acid-free paper responsibly manufactured from sustainable forestry in which at least two trees are planted for each one used for paper production.

CONTENTS

LIST OF CONTRIBUTORS

Adams, M.E.
Departments of Entomology and Neuroscience, University of California Riverside, Riverside, CA 92521, USA

Anglister, J.
Department of Life Sciences, Ben-Gurion University of the Negev, Beer Sheva 84105, Israel

Becerril B.
Institute of Biotechnology, National Autonomous University of Mexico, Avenida Universidad, 2001, Apartado Postal 510–3, Cuernavaca 62210, Mexico (*baltazar@ibt.unam.mx*)

Benoit, E.
Institut Fédératif de Neurobiologie Alfred Fessard, Laboratoire de Neurobiologie Cellulaire et Moléculaire, C.N.R.S., UPR 9040, 1 avenue de la Terrasse, Bâtiments 32–33, 91198 Gif-sur-Yvette cedex, France (*benoit@nbcm.cnrs-gif.fr*)

Bertrand, D.
Department of Physiology, C.M.U., 1 rue Michel-Servet, CH-1211 Geneva 4, Switzerland (*betrand@ibm.unge.ch*)

Bon, C.
Unité des Venins, Institut Pasteur, 25 rue du Docteur Roux, 75724 Paris cedex 15, France (*cbon@pasteur.fr*)

Brier, T.J.
Division of Molecular Toxicology, School of Life and Environmental Sciences, The University of Nottingham, Nottingham NG7 2RD, United Kingdom

Bulaj, G.
Department of Biology, University of Utah, Salt Lake City, Utah, USA

Calvete, J.J.
Istituto de Biomedicine, CSIC, Valencia, Spain

Camargo, A.C.M.
Instituto Butantan, São Paulo, Brazil

Chijiwa, T.
Department of Life Science, Faculty of Engineering, Sojo University, Kumamoto 860, Japan

Chippaux, J.-P.
Institut de Recherche pour le Développement, B.P. 1386, Dakar, Senegal (*Jean-Philippe.Chippaux@ird.sn*)

Corona, M.
Institute of Biotechnology, National Autonomous University of Mexico, Avenida Universidad, 2001, Apartado Postal 510–3, Cuernavaca 62210, Mexico

De Waard, M.
INSERM EMI 9931, CEA/DBMS, 17 rue des Martyrs, 38054 Grenoble, France

Delepierre, M.
Unité RMN des Biomolécules, CNRS URA 2185, Institut Pasteur, 28 rue du Docteur Roux, 75724 Paris cedex 15, France (*murield@pasteur.fr*)

Domont, G.B.
Departamento de Bioquimica, Instituto de Quimica, Universidade Federal do Rio de Janeiro, 21.949–900 Rio de Janeiro, RJ, Brazil (*gilberto@iq.ufrj.br*)

Favreau, P.
Atheris Laboratories, case postale 314, CH-1233 Bernex-Geneva, Switzerland (*philippe.farreau@atheris.ch*)

Fox, J.W.
Biomolecular Research Facility, University of Virginia Health Systems, Charlottesville, Virginia, USA (*jwf8x@virginia.edu*)

Froy, O.
Department of Plant Sciences, George S. Wise Faculty of Life Sciences, Tel-Aviv University, Ramat-Aviv 69978, Tel-Aviv, Israel

Gasparini, S.
Département d'Ingénierie et d'Etudes des Protéines, CEA Saclay, 91191 Gif-sur-Yvette cedex, France (*sylvaine.gasparini@cea.fr*)

Gilles, N.
Département d'Ingénierie et d'Etudes des Protéines, CEA Saclay, 91191 Gif-sur-Yvette cedex, France (*nicolas.gilles@cea.fr*)

Gordon, D.
Department of Plant Sciences, Tel-Aviv University, Ramat-Aviv, Tel-Aviv 69978, Israel (*dgordon@post.tau.ac.il*)

Goyffon, M.
L.E.R.A.I., Muséum National d'Histoire Naturelle, 57 rue Cuvier, 75724 Paris cedex 15, France (*mgoyffon@mnhn.fr*)

Gubenšek, F.
Department of Biochemistry and Molecular Biology, Jozef Stefan Institute, 1000 Ljubljana, Slovenia (*franc.gubensek@ijs.si*)

Gurevitz, M.
Department of Plant Sciences, George S. Wise Faculty of Life Sciences, Tel-Aviv University, Ramat-Aviv 69978, Tel-Aviv, Israel (*mamgur@post.tau.ac.il*)

Harvey, A.L.
Strathclyde Institute for Drug Research and Department of Physiology and Pharmacology, University of Strathclyde, 27 Taylor Street, Glasgow G4 ONR, United Kingdom (*sidr@strath.ac.uk*)

Hattori, S.
Institute of Medical Science, University of Tokyo, Oshima-gun, Kagoshima 894, Japan

Heinemann, S.H.
Research Unit Molecular and Cellular Biophysics at the Friedrich Schiller University Jena, Drackendorfer St. 1, D-07747, Jena, Germany

Holmes, M.J.
Tropical Marine Science Institute, National University of Singapore, 119223 Singapore

Hotze, E.
Departement of Microbiology and Immunology, Biomedical Research Center, The University of Oklahoma Health Sciences Center, Oklahoma City, OK 73104, USA

Imperial, J.S.
Department of Biology, University of Utah, Salt Lake City, Utha, USA

Joseph, J.S.
Bioinformatics Centre, Institute of Molecular and Cell Biology, 30 Medical Drive, Singapore 117609 (*medp7015@nus.edu.sg*)

Kallen, R.G.
Department of Biochemistry and Biophysics, University of Penn, 422 Curie Blvd, Philadelphia, PA 19104-6059, USA

Kamiguti, A.S.
Department of Haematology, University of Liverpool, Liverpool, United Kingdom (*aurakami@liv.ac.uk*)

Karbat, I.
Department of Plant Sciences, George S. Wise Faculty of Life Sciences, Tel-Aviv University, Ramat-Aviv 69978, Tel-Aviv, Israel

Kem, W.R.
Department of Pharmacology and Therapeutics, University of Florida College of Medicine, Gainesville, FL 32610-0267, USA (*Kem@pharmacology.ufl.edu*)

Kini, R.M.
Department of Biological Sciences, Faculty of Science, Department of Biochemistry, Faculty of Medicine, National University of Singapore, 119260 Singapore (*dbskinim@nus.edu.sg*)

Kordiš, D.
Department of Biochemistry and Molecular Biology, Jozef Stefan Institute, 1000 Ljubljana, Slovenia (*dusan.kordis@ijs.si*)

Križaj, I.
Department of Biochemistry and Molecular Biology, Jozef Stefan Institute, 1000 Ljubljana, Slovenia (*igor.krizaj@ijs.si*)

Kumar, T.K.S.
Department of Chemistry, National Tsing Hua University, Hsinchu, Taiwan (*skumar@ms.nthu.edu.tw*)

Leduc, M.
Unité des Venins, Institut Pasteur, 25 rue du Docteur Roux, 75724 Paris cedex 15, France

Lewis, R.
Institute for Molecular Bioscience, The University of Queensland, Brisbane, QLD 4072, Australia (*R.Lewis@mailbox.uq.oz.au*)

Lotan, I.
Tel-Aviv University, Sackler School of Medicine, Department of Physiology and Pharmacology, Ramat Aviv, Israel

Malany, S.
Department of Pharmacology, 0636, University of California at San Diego, La Jolla, CA 92093, USA (Present address: Max Planck Institute, Heidelberg, Germany)

Marcinkiewicz, C.
Sol Sherry Thrombosis Research Center, Temple University School of Medicine, Philadelphia, PA, USA

McLane, M.A.
Department of Medical Technology, University of Delaware, McKinly Lab 057, Newark, DE 19716, USA (*mclane@udel.edu*)

Mellor, I.R.
Division of Molecular Toxicology, School of Life and Environmental Sciences, The University of Nottingham, NG7 2RD, United Kingdom

Ménez, A.
Département d'Ingénierie et d'Etudes des Protéines, CEA Saclay, 91191 Gif-sur-Yvette cedex, France (*andre.menez@cea.fr*)

Merino, E.
Institute of Biotechnology, National Autonomous University of Mexico, Avenida Universidad, 2001, Apartado Postal 510–3, Cuernavaca 62210 Mexico

Molgó, J.
Institut Fédératif de Neurobiologie Alfred Fessard, Laboratoire de Neurobiologie Cellulaire et Moléculaire, C.N.R.S., UPR 9040, 1 avenue de la Terrasse, Bâtiments 32–33, 91198 Gif-sur-Yvette cedex, France (*Jordi.Molgo@nbcm.cnrs-gif.fr*)

Molles, B. Department of Pharmacology, 0636, University of California at San Diego, La Jolla, CA 92093, USA (*bmolles@ucsd.edu*)

Montecucco, C.
Centro CNR Biomembrane and Dipartimento di Scienze Biomediche, Università di Padova, Via Colombo 3, 35100 Padova, Italy (*cesare@civ.bio.unipd.it*)

Nicholson, G.M.
Department of Health Sciences, University of Technology, Sydney, Broadway NSW 2007, Australia (*Graham.Nicholson@uts.edu.ac*)

Niewiarowski, S.
Department of Physiology, Temple University School of Medicine, Philadelphia, PA, USA

Norton, R.S.
The Walter and Eliza Hall Institute of Medical Research, Parkville 3050, Australia (*ray.norton@bioresi.com.au*)

Oda-Ueda, N.
Department of Applied Life Science, Faculty of Engineering, Sojo University, Kumamoto 860, Japan (*naoko@bio.kumamoto-it.ac.jp*)

Ogawa, T.
Department of Biochemistry, School of Agricultural Science, Tohoku University, Aoba-ku; Sendai 981, Japan (*ogawa@biochem.tohoku.ac.jp*)

Ohno, M.
Department of Applied Life Science, Faculty of Engineering, Sojo University, Kumamoto 860, Japan (*ohno@ed.kumamoto-it.ac.jp*)

Olivera, B.M.
Department of Biology, University of Utah, Salt Lake City, Utah, USA (*oliveralab@bioscience.utah.edu*)

Osaka, H.
Department of Pharmacology, 0636, University of California at San Diego, La Jolla, CA 92093, USA (Present address: Yokohama City University, Yokohama, Japan)

Pelhate, M.
Laboratoire de Neurophysiologie, Faculté de Médecine, Université d'Angers, 49045 Angers, France (*marcel.pelhate@univ-angers.fr*)

Perales, J.
Departamento de Fisiologia e Farmacodinâmica, IOC, FIOCRUZ, Av. Brazil 4365, 21.045-900, Rio de Janeiro, RJ, Brazil (*jperales@gene.dbbm.fiocruz.br*)

Possani, L.D.
Institute of Biotechnology, National Autonomous University of Mexico, Avenida Universidad, 2001, Apartado Postal 510–3, Cuernavaca 62210 Mexico (*possani@ibt.unam.mx*)

Rao, V.S.
Department of Biological Sciences, Faculty of Science, Department of Biochemistry, Faculty of Medicine, National University of Singapore, 119260 Singapore (*scip9063@nus.edu.sg*)

Rochat, H.
CNRS UMR 6560, IFR Jean Roche, Faculté de Médecine Nord, Boulevard Pierre Dramard, 13916 Marseille cedex 20, France (*rochat.h@jean-roche.univ-mrs.fr*)

Rossetto, O.
Centro CNR Biomembrane and Dipartimento di Scienze Biomediche, Università di Padova, Via Colombo 3, 35100 Padova, Italy (*rosetto@civ.bio.unipd.it*)

Sabatier, J.-M.
CNRS UMR 6560, IFR Jean Roche, Faculté de Médecine Nord, Boulevard Pierre Dramard, 13916 Marseille cedex 20, France (*sabatier.jm@jean-roche.univ-mrs.fr*)

Sauviat, M.-P.
Unité INSERM 451, 91761 Palaiseau cedex, France (*sauviat@enstag.ensta.fr*)

Serrano, S.M.T.
Biomolecular Research Facility, University of Virginia Health Systems, Charlottesville, Virginia, USA

Servent, D.
Département d'Ingénierie et d'Etudes des Protéines, CEA Saclay, 91191 Gif-sur-Yvette cedex, France (*denis.servent@cea.fr*)

Shaanan, B.
Department of Life Sciences, Ben-Gurion University of the Negev, Beer-Sheva 84105, Israel

Shannon, J.D.
Biomolecular Research Facility, University of Virginia Health Systems, Charlottesville, Virginia, USA

Sherman, N.
Biomolecular Research Facility, University of Virginia Health Systems, Charlottesville, Virginia, USA

Shichor, I.
Tel-Aviv University, Sackler School of Medicine, Department of Physiology and Pharmacology, Ramat Aviv, Israel

Srisailam, S.
Department of Chemistry, National Tsing Hua University, Hsinchu, Taiwan

Stefansson, B.
Biomolecular Research Facility, University of Virginia Health Systems, Charlottesville, Virginia, USA

Stöcklin, R.
Atheris Laboratories, case postale 314, CH-1233 Bernex-Geneva, Switzerland (*reto.stocklin@atheris.ch*)

Taylor, P.
Department of Pharmacology, 0636, University of California at San Diego, La Jolla, CA 92093, USA (*pwtaylor@ucsd.edu*)

Theakston, R.D.G.
Liverpool School of Tropical Medicine, University of Liverpool, Liverpool, United Kingdom (*R.D.G.Theakston@liverpool.ac.uk*)

Turkov, M.
Department of Plant Sciences, George S. Wise Faculty of Life Sciences, Tel-Aviv University, Ramat-Aviv 69978, Tel-Aviv, Israel

Tweten, R.K.
Department of Microbiology and Immunology, Biomedical Research Center, The University of Oklahoma Health Sciences Center, Oklahoma City, OK 73104, USA (*rod-tweten@ouhsc.edu*)

Usherwood, P.N.R.
Division of Molecular Toxicology, School of Life and Environmental Sciences, The University of Nottingham, Nottingham NG7 2RD, United Kingdom (*peter.usherwood@nottingham.ac.uk*)

Vethanayagam, R.R.
Department of Chemistry, National Tsing Hua University, Hsinchu, Taiwan

Wierzbicka-Patynowski, I.
Department of Medical Technology, University of Delaware, Newark, DE, USA

Wilunsky, R.
Department of Plant Sciences, George S. Wise Faculty of Life Sciences, Tel-Aviv University, Ramat-Aviv 69978, Tel-Aviv, Israel

Wisner, A.
Unité des Venins, Institut Pasteur, 25 rue du Docteur Roux, 75724 Paris cedex 15, France (*awisner@pasteur.fr*)

Yu, C.
Department of Chemistry, National Tsing Hua University, Hsinchu, Taiwan (*cyu@mx.edu.tw*)

Zilberberg, N.
Department of Plant Sciences, George S. Wise Faculty of Life Sciences, Tel-Aviv University, Ramat-Aviv 69978, Tel-Aviv, Israel

PREFACE

Why are toxins and venoms such fascinating compounds? Over the years, answers to this question have naturally reflected the preoccupations and the sociohistorical context of the enquirer. Francesco Redi saw them as a means of fighting against irrationalism at a time, the 17th century, when mysterious natural actions frequently received metaphysical explanations. Redi believed that vipers were not dangerous by virtue of the 'irritated spirits they could emit', as commonly claimed at the time, but simply because they produced a 'toxic' juice, whose administration could kill. What intrigued Felice Fontana, in the 18th century, and many later scientists, was how a venom kills and what it contains. Fontana was only too aware of the magnitude of the task in answering such questions. In the second half of the 19th century, the physiologist Claude Bernard considered that poisons, including venoms and toxins, specifically and efficiently affect the functioning of essential physiological processes. He saw poisons as extraordinary 'dissecting' instruments to be used to uncover the constitutive principles of physiological processes. Bernard's simple and brilliant reasoning is still largely exploited and its application has led to vital discoveries in neuroscience. In the late 19th century, many other gifted scientists, like Albert Calmette, considered venoms and toxins interesting because they trigger a specific immune response in host animals. Since then, serotherapy and vaccination have constituted major therapeutic approaches to the protection of humans against toxins from microbes and animals. By the late 1960s, the application of separation techniques to proteins finally began to yield answers to the questions raised by Felice Fontana two centuries before. This intense interest in toxinology over the last three centuries shows no sign of waning, and in this book we consider a whole range of fascinating perspectives.

Perspective comes from *perspectum*, a Latin word meaning 'to look through'. Can established knowledge be used to anticipate developments in a field of scientific endeavour? This is an absorbing but most difficult challenge, as speculation must be rigorously excluded from the analysis. So how should we proceed? Our illustrious predecessors pointed the way by posing crystal-clear questions based on the most precise acquired knowledge. This was the exacting task undertaken by the authors of this book. Through twenty-seven chapters, they have not only reviewed the most recent major advances in our understanding of toxins from venomous animals and microbes, but have also addressed intriguing unsolved questions. Will proteomics help us to discover more toxins? What are the intimate modes of action of toxins on ion channels, receptors, enzymes, cell membranes? What is the molecular basis of specificity and more generally of pleiotropism of toxins? What else can be learned about toxin tertiary structures and dynamics? How can we enhance our understanding of target architecture and complex toxin-target structures? Will structural genomics help clarify these questions? Are toxins subject to accelerated evolution? Can they be exploited as models for the design of new drugs? What does the future hold for the treatment of snake bites? These are just some of the many questions that have been addressed in this book. Complete answers may not always be readily apparent, but the attendant reasoning shows how toxinology has become an important area of biochemistry, directly associated with advances in cellular microbiology, molecular pharmacology, molecular physiology, cell biology, protein engineering, and other disciplines.

I would like to thank all the authors for their readiness to accept this challenge of looking into the future of toxinology. My thanks also go to Florence Izabelle for her constant and most efficient collaboration.

André Ménez

Part I Toxins from Microorganisms

1 Bacterial Toxins with Metalloprotease Activity

ORNELLA ROSSETTO and CESARE MONTECUCCO*

1.1 INTRODUCTION

Metalloproteases are hydrolytic enzymes characterised by an active site containing a metal atom, usually zinc. They include amino-peptidases, carboxy-peptidases and endopeptidases depending on whether they remove an N- or a C-terminal residue or cleave internal peptide bonds of the protein substrate. Zinc-dependent endopeptidases are characterised by the presence of a zinc-binding motif consisting of His-Glu-X-X-His [1]. Hundreds of different metalloproteases are produced by microorganisms and are involved in their metabolic activities or are released outside to cleave substrates not elaborated by the microorganism itself. The present chapter will only deal with bacterial metalloproteases which act on a specific cellular or tissue target of the mammalian bacterial host.

The recent determination of their primary sequence has led to the discovery of the metalloproteolytic activity of the bacterial toxins responsible for tetanus, botulism and anthrax. The protease domain of these toxins enters into the cytosol where it displays a zinc-dependent endopeptidase activity of remarkable specificity. Tetanus (TeNT) and botulinum toxins (BoNTs) cleave three protein components of the neuroexocytosis machinery leading to the blockade of neurotransmitter release and consequent paralysis. BoNTs are increasingly used in medicine for the treatment of human diseases characterised by hyperfunction of cholinergic terminals.

The lethal factor of *Bacillus anthracis* is specific for the MAP kinase-kinases that are cleaved within their amino-terminus. In this case, however, such a specific biochemical lesion has not yet been correlated with the pathogenesis of anthrax.

Fragilysin (BFT) is produced by toxigenic strains of the intestinal pathogen *Bacteroides fragilis* and attacks E-cadherin.

---

* Corresponding author

Perspectives in Molecular Toxinology
Edited by A Ménez © 2002 John Wiley & Sons, Ltd

1.2 TETANUS AND BOTULINUM NEUROTOXINS

1.2.1 INTRODUCTION

Tetanus neurotoxin (TeNT) and botulinum neurotoxin (BoNT) were identified as the sole cause of tetanus and botulism, respectively, a little over a century ago, after the discovery of the anaerobic and spore-forming bacteria of the genus *Clostridium* [2–4]. There are seven types of BoNT (indicated with letters from A to G) which differ in antigenicity and biochemical activity [5, 6]. BoNTs bind to and enter peripheral cholinergic terminals, causing a sustained block of acetylcholine (ACh) release, with ensuing flaccid paralysis and autonomic symptoms. Tetanus neurotoxin (TeNT) acts on the CNS and blocks neurotransmitter release at the inhibitory interneurons of the spinal cord, resulting in a frequently lethal spastic paralysis. Despite the opposite clinical symptoms of tetanus and botulism, the neurotoxins affect the same neuronal function: neuroexocytosis [7, 8].

1.2.2 GENETICS AND STRUCTURE

In *C. tetani* and in *C. botulinum G*, the neurotoxin genes are contained within large plasmids, whereas in *C. botulinum* A, B, E and F the neurotoxin genes have a chromosomal localisation and in *C. botulinum* C and D toxins are encoded by bacteriophages. Usually one bacterium harbours one toxin gene, but several cases of multiple toxin genes have been reported [9]. These genes do not contain a secretion signal sequence and the protein neurotoxins are released by bacterial autolysis as single polypeptide chains of 150 kDa, which are later activated by a specific proteolytic cleavage within a loop subtended by a highly conserved disulphide bridge [10]. The heavy chain (H, 100 kDa) and the light chain (L, 50 kDa) remain associated via noncovalent interactions and by the conserved interchain S–S bond, whose integrity is essential for neurotoxicity, but which has to be reduced to allow the display of the metalloproteolytic activity of the L chain in the cytosol [11] (Figure 1.1A).

The crystallographic structures of BoNT/A and BoNT/B and of the C-terminal part of the heavy chain of TeNT revealed that the 50 kDa receptor binding domain, termed Hc, consists of two sub-domains [12–14]. The N-terminal part of Hc (HcN) consists of sixteen β-strands and four α-helices arranged in a jelly roll motif, closely similar to that of carbohydrate-binding proteins of the legume lectin family [14, 15]. The amino acid sequence of this sub-domain is highly conserved among BoNTs and TeNT, suggesting a closely similar three-dimensional structure. In contrast, the sequence of the C-terminal part of Hc (HcC) is poorly conserved, but folds similarly to proteins of the trypsin inhibitor family. On the basis of experiments performed with TeNT, it was suggested that HcC plays a major role in neurospecific binding [16].

(a)

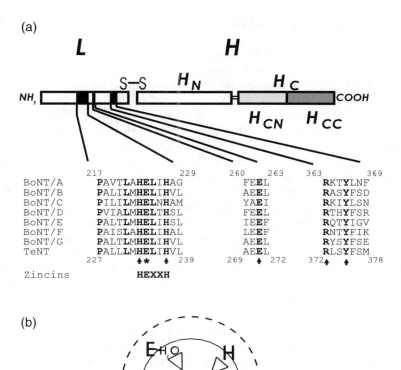

Figure 1.1 The three-functional domain structure of CNTs. a) Schematic structure of activated di-chain CNTs. The neurotoxin is composed of two polypeptide chains held together by a single disulphide bridge. The C-terminal portion of the heavy chain (H, 100 kDa) is responsible for neurospecific binding (domain H_C) while the N-terminus (H_N) is implicated in the translocation of the light chain in cytosol and pore formation. Structurally H_C can be further subdivided into two portions of 25 kDa, HcN and HcC. The light chain (L, 50 kDa) is a zinc-endopeptidase responsible for the intracellular activity of CNTs. The segments presenting high homology between different serotypes are in black. A short α-helix (217–229 in BoNT/A and 227–239 in TeNT), in the central part of the L chain, shows the highest homology and contains the zinc-binding motif of metallo-endopeptidases. Amino acids involved in the coordination of zinc or in the hydrolysis of the substrate are indicated by an arrow. The glutamic acid coordinating a water molecule responsible for the target hydrolysis is indicated by an asterisk. b) The active site architecture shows a primary sphere of zinc coordinating residues (circle) and a secondary layer of residues, not as close to the zinc centre (broken circle)

The N-terminal part of the heavy chain (H_N) features two ~ 100 Å-long antiparallel α-helices, similar to those of the membrane interacting proteins colicins and influenza hemagglutinin [14, 15]. The H_Ns of the CNTs are highly homologous and their predicted secondary structures are also highly similar, in agreement with their proposed role in transmembrane translocation of the L chain [15, 17].

The L chain is a metalloprotease with little protein–protein interaction with the adjacent translocation domain (H_N), which is in turn linked to the receptor-binding domain. At the centre of the long cleft-shaped active site there is a zinc atom coordinated via the two histidines and the glutamic residues of the zinc-binding motif, and by Glu262 in BoNT/A and Glu268 in BoNT/B, a residue conserved among clostridial neurotoxins which corresponds to Glu271 of TeNT (Figure 1.1A). The Glu residue of the motif is particularly important because it coordinates the water molecule which actually performs the hydrolytic reaction of proteolysis. Its mutation leads to complete inactivation of these neurotoxins [18]. The critical role of Glu271 of TeNT and Glu262 in BoNT/A has been shown to be that of providing a negatively charged carboxylate moiety ([19] and in preparation). This active site architecture is similar to that of thermolysin and identifies a primary sphere of residues essential to the catalytic function, which coincides with the zinc coordinating residues. In addition, it appears that a secondary layer of residues, less close to the zinc centre, is present at the active site of clostridial neurotoxins (Figure 1.1B). Among these residues, Arg363 and Tyr366 in BoNT/A could play a role in the catalytic activity of this family of metalloproteases. In particular, Tyr366 in BoNT/A (corresponding to Tyr373 in BoNT/B and to Tyr375 in TeNT) point its phenolic ring inside the cleft-shaped active site of the toxin [14, 15]. The mutation of Tyr375 with an alanine inactivates the TeNT L chain, clearly indicating that this residue plays a critical role in the hydrolysis of the substrate [19]. It has been proposed that Tyr373 of BoNT/B assists the hydrolysis reaction by donating a proton to the amide nitrogen of VAMP Phe77 which, together with bound water molecules, stabilises the leaving group [20].

The active site of the L chain faces the H chain in the unreduced toxin, accounting for its lack of proteolytic activity, and becomes accessible to the substrate upon reduction of the interchain disulphide bridge. Their proteolytic activity is zinc-dependent and heavy metal chelators such as ortho-phenantroline, which remove bound zinc, generate inactive apo-neurotoxins, but the active site metal atom can be reacquired upon incubation of apo-toxin in zinc-containing buffers [11]. The biochemical and structural properties of clostridial neurotoxins define them as a distinct group of metalloproteases, whose origin cannot at present be traced to any of the known families of metalloprotease [11].

Such structural organisation of the CNTs has been shaped by evolution to fulfil the requirements of their mode of neuron intoxication which consists of four steps [17, 21]: 1) binding, 2) internalisation, 3) membrane translocation, and 4) proteolytic cleavage of their substrates (Figure 1.2).

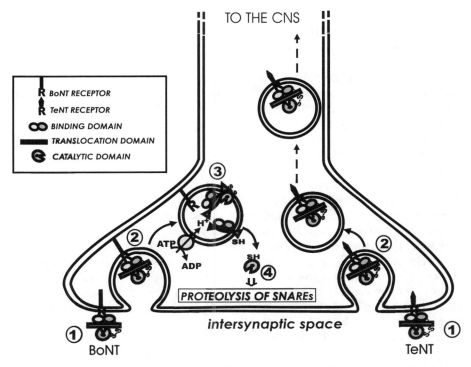

Figure 1.2 Entry of BoNTs and TeNT inside nerve terminals. ① BoNTs and TeNT bind to the presynaptic membrane at as yet unidentified receptors of peripheral nerve terminals. ② The protein receptor of TeNT would be responsible for its inclusion in an endocytic vesicle that moves in a retrograde direction all along the axon to the inhibitory interneurons of the spinal cord (CNS), whereas BoNT protein receptors would guide them inside vesicles that acidify within the NMJ. ③ At low pH, BoNTs and TeNT change conformation, insert into the lipid bilayer of the vesicle membrane and translocate the L chain into the cytosol of peripheral and central neurons respectively. ④ Inside the cytosol the L chain catalyses the proteolysis of one of the three SNARE proteins

1.2.3   NEURONAL INTOXICATION

From the site of production or adsorption (intestine or wounds), BoNTs and TeNT diffuse in the body fluids, up to the presynaptic membrane of cholinergic terminals where they bind very specifically. The Hc domain plays a major role in neurospecific binding [22], but additional regions may be involved in determining the remarkable specificity for cholinergic terminals of CNTs.

Identification of the presynaptic receptor(s) of CNTs has been attempted by several investigators. Polysialogangliosides are certainly involved [23–25] together with as yet unidentified proteins of the presynaptic membrane [26]. The presence of both lectin-like and protein binding sub-domains in the Hc domain supports the suggestion that CNTs bind strongly and specifically to the presynaptic membrane because they display multiple interactions with sugar- and protein-binding sites [14, 15]. Recently, BoNT/B was shown to bind strongly to the synaptic vesicle protein synaptotagmin II only in the presence of poly-sialogangliosides [24], but its role *in vivo* remains to be established. Identification of the receptors for the various CNTs will constitute a major advance in the understanding of the mechanism of neuron intoxication and help to improve current therapeutic protocols employing BoNT to treat human syndromes of hyperfunction of cholinergic terminals and excessive muscle contraction.

As depicted in Figure 1.2, the L chains of CNTs block neuroexocytosis by acting in the cytosol and they reach this cell compartment following endocytosis and membrane translocation. They are internalised inside acidic cellular com-partments via a temperature- and energy-dependent process [27]. Nerve stimu-lation facilitates intoxication by CNTs [28] and a close link exists between stimulus-contraction coupling and endocytosis at nerve terminals [29]. Hence endocytosis and other factors such as nerve stimulation-dependent proteolytic activity in the cytosol [30] may partly account for this effect, which is potentially very relevant for the development of novel protocols of therapy employing BoNT. The protein receptor of TeNT would be responsible for its inclusion in an endocytic vesicle that moves in a retrograde direction all along and inside the axon, whereas BoNTs' protein receptors would guide them inside vesicles that acidify within the NMJ. The TeNT-carrying vesicles reach the cell body in the spinal cord and then move to dendritic terminals to release the toxin in the intersynaptic space. TeNT equilibrates between pre- and post-synaptic mem-branes and then binds and enters the inhibitory interneurons of the spinal cord via synaptic vesicle endocytosis.

To reach the cytosol the L chain has to cross the hydrophobic barrier of the vesicle membrane (Figure 1.2) and the acidity of the lumen is essential for such a movement. CNTs have to be exposed to a low pH step for nerve intoxication to occur [31]. Acidic pH does not induce a direct activation of the toxin via a structural change, but is required in the process of transmembrane transloca-tion of the L chain itself. CNTs undergo a low pH-driven conformational change from a water-soluble 'neutral' structure to an 'acid' structure character-ised by the surface exposure of hydrophobic patches, which lead the H and L chains in the hydrocarbon core of the lipid bilayer [32]. Following the low pH-induced membrane insertion, BoNTs and TeNT form transmembrane ion channels in planar lipid bilayers of low conductance [33, 34]. There is a general consensus that the toxin channels participate in the process of transmembrane translocation of the L domain, from the vesicle membrane to the nerve terminal cytosol, but there is no agreement on how this process may take place. The BoNT translocation domain is different from those of other pore-forming toxins since the long pair of α-helices, with their triple helix bundle at either end, resemble more some coiled coil viral proteins [14, 15] which do not translocate through pores but change structure at low pH and insert into membranes. It has been proposed that the L chain translocates across the vesicle membrane within a channel opened laterally to lipids, rather than inside a wholly proteinaceous pore [21], accounting for the fact that the L chain does contact the fatty acid chains of lipids during translocation [33]. The H chain is suggested to form a transmembrane hydrophilic cleft that nests in the passage of the partially unfolded L chain with its hydrophobic segments facing the lipids. Facing the cytosolic neutral pH, the L chain refolds and regains its water-soluble neutral conformation. This model is also supported by the finding that the protein-translocating channel of the endoplasmic reticulum has been shown to be open laterally to lipids.

Once in the cytosol, CNTs exploit their catalytic activity. BoNTs and TeNT are remarkably specific proteases that recognise and cleave only three proteins, the so-called SNARE proteins, which form the core of the neuroexocytosis machinery [17, 26]. TeNT, BoNT/B, /D, /F and /G cleave VAMP, at different single peptide bonds [17, 35]; BoNT/A and /E cleave SNAP-25 at different sites within the COOH-terminus whereas BoNT/C cleaves both syntaxin and SNAP-25 [36–38]. Strikingly, TeNT and BoNT/B cleave VAMP at the same peptide bond (Gln76–Phe77) and yet, when injected into the animal, they cause the opposite symptoms of tetanus and botulism, respectively [35], conclusively demonstrating that the different symptoms of the two diseases derive from different sites of intoxication rather than from a different molecular mechanism of action.

VAMP, SNAP-25 and syntaxin form a heterotrimeric coil-coiled complex, termed the SNARE complex, which induces the juxtaposition of vesicle to the target membrane [39] and is involved in their fusion [40]. VAMP is a family of vesicular SNAREs with a short C-terminal tail facing the vesicle lumen, a single transmembrane domain and the remaining N-terminal part exposed to the cytosol. Different VAMP isoforms are located on different cell vesicles and contribute to address each vesicle to its appropriate target membrane with which it will fuse. VAMP-1 and -2 are the isoforms mainly involved in the binding and fusion of neurotransmitter-containing synaptic vesicles with the presynaptic membrane (neuroexocytosis). Syntaxin is anchored to target membranes via a C-terminal hydrophobic tail. Of the many syntaxin isoforms presently known, syntaxin 1A, 1B and 2 are the isoforms mainly involved in neuroexocytosis. SNAP-25 (few isoforms) are 25 kDa SNARE proteins bound to the target membrane via fatty acids covalently linked to cysteine residues present in the middle of the polypeptide chain.

The proteolysis of one SNARE protein prevents the formation of the complex and consequently the release of the neurotransmitter. The SNARE complex is insensitive to CNT proteolysis [41], as expected on the basis of the fact that proteases are known to attack predominantly unstructured exposed loops.

The molecular basis of the specificity of the metalloprotease activity of the clostridial neurotoxins for the three SNAREs is only partially known. The sequences flanking the cleavage sites of the three CNT substrates do not show a conserved pattern that could account for such specificity. Experimental evidence indicates the involvement of a nine-residue-long motif, termed hereafter SNARE motif and characterised by three carboxylate residues alternated with hydrophobic and hydrophilic residues [42–44]. The motif is present in two copies (V1 and V2) in VAMP and syntaxin and four copies in SNAP-25. The various CNTs differ with respect to the specific interaction with the SNARE motif [26]. The findings that only protein segments including at least one SNARE motif are cleaved by the toxins and that the motif is exposed at the protein surface [42, 43, 45] clearly indicate the involvement of the SNARE motif in the specificity of action of botulinum neurotoxins. Moreover, different SNARE isoforms coexist within the same cell [46], but only some of them are susceptible to proteolysis by the CNTs and it has been shown that resistance is associated with mutations in SNARE motifs or at the cleavage site [17].

1.3 THE ANTHRAX LETHAL FACTOR

1.3.1 INTRODUCTION

Anthrax is a disease of animals and humans, caused by toxigenic strains of *Bacillus anthracis*, a Gram positive spore-forming bacterium, which secretes three distinct proteins, acting in binary combinations [47]. They are: the protective antigen (PA) which elicits a protective immune response against anthrax [48], the lethal factor (LF) and the oedema factor (EF). The three proteins are encoded by genes included in a large plasmid (pXO1) and are synthesised and secreted independently. The injection of PA+LF (LeTx, lethal toxin) causes a rapid death of laboratory animals [49]. LeTx lyses some cell lines and primary cultures of murine macrophages [50, 51] whereas intradermal injection of PA+EF (EdTx, oedema toxin) produces oedema in the skin [47]. Separately, none of these proteins is toxic. It is now well established that they represent a unique variation in the A-B toxin pattern. PA is the common cell-binding domain (B) which mediates the entry into the cytosol of two different enzyme domains (A): EF and LF, which elicit cell damage. EF is a calmodulin-dependent adenylate cyclase [52] whereas LF is a zinc-binding protein which includes the HEXXH motif and acts in the cytosol via a metalloproteolytic activity [53, 54].

1.3.2 THE PROTECTIVE ANTIGEN PA AND TOXIN INTERNALISATION AND TRANSLOCATION

The mature protective antigen (PA83) is a 735-amino-acid protein (82 kDa) and the crystal structure shows that it is a long, flat protein, rich in β-sheet structure [55]. The role of PA is to cause binding of LF and EF to the cell surface, so that they will be internalised by endocytosis, and to provide a membrane channel for their translocation from the endosome to the cytosol (Figure 1.3). The first step corresponds to the highly specific binding of PA to the cell surface receptor. The receptor of PA, partially proteinaceous, is present on many cell types [56] and is still unidentified. Upon cell binding, PA is cleaved by furin or furin-like proteases and the N-terminal 20 kDa fragment (PA20) is released [57]. The proteolytic activation of PA83 is essential since it exposes the binding site for EF and LF and allows the oligomerisation of the remaining 63 kDa fragment (PA63) in a ring-shaped heptamer. The heptamer then binds LF and EF competitively. Some data support the hypothesis that each PA63 monomer binds one EF or LF, suggesting the binding of seven molecules of LF and/or EF per heptamer. Formation of the heptamer and fixation of EF and/or LF is then followed by the internalisation of this hetero-oligomeric complex. Oligomerisation triggers receptor-mediated endocytosis of receptor-bound PA63 [58] and both EdTx and LeTx require passage through an acidic vesicle to enter the cytosol, because inhibitors of endosomal acidification or of endocytosis prevent toxicity [57, 59]. At acidic pH the PA63-heptamer prepore converts to an active pore and it is assumed that LF/EF translocate through the lumen of the pore, but little is known of how this process occurs. Several data suggest that EF and LF participate actively in their translocation. The two

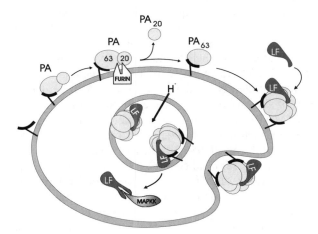

Figure 1.3 Schematic representation of anthrax LF toxin action. Upon cell binding, PA is cleaved by furin and the N-terminal 20 kDa fragment (PA20) is released. The proteolytic activation of PA83 is essential since it exposes the binding site for LF (or EF) and allows the oligomerization of the remaining 63 kDa fragment (PA63) in a ring-shaped heptamer. Formation of the heptamer and fixation of LF is then followed by the internalization of this hetero-oligomeric complex through an acidic vesicle. At acidic pH the PA63-heptamer prepore converts to an active pore and LF translocates through the lumen of the pore to the cytosol where it catalyses the proteolysis of the MAP kinase kinases (MAPKK) (modified from [57])

components interact with lipid bilayers in a pH-dependent manner [54] and this interaction, optimal at acidic pH, is irreversible for EF and reversible for LF. Moreover, recent experiments on CHO-K1 cells, show that LF is completely translocated into the cytoplasm whereas EF remains membrane-associated [60]. The two components may be translocated in a partially unfolded state [61, 62].

1.3.3 THE METALLOPROTEOLYTIC ACTIVITY OF THE LETHAL FACTOR LF

The mature protein contains 776 residues with an apparent molecular mass of 85 kDa. The molecule can be divided into three parts. The aminoterminal part of LF (LF254) is involved in the binding to PA and has a substantial similarity with the aminoterminal part of EF [63]. The central part of the molecule (307–383) contains a series of four imperfect repeats rich in glutamate, and deletions in this region render the protein unstable and inactive [64]. The catalytic domain resides in the C-terminal part and the recognition that LF contains the consensus sequence of zinc metalloprotease [53] started a process that eventually led to the identification of its catalytic activity. Substitution of Ala for H686, E687 or H690 in the sequence $_{686}$HEFGH$_{690}$ abolishes the binding of zinc to LF and its toxicity on macrophage cell lines and on Fisher rat 344 [53, 65]. From these results it became quite obvious that, like the clostridial botulinum and tetanus toxins, LF is a zinc protease, and this hypothesis was supported by the discovery that inhibitors of zinc-dependent aminopeptidases (e.g. bestatin, aromatic amino acid amides and hydroxamates) protect macrophages from LF [53, 66]. However, searches for a cellular substrate continued until recently. Independently, and by two totally different experimental approaches, LF was shown to cleave the aminoterminus of mitogen-activated protein kinases (MAPKKs) Mek1 and Mek2 [67, 68]. One group identified Mek1 and Mek2 as metalloproteolytic substrate of LF following the finding that the MAPKK inhibitor PD98059 and LF gave similar profiles of toxicity with respect to a series of tumoral cell lines [67]. The other group identified the same cytosolic substrates of LF metalloproteolytic activity by a yeast two-hybrid technique, using as bait an LF mutated at the glutamic acid 687 (LF^{E687A}) of the zinc-binding motif to screen a Hela c-DNA library. The MAPK pathway relays environmental signals to the transcriptional machinery in the nucleus and thus modulates gene expression via a burst of protein phosphorylation. Seven different MAPKKs are known, composing three distinct MAPK cascades [69]. Recently, it has been reported that LF cleaves all the MAPKKs except MAPKK5 and a consensus motif for the cleavage site was identified [70, 71]. Cleavage invariably occurs within the N-terminal proline-rich region preceding the kinase domain, and the release of the N-terminal part of MAPKKs is accompanied by phosphorylation of the MAP kinases (MAPK) ERK in cultured macrophages [68]. Phosphorylated ERK is the nuclear active form of the protein and it is possible that in macrophages phosphorylated ERK starts a pathway leading to their death. Although the signaling pathways involving MAPKKs play a crucial role in the activation of macrophages and are directly involved in the production of cytokines such as TNF, IL-1 and IL-6, the link between the cleavage of MAPKKs by LF, macrophage lysis and pathogenesis remains unclear. Release of proinflammatory cytokines potentially elaborated by macrophages upon treatment with LeTx could account for shock, and mice depleted of macrophages have been reported to become insensitive to LeTx challenge [72]. However, there are conflicting data concerning the modulation of TNF and IL-1 by LeTx. Whereas it was reported that sublytic doses of LeTx induced production of these cytokines [72], a lack of effect and an LeTx inhibition of the production of NO and TNF induced by LPS were reported [70, 73]. Moreover, macrophage cell lines resistant to the lytic effect of LeTx, and peritoneal macrophages isolated from mouse strains insensitive to LeTx, are still sensitive to the protease activity of LF on their MAPKKs. The lack of correlation seems to indicate that there are other cytosolic targets of LF involved in cytotoxicity.

1.4   FRAGILYSINS FROM *BACTEROIDES FRAGILIS*

1.4.1   INTRODUCTION

Bacteroides fragilis spp. are included in the normal commensal intestinal flora of the majority of adults ($< 0.5\%$ of the total), but have recently been identified as anaerobic bacteria in isolates from clinical specimens, bloodstream infections and abdominal abscesses [74, 75]. Toxigenic *B. fragilis* strains, termed enterotoxigenic *B. fragilis* (ETBF), were first identified during studies of an epidemic of diarrheal disease in lambs, and it was later found in many other animal species [76, 77]. A 20 kDa fraction of culture filtrates of ETBF stimulated a striking secretory response in lamb ligated intestinal segments [78], suggesting the presence of a secreted toxin, termed fragilysin and abbreviated BFT [79]. In different countries, the association of *B. fragilis* with human diarrheal diseases is well established [80, 81, 82] , though we have yet to assess the link between severity of the disease and expression and release of BFT. Recently, an association between clinically active inflammatory bowel disease and infection with toxigenic *B. fragilis* was suggested [83].

1.4.2   GENETICS AND STRUCTURE OF FRAGILYSIN (BFT)

Three highly homologous genes encoding for BFTs have been identified so far [84–88] and were termed *bft-1* (from strain VPI 13784), *bft-2* (from strain 86-5443-2-2) and *bft-3* (from strain Korea 419) [79, 86, 89]. These genes are included in a 6-kb region, present only in enteropathogenic strains of *B. fragilis* [84], which has several characteristics of a pathogenicity island [90]. This chromosomal region also includes two opening frames (ORF-A and ORF-B) in addition to *bft-2* [91]. ORF-A is small (375 bp) and has no significant similarity to any protein sequence in the database. ORF-B encodes a predicted 44.4 kDa protein including a HExxH metalloprotease motif similar to those contained in the BFTs, a lipoprotein signal peptide and a nucleotide-triphosphate-binding site motif [92]. This protein has been termed metalloprotease II (MP-II) and neither its biologic activity nor its role in pathogenesis are known [91].

The BFT genes consist of one open reading frame encoding for a protein of nearly four hundred residues [84, 85]. The analysis of their sequences suggests that BFT is synthesised as a pre-pro-protein peptide and that it belongs to the intramolecular chaperone protease family [79, 93]. As depicted in Figure 1.4, this protein is composed of three consecutive parts: an 18 residues long signal sequence precedes a 193 residues long pro-portion which is essential for the correct folding of the 186 residues long enzymatic domain [79, 94]. Differences must exist in the intracellular synthesis, processing and secretion of BFTs among the various enteropathogenic *B. fragilis* strains, to account for a large variability in the net amount of secreted toxin [95–99], but this issue remains to be investigated in detail.

Unlike clostridial neurotoxins, whose structural and biochemical properties define a unique novel family of metalloproteases [19, 100], the analysis of the BFT genes indicates that they belong to the metzincins group of metalloproteases, characterised by the common presence of the motif HexxHxxGxxH and by the presence of a conserved methionine near the C-terminus [101, 102]. The catalytic zinc atom is coordinated by three conserved histidines and the water-binding glutamate of the motif. BFTs have both autoproteolytic activity and *in vitro* proteolytic activity for substrates such as actin, gelatin, casein and azocoll

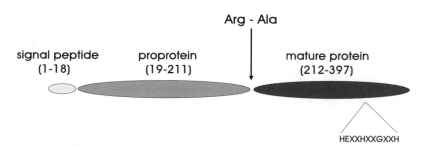

Figure 1.4   Schematic structure of BFT. The predicted amino acid sequences of the *bft-1*, *bft-2* and *bft-3* genes suggest that the encoded proteins are pre-proprotein toxins. Cleavage between Arg and Ala residues appears to release the mature BFT protein. Each protein contains an HEXXHXXGXXH motif (and a conserved methionine near the C-terminus) suggesting that they belong to the metzincins group of metalloproteases

[88]. The *bft-2* gene additionally includes a 20-residue-long COOH-terminal amphipathic segment [84], which has been suggested to mediate the oligomer-isation and the membrane insertion of the protein toxin with creation of an ion channel [79]. This hypothesis was put forward to account for the fact that BFT-2 exhibits greater biologic activity than BFT-1 or BFT-3 when tested on HT29/C1 cells [79]. An alternative explanation is that the COOH-terminal segment of BFT-2 mediates the association of BFT-2 with the cell surface leading to a much more efficient cleavage of cell surface proteins, due to the presence of both toxin and substrate on the two-dimensional solvent constituted by the plasma membrane. The virulence of ETBF strains containing two copies of the *bft* and *mpII* genes in animal models of ETBF disease is not known at present.

Though BFT causes a morphological change in HT29/C1 cells, a human colonic carcinoma cell line [96], the most appropriate *in vitro* system to test for the activity of BFT consists of polarised epithelial cell monolayers grown on filters, which form tight junctions and develop a trans-epithelial resistance which is a sensitive measure of the tight junctional sealing [103, 104]. The three BFT isoforms have been purified and their *in vitro* activity was assayed on HT29/C1 cells. Their order of potency is BFT-2 > BFT-1 > BFT-3 [79]. The half-maximal concentration of BFT-2 altering HT29/C1 cell morphology is approximately 12.5 pM (measured at three hours, 0.5 pM at eighteen hours), whereas it is approximately 1 nM in polarised T84 monolayers [105, 106]. Such a difference could be due to a different degree of membrane association of BFT-2 with the two cell lines.

1.4.3   THE METALLOPROTEOLYTIC ACTIVITY OF FRAGILYSIN (BFT)

In analogy with tetanus, botulism and anthrax, one is tempted to suggest that BFT is an essential virulence factor. However, wild-type ETBF strains and isogenic mutants of the same strain differing only in the in-frame deletion of the *bft* gene have not yet been compared for their virulence. Moreover, current knowledge of the pathogenicity of other gastrointestinal pathogens suggests a word of caution in attributing a predominant importance to single virulence factors. On the other hand, the activity of purified BFT is well documented at both the physiological and biochemical levels [79] and BFT promises to be a valuable addition for cell biologists and physiologists studying epithelia. In ligated intestinal loops isolated from different animal species, BFT stimulates dose-dependent secretion of fluids containing sodium, chloride and proteins [95]. At higher doses, BFT causes a hemorrhagic inflammation. Accordingly, BFT stimulates T84 cells to secrete the pro-inflammatory cytokine IL-8 and TGF-β, which promote the repair of 'wounded' epithelium in a dose-dependent manner [107]. At the single cell level, BFT induces time- and concentration-dependent changes in the structure of filamentous actin, without altering the total F-actin content of the cells. In parallel, BFT stimulates a rapid and sustained increase in the volume of HT29/C1 cells [108]. However, the main activity of BFT is the decrease in the trans-epithelial resistance of polarised monolayers of epithelial cells in a dose- and time-dependent manner [106, 109, 110]. Electron microscopic analysis of BFT treated monolayers of T84 cells shows a decrease in microvilli and an effacement of some tight junctions and of the zonula adherens [105]. BFT is more active on the basolateral than on the apical domains of polarised (i.e. tight junction-forming) epithelial cells. T84 cells treated for 48 hours or more even with low concentrations of BFT undergo cell death with features of apoptosis. Unlike clostridial neurotoxins and LF, BFT does not appear to enter cells, but acts on the cell surface on selected targets, which include the tight junctions. Of the tight junctional proteins tested, only E-cadherin was cleaved by BFT, and cleavage could be detected within one minute from toxin addition in close agreement with the rapidity of BFT action. BFT cleavage of E-cadherin takes place in two steps: the extracellular domain of E-cadherin is degraded in an ATP-independent manner, followed by the degradation of its intracellular domain in an ATP-dependent manner, most likely mediated by intracellular proteases.

BFT is the first bacterial toxin known to remodel the intestinal epithelial cytoskeleton and F-actin architecture via cleavage of a cell surface molecule and represents the prototype of a novel class of bacterial toxins, while known actinomorphic bacterial toxins require cell internalisation and determine a covalent modification of cellular substrates [111].

These data suggested the following model for the pathogenesis of *B. fragilis*-induced intestinal secretion [79]. The toxigenic bacterial strains are presumed to attach to the apical membrane of intestinal epithelial cells and secrete BFT, which may diffuse through the zonula occludens to reach and cleave E-cadherin. Cleavage of the extracellular domain of E-cadherin is followed by loss of the intracellular domain of this molecule. Because the intracellular domain of E-cadherin is tethered to the apical network of F-actin in the epithelial cell, loss of these protein–protein interactions may precipitate focal morphological changes in the apical cellular cytoskeleton with loss of microvilli and decreased barrier function which allows the delivery of BFT to the basolateral membranes of the intestinal epithelial cells. Here, further cleavage of E-cadherin may increase the apical morphological changes initially stimulated by BFT. These dramatic changes in the apical membrane of epithelial cells caused by BFT are hypothe-sised to alter the function of one or more ion transporters, resulting in net intestinal secretion. Concomitantly, protein synthesis is stimulated in the intes-tinal epithelial cells resulting in secretion of the proinflammatory cytokine IL-8. This molecule initiates the recruitment of polymorphonuclear leukocytes to the intestinal submucosa. This resulting inflammatory response is predicted to contribute to the intestinal secretory response. However, it is not yet known if ETBF stimulate intestinal inflammation in humans, and several aspects of this model still require direct experimental testing.

ACKNOWLEDGEMENTS

Work performed in the authors' laboratory is supported by Telethon-Italia grant 1068 and MURST grant 990698133.

REFERENCES

[1] Rawlings, N.D. and Barrett, A.J. (1995) Evolutionary families of metallopeptidases. *Methods Enzymol.* **248**, 183–228.

[2] Emergem, Ev. (1897) Ueber einen neuen anaeroben Bacillus und seine Beziehungen zum Botulism. *Z. Hyg. Infekt.* **26**, 1–56.

[3] Faber, K. (1890) Die Pathogenie des Tetanus. *Berl. Klin. Wochenschr.* **27**, 717–20.

[4] Tizzoni, G. and Cattani, G. (1890) *Uber das Tetanusgift. Zentralbl. Bakt.* **8**, 69–73.

[5] Smith, L.D. and Sugiyama, H. (1988) *Botulism: the organism, its toxins, the disease.* In: Springfield I, ed. C.C. Thomas Publisher.

[6] Hatheway, C.L. (1995) Botulism: the present status of the disease. *Curr. Top. Microbiol. Immunol.* **195**, 55–75.

[7] Burgen, A.S.V., Dickens, F. and Zatman, L.J. (1949) The action of botulinum toxin on the neuro-muscular junction. *J. Physiol. (London)* **109**, 10–24.

[8] Brooks, V.B., Curtis, D.R. and Eccles, J.C. (1955) Mode of action of tetanus toxin. *Nature* **175**, 120–1.

[9] Popoff, M.R. and Marvaud, J.C. (1999) Structural and genomic features of clostridial neurotoxins. In: Alouf JE, J.H. Freer, eds. *The comprehensive sourcebook of bacterial protein toxins.* 2nd ed. London: Academic Press, 174–201.

[10] DasGupta, B.R. (1994) Structures of botulinum neurotoxin, its functional domains, and perspectives on the crystalline type A toxin. In: Jankovic J, Hallett M, eds. *Therapy with Botulinum toxin.* New York: Marcel Dekker, 15–39.

[11] Schiavo, G. and Montecucco, C. (1995) Tetanus and botulism neurotoxins: isolation and assay. *Methods Enzymol.* **248**, 643–52.

[12] Umland, T.C., Wingert, L.M., Swaminathan, S., Furey, W.F., Schmidt, J.J. and Sax, M. (1997) Structure of the receptor binding fragment HC of tetanus neurotoxin. *Nat. Struct. Biol.* **4**, 788–92.

[13] Lacy, D.B., Tepp, W., Cohen, A.C., DasGupta, B.R. and Stevens, R.C. (1998) Crystal structure of botulinum neurotoxin type A and implications for toxicity. *Nature Struct. Biol.* **5**, 898–902.

[14] Swaminathan, S. and Eswaramoorthy, S. (2000) Structural analysis of the catalytic and binding sites of Clostridium botulinum neurotoxin B. *Nat. Struct. Biol.* **7**(8), 693–9.

[15] Lacy, D.B. and Stevens, R.C. (1999) Sequence homology and structural analysis of the clostridial neurotoxins. *J. Mol. Biol.* **291**(5), 1091–104.

[16] Halpern, J.L. and Loftus, A. (1993) Characterization of the receptor-binding domain of tetanus toxin. *J. Biol. Chem.* **268**(15), 11188–92.

[17] Schiavo, G., Matteoli, M. and Montecucco, C. (2000) Neurotoxins affecting neuroexocytosis. *Physiol. Rev.* **80**(2), 717–66.

[18] Li, L., Binz, T., Niemann, H. and Singh, B.R. (2000) Probing the mechanistic role of glutamate residue in the zinc-binding motif of type A botulinum neurotoxin light chain. *Biochemistry* **39**(9), 2399–405.

[19] Rossetto, O., Caccin, P., Rigoni, M., Tonello, F., Bortoletto, N., Stevens, R.C. and Montecucco, C. (2001a) Active-site mutagenesis of tetanus neurotoxin implicates TYR-375 and GLU-271 in metalloproteolytic activity. *Toxicon* **39**, 1151–9.

[20] Hanson, M.A. and Stevens, R.C. (2000) Cocrystal structure of synaptobrevin-II bound to botulinum neurotoxin type B at 2.0 Å resolution. *Nature Struct. Biol.* **7**, 687–92.

[21] Montecucco, C., Papini, E. and Schiavo, G. (1994) Bacterial protein toxins penetrate cells via a four-step mechanism. *FEBS Lett.* **346**(1), 92–8.

[22] Lalli, G., Herreros, J., Osborne, S.L., Montecucco, C., Rossetto, O. and Schiavo, G. (1999) Functional characterisation of tetanus and botulinum neurotoxins binding domains. *J. Cell. Sci.* **112**, 2715–24.

[23] Habermann, E. and Dreyer, F. (1986) Clostridial neurotoxins: handling and action at the cellular and molecular level. *Curr. Top. Microbiol. Immunol.* **129**, 93–179.

[24] Nishiki, T., Tokuyama, Y., Kamata, Y. *et al.* (1996) The high-affinity binding of Clostridium botulinum type B neurotoxin to synaptotagmin II associated with gangliosides GT1b/GD1a. *FEBS Lett.* **378**(3), 253–7.

[25] Halpern, J.L. and Neale, E.A. (1995) Neurospecific binding, internalization, and retrograde axonal transport. *Curr. Top. Microbiol. Immunol.* **195**, 221–41.

[26] Rossetto, O., Seveso, M., Caccin, P., Schiavo, G. and Montecucco, C. (2001b) Tetanus and botulinum neurotoxins: turning bad guys into good by research. *Toxicon* **39**, 27–41.

[27] Dolly, J.O., Black, J., Williams, R.S. and Melling, J. (1984) Acceptors for botulinum neurotoxin reside on motor nerve terminals and mediate its internalization. *Nature* **307**, 457–60.

[28] Hughes, R. and Whaler, B.C. (1962) Influence of nerve-endings activity and of drugs on the rate of paralysis of rat diaphragm preparations by *Clostridium botulinum* type A toxin. *J. Physiol. (London)* **160**, 221–33.

[29] Cremona, O. and De Camilli, P. (1997) Synaptic vesicle endocytosis. *Curr. Opin. Neurobiol.* **7**(3), 323–30.

[30] Hua, S.Y. and Charlton, M.P. (1999) Activity-dependent changes in partial VAMP complexes during neurotransmitter release. *Nat. Neurosci.* **2**(12), 1078–83.

[31] Simpson, L.L., Coffield, J.A. and Bakry, N. (1994) Inhibition of vacuolar adenosine triphosphatase antagonizes the effects of clostridial neurotoxins but not phospholipase A2 neurotoxins. *J. Pharmacol. Exp. Ther.* **269**(1), 256–62.

[32] Montecucco, C., Schiavo, G. and DasGupta, B.R. (1989) Effect of pH on the interaction of botulinum neurotoxins A, B and E with liposomes. *Biochem. J.* **259**(1), 47–53.

[33] Hoch, D.H., Romero Mira, M., Ehrlich, B.E., Finkelstein, A., DasGupta, B.R. and Simpson, L.L. (1985) Channels formed by botulinum, tetanus, and diphtheria toxins in planar lipid bilayers: relevance to translocation of proteins across membranes. *Proc. Natl. Acad. Sci. USA* **82**(6), 1692–6.

[34] Boquet, P. and Duflot, E. (1982) Tetanus toxin fragment forms channels in lipid vesicles at low pH. *Proc. Natl. Acad. Sci. USA* **79**(24), 7614–8.

[35] Schiavo, G., Benfenati, F., Poulain, B. *et al.* (1992) Tetanus and botulinum-B neurotoxins block neurotransmitter release by proteolytic cleavage of synaptobrevin. *Nature* **359**, 832–5.

[36] Blasi, J., Chapman, E.R., Link, E. *et al.* (1993a) Botulinum neurotoxin A selectively cleaves the synaptic protein SNAP-25. *Nature* **365**, 160–3.

[37] Blasi, J., Chapman, E.R., Yamasaki, S., Binz, T., Niemann, H. and Jahn, R. (1993b) Botulinum neurotoxin C1 blocks neurotransmitter release by means of cleaving HPC-1/syntaxin. *Embo J.* **12**, 4821–8.

[38] Schiavo, G., Santucci, A., DasGupta, B.R. *et al.* (1993) Botulinum neurotoxins serotypes A and E cleave SNAP-25 at distinct COOH-terminal peptide bonds. *FEBS Lett.* **335**(1), 99–103.

[39] Sutton, R.B., Fasshauer, D., Jahn, R. and Brunger, A.T. (1998) Crystal structure of a SNARE complex involved in synaptic. *Nature* **395**, 347–53.

[40] Chen, Y.A., Scales, S.J., Patel, S.M., Doung, Y.C. and Scheller, R.H. (1999) SNARE complex formation is triggered by Ca^{2+} and drives membrane fusion. *Cell* **97**(2), 165–74.

[41] Hayashi, T., McMahon, H., Yamasaki, S. *et al.* (1994) Synaptic vesicle membrane fusion complex: action of clostridial neurotoxins on assembly. *EMBO J.* **13**(21), 5051–61.

[42] Rossetto, O., Schiavo, G., Montecucco, C. *et al.* (1994) SNARE motif and neurotoxins. *Nature* **372**, 415–6.

[43] Pellizzari, R., Rossetto, O., Lozzi, L. *et al.* (1996) Structural determinants of the specificity for synaptic vesicle-associated membrane protein/synaptobrevin of tetanus and botulinum type B and G neurotoxins. *J. Biol. Chem.* **271**(34), 20353–8.

[44] Vaidyanathan, V.V., Yoshino, K., Jahnz, M. *et al.* (1999) Proteolysis of SNAP-25 isoforms by botulinum neurotoxin types A, C, and E: domains and amino acid residues controlling the formation of enzyme-substrate complexes and cleavage. *J. Neurochem.* **72**(1), 327–37.

[45] Washbourne, P., Pellizzari, R., Baldini, G., Wilson, M.C. and Montecucco, C. (1997) Botulinum neurotoxin type-A and type-E require the SNARE motif in SNAP-25 for proteolysis. *FEBS Lett.* **418**, 1–5.

[46] Bock, J.B. and Scheller, R.H. (1999) SNARE proteins mediate lipid bilayer fusion. *Proc. Natl. Acad. Sci. USA* **96**(22), 12227–9.

[47] Stanley, J.L. and Smith, H. (1961) Purification of factor I and recognition of a third factor of the anthrax toxin. *J. Gen. Microbiol.* **26**, 49–66.

[48] Gladstone, G.P. (1946) Immunity of anthrax: protective antigen present in cell-free culture filtrates. *Br. J. Exp. Pathol.* **27**, 393–410.

[49] Ezzell, J.W., Ivins, B.E. and Leppla, S.H. (1984) Immunoelectrophoretic analysis, toxicity, and kinetics of in vitro production of the protective antigen and lethal factor components of Bacillus anthracis toxin. *Infect. Immun.* **45**, 761–7.

[50] Beall, F.A., Taylor, M.J. and Thorne, C.B. (1962) Rapid lethal effects in rats of a third component found upon fractionating the toxin of *Bacillus anthracis*. *J. Bacteriol.* **83**, 1274–80.

[51] Friedlander, A.M. (1986) Macrophages are sensitive to anthrax lethal toxin through an acid-dependent process. *J. Biol. Chem.* **261**, 7123–6.

[52] Leppla, S.H. (1982) Anthrax toxin edema factor: a bacterial adenylate cyclase that increases cyclic AMP concentrations in eukaryotic cells. *Proc. Natl. Acad. Sci. USA* **79**, 3162–6.

[53] Klimpel, K.R., Arora, N. and Leppla, S.H. (1994) Anthrax toxin lethal factor contains a zinc metalloprotease consensus sequence which is required for lethal toxin activity. *Mol. Microbiol.* **13**, 1093–100.

[54] Kochi, S.K., Schiavo, G., Mock, M. and Montecucco, C. (1994) Zinc content of the *Bacillus anthracis* lethal factor. *FEMS Microbiol. Lett.* **124**, 343–8.

[55] Petosa, C., Collier, R.J., Klimpel, K.R., Leppla, S.H. and Liddington, R.C. (1997) Crystal structure of the anthrax toxin protective antigen. *Nature* **385**, 833–8.

[56] Escuyer, V. and Collier, R.J. (1991) Anthrax protective antigen interacts with a specific receptor on the surface of CHO-K1 cells. *Infect. Immun.* **59**, 3381–6.

[57] Leppla, S. (1995) Anthrax toxins. In Bacterial Toxins and Virulence Factors in Disease, ed. J Moss, B Iglewski, M Vaughan, AT Tu, *Handbook of Natural Toxins*, New York, Basel, Hong Kong. **8**, 543–72.

[58] Beauregard, K., Collier, R.J. and Swanson, J.A. (2000) Proteolytic activation of receptor-bound anthrax protective antigen on macrophages promotes its internalization. *Cell. Microbiol.* **2**, 251–8.

[59] Menard, A., Altendorf, K., Breves, D., Mock, M. and Montecucco, C. (1996a) The vacuolar ATPase proton pump is required for the cytotoxicity of Bacillus anthracis lethal toxin. *FEBS Lett.* **386**, 161–4.

[60] Guidi-Rontani, C., Weber-Levy, M., Mock, M. and Cabiaux, V. (2000) Transloca-
     tion of *Bacillus anthracis* lethal and oedema factors across endosome membranes.
     Cell. Microbiol. **2**, 259–64.
[61] Wang, X.M., Mock, M., Ruysschaert, J-M. and Cabiaux, V. (1996) Secondary
     structure of anthrax lethal toxin proteins and their interaction with large unilamel-
     lar vesicles: a Fourier-transform infrared spectroscopy approach. *Biochemistry* **35**,
     14939–46.
[62] Wesche, J., Elliott, J.L., Falnes, P.O., Olsnes, S. and Collier, R.J. (1998) Character-
     ization of membrane translocation by anthrax protective antigen. *Biochemistry* **37**,
     15737–46.
[63] Arora, N. and Leppla, S.H. (1993) Residues 1–254 of anthrax toxin lethal factor are
     sufficient to cause cellular uptake of fused polypeptides. *J. Biol. Chem.* **268**, 3334–
     41.
[64] Quinn, C.P., Singh, Y., Klimpel, K.R. and Leppla, S.H. (1991) Functional mapping
     of anthrax toxin lethal factor by in-frame insertion mutagenesis. *J. Biol. Chem.* **266**,
     20124–30.
[65] Brossier, F., Weber-Levy, M., Mock, M. and Sirard, J-C. (2000) Role of toxin
     functional domains in anthrax pathogenesis. *Infect. Immun.* **68**, 1781–6.
[66] Menard, A., Papini, E., Mock, M. and Montecucco, C. (1996b) The cytotoxic
     activity of Bacillus anthracis lethal factor is inhibited by leukotriene A4 hydrolase
     and metallopeptidase inhibitors. *Biochem. J.* **320**, 687–91.
[67] Duesbery, N.S., Webb, C.P., Leppla, S.H., Gordon, V.M., Klimpel, K.R., Cope-
     land, T.D., Ahn, N.G., Oskarsson, M.K., Fukasawa, K., Paull, K.D. and Vande
     Woude, G.F. (1998) Proteolytic inactivation of MAP-Kinase-Kinase by anthrax
     lethal factor. *Science* **280**, 734–7.
[68] Vitale, G., Pellizzari, R., Recchi, C., Napolitani, G., Mock, M. and Montecucco, C.
     (1998) Anthrax lethal factor cleaves the N-terminus of MAPKKs and induces
     Tyrosine/Threonine phosphorylation of MAPKs in cultured macrophages. *Bio-
     chem. Biophys. Res. Commun.* **248**, 706–11.
[69] Widmann, C., Gibson, S., Jarpe, M.B. and Johnson, G.L. (1999) Mitogen-activated
     protein kinase: conservation of a three-kinase module from yeast to human. *Phy-
     siol. Rev.* **79**, 143–80.
[70] Pellizzari, R., Guidi-Rontani, C., Vitale, G., Mock, M. and Montecucco, C. (1999)
     Anthrax lethal factor cleaves MKK3 in macrophages and inhibits the LPS/IFNγ-
     induced release of NO and TNFα. *FEBS Lett.* **462**, 199–204.
[71] Vitale, G., Bernardi, L., Napolitani, G., Mock, M. and Montecucco, C. (2000)
     Susceptibility of mitogen-activated protein kinase kinase family members to prote-
     olysis by anthrax lethal factor. *J. Biochem.* **15**, 739–45.
[72] Hanna, P.C., Acosta, D. and Collier, R.J. (1993) On the role of macrophages in
     anthrax. *Proc. Natl. Acad. Sci. USA* **90**, 10198–201.
[73] Erwin, J.L., Little, S.F., Friedlander, A.M., DaSilva, L., Bavari, S. and Chanh, T.C.
     (2000) Macrophage-derived cell lines do not express pro-inflammatory cytokines
     after exposure to anthrax lethal toxin. Presented at *Am. Soc. Microbiol., Gen.
     Meeting, 100th*, Los Angeles.
[74] Polk, F.B. and Kasper, D.L. (1996) *Bacteroides fragilis* Subspecies in Clinical
     Isolates. *Ann. Intern. Med.* **86**, 569–71.
[75] Nguyen, M.H., Yu, V.L., Morris, A.J., McDermott, L., Wagener, M.W., Harrell,
     L. and Snydman, D.R. (2000) Antimicrobial resistance and clinical outcome of
     Bacteroides bacteremia: findings of a multicenter prospective observational trial.
     Clin. Infect. Dis. **30**, 870–6.
[76] Obiso, R.J. Jr., Lyerly, D.M., Van Tassell, R.L. and Wilkins, T.D. (1995) Proteo-
     lytic activity of the *Bacteroides fragilis* enterotoxin causes fluid secretion and
     intestinal damage *in vivo*. *Infect. Immun.* **63**, 3820–6.

[77] Sears, C., Myers, L.L., Lazenby, A. and Van Tassell, R.L. (1995) Enterotoxigenic *Bacteroides fragilis*. *Clin. Infec. Dis.* **20**(Suppl 2), S142–8.

[78] Myers, L.L., Shoop, D.S., Firehammer, B.D. and Border, M.M. (1985) Association of enterotoxigenic *Bacteroides fragilis* with diarrheal disease in calves. *J. Infect. Dis.* **152**, 1344–7.

[79] Sears, C.L. (2001) The toxins of bacteroides fragilis. *Toxicon*, **39**, 1737–1746.

[80] Sack, R.B., Albert, M.J., Alam, K., Neogi, P.K.B. and Akbar, M.S. (1994) Isolation of enterotoxigenic *Bacteroides fragilis* from Bangladeshi children with diarrhea: a case-control study. *J. Clin. Microbiol.* **32**, 960–3.

[81] San Joaquin, V.H., Griffis, J.C., Lee, C. and Sears, C.L. (1995) Association of *Bacteroides fragilis* with childhood diarrhea. *Scand. J. Infect. Dis.* **27**, 211–15.

[82] Pantosti, A., Menozzi, M.G., Frate, A., Sanfilippo, L., D'Ambrosio, F. and Malpeli, M. (1997) Detection of enterotoxigenic *Bacteroides fragilis* and its toxin in stool samples from adults and children in Italy. *Clin. Infect. Dis.* **24**, 12–16.

[83] Prindiville, T.P., Sheikh, R.A., Cohen, S.H., Tang, Y.J., Cantrell, M.C. and Silva, J. Jr. (2000) *Bacteroides fragilis* enterotoxin gene sequences in patients with inflammatory bowel disease. *Emerg. Infect. Dis.* **6**, 171–4.

[84] Franco, A.A., Mundy, L.M., Trucksis, M., Wu, S., Kaper, J.B. and Sears, C.L. (1997) Cloning and characterization of the *Bacteroides fragilis* metalloprotease toxin gene. *Infec. Immun.* **65**, 1007–13.

[85] Kling, J.J., Wright, R.L., Moncrief, J.S. and Wilkens, T.D. (1997) Cloning and characterization of the gene for the metalloprotease enterotoxing of *Bacteroides fragilis*. *FEMS Microbiol. Letters* **146**, 279–84.

[86] Kato, N., Liu, C.X., Kato, H., Watanabe, K., Tanaka, Y., Yamamoto, T., Suzuki, K. and Ueno, K. (2000) A new subtype of the metalloprotease toxin gene and the incidence of the three *bft* subtypes among *Bacteroides fragilis* isolates in Japan. *FEMS Microbiol. Lett.* **182**, 171–6.

[87] Moncrief, J.S., Obiso, R., Barroso, L.A., Kling, J.J., Wright, R.L., Van Tassell, R.L., Lyerly, D.M. and Wilkins, T.D. (1995) The enterotoxin of *Bacteroides fragilis* is a metalloprotease. *Infect. Immun.* **63**, 175–81.

[88] Chung, G.T., Franco, A.A., Wu, S., Rhie, G.E., Cheng, R., Oh, H.B. and Sears, C.L. (1999) Identification of a third metalloprotease toxin gene in extraintestinal isolates of *Bacteroides fragilis*. *Infect. Immun.* **67**, 4945–9.

[89] d'Abusco, A.S., Del Grosso, M., Censini, S., Covacci, A. and Pantosti, A. (2000) The alleles of the bft gene are distributed differently among enterotoxigenic *Bacteroides fragilis* strains from human sources and can be present in double copies. *J. Clin. Microbiol.* **38**, 607–12.

[90] Hacker, J. and Kaper, J.B. (2000) Pathogenicity islands and the evolution of microbes. *Annu. Rev. Microbiol.* **54**, 641–79

[91] Franco, A.A., Cheng, R.K., Chung, G.T., Wu, S., Oh, H.B. and Sears, C.L. (1999) Molecular evolution of the pathogenicity island of enterotoxigenic *Bacteroides fragilis* strains. *J. Bacteriol.* **181**, 6623–33.

[92] Moncrief, J.S., Duncan, A.J., Wright, R.L., Barroso, L.A. and Wilkins, T.D. (1998) Molecular Characterization of the Fragilysin Pathogenicity Islet of Enterotoxigenic *Bacteroides fragilis*. *Infec. Immun.* **66**, 1735–9.

[93] Inouye, M. (1991) Intramolecular chaperone: the role of the pro-peptide in protein folding. *Enzyme* **45**, 314–21.

[94] McIver, K.S., Kessler, E., Olson, J.C. and Ohman, D.E. (1995) The elastase propeptide functions as an intramolecular chaperone required for elastase activity and secretion in *Pseudomonas aeruginosa*. *Molec. Microbiol.* **18**, 877–89.

[95] Myers, L.L., Shoop, D.S., Stackhouse, L.L., Newman, F.S., Flaherty, R.J., Letson, G.W. and Sack, R.B. (1987) Isolation of enterotoxigenic *Bacteroides fragilis* from humans with diarrhea. *J. Clin. Microbiol.* **25**, 2330–3.

[96] Weikel, C.S., Grieco, F.D., Reuben, J., Myers, L.L. and Sack, R.B. (1992) Human colonic epithelial cells, HT29/C$_1$, treated with crude *Bacteroides fragilis* enterotoxin dramatically alter their morphology. *Infect. Immun.* **60**, 321–7.

[97] Mundy, L.M. and Sears, C.L. (1996) Detection of toxin production by *Bacteroides fragilis*: assay development and screening of extraintestinal clinical isolates. *Clin. Infect. Dis.* **23**, 269–76.

[98] Van Tassell, R.L., Lyerly, D.M. and Wilkins, T.D. (1994a) Characterization of enterotoxigenic *Bacteroides fragilis* by a toxin-specific enzyme-linked immunosorbent assay. *Clin. Diag. Lab. Immunol.* **1**, 578–84.

[99] Van Tassell, R.L., Lyerly, D.M. and Wilkins, T.D. (1994b) Production of antisera against the enterotoxin of *Bacteroides fragilis* and their use in a cytotoxicity neutralization assay of HT-29 cells. *Clin. Diag. Lab. Immunol.* **1**, 473–6.

[100] Montecucco, C. and Schiavo, G. (1993) Tetanus and botulism neurotoxins: a new group of zinc proteases. *Trends Biochem. Sci.* **18**, 324–7.

[101] Bode, W., Gomis-Ruth, F.X. and Stockler, W. (1993) Astacins, serralysins, snake venom and matrix metalloproteinases exhibit identical zinc-binding environments (HEXXHXXGXXH and Met- turn) and topologies and should be grouped into a common family, the 'metzincins'. *FEBS Lett.* **331**, 134–40.

[102] Obiso, R.J. Jr., Bevan, D. and Wilkins, T.D. (1997) Molecular modeling and analysis of fragilysin, the *Bacteroides fragilis* toxin. *Clin. Infect. Dis.* **25**, S153–5.

[103] Eaton, S. and Simons, K. (1995) Apical, basal, and lateral cues for epithelial polarization. *Cell* **82**, 5–8.

[104] Kraehenbuhl, J.P. and Neutra, M.R. (1992) Molecular and cellular basis of immune protection of mucosal surfaces. *Physiol. Rev.* **72**, 853–79.

[105] Chambers, F.G., Koshy, S.S., Saidi, R.F., Clark, D.P., Moore, R.D. and Sears, C.L. (1997) *Bacteroides fragilis* toxin exhibits polar activity on monolayers of human intestinal epithelial cells (T84 Cells) *in vitro. Infec. Immun.* **65**, 3561–70.

[106] Saidi, R.F. and Sears, C.L. (1996) *Bacteroides fragilis* toxin rapidly intoxicates human intestinal epithelial cells (HT29/C$_1$) *in vitro. Infec. Immun.* **64**, 5029–34.

[107] Sanfilippo, L., Li, C.K., Seth, R., Balwin, T.J., Menozzi, M.G. and Mahida, Y.R. (2000) *Bacteroides fragilis* enterotoxin induces the expression of IL-8 and transforming growth factor-beta (TGF-beta) by human colonic epithelial cells. *Clin. Exp. Immunol.* **119**, 456–63.

[108] Koshy, S.S., Montrose, M.H. and Sears, C.L. (1996) Human intestinal epithelial cells swell and demonstrate actin rearrangement in response to the metalloprotease toxin of *Bacteroides fragilis. Infec. Immun.* **64**, 5022–8

[109] Wells, C.L., Van De Westerlo, E.M.A., Jechorek, R.P., Feltis, B.A., Wilkins, T.D. and Erlandsen, S.L. (1996) *Bacteroides fragilis* Enterotoxin Modulates Epithelial Permeability and Bacterial Internalization by HT-29 Enterocytes. *Gastroenterol.* **110**, 1429–37.

[110] Obiso, R.J. Jr., Azghani, A.O. and Wilkins, T.D. (1997) The *Bacteroides fragilis* toxin fragilysin disrupts the paracellular barrier of epithelial cells. *Infec. Immun.* **65**, 1431–9.

[111] Lerm, M., Schmidt, G. and Aktories, K. (2000) Bacterial protein toxins targeting rho GTPases. *FEMS Microbiol. Lett.* **188**, 1–6.

2 The Cholesterol-dependent Cytolysins: Current Perspectives on Membrane Assembly and Insertion

EILEEN HOTZE and RODNEY K. TWETEN*

2.1 INTRODUCTION

Cholesterol-dependent cytolysins (CDCs) are a group of pore-forming toxins which, as the name implies, act exclusively on membranes containing cholesterol. These toxins are produced by at least 23 species of Gram positive bacteria including species of *Streptococcus, Bacillus, Clostridium, Listeria*, and *Arcanobacterium*. With the exception of pneumolysin from *Streptococcus pneumoniae*, all identified CDCs are secreted into the extra-cellular environment, presumably by a type II secretion system since the members of this family which have been sequenced all contain a typical type II signal peptide.

The mechanism of action of the CDCs involves a complex series of events that begins with the monomer recognising and binding to cholesterol-containing membranes. How cholesterol may function in this recognition event is described in more detail in the subsequent section. Once bound to the membrane, the monomers are thought to diffuse laterally and then interact with one another to form large homo-oligomeric pore-forming structures of up to 50 monomers. These large oligomeric complexes of approximately 25 nm can be readily identified as ring- or arc-shaped structures by electron microscopy [1–5]. The interaction of the monomers and formation of the oligomeric structure creates a large amphipathic β-sheet which inserts into the membrane and forms the transmembrane channel. The role of these toxins in pathogenesis also seems to be evolving from simple cytolysins to more complex roles that suggest that the large channels formed by the CDCs can be used in various ways. Although significant progress has been made in understanding the multiple steps involved in the pathway of cytolytic activity of the CDCs, many aspects of this process remain unclear, as do the functional roles that these toxins play in pathogenesis.

The nomenclature used for this family of pore-forming toxins has been undergoing considerable variation recently. These toxins were originally given

---

* Corresponding author

Perspectives in Molecular Toxinology
Edited by A Ménez © 2002 John Wiley & Sons, Ltd

the designation of 'oxygen-labile' or 'thiol-activated cytolysins' since crude preparations over time lost haemolytic activity that could be restored by the addition of a reducing agent. Recently, this toxin family has been termed the cholesterol-binding cytolysins (CBC) [6] since several studies have suggested that these toxins bind cholesterol directly. However, as discussed in detail below, this hypothesis is under some scrutiny. Therefore, we have used the term cholesterol-dependent cytolysin (CDC) [5, 7, 8] reflecting the absolute requirement of cholesterol for the full activity of these toxins, but without implying a specific function of the cholesterol.

A more extensive review of the general characteristics of this group of toxins has recently been published [6], and therefore this review will specifically discuss recent advances towards understanding the cytolytic mechanism of the CDCs and the functional role of these proteins in pathogenesis.

2.2 CDC STRUCTURE

The primary structures of the members of the CDC family of toxins display significant homology, they exhibit 40–70% identity at the primary level and hence are likely to exhibit certain common functional features. Sequence analysis of the known members of the CDC family revealed a single conserved cysteine, with the exception of pyolysin and intermedilysin. The lone cysteine residue is located in an undecapeptide near the C-terminus of the molecule which is the most highly conserved region found in the CDCs. This undecapeptide is critical for biological activity, as it is relatively intolerant to mutation [9]. It appears that the sulphhydryl group is not required for cytolytic activity [10, 11] and alanine has been found in place of the cysteine in the CDCs pyolysin [12] and intermedilysin [13]. However, chemical modification of the unique cysteine residue in those toxins that contain it results in loss of activity indicating that the cysteine is located at a functionally important site [14].

Although the crystal structure of the membrane-bound oligomer of a CDC has not been solved, the crystal structure of the water-soluble monomeric form of the CDC perfringolysin O (PFO) has been solved and serves as a model for the rest of the CDC family [15]. The crystal structure reveals an elongated molecule (11.5 nm long) rich in β-sheet structure (40%) (Figure 2.1). PFO comprises four discontinuous domains: domain 1 (residues 37–53, 90–178, 229–274, 350–373), domain 2 (residues 54–89, 374–390), domain 3 (residues 179–228, 275–349) and domain 4 (residues 391–500). The role of each domain in pore formation by the CDCs will be discussed in detail below.

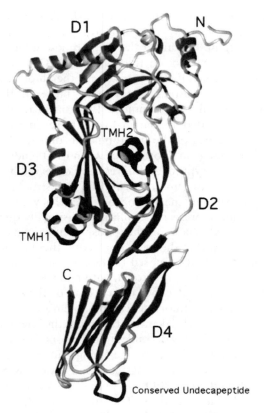

Figure 2.1 Ribbon diagram of the crystal structure of PFO. Shown are the locations of the 4 domains of PFO (D1–D4), the α-helices that ultimately form the two transmembrane β-hairpins (TMH1 and TMH2, in black), the highly conserved undecapeptide (black) and the carboxy (C) and amino (N) termini

2.3   MEMBRANE RECOGNITION

Cellular recognition and the role of cholesterol constitute a conundrum with the CDCs. There is strong evidence that the CDCs bind cholesterol, yet there are tantalising reports that cholesterol may not be involved in the initial binding event. Does cholesterol form the receptor or part of the receptor, or is its role not associated with membrane recognition?

2.3.1   DOMAIN 4

PFO cleaved by trypsin after amino acid residue lysine 304 yields two fragments: T1, which encompasses residues from the amino terminus to residue 304;

and T2, which contains residues 305 to the carboxy terminus and primarily comprises domain 4. T2 was shown to bind to erythrocyte membranes [16, 17] suggesting that this region contained the membrane-binding domain. In addition, Nakamura et al. [18] showed that binding PFO to vesicles containing cholesterol resulted in a significant increase in the fluorescence emission of the PFO tryptophans, suggesting that one or more of these tryptophans was moving into a less polar environment. PFO has a total of seven tryptophans, six of which are located in domain 4 with the seventh residing in domain 1. The intrinsic fluorescence of the tryptophan in domain 1 does not significantly affect the total change in fluorescence when PFO binds a membrane [18]. Therefore, the increase in tryptophan fluorescence observed by Nakamura et al. could be correlated with an environmental and/or structural change in the domain 4 tryptophans upon membrane binding. They further showed that some of the tryptophan fluorescence could be collisionally quenched by bromine attached to the acyl chain of the phospholipids. Since the collisional quencher was restricted within the membrane core, these results indicated that domain 4 was at least partially exposed to the nonpolar bilayer core. Three of the six tryptophans in domain 4 are present within the conserved undecapeptide of PFO, although at present it is not clear which of these tryptophans is/are responsible for the change in fluorescence upon membrane binding.

More recently Heuck et al. [19] demonstrated that the interaction of PFO with liposomal membranes was highly dependent on the concentration of cholesterol. They monitored membrane binding by the changes in tryptophan fluorescence of PFO when incubated with liposomes containing various concentrations of cholesterol. Efficient binding of PFO was only observed at cholesterol concentrations that exceeded 45 mole-percent. They also demonstrated that the tryptophan fluorescence could be quenched by collisional quenchers located near the surface of the membrane bilayer, but not by those located near the bilayer core. These results showed that domain 4 does not penetrate deeply into the bilayer and likely resides near the membrane surface.

2.3.2 CELLULAR RECOGNITION

The absolute requirement of cholesterol for CDC activity has been well documented, but the exact role of cholesterol in this process remains unclear. It has been shown that the haemolytic activity of PFO and other CDCs including tetanolysin, pneumolysin and streptolysin O (SLO) is inactivated when the toxin is pre-incubated with cholesterol and other sterols (reviewed in [6]). It is thought that the toxin binds the cholesterol and is unable to interact with the cell surface. Jacobs et al. [20] used immunoblot analyses and flow cytometry to demonstrate that pre-incubation with cholesterol did not influence binding of the listeriolysin–cholesterol complex to red blood cells, eukaryotic cells or artificial membranes. Enzymatic treatment of the membrane-bound LLO with cholesterol oxidase restored lytic activity. They determined that the inhibition of lytic activity by cholesterol was not due to decreased binding of LLO to target membranes, but rather to interference with a subsequent step leading to polymerisation of the toxin. However, several studies have also demonstrated that cholesterol is specifically bound by the CDC PFO [16, 21, 22] which clearly shows that whatever the role of cholesterol PFO can specifically recognise and bind to it. These two observations are not mutually exclusive, and it is entirely possible that PFO and other CDCs do specifically bind cholesterol, but not as part of the initial cellular recognition event.

Heuck et al. [19] demonstrated that both the change in tryptophan fluorescence by the domain 4 tryptophans of PFO and the insertion of the domain 3 β-hairpins into the membrane required liposomes containing greater than 45 mole-percent cholesterol. Binding of PFO to the liposomes is undetectable at less than 45 mole-percent but increases to 100 % at 50 mole-percent (the balance of the membrane composition is phosphatidylcholine). It would seem that if cholesterol functioned as a simple receptor then such a sharp transition between no binding and binding of PFO would not be observed. Whether cellular membranes must contain this level of cholesterol for full CDC activity is unclear. However, the fundamental question of whether cholesterol functions as a receptor for the CDCs, or for only some of the CDCs, or in a post-binding event as suggested by the results of Jacobs et al. [20] remains unanswered. The fact that such high levels are required and that such a sharp transition in binding occurs in liposomes as cholesterol concentration goes from 45 to 50 mole-percent suggests that the structure of cholesterol in the membrane plays a role in this interaction. Recently, Waheed et al. [23] reported that an inactive derivative of PFO, generated by cleavage with protease, specifically bound to lipid rafts in cells, structures that are well known to be enriched in cholesterol. However, as described below, there is some evidence that cholesterol cannot be the sole determinant of membrane recognition for at least one member of the CDC family.

Due to the absolute requirement of cholesterol for activity, it is has been generally thought that cholesterol could function as the cellular receptor for the CDCs. As mentioned above, this concept was reinforced by the observation that PFO can bind specifically to cholesterol among the chloroform–methanol extracted lipids from human and sheep erythrocytes [16, 21, 22]. However, the cellular specificity exhibited by intermedilysin (ILY), one of the more recent additions to the CDC family, indicates that cholesterol may not be the prime determinant in cellular recognition for this toxin. ILY has been shown to exhibit significant cytolytic activity only on human erythrocytes; erythrocytes from other mammals such as horse, sheep, rabbit, etc. are completely resistant to its cytolytic activity [13]. Therefore, it seems unlikely that cholesterol could be the sole determinant for the cellular specificity of ILY since other CDCs can lyse animal and human cells efficiently. Furthermore, treatment of the human erythrocytes with trypsin significantly reduced the activity of ILY demonstrating that there may be a proteinaceous component involved in cellular recognition by ILY. The level of specificity exhibited by ILY has not been previously reported for a CDC and so ILY offers a unique opportunity to investigate cellular recognition by these toxins. Whether the other CDCs have specific cellular receptors remains an intriguing question.

2.4 FORMATION OF OLIGOMERIC COMPLEXES AND INSERTION OF THE TRANSMEMBRANE DOMAINS

2.4.1 IDENTIFICATION OF THE TRANSMEMBRANE DOMAIN OF PFO

Based on the crystal structure, domain 4 of PFO (residues 391–500) is 29 Å in length and contains the highly conserved 11-amino-acid sequence present in all the CDCs. This domain is long enough to span the bilayer and has a high proportion of hydrophobic residues in the undecapeptide loop at its tip (Figure 2.1). This observation coupled with the observation by Nakamura *et al.* [18] that some of the domain 4 tryptophans were exposed to the lipid led Rossjohn *et al.* to propose that this domain may form the membrane-penetrating domain of PFO [15]. However, subsequent studies demonstrated that the transmembrane domain of this toxin is located elsewhere and that domain 4 does not significantly penetrate the bilayer.

Shepard *et al.* [7] used cysteine scanning mutagenesis and several complementary fluorescent techniques that used the environmentally-sensitive, sulphhydryl-specific fluorophore N,N'-dimethyl-N-(iodoacetyl)-N'-(7-nitrobenz-2-oxa-1,3-diazolyl)ethylenediamine (NBD) to demonstrate that amino acid residues S190–N217 of domain 3 (Figure 2.1) could insert into and span the bilayer. Soon after, Shatursky *et al.* [8], using the same methods, identified a second membrane-spanning β-hairpin in domain 3 that comprised residues K288–D311. They further demonstrated that each of the two regions of domain 3 formed amphipathic β-hairpins that completely spanned the lipid bilayer. Remarkably, both of the transmembrane β-hairpins (TMHs) identified in these studies were shown to undergo an α-helical to β-strand transition which required the conversion of a total of six short α-helices in the monomer into the two transmembrane β-hairpins. These six α-helices comprise 11% of the primary structure of PFO and therefore represent a major change in secondary structure of domain 3. The results of these studies also suggested that domain 3 must swing out from its interface with domains 1 and 2 in order to extend these transmembrane β-sheets into the membrane.

Heuck *et al.* [19] demonstrated that the domain 4 tryptophans were not quenched with nitroxide probes restricted to the core of the bilayer membrane, which showed that domain 4 did not significantly penetrate the bilayer. However, they also showed that domain 4 and the domain 3 transmembrane β-hairpins appear to be conformationally coupled. Mutations in the transmembrane β-hairpins that affected the rate of their insertion also had a similar effect on the rate of change in the domain 4 tryptophan fluorescence. Using the disulphide-locked mutant of Hotze et al. [24], which cannot insert either of its transmembrane β-hairpins (described below), they also showed that the interaction of domain 4 alone with the membrane could not induce pore formation. From these data it was clear that domain 4 interacts with the membrane prior to the insertion of the domain 3 transmembrane β-hairpins.

2.4.2   OLIGOMER ASSEMBLY AND INSERTION

One of the distinguishing characteristics of all CDCs is the formation of large homo-oligomeric structures (comprising an estimated 50 monomers) on sterol-containing membranes [4, 25]. These ring- or arc-shaped structures can be easily visualised by electron microscopy [1–5]. However, the large size of the oligomer (approximately 1.5–2.5 mDa) prevents it from being resolved by SDS-PAGE (although for many years it was believed that SDS could dissociate the CDC oligomer). Therefore, an agarose-based SDS electrophoresis system was adapted by Shepard et al. [5] for separation and visualisation of the SDS resistant PFO oligomer. These studies showed that the PFO oligomer complex generated on liposomes comprised a large, relatively homogeneous SDS-resistant complex rather than a wide distribution of oligomer sizes.

Although there is no doubt of the existence of the CDC oligomer complex in the pathway of CDC pore formation, the mechanism of oligomer formation and the timing of the insertion of the transmembrane hairpins remain controversial. Two current models have been proposed to explain the formation of the pore complex. The first model is based on the mechanism described for other pore-forming toxins such as *Clostridium septicum* α-toxin [26], *Staphylococcus aureus* α-haemolysin [27, 28], and the protective antigen of anthrax toxin [29]. These toxins have been shown to form oligomeric prepore complexes prior to the formation of the pore. The function of the prepore complex has not been rigorously defined, but it likely organises the transmembrane domains so that they insert into the membrane in a concerted fashion. This concept was adopted by Rossjohn et al. [15] who predicted that PFO would follow this model and form a prepore complex prior to insertion of the membrane-spanning domain. One conceptual problem was the fact that the prepore model was based on pore-forming toxins whose prepore complexes are much smaller than that of PFO and are comprised of only up to seven monomers. The oligomer complex of PFO had been predicted to be up to 50 monomers [25] with each contributing two transmembrane hairpins to the β-barrel [8]. Therefore, the organisation of up to 100 transmembrane β-hairpins would have to occur in the CDC prepore complex, a somewhat daunting task that has sparked some controversy.

Based on the size of the CDC oligomer and its apparent heterogeneity Palmer et al. [30] suggested that the prepore-based mechanism might not adequately describe some of the features of the CDC oligomeric complex. It was suggested that the control and coordinate insertion of the large numbers of transmembrane β-hairpins in the CDC from a prepore oligomer was virtually impossible and energetically unfavourable. They proposed a continuous growth model in which a complex of SLO, as small as a dimer, would initially insert into the membrane and form a small pore. This small inserted oligomer complex then would act as a nucleation site for the addition of other soluble monomers. In this model, as each monomer is added it is predicted to bind simultaneously to the end of the growing oligomer and insert its transmembrane domain. Therefore, the overall pore size would enlarge as the oligomer grew in size. The defining principle of this model is that the processes of oligomerisation and insertion are intimately linked: one cannot occur in the absence of the other. Palmer *et al.* [30] described a mutant of SLO in which threonine 250 was converted to a cysteine and which upon derivatisation with the sulphhydryl-modifying reagent *N*-(iodoacetaminoethyl)-1-naphthylamine-5-sulphonic acid (IAEDANS) resulted in the apparent loss of the ability of SLO to oligomerise on the membrane, but which still bound erythrocytes. The results of these studies indicated that the non-lytic phenotype of the mutant was related to a defect in the ability to oligomerise with itself. However, they suggest that this mutant was capable of incorporating into an oligomer of native SLO and affecting the growth of the oligomer. Osmotic protection experiments as well as marker release studies indicated that the mixed oligomers of mutant and native SLO formed pores of reduced size. The smaller pore size presumably resulted from a smaller average size of the oligomer as the ratio of mutant to native toxin increased, although direct quantification of a reduction in size of the complex was not reported.

Several recent studies with the CDC PFO have provided evidence supporting the prepore model. Shepard *et al.* [5] sought to differentiate between the two models by determining if oligomerisation of PFO into a prepore complex could occur in the absence of significant insertion of the two transmembrane hairpins. Their results demonstrated that the PFO oligomeric complex could form on cholesterol-containing liposomes at 4° C in the absence of significant membrane insertion of the β-hairpins [5]. In addition, they also showed that when PFO was applied to a planar lipid bilayer it induced large and discrete stepwise changes in conductance, which was consistent with the insertion of a large, pre-assembled pore complex into the bilayer.

Hotze *et al.* provided further evidence for the existence of the prepore complex as an intermediate in the cytolytic pathway [24]. In their studies a disulphide bridge was engineered between glycine 57 and serine 190 of PFO. This disulphide bridge locked the TMH 1 of PFO to domain 2 and abolished haemolytic activity of the oxidised form of this mutant. The oxidised form of this mutant was haemolytically inactive whereas full haemolytic activity was recovered upon addition of reducing agent. If reducing reagent was added after the oxidised mutant had been pre-incubated with the erythrocytes, thus allowing prepore oligomer formation, haemolysis proceeded at a rate seven times faster then the rate of wild-type PFO. This dramatic increase in the rate of haemolysis suggests that the PFO had advanced to a stage in the haemolytic pathway that primed it for insertion of the transmembrane hairpins. Fluorescence-based experiments indicated that in the disulphide-locked state neither of the TMHs could insert into the bilayer, despite the fact that only TMH 1 was involved in the disulphide bridge. It was also shown that if the disulphide bond was reduced after the $PFO^{C57-C190}$ was converted to the prepore complex, the membrane insertion of the TMHs was also very rapid, suggesting that the formation of the prepore facilitated the insertion of the transmembrane β-sheet. Therefore, instead of posing an energetic barrier to membrane insertion, as suggested by Palmer *et al.* [30], the insertion of the β-sheet from a prepore complex actually facilitates the insertion process. SDS-agarose gel and fluorescence resonance energy transfer (FRET) analyses of the disulphide-trapped mutant revealed that the mutant was as capable of forming oligomers as was wild-type PFO. Electron microscopy of the rings formed by the disulphide-trapped mutant also showed the typical ring and arc structures indistinguishable from the pre-reduced $PFO^{C190-C57}$. In addition, a combination of spectroscopic experiments by Heuck *et al.* [19] using this disulphide-locked mutant, demonstrated that domain 4 interacts with the membrane prior to domain 3 and that domain 4 can move into its membrane-bound state before TMH insertion begins.

These results suggest the following pathway for the insertion of the CDCs. First, domain 4 recognises the membrane and is anchored to the surface via this domain. At the time of binding, or shortly thereafter, one or more of the domain 4 tryptophans move into a nonpolar environment near the surface of the membrane. Some conformational modification of the monomer may occur at this time and enable it to begin oligomerising with other bound CDC monomers to form the prepore complex. Once the prepore complex reaches a sufficient size it inserts its transmembrane β-sheet into the membrane forming the pore. The trigger for conversion of the prepore to pore complex may be linked to the thermodynamics of inserting an amphipathic β-sheet into the membrane. Perhaps only when a sufficient number of monomers have assembled into the prepore is there sufficient free energy, perhaps derived from structural changes in each monomer, to drive the membrane insertion of the β-sheet.

2.5   THE ROLE OF THE CDCs IN BACTERIAL PATHOGENESIS

The CDCs have always been considered cytolytic proteins, but in view of recent studies on the *in vivo* roles of the CDCs it is no longer clear that the primary function of these proteins is cell lysis. There is mounting evidence that indicates that the CDC pore may be used in other ways and that direct cell killing by

CDC-dependent lysis may not be the primary function of these toxins in many of the pathogens which produce them.

2.5.1   LISTERIOLYSIN O

The earliest studies that directly addressed the role of a CDC in the pathogenesis of a bacterial cell were carried out by Portnoy *et al.* [31] who demonstrated that listeriolysin O (LLO) was necessary for the escape of *Listeria monocytogenes* from the endocytic vacuole. The role of LLO in endosomal escape was subsequently elegantly demonstrated in a report by Bielecki *et al.* [32] in which they showed that *Bacillus subtilis* expressing the LLO gene could also escape from an macrophage endosome. Presumably, the LLO disrupted the endosome by the formation of its large membrane pores, although the specific process of membrane disruption is not clear since an endosome cannot 'lyse' within the cell. Possibly, since all of the LLO is expressed within the vacuole, the endosomal membrane is simply destabilised to a point that the bacterial cell can escape the fragmented membrane.

LLO seems well suited to this task since it has been shown to be active in pore formation at pH 5–6 (the pH of the endosome), but is significantly less active at neutral pH [33]. It is likely that decreasing the pore-forming activity of LLO once the bacterial cell escapes the phagosome is important to the intracellular survival and replication of *L. monocytogenes*. The lysis of the host cell would be counter-productive to the life cycle of the *Listeria* which replicates within the protected environment of the cell cytoplasm. Substitution of LLO with PFO, which is active at both 5.5 and neutral pH, allows the bacterial cell to escape the phagosome, but does not facilitate its cytoplasmic replication since PFO retains full activity at neutral pH and lyses the eukaryotic cell [34].

The relatively low activity of LLO at neutral pH is apparently not sufficient for intracellular survival of *Listeria*. Recently, Decatur and Portnoy [35] showed that LLO also contains a PEST (P, Pro; E, Glu; S, Ser; T, Thr) sequence near its amino terminus. PEST sequences normally target a protein for degradation in the eukaryotic cell cytoplasm [36]. Mutants of LLO containing an altered PEST sequence could facilitate the escape of the bacteria from the phagosome, but they did not replicate in the cytoplasm due to the destruction of the host cell membrane by LLO. Although LLO is significantly less active at neutral pH, presumably it is still sufficiently active to destroy the integrity of the host cell membrane. Therefore, the elimination of any residual LLO by the PEST-dependent degradation pathway may be necessary to ensure that residual LLO does not kill the host cell. One could envision a scenario in which once the LLO has facilitated the escape of the bacterial cell from the endosome the shift in pH from 5–6 to neutral pH slows the cytolytic activity and may provide sufficient time for the remaining active LLO to be targeted for degradation via its PEST sequence. Whether there are other functions for the PEST sequence in LLO remains to be elucidated.

2.5.2 PNEUMOLYSIN

Pneumolysin is produced by *Streptococcus pneumoniae* and has been shown
to be an important virulence factor in the development of *S. pneumoniae* lobar
pneumoniae, although its function in pathogenesis is incompletely understood
[37]. Pneumolysin is unique among the CDCs in that it is not actively secreted
by the cell as are the other CDCs and, in addition to its cytolytic activity,
it can also activate the classical pathway of complement [38]. The region of
pneumolysin involved in complement activation exhibits some similarity
with human C-reactive protein that can also activate the classical pathway
of complement. The roles of the cytolytic activity and complement activation
of pneumolysin in pneumonia appear to be distinct. The cytolytic activity
appears to be important in the tissue destruction in the lung whereas the
complement fixing activity seems to be important in bacterial survival within
the host [38].

2.5.3 PERFRINGOLYSIN O

The role of PFO during the development of *Clostridium perfringens* gas gangrene does not appear to be primarily a cytolytic role in view of recent studies
by Rood and colleagues. PFO is produced by type A *Clostridium perfringens*
strains which are the most frequent cause of traumatic gas gangrene in humans.
Gangrene is distinguished by two characteristics: massive tissue destruction
within the muscle and vascular leukostasis that prevents the infiltration of
polymorphonuclear leukocytes (PMNLs) into the necrotic tissue. *In vitro* studies have indicated that PFO has a direct effect on PMNL physiology [39, 40, 41]
and since it is clearly cytolytic it was not unreasonable to assume that it was also
involved in the destruction of tissue during the infection. However, using gene
knockouts of PFO or α-toxin in *C. perfringens*, and using the mouse model for
gangrene, it has been shown that the α-toxin (phospholipase/sphingomyelinase
C) is largely responsible for tissue destruction (PFO contributes to necrosis, but
to a significantly lesser extent than α-toxin), whereas PFO seems to be primarily, though not solely, responsible for vascular leukostasis [42, 43]. Therefore,
contrary to the expectation that PFO would contribute significantly to the
cytolytic tissue destruction during the development of gangrene, it instead
appears that a key role for PFO may be the modulation of the cellular immune
response.

2.5.4 STREPTOLYSIN O

Caparon and colleagues recently reported an intriguing function for the CDC
streptolysin O which is produced by *Streptococcus pyogenes*. *S. pyogenes* causes

several types of skin infections and they have shown that *S. pyogenes* binds to keratinocytes via one or more adhesions [44]. Madden *et al.* [45] made the remarkable observation that while attached to the keratinocytes *S. pyogenes* establishes a functional translocation channel between the streptococcal cell and the cytoplasm of the keratinocyte. They have identified at least one protein that is produced by *S. pyogenes*, an NAD glycohydrolase (SPN), which is translocated from the bacterial cell to the cytoplasm of the keratinocyte. This process appears to be directed since the majority of the SPN is translocated into the cytoplasm of the keratinocyte. Since SPN is secreted by a type II secretion system across the cell membrane of the bacterial cell it is presumably translocated from the bacterial cell surface to the cytoplasm of the keratinocyte via the SLO channel. Since it appears that this system can facilitate the direct transfer of proteins from the gram positive bacterial cell to the eukaryotic cell, Madden *et al.* [45] have suggested that it may be the functional equivalent of the gram negative bacterial type III secretion system in a gram positive organism. It is likely that this system will yield many more interesting aspects of the translocation process in this organism since it requires a certain degree of coupling between the bacterial cell and the SLO translocation channel. Furthermore, these data suggest that the secretion of the SPN must be localised on the bacterial cell membrane in the vicinity of the SLO pore and that the SLO pore may form a tight junction with this region of the bacterial membrane.

2.6   SUMMARY AND PERSPECTIVES

Over the past five years numerous advances have significantly increased our understanding of the process of pore formation by the CDCs. Although these studies have provided several key insights into the mechanism of pore formation by the CDCs, these and other recent studies have resulted in more questions about this process. Future studies should reveal even more intriguing aspects of the CDC mechanism, particularly regarding the role of cholesterol, cellular specificity of the CDCs, the forces that drive membrane insertion and what factors affect the size of an insertion-competent prepore. Furthermore, the *in vivo* studies with SLO, PFO and LLO all suggest that the large channel formed by the CDCs can be used for processes other than cell lysis. The recent discovery that the streptolysin O pore may function as an analogue to a type III secretion channel for a Gram positive bacterium demonstrates that the large pore generated by the CDCs can be used by bacteria in unexpected ways. It is clear that the CDCs have evolved to fulfil a variety of functions for Gram positive bacterial pathogens and it is likely that, with time, investigators may discover that bacteria have other uses for these functionally adaptable proteins.

REFERENCES

[1] Smith, L.D.S. (1975) *The Pathogenic Anaerobic Bacteria.* Springfield, Illinois, Charles C. Thomas.

[2] Mitsui, K., T. Sekiya, *et al.* (1979) 'Alteration of human erythrocyte plasma membranes by perfringolysin O as revealed by freeze-fracture electron microscopy Studies on *Clostridium perfringens* exotoxins V.' *Biochim. Biophys. Acta* **554**, 68–75.

[3] Mitsui, K., T. Sekiya, *et al.* (1979) 'Ring formation of perfringolysin O as revealed by negative stain electron microscopy.' *Biochim. Biophys. Acta* **558**, 307–13.

[4] Bhakdi, S., J. Tranum-Jensen, *et al.* (1984). Structure of streptolysin-O in target membranes. *Bacterial Protein Toxins.* J.E. Alouf, F.J. Fehrenbach, J.H. Freer and J. Jeljaszewicz. London, Academic Press. **24**, 173–80.

[5] Shepard, L.A., O. Shatursky, *et al.* (2000). 'The mechanism of assembly and insertion of the membrane complex of the cholesterol-dependent cytolysin perfringolysin O: Formation of a large prepore complex.' *Biochemistry* **39**, 10284–93.

[6] Alouf, J.E. (1999) Introduction to the family of the structurally related cholesterol-binding cytolysins ('sulfhydryl-activated toxins'). *Bacterial Toxins: A Comprehensive Sourcebook.* J. Alouf and J. Freer. London, Academic Press, 443–56.

[7] Shepard, L.A., A.P. Heuck, *et al.* (1998) 'Identification of a membrane-spanning domain of the thiol-activated pore-forming toxin *Clostridium perfringens* perfringolysin O: an α-helical to β-sheet transition identified by fluorescence spectroscopy.' *Biochemistry* **37**, 14563–74.

[8] Shatursky, O., A.P. Heuck, *et al.* (1999) 'The mechanism of membrane insertion for a cholesterol dependent cytolysin: A novel paradigm for pore-forming toxins.' *Cell* **99**, 293–9.

[9] Sekino-Suzuki, N., M. Nakamura, *et al.* (1996) 'Contribution of individual tryptophan residues to the structure and activity of theta-toxin (perfringolysin o), a cholesterol-binding cytolysin.' *Eur. J. Biochem.* **241**, 941–7.

[10] Pinkney, M., E. Beachey, *et al.* (1989) 'The thiol-activated toxin streptolysin O does not require a thiol group for activity.' *Infect. Immun.* **57**, 2553–8.

[11] Saunders, K.F., T.J. Mitchell, *et al.* (1989) Pneumolysin, the thiol-activated toxin of *Streptococcus pneumoniae*, does not require a thiol group for in vitro activity. *Infection and Immunity.* **57**, 2547–52.

[12] Billington, S.J., B.H. Jost, *et al.* (1997) The *Arcanobacterium* (Actinomyces) *pyogenes* hemolysin, pyolysin, is a novel member of the thiol-activated cytolysin family *J. Bacteriol.* **179**, 6100–6.

[13] Nagamune, H., C. Ohnishi, *et al.* (1996) 'Intermedilysin, a novel cytotoxin specific for human cells secreted by *Streptococcus intermedius* UNS46 isolated from a human liver abscess.' *Infect. Immun.* **64**, 3093–100.

[14] Iwamoto, M., Y. Ohno-Iwashita, *et al.* (1987) 'Role of the essential thiol group in the thiol-activated cytolysin from *Clostridium perfringens.*' *Eur. J. Biochem.* **167**, 425–30.

[15] Rossjohn, J., S.C. Feil, *et al.* (1997) 'Structure of a cholesterol-binding thiol-activated cytolysin and a model of its membrane form.' *Cell* **89**, 685–92.

[16] Iwamoto, M., Y. Ohno-Iwashita, *et al.* (1990) 'Effect of isolated C-terminal fragment of theta-toxin (perfringolysin-O) on toxin assembly and membrane lysis.' *Eur. J. Biochem.* **194**, 25–31.

[17] Tweten, R.K., R.W. Harris, *et al.* (1991) Isolation of a tryptic fragment from *Clostridium perfringens* theta-toxin that contains sites for membrane binding and self-aggregation *J. Biol. Chemistry.* **266**, 12449–54.

[18] Nakamura, M., N. Sekino, *et al.* (1995) 'Interaction of theta-toxin (perfringolysin O), a cholesterol-binding cytolysin, with liposomal membranes: change in the aromatic side chains upon binding and insertion.' *Biochemistry* **34**, 6513–20.

[19] Heuck, A.P., E. Hotze, *et al.* (2000) 'Mechanism of membrane insertion of a multimeric β-barrel protein: Perfringolysin O creates a pore using ordered and coupled conformational changes.' *Molec. Cell* **6**, 1233–42.

[20] Jacobs, T., A. Darji, *et al.* (1998) 'Listeriolysin O: cholesterol inhibits cytolysis but not binding to cellular membranes.' *Mol. Microbiol.* **28**, 1081–9.

[21] Ohno-Iwashita, Y., M. Iwamoto, *et al.* (1988) 'Protease nicked θ-toxin of *Clostridium perfringens*, a new membrane probe with no cytolytic effect, reveals two classes of cholesterol as toxin-binding sites on sheep erythrocytes.' *Eur. J. Biochem.* **176**, 95–101.

[22] Ohno-Iwashita, Y., M. Iwamoto, *et al.* (1991) 'A cytolysin, theta-toxin, preferentially binds to membrane cholesterol surrounded by phospholipids with 18-carbon hydrocarbon chains in cholesterol-rich region.' *J. Biochem. (Tokyo)* **110**, 369–75.

[23] Waheed, A.A., Y. Shimada, *et al.* (2001) 'Selective binding of perfringolysin O derivative to cholesterol-rich membrane microdomains (rafts).' *Proc. Natl. Acad. Sci. USA* **17**, 17.

[24] Hotze, E.M., E.M. Wilson-Kubalek, *et al.* (2001) 'Arresting pore formation of a cholesterol-dependent cytolysin by disulfide trapping synchronizes the insertion of the transmembrane β-sheet from a prepore intermediate.' *J. Biol. Chem.* **276**, 8261–8.

[25] Olofsson, A., H. Hebert, *et al.* (1993) 'The projection structure of perfringolysin-O (*Clostridium Perfringens* theta-toxin).' *FEBS Lett.* **319**, 125–7.

[26] Sellman, B.R., B.L. Kagan, *et al.* (1997) 'Generation of a membrane-bound, oligomerized pre-pore complex is necessary for pore formation by *Clostridium septicum* alpha toxin.' *Molec. Microbiol.* **23**, 551–8.

[27] Walker, B., Braha, O. *et al.* (1995) 'An intermediate in the assembly of a pore-forming protein trapped with a genetically-engineered switch'. *Chem. Biol.* **2**, 99–105.

[28] Fang, Y., S. Cheley, *et al.* (1997) 'The heptameric prepore of a staphylococcal alpha-hemolysin mutant in lipid bilayers imaged by atomic force microscopy.' *Biochemistry* **36**, 9518–22.

[29] Miller, C.J., J.L. Elliot, *et al.* (1999) 'Anthrax protective antigen: prepore-to-pore conversion.' *Biochemistry* **38**, 10432–41.

[30] Palmer, M., R. Harris, *et al.* (1998) 'Assembly mechanism of the oligomeric streptolysin O pore: the early membrane lesion is lined by a free edge of the lipid membrane and is extended gradually during oligomerization.' *EMBO J.* **17**, 1598–1605.

[31] Portnoy, D., P.S. Jacks, *et al.* (1988) 'The role of hemolysin for intracellular growth of *Listeria monocytogenes*.' *J. Exp. Med.* **167**, 1459–71.

[32] Bielecki, J., P. Youngman, *et al.* (1990) '*Bacillus subtilis* expressing a haemolysin gene from *Listeria monocytogenes* can grow in mammalian cells.' *Nature* **345**, 175–6.

[33] Geoffroy, C., J.L. Gaillard, *et al.* (1987) 'Purification, characterization, and toxicity of the sulfhydryl-activated hemolysin listeriolysin O from *Listeria monocytogenes*.' *Infect. Immun.* **55**, 1641–6.

[34] Jones, S. and D.A. Portnoy (1994) 'Characterization of *Listeria monocytogenes* pathogenesis in a strain expressing perfringolysin O in place of listeriolysin O.' *Infect. Immun.* **62**, 5608–13.

[35] Decatur, A.L. and D.A. Portnoy (2000) 'A PEST-like sequence in listeriolysin O essential for *Listeria monocytogenes* pathogenicity.' *Science* **290**, 992–5.

[36] Rechsteiner, M. and S.W. Rogers (1996) 'PEST sequences and regulation by proteolysis.' *Trends Biochem. Sci.* **21**, 267–71.

[37] Rubins, J.B. and E.N. Janoff (1998) 'Pneumolysin – a multifunctional pneumococcal virulence factor.' *J. Lab. Clin. Med.* **131**, 21–7.

[38] Paton, J.C., K.B. Rowan, *et al.* (1984) 'Activation of human complement by the pneumococcal toxin pneumolysin.' *Infect. Immun.* **43**, 1085–7.

[39] Stevens, D.L., J. Mitten, *et al.* (1987) 'Effects of alpha and theta toxins from *Clostridium perfringens* on human polymorphonuclear leukocytes.' *J. Infect. Dis.* **156**, 324–33.

[40] Bryant, A.E., R. Bergstrom, *et al.* (1993) '*Clostridium perfringens* invasiveness is enhanced by effects of theta toxin upon PMNL structure and function: the roles of leukocytotoxicity and expression of CD11/CD18 adherence glycoprotein.' *FEMS Immunol Med Microbiol* **7**, 321–36.

[41] Stevens, D.L. and A.E. Bryant (1993) 'Role of theta-toxin, a sulfhydryl-activated cytolysin, in the pathogenesis of clostridial gas gangrene.' *Clin. Infect. Dis.* **16**, S195–9.

[42] Awad, M.M., A.E. Bryant, *et al.* (1995) 'Virulence studies on chromosomal alpha-toxin and theta-toxin mutants constructed by allelic exchange provide genetic evidence for the essential role of alpha-toxin in *Clostridium perfringens*-mediated gas gangrene.' *Mol Microbiol* **15**, 191–202.

[43] Ellemor, D.M., R.N. Baird, *et al.* (1999) 'Use of genetically manipulated strains of *Clostridium perfringens* reveals that both alpha-toxin and theta-toxin are required for vascular leukostasis to occur in experimental gas gangrene.' *Infect Immun* **67**, 4902–7.

[44] Okada, N., A.P. Pentland, *et al.* (1994) 'M protein and protein F act as important determinants of cell-specific tropism of *Streptococcus pyogenes* in skin tissue.' *J. Clin. Invest.* **94**, 965–77.

[45] Madden, J.C., N. Ruiz, *et al.* (2001) 'Cytolysin-mediated translocation (CMT): a functional equivalent of type III secretion in gram-positive bacteria.' *Cell* **104**, 143–52.

3   Toxin-producing Dinoflagellates

MICHAEL JAMES HOLMES and RICHARD LEWIS*

Dinoflagellates are a group of microscopic, mostly unicellular algae (protists) that can be found from tropical to polar seas. Most species are planktonic and marine, forming an important component of the phytoplankton in the world's oceans. They have a fossil record stretching back only to the Mesozoic; however, biochemical evidence suggests that ancestors of dinoflagellates swam in early Cambrian seas some 500 million years ago [1, 2]. There are about 2000 living species of dinoflagellates with approximately half of these being photosynthetic [3]. All toxin-producing species, with the exception of *Phalacroma rotundatum* and *Pfiesteria piscicida*, are photosynthetic or mixotrophic. However, *Pfiesteria piscicida* can function as a mixotroph using cleptochloroplasts [4].

Only a small number of dinoflagellate species produce toxins that kill aquatic life or accumulate through food chains to cause human poisoning. Toxin-producing species are still being discovered, especially with the increased research emphasis in the last 20 years on benthic dinoflagellates. Anderson and Lobel [5] observed that the proportion of toxic species is greater among free-living, benthic species than planktonic forms. This generalisation is still valid, although, many of the recently described species have not been tested for toxicity. In this review we have listed the marine dinoflagellates known to produce toxins that are lethal to mammals, concentrating on benthic species and those species that produce toxins that cause human poisoning. Compiling a list of toxin-producing dinoflagellates is not a simple process [6]. Firstly, it begs the question; what is a toxin? Figure 3.1 shows the structures of toxins of dinoflagellate origin, or likely dinoflagellate origin, that are known to reach levels that cause human illness. There are also increasing numbers of bioactive compounds being discovered from dinoflagellates that are toxic only when injected into animals, or tested *in vitro*. Bioactive compounds produced by dinoflagellates that have not (yet) been shown to cause any clear adverse health effects in humans are shown in Figure 3.2.

Another growing problem in compiling a list of toxin-producing dinoflagellates is the frequent name changes that have occurred for many species. This greatly complicates the literature for non-taxonomists (most toxinologists). For example, the continual re-evaluation of detailed thecal plate morphology in

---

* Corresponding author

Perspectives in Molecular Toxinology
Edited by A Ménez © 2002 John Wiley & Sons, Ltd

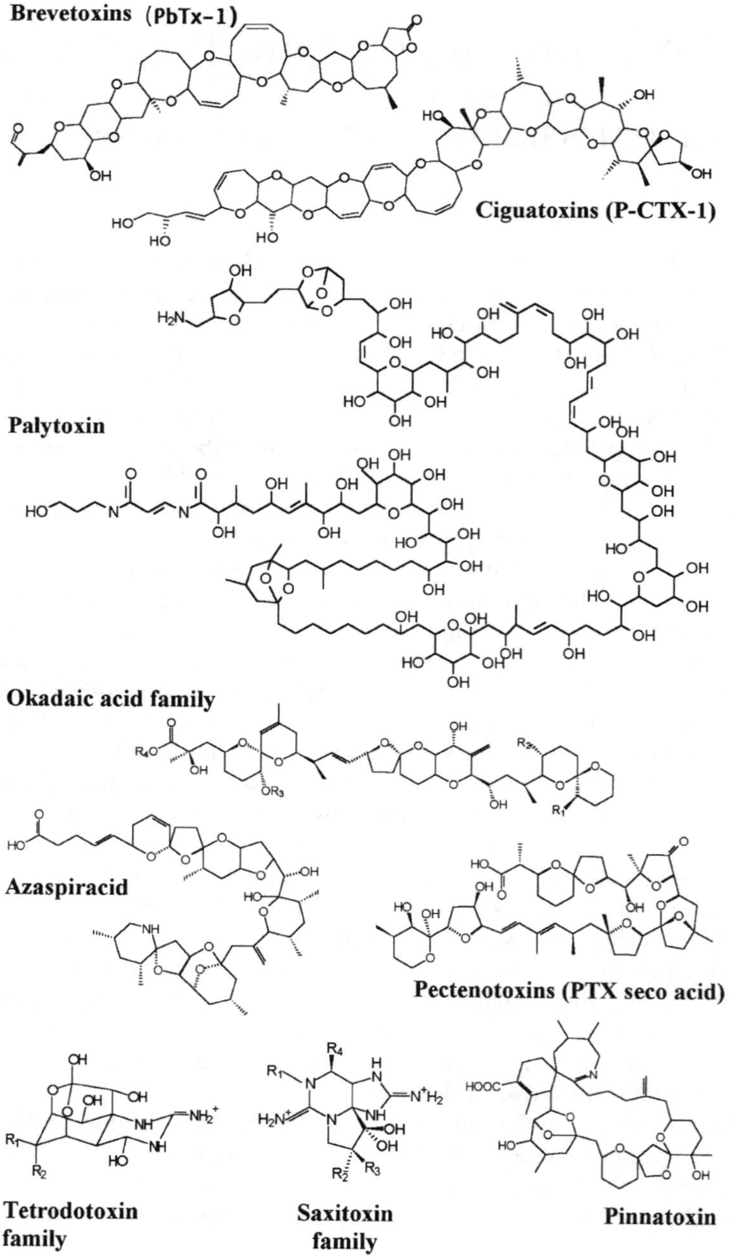

Figure 3.1   Major families of dinoflagellate toxins with identified health risks to humans. The origins of the azaspiracids and pinnatoxins are not known, but may be dinoflagellates based on chemistry

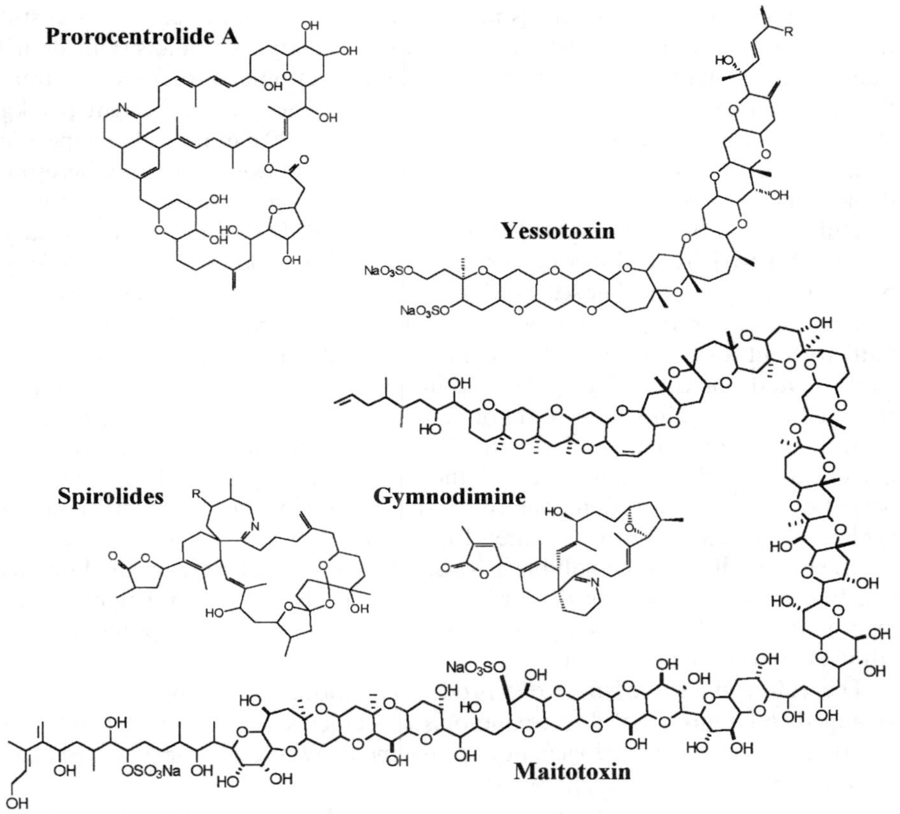

Figure 3.2 Dinoflagellate toxins with minimal or uncertain health effects on humans

Alexandrium has left a legacy of confusing taxonomic designations not easy for newcomers to the field to comprehend [7]. This difficulty arises because the species concept is not easy to apply to dinoflagellates (see [8, 9]). Historically they have been defined in terms of 'morphospecies' but this concept may not always be sufficient to delineate cryptic species (genospecies) or alternatively may reflect morphological plasticity that cannot be reconciled with genetic differences. For example, some currently valid species are suspected to be conspecific (e.g. *Alexandrium tamarense* and *A. fundyense*) and toxin production can vary between strains (infraspecific variation as in *A. tamarense*) [7, 10]. Caution is therefore necessary in interpreting the following species list and the reader is referred to [2, 3, 9, 11, 12] for more information on species definitions and taxonomy of dinoflagellates.

For known toxins, chemical, immunological, pharmacological or instrumental methods are usually preferred over animal bioassays because of their higher precision and sensitivity, as well as for ethical reasons. However, for new or poorly characterised toxins this is not always possible, and mice are often still required as test animals to determine potential toxicity to humans. Our definition of a likely mammalian toxin is based upon injection of a fraction or extract that is lethal to mice at a maximum dose of 1 g dried extract weight per kg mouse body weight (i.e. 1 g.kg^{-1} or 20 mg extract per 20 g mouse). Non-specific lethality (e.g. from non-toxic lipids) can occur at even lower doses and therefore the presence of toxicity, especially of crude extracts, has to be interpreted with careful observation of the signs displayed by animals [13]. Caution is necessary to interpret toxicity at doses higher than 1 g.kg^{-1} and descriptions of toxicity based upon injection of doses as high as 7 g.kg^{-1} are not surprising and not informative. Literature that reports toxicity of new toxins as 'cells per lethal unit' without also specifying the weight of extract injected are also of limited value. In addition to specifying the dose injected, the following details should be reported in any toxin study: route of administration (usually intraperitoneally = i.p.), the strain of mice used (usually an outbred strain such as Swiss or Quackenbush), the weight range of the mice used (often 18–21 g), the time period of observation (often over 24 – 48 hours for slow-acting toxins) and the vehicle of injection, such as water for water-soluble toxins or a non-toxic detergent or oil for lipid-soluble toxins. The likely oral potency to humans can be further assessed by dosing extracts perorally, as a number of toxins (e.g. maitotoxins) and toxic lipids often have significantly lower potency when administered by the oral route.

The following list of 62 toxin-producing dinoflagellate species has been arranged to be most useful to toxinologists (see below) so it does not follow a strictly taxonomic (or phylogenetic) arrangement. Where appropriate, the classification of Fensome et al. [2, 9] is used.

(a) The species are first separated into those whose motile (flagellated) form is predominantly planktonic or benthic. These are the two major habits of toxin-producing dinoflagellates.

(b) Under these two categories, species are then subdivided into unarmoured (athectae) and armoured (thecate) forms. This is a useful characteristic that has been traditionally used to separate dinoflagellates at the light microscope level. However, ultra-structure studies have shown that species previously thought devoid of armour can have thin optically undetectable thecal plates.

(c) They are then grouped on the basis of Order and Family, corresponding to the major body forms within the dinoflagellates (e.g. dinophysoids, gymnodinoids, prorocentroids etc.).

(d) Under Family, the genera and species are arranged alphabetically. Species producing the same structural class of toxins are grouped together followed by species within the same genus that produce other toxins. Species producing poorly characterised toxins or non-mammalian toxins are listed last.

At the end of the list, the species are cross-referenced to the major toxin classes produced by dinoflagellates (Table 3.1).

Table 3.1 Origin, potency, regulatory level, and pharmacology of major classes of dinoflagellate toxins.

Class of toxin	Origin[a]	Maximum potency[b] (i.p. mice, μg/kg)	Regulatory action limit for seafood safety (in meat)	Mode of action
Saxitoxins	*Gymnodinium catenatum* (1), *Pyrodinium bahamense* var. *compressum* (34), *Alexandrium* spp. (20–31)	6 (STX)	80 μg STX equivalents/100 g (Philippines 40 μg STX equivalents/100 g to protect children)	Inhibit sodium channels (site 1)
Tetrodotoxins	*Alexandrium tamarense* (29)	10 (TTX)	None enacted (human poisoning mainly from accidental consumption)	Inhibit sodium channels (site 1)
Ciguatoxins	*Gambierdiscus* spp. (42?, 43, 45, 46)	0.25 (P-CTX-1)	None enacted yet (estimated at $\geq$ 0.05 ppb)	Activate sodium channels (site 5)
Brevetoxins	*Karenia brevis* (2)	100 (PbTx-1)	USA $\geq$ 20 mouse units/100 g (may need revision (Dickey *et al.* 1999))	Activate sodium channels (site 5)
Okadaic acids	*Dinophysis* spp. (10–18), *Prorocentrum* spp. (53–58)	160 (DTX-1)	Variable, generally 20 μg okadaic acid equivalents/100 g. Europe to set new limit of 16 μg/100 g	Inhibit protein phosphatases, especially PP1 and PP2A
Palytoxin	*Ostreopsis siamensis* (52)	0.17 (Palytoxin)	None enacted yet	Creates non-selective cation channel at the Na$^+$/K$^+$ ATPase
Pectenotoxins	*Dinophysis* spp. (11, 13, 14)	250 (PTX-1)	None enacted yet, PTX2-SA causes human poisoning	Hepatotoxins
Prorocentrolides	*Prorocentrum* spp. (57–58)	400 (Prorocentrolide A)	None enacted. No evidence for human poisoning	?

continues overleaf

Table 3.1 (continued)

Class of toxin	Origin[a]	Maximum potency[b] (i.p. mice, μg/kg)	Regulatory action limit for seafood safety (in meat)	Mode of action
Yessotoxins	*Lingulodinium polyedrum* (35), *Protoceratium reticulatum* (36)	100 (YTX)	None enacted. Involvement in human poisoning not clear	?
Maitotoxins	*Gambierdiscus* spp.? (42–47?)	0.05 (MTX-1)	None enacted. Involvement in human poisoning (if any) restricted to consumption of certain herbivorous fish	Create a nonselective cation channel
Spirolides	*Alexandrium ostenfeldii* (28)	?	None enacted. Involvement in human poisoning not clear	?
Gymnodimines	'*Gymnodinium*' sp. (?)	450 (Gymnodimine)	None enacted. Levels detected in shellfish not poisonous to humans	?

[a] Numbers in parentheses refer to the species number in the species list
[b] Abbreviation in parentheses refers to the actual toxin associated with this LD_{50}. In some cases, this most potent form is not produced by the dinoflagellate, with only less toxic analogues or precursors being synthesised by the dinoflagellate

3.1 REVIEW OF TOXIN-PRODUCING DINOFLAGELLATES

3.1.1 DINOFLAGELLATES WITH MOTILE CELL PREDOMINANTLY PLANKTONIC

Unarmoured Dinoflagellates

Gymnodinoids (Order Gymnodiniales, Family Gymnodiniaceae but see [14])

1. *Gymnodinium catenatum* Graham 1943: Paralytic shellfish poisoning (PSP) toxins collectively know as 'saxitoxins' [15]. There are currently 21 structural analogues of saxitoxin comprising the known PSP toxins; although, this number is expected to increase soon [16]. These 21 toxins can be separated into four structural families depending upon the chemical substitutions; six carbamate toxins (including the parent structure, saxitoxin), six N-sulfocarbamoyl toxins, six decarbamoyl toxins and three deoxy-decarbamoyl toxins [17–19]. The PSP-toxin-producing dinoflagellates (*G. catenatum, Pyrodinium bahamense* var. *compressum* and *Alexandrium* spp.) do not produce all 21 toxins but only a subset of toxins that can vary between species/strains. The first reports of *G. catenatum* causing PSP came from Spain, the west coast of Mexico, Southern Japan and Southern Australia but the dinoflagellate has since been found in South America and the tropical Indo-West Pacific [20]. To date, the PSP toxin profile of all *G. catenatum* isolates has been dominated by the less-potent N-sulfocarbamoyl toxins [19].
2. *Karenia brevis* (Davis) G. Hansen and Moestrup 2000: (synonyms: *Gymnodinium breve, Ptychodiscus brevis*) Produces brevetoxins (PbTX or BTX) [21–24] of which there are two structural families based on 'A' (e.g. PbTx-2) and 'B' (e.g. PbTx-1) polyether ring systems. Structures for many PbTx analogues (13) have been determined from dinoflagellates and shellfish from the USA [24] and New Zealand [25] and recent research suggests that this number will soon increase [26, 27]. To date, nine analogues (BTX-1 to -3 and -5 to -10) have been extracted from cultures of *K. brevis* from the USA [28]. Shellfish accumulate and metabolise these toxins [26] to cause neurotoxic shellfish poisoning (NSP). The brevetoxins also kill fish [29].

Toxic Gymnodinoids with poorly characterised toxins
3. *Cochlodinium polykrikoides* Margalef 1961: (synonym: *C. heterolobatum*) Kills fish [30] possibly through production of superoxide radicals [31].
4. *'Gymnodinium' pulchellum* J. Larsen 1994: (Taxonomic position not clarified see [14]) Fish killer (Onue and Nozawa [32] as *Gymnodinium* type '84 K, see [33, 34]).
5. *Karenia brevisulcata* (Chang) G. Hansen and Moestrup 2000: (synonym: *Gymnodinium brevisulcatum*) Kills fish and invertebrates, also causes respiratory symptoms in humans in New Zealand [35].

6. *Karenia digitata* Yang, Takayama, Matsuoka and Hodgkiss 2000: Fish kills in east Asia [36].
7. *Karenia mikimotoi* (Miyake and Kominami ex Odata) G. Hansen and Moestrup 2000: (synonyms: *Gymnodinium mikimotoi, G. nagasakiense*) Fish killer, galactolipids [37].
8. *Karlodinium micrum* (Leadbeater and Dodge) J. Larsen 2000: (synonyms: *Woloszynskia micra, Gymnodinium micrum, Gymnodinium galatheanum, Gyrodinium galatheanum*) Fish killer [38].
9. *Karlodinium veneficum* (Ballantine) J. Larsen 2000: (synonym: *Gymnodinium veneficum*) Kills fish and invertebrates [8].

Gymnodinium aureolum (Hulbert) G. Hansen 2000 (synonym *Gyrodinium aureolum*) is a dinoflagellate described in the literature as a fish-killing species but this may be attributable to *Karenia mikimotoi*. There are also a number of new 'Gymnodinoid' species from New Zealand recently shown to produce analogues of brevetoxin or gymnodimine [39–41]. Shellfish accumulate (and metabolise?) these brevetoxin analogues to cause NSP analogous to *K. brevis* in the Gulf of Mexico [25, 27, 29, 42]. Shellfish can also accumulate gymnodimines but these are believed to be at concentrations non-toxic to humans [43].

Armoured Dinoflagellates

Dinophysoids (Order Dinophysiales, Family Dinophysiaceae)

Genus *Dinophysis* Ehrenberg 1839 and *Phalacroma* Stein 1883 (some authorities merge *Phalacroma* with *Dinophysis* see [12]). Species of *Dinophysis* are the origin of okadaic acid and its analogues, the major diarrhetic shellfish poisoning (DSP) toxins that contaminate shellfish around the world [44, 45]. Pectenotoxin-2 (PTX-2) is a hepatotoxin also produced by some *Dinophysis* species [46] but as yet there is no evidence this toxin causes human poisoning. PTX-2 can be metabolised in shellfish to a range of pectenotoxins [47]. However, the closely related PTX-2 seco acids do cause human poisoning [48]. Not all authorities consider the pectenotoxins (or yessotoxin) should be classified as part of the DSP family of toxins [49].

Production of toxins by *Dinophysis* species is determined from analysis of wild cells because these species cannot yet be grown in culture. Because these studies often extract phytoplankton net samples containing mixtures of species, there is the possibility of wrongly attributing toxin production. A recent hypothesis by Imai and Nishitani [50] that *D. acuminata* and *D. fortii* are intrinsically non-toxic and only accumulate DSP toxins from ingestion of toxic picophytoplankton should be investigated. This (or strain differences) may be an explanation of the considerable variability of toxicity among *Dinophysis* spp. [51].

10. *Dinophysis acuminata* Claparède and Lachmann1859 (synonyms: *D. borealis, D. lachmanii, D. boehmi*): Produces okadaic acid [46] and 35-methyl okadaic acid also known as dinophysistoxin-1 (DTX-1) [52], although, okadaic acid generally dominates [53, 54].

11. *Dinophysis acuta* Ehrenberg 1839: Produces okadaic acid and DTX-1 [46] and isomers of okadaic acid called DTX-2 [55] and DTX-2C [56]. Also pectenotoxin-2-seco acid (PTX2SA) and 7-*epi*-PTX2SA [57]. DTX-2 can sometimes be the dominant toxin [58].

12. *Dinophysis caudata* Saville-Kent 1881 (synonym *D. homunculis*): Only low concentrations of okadaic acid so far detected [59].

13. *Dinophysis fortii* Pavillard 1923: Produces DTX-1 [44], okadaic acid and PTX-2 [46].

14. *Dinophysis norvegica* Claparède and Lachmann1859: Okadaic acid and DTX-1 [46], PTX-2 and PTX2SA [60].

15. *Dinophysis sacculus* Stein 1883: Probable producer of okadaic acid [61]. This species is morphologically similar to *D. acuminata* [62].

16. *Dinophysis tripos* Gourret 1883: DTX-1 [46], morphologically similar to *D. caudata* but DTX-1 not detected from the latter species [59].

17. *Phalacroma mitra* Schütt 1895: DTX-1 [46].

18. *Phalacroma rotundatum* (Claparède and Lachmann) Kofoid and Michener 1911 (synonym: *D. whittingae*): DTX-1 [46], although Cembella [63] found no DSP toxins from North American cells. Wright and Cembella [51] consider the original report suspect, as this is the only facultative heterotrophic dinophysoid species that produces toxins.

Species with uncharacterised toxins
19. *Dinophysis miles* Cleve 1900: Philippines cells reported toxic [64].

Gonyaulacoids (Order Gonyaulacales, Family Goniodomaceae)

Genus *Alexandrium* Halim (synonyms: *Protogonyaulax, Gessnerium, Gonyaulax*) revised by Balech [65] but see cautions by Taylor and Fukuyo [66], and Scholin [7]. *Alexandrium* species are the origin of most PSP contamination of shellfish around the world [45, 67]. The taxonomy of this group is confusing and requires considerable expertise. A number of currently valid species are thought to be conspecific with the relatedness of isolates often dependent upon geographical proximity. For example, Scholin [7] suggests that some *A. tamarense, A. catenella* and *A. fundyense* from North America share a high degree of genetic similarity despite their morphological differences, whereas, some *A. tamarense* from Europe and North America can be morphologically identical but genetically distinct.

Species that produce paralytic shellfish poison (PSP) toxins. (Note: we have not reported the specific PSP toxins produced by each of these species.)

20. *Alexandrium acatenella* (Whedon and Kofoid) Balech 1985 (synonym: *Gonyaulax acatenella*).
21. *Alexandrium andersonii* Balech 1990: Reported as non-toxic from the east coast of North America [68, 69] but isolates from Italy recently reported to produce PSP toxins [70].
22. *Alexandrium angustitabulatum* Taylor 1995.
23. *Alexandrium catenella* (Whedon and Kofoid) Balech 1985 (synonym: *Gonyaulax catenella, Gessnerium acatenellum, Protogonyaulax acatenella*).
24. *Alexandrium cohorticula* (Balech) Balech 1985 (synonyms: *Gonyaulax cohorticula, Gessnerium cohorticula, Protogonyaulax cohorticula*): Reportedly a PSP producer from South East Asia [71, 72] but Taylor *et al.* [64] suggests caution as this species is very similar to *A. tamiyavanichi*.
25. *Alexandrium fundyense* Balech 1985 (possibly conspecific with *A. tamarense*).
26. *Alexandrium lusitanicum* Balech 1985 (possibly conspecific with *A. minutum*).
27. *Alexandrium minutum* Halim 1960 (synonym *A. ibericum*).
28. *Alexandrium ostenfeldii* (Paulsen) Balech and Tangen 1985 (synonyms: *Goniodoma ostenfeldii, Goniaulax tamarensis, Goniaulax ostenfeldii, Heteraulacus ostenfeldi, Gonyaulax globosa, Gonyaulax trygvei, Protogonyaulax globosa, Gessnerium ostenfeldii, Triadinium ostenfeldii*): A weak producer of PSP toxins [73]. The species also produces 'fast-acting toxins' called spirolides. A range of spirolides (e.g. A, C, D and 13-desmethyl C) has been isolated from *A. ostenfeldii* with additional analogues isolated from Nova Scotia shellfish [74, 75]. The potential for spirolides to cause human poisoning is not known although they are structurally similar to the pinnatoxins that have caused major shellfish poisoning events in Japan and China [76].
29. *Alexandrium tamarense* (Lebour) Balech 1992 (synonyms: *Gonyaulax tamarensis* var. *excavata, Gonyaulax excavata, Gessnerium tamarensis, Protogonyaulax tamarensis, Alexandrium excavatum*): Toxic and non-toxic strains exist [7]. Also suggested to be able to produce tetrodotoxin [77].
30. *Alexandrium tamiyavanichi* Balech 1994 (synonym: *Protogonyaulax cohorticula*).

Possible PSP producer:
31. *Alexandrium fraterculis* (Balech) Balech 1985 (synonyms: *Gonyaulax fratercula, Gessnerium fraterculum, Protogonyaulax fratercula*): Noguchi *et al.* [78] found this species to be non-toxic but Balech [65] reported its coincidence with toxic mussels in Uruguay when no other suspect organisms were present.

Alexandrium spp suggested toxic in the absence of PSP toxin production:

32. *Alexandrium monilatum* (Howell) F.J.R. Taylor 1979 (synonyms: *Gonyaulax monilata, Gessnerium monilata, G. mochimaensis, Pyrodinium monilatum*): Fish killer but toxin(s) poorly characterised [79].

33. *Alexandrium pseudogonyaulax* (Biecheler) Horiguchi ex Kita and Fukuyo 1992 (synonym: *Goniodoma pseudogoniaulax*): Produces the macrolide goniodomin-A [80]. Murakami *et al.* [81] subsequently refer to this species as *A. hiranoi* Kita and Fukuyo 1988 and consider *Goniodoma pseudogoniaulax* a junior synonym but Balech [65] considers *A. pseudogonyaulax* and *A. hiranoi* are separate species.

Other species:

34. *Pyrodinium bahamense* var. *compressum* (Böhm) Steidinger, Testor and Taylor 1980: Produces PSP toxins that contaminate shellfish and clupeid fish in South East Asia and the west coast of Central America [82, 83]. Western Pacific Ocean isolates of this species produce mainly gonyautoxin 5, saxitoxin and/or neosaxitoxin [84, 85]. A nontoxic, bioluminescent Atlantic Ocean variety exists called *P. bahamense* var. *bahamense* [12].

Gonyaulacoids (Order Gonyaulacales, Family Gonyaulacaceae)

35. *Lingulodinium polyedrum* (Stein) Dodge 1989: (synonym: *Gonyaulax polyedra*): Produces yessotoxin (YTX) and homoYTX [86, 87]. These toxins are sometimes classified with the toxins that cause diarrhetic shellfish poisoning (DSP). However, their involvement in human poisoning is not clear and they do not cause diarrhea in experimental animals [49, 88].

36. *Protoceratium reticulatum* (Claparède and Lachmann) Bütschli 1885: (synonyms: *P. aceros, Gonyaulax grindleyi*): YTX [89].

Peridinoids (Order Peridinales, Family Pfiesteriaceae) Phantom dinoflagellates (Note: Steidinger *et al.* [4] place these organisms in the Order Dinamoebales)

Genus *Pfiesteria* Steidinger and Burkholder

37. *Pfiesteria piscicida* Steidinger and Burkholder 1996: Kills fish [90] and invertebrates [91] and also linked to contact poisoning and aerosolised toxins harmful to humans [92]. No evidence for accumulation of toxins into seafood.

There are a number of morphologically similar species known to exist and a second toxic species *Pfiesteria shumwayae* sp. nov. is being described [93].

Prorocentroids (Order Prorocentrales, Family Prorocentraceae)

Genus *Prorocentrum* Ehrenberg (synonym *Exuviaella*). Most toxin-producing species of *Prorocentrum* are benthic. The only confirmed toxin-producing planktonic species is *P. cordatum*. The toxicity of *P. balticum* is uncertain [12].
38.   *Prorocentrum cordatum* (Ostenfeld) Dodge 1901: see [94] (synonyms: *P. minimum, P. mariae-lebouriae*): Neurotoxicity [95] and fish kills.

3.1.2   DINOFLAGELLATES WITH MOTILE CELL PREDOMINANTLY BENTHIC

Unarmoured Dinoflagellates

Gymnodinoids (Order Gymnodiniales, Family Gymnodiniaceae but see [14])

Genus *Amphidinium* Claparède and Lachmann. A polyphyletic genus [14]. Species in this genus are symbiotic, benthic or planktonic or move between the latter two environments. A range of macrolide structures and analogues (e.g. amphidinolides A–T, amphidinin A, caribenolide I, luteophanols, colopsinol) have been isolated from unidentified *Amphidinium*.
39.   *Amphidinium carterae* Hulbert 1957: Produces cytotoxic/hemolytic compounds [64, 96].
40.   *Amphidinium operculatum* Claparède and Lachmann 1859: (synonym: *A. klebsii*) Produces a range of potent antifungal and haemolytic compounds called amphidinols, e.g. amphidinol 1 [97].

Armoured Dinoflagellates

Gonyaulacoids (Order Gonyaulacales, Family Goniodomaceae)

Genus *Coolia* Meunier. Two species, *Coolia areolata* Ten-Hage, Turquet, Quod and Couté 2001 and *C. tropicalis* Faust 1995 have not yet been tested for toxicity.
41.   *Coolia monotis* Meunier 1919. Most strains of this species are apparently non-toxic [98, 99] but a neurotoxin called cooliatoxin was purified from an Australian strain [100]. The reference by Nakajima *et al.* [99] has often been used as evidence for toxicity of this species but this paper only describes a minor haemolytic activity. Extracts and liquid–liquid partitions (see [100]) from a Singapore clone of *C. monotis* were not lethal to mice at $1 \, \mathrm{g.kg^{-1}}$, and electrospray mass spectrometry analysis of extracts partially purified on a silica gel chromatography column did not detect any cooliatoxin (Holmes, unpublished result). The toxic Australian clone

extracted by Holmes *et al.* [100] had identical plate morphology and architecture as *C. monotis*, but the cells were relatively large and possibly more spherical (less anterior–posterior compression) than has been described for this species [101, 102].

Genus *Gambierdiscus* Adachi and Fukuyo

42. *Gambierdiscus toxicus* Adachi and Fukuyo 1979: First linked to ciguatera by Yasumoto *et al.* [103, 104]. Produces ciguatoxin (CTX) analogues but this varies with the strain [105, 106; but see comments below on recently described *Gambierdiscus* species). The CTXs produced by this dinoflagellate are accumulated and biotransformed through marine food chains into fishes to cause ciguatera [107]. Ciguatoxins produced by Pacific Ocean strains of '*G. toxicus*' include CTX-3C [108] and CTX-4A, also known as scaritoxin [109], both of which can be accumulated into fish. The toxin originally thought to be produced by '*G. toxicus*' (CTX-4B, initially coded GT4B, [110]) is formed after epimerisation of CTX-4A. CTX-4A (also known as P-CTX-4A) is the principal progenitor of the major ciguatoxins found in Pacific Ocean fishes. In terms of total toxicity, the major toxin in ciguateric fishes from the Pacific is P-CTX-1 [111–113] and from the Caribbean is C-CTX-1 [114]. These toxins have not been found in dinoflagellates. P-CTX-1 is formed through oxidative metabolism in fishes of CTX-4A via CTX-4B and CTX-3 or via CTX-2 and CTX-3 [112, 115]. P-CTX-2 and P-CTX-3 are also common in fish but are less toxic than P-CTX-1 [112, 116]. The origin of the precursor(s) of C-CTX-1 is presumed to be Caribbean strains of *Gambierdiscus* but this is yet to be proven. Yasumoto *et al* [117] recently described structural elucidation of 16 Pacific ciguatoxin congeners only available in microgram quantities, including 5 produced by *Gambierdiscus*. The contribution of these minor ciguatoxins to the ciguatera disease is unclear, although they may explain some of the variability of disease symptoms. Gambierol is another toxic CTX congener [118] produced by *Gambierdiscus* but not yet found in fishes.

Gambierdiscus toxicus also produces maitotoxin, the most potent marine toxin known [119]. Maitotoxin was first isolated from surgeon-fishes [120], but only from the liver and gut/stomach contents [121]. There is no evidence for the accumulation of maitotoxin into the flesh of fish to cause human poisoning, although it has often been cited as part of the ciguatera syndrome. Without accumulation into the flesh of fish, the only possible involvement of maitotoxin in human poisoning would be in societies that eat the viscera of herbivorous fishes. Holmes *et al.* [122] found that different strains of the dinoflagellate produce different maitotoxins; although, this may be a function of different species (see comments below).

Gambieric acids are non-toxic, potent antifungal compounds also isolated from '*G. toxicus*' [123]

All of the toxin chemistry discussed above was carried out before the discovery of the morphologically similar species *G. pacificus*, *G. australes* and *G. polynesiensis* by Chinain *et al.* [124], and could reflect isolation of toxins from one or more of these similar species. The specific link between CTX production and *G. toxicus* is, once again, open to speculation. The identity of the *Gambierdiscus* strain used for isolation of CTX-4A, CTX3C and maitotoxin (MTX-1) by Yasumoto's group especially needs clarification. Chinain *et al.* [124] found production of ciguatoxins by cultures of *G. australes*, *G. pacificus* and *G. polynesiensis* but not *G. toxicus*. *Gambierdiscus belizeanus* is another recently discovered species but its aerolated plate morphology is distinctive, especially under SEM, and this species is unlikely to have been confused with any of the other species. However, its involvement in ciguatera is still an open question. Similarly, *G. yasumotoi* is a recently described species but its round cell morphology easily distinguishes it by light spectroscopy from all the other anterior–posteriorly compressed species.

Holmes *et al.* [105] extracted 13 cultured strains of putative *G. toxicus* for toxins. Re-examination of a limited number of SEM micrographs from this work indicates that amongst the nine Australian clones were species of what now would be called *G. toxicus*, *G. australes* and *G. polynesiensis* with ciguatoxins found only from cultures we now tentatively identify as *G. polynesiensis*. Chinain *et al.* [124] also reported highest CTX production from this species. The putatively mono-sulphated maitotoxin (MTX-2) isolated by Holmes *et al.* [122] may have been isolated from *G. australes* and the small di-sulphated MTX-3 [125, 126] may have come from *G. polynesiensis*.

43. *Gambierdiscus australes* Faust and Chinain 1999: Ciguatoxin(s) and maitotoxin(s) [124]. A likely origin of ciguatera.
44. *Gambierdiscus belizeanus* Faust 1995: Produces toxin(s) that induce maitotoxin-like signs in mice (Holmes, unpublished observations). Involvement in ciguatera not proven.
45. *Gambierdiscus pacificus* Chinain and Faust 1999: Ciguatoxin(s) and maitotoxins(s) [124]. A likely origin of ciguatera.
46. *Gambierdiscus polynesiensis* Chinain and Faust 1999: Ciguatoxin(s) and maitotoxin(s) [124]. A likely origin of ciguatera, especially in the western Pacific Ocean.
47. *Gambierdiscus yasumotoi* Holmes 1998: Unidentified toxin inducing signs in mice different from ciguatoxins and most similar to maitotoxins [127]. However, two bioassay signs in mice differentiate the *G. yasumotoi*-toxin from known maitotoxins: (a) absence of cyanosis produced by *G. yasumotoi*-toxin; and (b) sub-lethal concentrations of this toxin can cause limb paralysis. In contrast, doses of maitotoxin that cause limb paralysis are inevitably fatal (Holmes, unpublished observation). No evidence for accumulation of this type of toxin in ciguateric fish.

Genus *Ostreopsis* Schmidt. A number of *Ostreopsis* species have been shown to produce compounds that are lethal to mice, although, most of these are not well characterised. Toxicity of the following benthic species has not been reported, presumably because they have not been analysed: *Ostreopsis belizeanus Faust 1999, O. caribbeanus Faust 1999, O. labens Faust and Morton 1995 and O. marinus Faust 1999.*

48. *Ostreopsis heptagona* Norris, Bomber and Balech 1985: Methanol extracts of this species were described as having 'limited toxicity' [128] or weak toxicity to mice [129].

49. *Ostreopsis lenticularis* Fukuyo 1981: Produces a number of toxins called ostreotoxins that are lethal to mice but most are not well characterised [130–133].

50. *Ostreopsis mascarenensis* Quod 1994: Crude methanol extracts lethal to mice [134].

51. *Ostreopsis ovata* Fukuyo 1981: A butanol-soluble fraction was lethal to mice [99] but the toxicity was so low ($> 2 \times 10^8$ cells per lethal 'mouse unit') that this species could be considered as nontoxic. Methanol extracts of a Singapore clone were not lethal to mice at a maximum dose of 1 g.kg^{-1} [135].

52. *Ostreopsis siamensis* Schmidt 1902: Produces a palytoxin analogue called ostreocin D [136]. Palytoxin analogues are the cause of clupeotoxism [137]. This is a separate disease to ciguatera that can be distinguished both on the basis of causative toxins and clinical examination. (The term clupeotoxism has also been applied to poisoning from eating clupeid fishes that have accumulated PSP toxins [66]).

Prorocentroids (Order Prorocentrales, Family Prorocentraceae)

Genus *Prorocentrum* Ehrenberg (synonym *Exuviaella*). Many new benthic species have been recently described. The genus has both planktonic and benthic species but most planktonic species do not produce toxins. Many of the benthic species produce okadaic acid and analogues of okadaic acid, the toxins responsible for diarrhetic shellfish poisoning (DSP). However, this does not necessarily indicate that the toxins produced by these species contaminate seafood species to cause human poisoning. This has to be proven or reasonable evidence provided to indicate the likelihood for each species. To date, this has only been shown for *P. lima* [138]. In addition, with many new species being described in recent years, there is the potential for toxin production being attributed to the wrong species, as occurred with okadaic acid being attributed to *P. concavum* when Dickey *et al.* [139] extracted what turned out to be a new species subsequently described by Faust [140] as *P. maculosum* [141]. So many new species have been described in the last 20 years that early reports of toxin production by *Prorocentrum* species need to be interpreted carefully with respect to the implicated species.

Benthic *Prorocentrum* spp. have also been implicated in ciguatera, although the evidence for this is mostly speculative. Gamboa *et al.* [142] reported finding okadaic acid from a Caribbean barracuda involved in causing ciguatera but did not quantify the toxin concentration. Okadaic acid and its analogues are poorly toxic compounds requiring ingestion of 30–50 μg of toxin to cause mild human poisoning [143–145]. Assuming an individual consumed a large meal of 200 g of fish fillet, poisoning would require that the fish-flesh have a concentration exceeding $\sim 0.2\,\mu\text{g.g}^{-1}$ or 0.2 ppm. In contrast, P-CTX-1 can cause ciguatera at concentrations greater than $0.1\,\text{ng.g}^{-1}$ or 0.1 ppb [115], a concentration three-orders of magnitude lower. It seems likely that lipid-soluble toxins like okadaic acid can accumulate in certain fish tissues but the extent of this accumulation is unknown. Lipid-soluble toxins are sometimes assumed to be capable of accumulating into body tissues but this generalisation should not be assumed for all marine animals. For example, okadaic acid and DXT-1 only accumulate to significant levels in the digestive tissues of mussels with the remaining tissues generally containing little toxin [146]. In the case of shellfish, this distinction is generally not important because apart from scallops, the soft tissues of most shellfish are eaten whole.

Toxicity has not been reported in the following benthic species, most of which have have not been analysed: *Prorocentrum caribbaeum* Faust 1993, *P. clipeus* Hoppenrath 2000, *P. foraminosum* Faust 1993, *P. formosum* Faust 1993, *P. panamensis* Grzebyk, Sako and Berland 1998, *P. reticulatum* Faust 1997, *P. ruetzlerianum* Faust 1990, *P. sabulosum* Faust 1994, *P. sculptile* Faust 1994 and *P. tropicalis* Faust 1997. The benthic species *P. emarginatum* Fukuyo 1981, *P. elegans* Faust 1993 and *P. norrisianum* Faust 1997 were recently reported to be non-toxic [147].

Species that produce okadaic acid or analogues
53. *Prorocentrum arenarium* Faust 1994: Indian Ocean strains produce okadaic acid and possibly low concentrations of DTX-1 and an isomer of okadaic acid called DTX-2 [148]. Non-toxic strains also reported [147].
54. *Prorocentrum belizeanum* Faust 1993: Okadaic acid and low concentrations of DTX-1 [147, 149].
55. *Prorocentrum faustiae* Morton 1998: Okadaic acid and DTX-1 [150].
56. *Prorocentrum hoffmannianum* Faust 1990: Okadaic acid and a fast-acting toxin [151] and hoffmanniolide, a non-toxic macrolide [152].
57. *Prorocentrum lima* (Ehrenberg) Dodge (Faust 1991): Okadaic acid [153], DTX-2 [154], DTX-1 [46] and an isomer of DTX-1 called DTX-1B [155]. Also a sulphated ester of okadaic acid (DTX-4) [156], diol esters of okadaic acid [154, 157], 2- and 7-deoxy-okadaic acid [158, 159], and a fast-acting toxin, probably prorocentrolide [160].
58. *Prorocentrum maculosum* Faust 1993: Okadaic acid ([139] but misidentified as *P. concavum* see [141]), diol esters of okadaic acid ([154]: cited as *P. concavum* but the culture strain extracted was the same as the strain isolated by Dickey *et al.* [139]; Quilliam, personal communication), sulphated esters of okadaic acid (DTX-5a, b) [161], and a fast-acting toxin called prorocentrolide B [162].

Species that produce toxins other than okadaic acid derivatives

59. *Prorocentrum borbonicum* Ten-Hage, Turquet, Quod, Puiseux-Dao and Coute 2000: Neurotoxicity but no DSP toxins found [163].
60. *Prorocentrum cassubicum* (Woloszynska) Dodge 1975: Uncharacterised toxins [164].
61. *Prorocentrum concavum* Fukuyo 1981: Extracts nontoxic to mice but cytotoxic and icthyotoxic fractions [165]. Reports misattributing production of okadaic acid [139] and diol esters of okadaic acid [154] were extracted from cultures of what is now thought to be *P. maculosum* [141].
62. *Prorocentrum mexicanum* Tafall 1942 (synonyms: *P. maximum*, *P. rhathymum*): Pacific Ocean strains nontoxic to mice and fish but weak haemolytic activity [99, 157]. An uncharacterised toxin lethal to mice found in Caribbean strains [166], cytotoxicity also reported [147].

A polar, lipid-soluble toxin called spiro-prorocentrimine was recently purified from a benthic, non-identified *Prorocentrum* species from Taiwan [167].

Others

The pinnatoxins are shellfish toxins that have caused human poisoning in Asia [74]. They are structurally similar to the spirolides produced by *Alexandrium ostenfeldii* [75] but the origin of the pinnatoxins is not known. Azaspiracids are a new class of marine toxins that accumulate in shellfish to cause azaspiracid shellfish poisoning (AZP) in humans [168]. The origin of these toxins, based upon their structure, is also thought to be a dinoflagellate.

REFERENCES

[1] Moldowan, J.M. and Talyzina, N.M. (1998) Biogeochemical evidence for dinoflagellate ancestors in the early Cambrian. *Science* **281**, 1168–70.
[2] Fensome, R.A., Saldarriaga, J.F. and Taylor, F.J.R. (1999) Dinoflagellate phylogeny revisited: reconciling morphological and molecular based phylogenies. *GRANA* **38**, 66–80.
[3] Taylor, F.J.R. (ed.) (1987) *The Biology of Dinoflagellates*. Blackwell Scientific, Oxford.
[4] Steidinger, K.A., Burkholder, J.A., Glasgow, H.B., Hobbs, C.W., Garrett, J.K., Truby, E.W., Noga, E.J. and Smith, S.A. (1996) *Pfiesteria piscicida* gen. et sp. nov. (Pfiesteriacea fam. nov.). a new toxic dinoflagellate with a complex life cycle and behavior. *J. Phycol.* **32**, 157–64.
[5] Anderson, D.M. and Lobel, P.S. (1987) The continuing enigma of ciguatera. *Biol. Bull.* **172**, 89–107.
[6] Sournia, A. (1995) Red tide and toxic marine phytoplankton of the world ocean: an inquiry into biodiversity. In: *Harmful Marine Algal Blooms*. Lassus P, Arzul G, Erard E, Gentien P and Marcaillou C (eds), pp. 103–112, Lavoisier, New York.

[7] Scholin, C.A. (1998) Morphological, genetic, and biogeographical relationships of the toxic dinoflagellates *Alexandrium tamarense*, *A catenella* and *A. fundyense*. In: *Physiological Ecology of Harmful Algal Blooms*. Anderson DM, Cembella and Hallegraeff GM (eds), pp. 13–27, Springer, Berlin.

[8] Taylor, F.J.R. (1985) The taxonomy and relationships of red tide dinoflagellates. In: *Toxic Dinoflagellates*. Anderson DM, White AW and Baden DG (eds), pp. 11–26, Elsevier, New York.

[9] Fensome, R.A., Taylor, F.J.R., Norris, G., Sarjeant, W.A.S., Wharton, D.I. and Williams, G.L. (1993) A classification of living and fossil dinoflagellates. *Micropaleontology*, Special publication number 7, 351 pp.

[10] Costas, E., Zardoya, R., Baustista, J., Garrido, A., Rojo, C. and López-Rodas, V. (1995) Morphospecies vs. geno-species in toxic marine dinoflagellates: an analysis of *Gymnodinium catenatum/Gyrodinium impudicum* and *Alexandrium minutum/A. lusitanicum* using antibodies, lectins, and gene sequences. *J. Phycol.* 31, 801–7.

[11] Taylor, F.J.R. (1993) The species problem and its impact on harmful phytoplankton studies. In: *Toxic Phytoplankton Blooms in the Sea*. Smayda TJ and Shimizu Y (eds), pp. 81–6, Elsevier, Amsterdam.

[12] Steidinger, K.A. and Tangen, K. (1997) Dinoflagellates. In: *Identifying Marine Phytoplankton*. Tomas CR (ed.), pp. 387–584, Academic Press, San Diego.

[13] Takagi, T., Hayashi, K. and Itabashi, Y. (1984) Toxic effect of free unsaturated fatty acids in the mouse assay of diarrhetic shellfish toxin by intraperitoneal injection. *Bull. Jap. Soc. Sci. Fish.* 50, 1413–1418.

[14] Daugbjerg, N., Hansen, G., Larsen, J. and Moestrup, Ø. (2000) Phylogeny of some of the major genera of dinoflagellates based on ultrastructure and partial LSU rDNA sequence data, including the erection of three new genera of unarmoured dinoflagellates. *Phycologia* 39, 302–17.

[15] Anderson, D.M., Sullivan, J.J. and Reguera, B. (1989) Paralytic shellfish poisoning in northwest Spain: the toxicity of the dinoflagellate *Gymnodinium catenatum*. *Toxicon* 27, 665–74.

[16] Negri, A.P., Bolch, C.J.S., Llewellyn, L.E. and Mendez, S. (2000) Paralytic shellfish toxins in *Gymnodinium catenatum* strains from six countries. In: *9th International Conference on Harmful Algal Blooms*, Hobart, Tasmania, 7–11 February 2000, conference abstracts p. 187.

[17] Bordner, J., Thiessen, W.E., Bates, H.A. and Rapoport, H. (1975) The structure of a crystalline derivative of saxitoxin. The structure of saxitoxin. *J. Am. Chem. Soc.* 97, 6008–6012.

[18] Shimizu, Y., Hsu, C-P., Fallon, W.E., Oshima, Y., Miura, I. and Nakanishi, K. (1978) Structure of neosaxitoxin. *J. Am. Chem. Soc.* 100, 6791–3.

[19] Oshima, Y., Blackburn, S.I. and Hallegraeff, G.M. (1993) Comparative study on paralytic shellfish toxin profiles of the dinoflagellate *Gymnodinium catenatum* from three different countries. *Mar. Biol.* 116, 471–6.

[20] Hallegraeff, G.M. and Fraga, S. (1998) Bloom dynamics of the toxic dinoflagellate *Gymnodinium catenatum*, with emphasis on Tasmanian and Spanish coastal waters. In: *Physiological Ecology of Harmful Algal Blooms*. Anderson DM, Cembella AD and Hallegraeff GM (eds), pp. 59–80, Springer, Berlin.

[21] Lin, Y-Y., Risk, M., Ray, M.S., Van Engen, D., Clardy, J., Golik, J., James, J.C. and Nakanishi, K. (1981) Isolation and structure of brevetoxin B from the 'red tide' dinoflagellate *Ptyochodiscus brevis*. *J. Am. Chem. Soc.* 103, 6773–5.

[22] Shimizu, Y., Chou, H-N., Bando, H., Van Duyne, G. and Clardy, J.C. (1986) Structure of brevetoxin A (GB-1), the most potent toxin in the Florida red tide organism *Gymnodinium breve* (*Ptychodiscus brevis*). *J. Am. Chem. Soc.* 108, 514–15.

[23] Pawlak, J., Tempesta, M.S., Golik, J., Zagorski, M.G., Lee, M.S., Nakanishi, K., Iwashita, T., Gross, M.L. and Tomer, K.B. (1987) Structure of brevetoxin A as constructed from NMR and MS data. *J. Am. Chem. Soc.* **109**, 1144–50.

[24] Baden, D.G. (1989) Brevetoxins: unique polyether dinoflagellate toxins. *FASEB J.* **3**, 1807–17.

[25] Ishida, H., Nozawa, A., Totoribe, K., Muramatsu, N., Nukaya, H., Tsuji, K., Yamaguchi, K., Yasumoto, T., Kaspar, H., Berkett, N. and Kosuge, T. (1995) Brevetoxin B1, a new polyether marine toxin from the New Zealand shellfish, *Austrovenus stutchburyi. Tetrahedron Lett.* **36**, 725–8.

[26] Dickey, R., Jester, E., Granade, R., Mowdy, D., Moncreiff, C., Rebarchik, D., Robi, M., Musser, S. and Poli, M. (1999) Monitoring brevetoxins during a *Gymnodinium breve* red tide: comparison of sodium channel specific cytotoxicity assay and mouse bioassay for determination of neurotoxic shellfish toxins in shellfish extracts. *Natural Toxins* **7**, 157–65.

[27] Morohashi, A., Satake, M., Naoki, H., Kaspar, F., Oshima, Y. and Yasumoto, T. (1999) Brevetoxin B4 isolated from Greenshell mussels *Perna canaliculis*, the major toxin involved in neurotoxic shellfish poisoning in New Zealand. *Natural Toxins* **7**, 45–8.

[28] Baden, D.G. and Adams, D.J. (2000) Brevetoxins: chemistry, mechanism of action and methods of detection. In: *Seafood and Freshwater Toxins: Pharmacology, Physiology and Detection.* Botana LM (ed.), pp. 505–32, Marcel Dekker, New York.

[29] Steidinger, K.A., Vargo, G.A., Tester, P.A. and Tomas, C.R. (1998) Bloom dynamics and physiology of *Gymnodinium breve* with emphasis on the Gulf of Mexico. In: *Physiological Ecology of Harmful Algal Blooms.* Anderson DM, Cembella AD and Hallegraeff GM (eds), pp. 133–53, Springer, Berlin.

[30] Kim, H.G. (1998) *Cochlodinium polykrikoides* blooms in Korean coastal waters and their mitigation. In: *Harmful Algae.* Reguera B, Blanco J, Fernández ML and Wyatt T (eds), pp. 227–8, Xunta de Galacia and International Oceanographic Commission of UNESCO.

[31] Kim, C.S., Lee, S.G., Lee, C.K., Kim, H.G. and Jung, J. (1999) Reactive oxygen species as causative agents in the ichthyotoxicity of the red tide dinoflagellate *Cochlodinium polykrikoides. J. Plankton Res.* **21**, 2105–15.

[32] Onoue, Y. and Nozawa, K. (1989) Separation of toxins from harmful red tides occurring along the coast of Kagoshima Prefecture. In: *Red Tides: Biology, Environmental Science, and Toxicology.* Okaichi T, Anderson DM and Nemoto T (eds), pp. 371–4, Elsevier, New York.

[33] Larsen, J. (1994) Unarmoured dinoflagellates from Australian waters I. The genus *Gymnodinium* (Gymnodiniales, Dinophyceae). *Phycologia* **33**, 24–33.

[34] Steidinger, K.A., Landsberg, J.H., Truby, E.W. and Roberts, B.S. (1998) First report of *Gymnodinium pulchellum* (Dinophyceae) in North America and associated fish kills in the Indian River, Florida. *J. Phycol.* **34**, 431–7.

[35] Chang, F.H. (1999) *Gymnodinium brevisulcatum* sp. nov. (Gymnodiniales, Dinophyceae), a new species isolated from the 1998 summer toxic bloom in Wellington Harbour, New Zealand. *Phycologia* **38**, 377–4.

[36] Yang, Z.B., Takayama, H., Matsuoka, K. and Hodgkiss, I.J. (2000) *Karenia digitata* sp. nov. (Gymnodiniales, Dinophyceae), a new harmful algal bloom species from the coastal waters of west Japan and Hong Kong. *Phycologia* **39**, 463–70.

[37] Yasumoto, T., Underdal, B., Aune, T., Hormazabal, V., Skulberg, O.M. and Oshima, Y. (1990) Screening of hemolytic and ichthyotoxic components of *Chrysochromulina polylepis* and *Gymnodinium aureolum* from Norwegian coastal waters. In: *Toxic Marine Phytoplankton.* Granéli E, Sundström L, Edler L and Anderson DM (eds), pp. 436–40, Elsevier, New York.

[38] Nielsen, M.V. (1993) Toxic effect of the marine dinoflagellate *Gymnodinium galatheanum* on juvenile cod *Gadus morhua*. *Mar. Ecol. Prog. Ser.* **95**, 273–7.

[39] Seki, T., Satake, M., Mackenzie, L., Kaspar, H. and Yasumoto, T. (1995) Gymnodimine, a new marine toxin of unprecedented structure isolated from New Zealand oysters and the dinoflagellate, *Gymnodinium* sp. *Tetrahedron Lett.* **36**, 7093–6.

[40] Haywood, A., Mackenzie, L., Garthwaite, I. and Towers, N. (1996) *Gymnodinium breve* 'look-alikes': three *Gymnodinium* isolates from New Zealand. In: *Harmful and Toxic Algal Blooms*. Yasumoto T, Oshima Y and Fukuyo Y (eds), pp. 227–30, International Oceanographic Commission of UNESCO, Paris.

[41] Miles, C.O., Wilkens, A.L., Stirling, D.J. and MacKenzie, A.L. (2000) New analogue of gymnodimine from a *Gymnodinium* species. *J. Agric. Food Chem.* **48**, 1373–6.

[42] MacKenzie, L., Rhodes, L., Till, D., Chang, F.H., Kaspar, H., Haywood, A., Kapa, J. and Walker, B. (1995) A *Gymnodinium* sp. bloom and contamination of shellfish with lipid soluble toxins in New Zealand, Jan–April 1993. In: *Harmful Marine Algal Blooms*. Lassus P, Azrul G, Erard E, Gentien P and Marcaillou C (eds), pp. 795–800, Lavoisier, New York.

[43] Mackenzie, L., Haywood, A., Adamson, J., Truman, P., Till, D., Seki, T., Satake, M. and Yasumoto, T. (1996) Gymnodimine contamination of shellfish in New Zealand. In: *Harmful and Toxic Algal Blooms*. Yasumoto T, Oshima Y and Fukuyo Y (eds), pp. 97–100, International Oceanographic Commission of UNESCO, Paris.

[44] Murata, M., Shimatani, M., Sugitani, H., Oshima, Y. and Yasumoto, T. (1982) Isolation and structural elucidation of the causative toxin of the diarrhetic shellfish poisoning. *Bull. Jap. Soc. Sci. Fish.* **48**, 549–52.

[45] Hallegraeff, G.M. (1995) Harmful algal blooms: a global overview. In: *Manual on Harmful Marine Microalgae*. Haellegraeff GM, Anderson DM and Cembella AD (eds), pp. 1–22, International Oceanographic Commission of UNESCO, Paris.

[46] Lee, J-S., Igarashi, T., Fraga, S., Dahl, E., Hovgaard, P. and Yasumoto, T. (1989) Determination of diarrhetic shellfish toxins in various dinoflagellate species. *J. App. Phycol.* **1**, 147–52.

[47] Yasumoto, T., Murata, M., Lee, J-S. and Torigoe, K. (1989) Polyether toxins produced by dinoflagellates. In: *Mycotoxins and Phycotoxins '88*. Natori S, Hasimoto K and Ueno Y (eds), pp. 375–82, Elsevier, Amsterdam.

[48] Quilliam, M., Eaglesham, G., Hallegraeff, G., Quaine, J., Curtis, J., Donald, R. and Nunez, P. (2000) Detection and identification of toxins associated with a shellfish poisoning incident in New South Wales, Australia. In: *9th International Conference on Harmful Algal Blooms*, Hobart, Tasmania, 7–11 February 2000, conference abstracts p 48.

[49] Fernández, M.L., Miguez, A., Cacho, E. and Marinez, A. (1996) Sanitary control of marine biotoxins in the European Union. National references laboratories network. In: *Harmful and Toxic Algal Blooms*. Yasumoto T, Oshima Y and Fukuyo Y (eds), pp. 11–14, International Oceanographic Commission of UNESCO, Paris.

[50] Imai, I. and Nishitani, G. (2000) Attachment of picophytoplankton to the cell surface of the toxic dinoflagellates *Dinophysis acuminata* and *D. fortii*. *Phycologia* **39**, 456–9.

[51] Wright, J.L.C. and Cembella, A.D. (1998) Ecophysiology and biosynthesis of polyether marine biotoxins. In: *Physiological Ecology of Harmful Algal Blooms*. Anderson DM, Cembella AD and Hallegraeff GM (eds), pp. 427–451, Springer, Berlin.

[52] Sato, S., Koike, K. and Kodama M (1996) Seasonal variation of okadaic acid and dinophysitoxin-1 in *Dinophysis* spp. in association with the toxicity of scallop. In: *Harmful and Toxic Algal Blooms*. Yasumoto T, Oshima Y and Fukuyo Y (eds), pp. 285–8, International Oceanographic Commission of UNESCO, Paris.

[53] Anderson, P., Hald, B. and Emsholm, H. (1996) Toxicity of *Dinophysis acuminata* in Danish coastal waters. In: *Harmful and Toxic Algal Blooms*. Yasumoto T, Oshima Y and Fukuyo Y (eds), pp. 281–4, International Oceanographic Commission of UNESCO, Paris.

[54] Johansson, N., Granéli, E., Yasumoto, T., Carlsson, P. and Legrand, C. (1996) Toxin production by *Dinophysis acuminata* and *D. acuta* cells grown under nitrogen sufficient and deficient conditions. In: *Harmful and Toxic Algal Blooms*. Yasumoto T, Oshima Y and Fukuyo Y (eds), pp. 285–8, International Oceanographic Commission of UNESCO, Paris.

[55] James, K.J., Bishop, A.G., Gillman, M., Kelly, S.S., Roden, C., Draisci, R., Lucentini, L., Giannetti, L. and Boria, P. (1997) Liquid chromatography with fluorimetric, mass spectrometric and tandem mass spectrometric detection for the investigation of the seafood-toxin-producing phytoplankton, *Dinophysis acuta*. *J. Chrom. A* **777**, 213–21.

[56] Draisci, R., Giannetti, L., Lucentini, L., Marchiafava, C., James, K.J., Bishop, A.G., Healy, B.M. and Kelly, S.S. (1998a) Isolation of a new okadaic acid analogue from phytoplankton implicated in diarrhetic shellfish poisoning. *J. Chrom. A* **798**, 137–45.

[57] Daiguji, M., Satake, M., James, K.J., Bishop, A., MacKenzie, L., Naoki, H. and Yasumoto, T. (1998) Structures of new pectenotoxin analogs, pectenotoxin-2 seco acid and 7-*epi*-pectenotoxin-2 seco acid, isolated from a dinoflagellate and Greenshell mussels. *Chemistry Letters* 653–4.

[58] Draisci, R., Lucentini, L., Giannetti, L., Marchiafava, C., James, K.J. and Kelly, S.S. (1998b) Diarrhoeic shellfish toxin profiles in phytoplankton and mussels from Italy and Ireland as determined by LC-MS and LC-MS-MS. In: *Harmful Algae*. Reguera B, Blanco J, Fernández ML and Wyatt T (eds), pp. 495–8, Xunta de Galicia and Intergovernmental Oceanographic Commission of UNESCO.

[59] Holmes, M.J., Teo, L.M., Lee, F.C. and Khoo, H.W. (1999) Persistent low concentrations of diarrhetic shellfish toxins in green mussels *Perna viridis* from the Johor Strait, Singapore: first record of diarrhetic shellfish toxins from South-East Asia. *Mar. Ecol. Prog. Ser.* **181**, 257–68.

[60] Goto, H., Igarashi, T., Watai, M., Yasumoto, T., Gomez, O.V., Valdivia, G.L., Noren, F., Gisselson, L-A. and Graneli, E. (2000) Worldwide occurrence of pectenotoxins and yessotoxins in shellfish and phytoplankton. In: *9th International Conference on Harmful Algal Blooms*, Hobart, Tasmania, 7–11 February 2000, conference abstracts p 20.

[61] Masselin, P., Lassus, P. and Bardouil, M. (1992) High performance liquid chromatography analysis of diarrhetic toxins in *Dinophysis* spp. from the French coast. *J. App. Phycol.* **4**, 385–9.

[62] Bravo, I., Reguera, B. and Fraga, S. (1995) Description of different morphotypes of *Dinophysis acuminata* complex in the Galacian Rias Baixas in 1991. In: *Harmful Marine Algal Blooms*. Lassus P, Arzul G, Erard E, Gentien P and Marcaillou C (eds), pp.21–6, Lavoisier, New York.

[63] Cembella, A.D. (1989) Occurrence of okadaic acid, a major diarrhetic shellfish toxin, in natural populations of *Dinophysis* spp. from the eastern coast of North America. *J. App. Phycol.* **1**, 307–10.

[64] Taylor, F.J.R., Fukuyo, Y. and Larsen, J. (1995) Taxonomy of harmful dinoflagellates. In: *Manual on Harmful Marine Microalgae*. Hallegraeff GM, Anderson DM and Cembella A (eds), pp. 283–317, IOC Manuals and guides No. 33, UNESCO, Paris.

[65] Balech, E. (1995) The genus *Alexandrium* Halim (Dinoflagellata). Sherkin Island Marine Station Special Publications.

[66] Taylor, F.J.R. and Fukuyo, Y. (1998) The neurotoxic dinoflagellate genus *Alexandrium* Halim: general introduction. In: *Physiological Ecology of Harmful Algal*

Blooms. Anderson DM, Cembella AD and Hallegraeff GM (eds), pp. 3–11, Springer, Berlin.

[67] Anderson, D.M., Cembella, A.D. and Hallegraeff, G.M. (eds) (1998) *Physiological Ecology of Harmful Algal Blooms.* Springer, Berlin.

[68] Anderson, D.M., Kulis, D.M., Sullivan, J.J., Hall, S. and Lee, C. (1990) Dynamics and physiology of saxitoxin production by the dinoflagellate *Alexandrium* spp. *Mar. Biol.* **104**, 511–24.

[69] Cembella, A.D. (1998) Ecophysiology and metabolism of paralytic shellfish toxins in marine microalgae. In: *Physiological Ecology of Harmful Algal Blooms.* Anderson DM, Cembella AD and Hallegraeff GM (eds), pp. 381–403, Springer, Berlin.

[70] Ciminiello, P., Fattorusso, E., Forino, M. and Montressor, M. (2000) Saxitoxin and neosaxitoxin as toxic principles of *Alexandrium andersoni* (Dinophyceae) from the Gulf of Naples, Italy. *Toxicon* **38**, 1871–7.

[71] Fukuyo, Y., Yoshida, K., Ogata, T., Ishimaru, T., Kodama, M., Pholpunthin, P., Wisessang, S., Phanichyakarn, V. and Piyakarnchana, T. (1989) Suspected causative dinoflagellates of paralytic shellfish poisoning in the Gulf of Thailand. In: *Red Tides: Biology, Environmental Science and Toxicology.* Okaichi T, Anderson DM and Nemoto T (eds), pp. 403–6, Elsevier, New York.

[72] Usup, G., Kulis, D.M. and Teen, L.P. (2000) PSP toxin profiles in the dinoflagellate *Alexandrium cohorticula* and toxic mussels from Sebatu, Malacca. In: *9th International Conference on Harmful Algal Blooms*, Hobart, Tasmania, 7–11 February 2000, conference abstracts p 242.

[73] Hansen, P.J., Cembella, A.D. and Moestrup, Ø. (1992) The marine dinoflagellate *Alexandrium ostenfeldii*: paralytic shellfish toxin concentration, composition, and toxicity to a tintinnid ciliate. *J. Phycol.* **28**, 597–603.

[74] Cembella, A.D., Lewis, N.I. and Quilliam, M.A. (2000) The marine dinoflagellate *Alexandrium ostenfeldii* (Dinophyceae) as the causative organism of spirolide shellfish toxins. *Phycologia* **39**, 67–74

[75] Hu, T.M., Burton, I.W., Cembella, A.D., Curtis, J.M., Quilliam, M.A., Walter, J. A. and Wright, J.L.C. (2001) Characterization of spirolides A, C, and 13–desmethyl C, new marine toxins isolated from toxic plankton and contaminated shellfish. *J. Nat. Prod.* **64**, 308–12.

[76] Chou, T., Haino, T., Kuramoto, M. and Uemura, D. (1996) Isolation and structure of pinnatoxin D, a new shellfish poison from the Okinawan bivalve *Pinna muricata. Tetrahedron Lett.* **37**, 4027–30.

[77] Kodama, M., Sato, S., Sakamoto, S. and Ogata, T. (1996) Occurrence of tetrodotoxin in *Alexandrium tamarense*, a causative dinoflagellate of paralytic shellfish poisoning. *Toxicon* **34**, 1101–5.

[78] Noguchi, T., Maruyama, J., Ikeda, T. and Hashimoto, K. (1985) *Protogonyaulax fratercula* as a nontoxic plankton. *Bull. Jap. Soc. Sci. Fish.* **51**, 1373.

[79] Anderson, D.M. (1998) Physiology and bloom dynamics of toxic *Alexandrium* species, with emphasis on life cycle transitions. In: *Physiological Ecology of Harmful Algal Blooms.* Anderson DM, Cembella AD and Hallegraeff GM (eds), pp. 29–48, Springer, Berlin.

[80] Murakami, M., Makabe, K., Yamaguchi, K., Konosu, S., Walchli, M.R. (1988) Goniodomin-A, a novel polyether macrolide from the dinoflagellate *Goniodoma pseudogoniaulax. Tetrahedron Lett.* **29**, 1149–52

[81] Murakami, M., Okita, Y., Matsuda, H., Okino, T. and Yamaguchi, K. (1998) From the dinoflagellate *Alexandrium hiranoi. Phytochemistry.* **48**, 85–8.

[82] Harada, T., Oshima, Y., Kamiya, H. and Yasumoto, T. (1982) Confirmation of paralytic shellfish toxins in the dinoflagellate *Pyrodinium bahamense* var. *compressa* and bivalves in Palau. *Bull. Jap. Soc. Sci. Fish.* **48**, 821–4.

[83] Hallegraeff, G.M. and Maclean, J.L. (eds) (1989) Biology, epidemiology and management of *Pyrodinium* red tides. *ICLARM Conf. Proc.* 21, Manila.

[84] Oshima, Y. (1989) Toxins in *Pyrodinium bahamense* var. *compressum* and infested marine organisms. In: *Biology, Epidemiology and Management of Pyrodinium Red Tides*. Hallegraeff GM and Maclean JL (eds), pp. 73–9, ICLARM Conf. Proc. 21, Manila.

[85] Usup, G., Kulis, D.M. and Anderson, D.M. (1994) Growth and toxin production of the toxic dinoflagellate *Pyrodinium bahamense* var. *compressum* in laboratory cultures. *Natural Toxins* 2, 254–62.

[86] Tubaro, A., Sidari, L., Loggia, R.D. and Yasumoto, T. (1998) Occurrence of yessotoxin-like toxins in phytoplankton and mussels from northern Adriatic Sea. In: *Harmful Algae*. Reguera B, Blanco J, Fernández ML and Wyatt T (eds), pp. 470–2, Xunta de Galacia and International Oceanographic Commission of UNESCO.

[87] Draisci, R., Ferretti, E., Palleschi, L., Marchiafava, C., Poletti, R., Milandri, A., Ceredi, A. and Pompei, M. (1999) High levels of yessotoxin in mussels and presence of yessotoxin and homoyessotoxin in dinoflagellates of the Adriatic Sea. *Toxicon* 37, 1187–93.

[88] Terao, K., Ito, E., Oarada, M., Murata, M. and Yasumoto, T. (1990) Histopathological studies on experimental marine toxin poisoning – 5. The effects in mice of yessotoxin isolated from *Patinopecten yessoensis* and of a desulfated derivative. *Toxicon* 28, 1095–104.

[89] Satake, M., MacKenzie, L. and Yasumoto, T. (1997a) Identification of *Protoceratium reticulatum* as the biogenic origin of yessotoxin. *Natural Toxins* 5, 164–7.

[90] Burkholder, J.M., Noga, E.J., Hobbs, C.H. and Glasgow, H.B. (1992) New 'phantom' dinoflagellate is the causative agent of major estuarine fish kills. *Nature* 358, 407–10.

[91] Burkholder, J.M., Glasgow, H.B. and Hobbs, C.W. (1995) Fish kills linked to a toxic ambush-predator dinoflagellate: distribution and environmental condition. *Mar. Ecol. Prog. Ser.* 124, 43–61.

[92] Glasgow, H.B., Burkholder, J.M., Schmechel, D.E., Tester, P.A. and Rublee, P.A. (1995) Insidious effects of a toxic estuarine dinoflagellate on fish survival and human health. *J. Toxicol. Environ. Health* 46, 501–22.

[93] Quilliam, M.A. (2001) Phycotoxins. *J. AOAC* 84, 194–201.

[94] Velikova, V. and Larsen, J. (1999) The *Prorocentrum cordatum*/*Prorocentrum minimum* taxonomic problem. *GRANA* 38, 108–12.

[95] Grzebyk, D., Denardou, A., Berland, B. and Pouchus, Y. (1997) Evidence of a new toxin in the red-tide dinoflagellate *Prorocentrum minimum*. *J. Plankton Res.* 19, 1111–24.

[96] Ammar, M., Diogène, G., Fessard, V. and Puiseux-Dao, S. (1996) Cytotoxic tests for evaluation of toxicity associated with carcinogenic potential of some microalgal toxins. In: *Harmful and Toxic Algal Blooms*. Yasumoto T, Oshima Y and Fukuyo Y (eds), pp. 477–8, International Oceanographic Commission of UNESCO, Paris.

[97] Satake, M., Murata, M., Yasumoto, T., Fujita, T. and Naoki, H. (1991) Amphidinol, a polyhydroxypolyene antifungal agent with an unprecedented structure, from a marine dinoflagellate, *Amphidinium klebsii*. *J. Am. Chem. Soc.* 113, 9859–61.

[98] Yasumoto, Y., Oshima, Y., Murakami, Y., Nakajima, I., Bagnis, R. and Fukuyo, Y. (1980) Toxicity of benthic dinoflagellates found in coral reefs. *Bull. Jap. Soc. Sci. Fish.* 46, 327–31.

[99] Nakajima, I., Oshima, Y. and Yasumoto, T. (1981) Toxicity of benthic dinoflagellates in Okinawa. *Bull. Jap. Soc. Sci. Fish.* 47, 1029–33.

[100] Holmes, M.J., Lewis, R.J., Jones, A. and Wong Hoy, A. (1995) Cooliatoxin, the first toxin from *Coolia monotis* (Dinophyceae). *Natural Toxins* **3**, 355–62.

[101] Fukuyo, Y. (1981) Taxonomical study on benthic dinoflagellates collected in coral reefs. *Bull. Jap. Soc. Sci. Fish.* **47**, 967–78.

[102] Dodge, J.D. (1982) *Marine dinoflagellates of the British Isles*. Her Majesty's Stationery Office, London.

[103] Yasumoto, T., Bagnis, R., Thevenin, S. and Garcon, M. (1977a) A survey of comparative toxicity in the food chain of ciguatera. *Bull. Jap. Soc. Sci. Fish.* **43**, 1015–19.

[104] Yasumoto, T., Nakajima, I., Bagnis, R. and Adachi, R. (1977b) Finding of a dinoflagellate as a likely culprit of ciguatera. *Bull. Jap. Soc. Sci. Fish.* **43**, 1021–26.

[105] Holmes, M.J., Lewis, R.J., Poli, M.A. and Gillespie, N.C. (1991) Strain dependent production of ciguatoxin precursors (gambiertoxins) by *Gambierdiscus toxicus* (Dinophyceae) in culture. *Toxicon* **29**, 761–75.

[106] Satake, M., Ishimaru, T., Legrand, A-M. and Yasumoto, T. (1993a) Isolation of a ciguatoxin analog from cultures of *Gambierdiscus toxicus*. In: *Toxic Phytoplankton Blooms in the Sea*. Smayda TJ and Shimizu Y (eds), pp. 575–9, Elsevier, Amsterdam.

[107] Randall, J.E. (1958) A review of ciguatera, tropical fish poisoning, with a tentative explanation of its cause. *Bull. Mar. Sci.* **8**, 236–67.

[108] Satake, M., Murata, M. and Yasumoto, T. (1993b) The structure of CTX3C, a ciguatoxin congener isolated from cultured *Gambierdiscus toxicus*. *Tetrahedron Lett.* **34**, 1975–8.

[109] Satake, M., Ishibashi, Y., Legrand, A-M. and Yasumoto, T. (1997b) Isolation and structure of ciguatoxin-4a, a new ciguatoxin precursor, from cultures of dinoflagellate *Gambierdiscus toxicus* and Parrotfish *Scarus gibbus*. *Biosci. Biotech. Biochem.* **60**, 2103–5.

[110] Murata, M., Legrand, A.M., Ishibashi, Y., Fukui, M. and Yasumoto, T. (1990) Structures and configurations of ciguatoxin from moral eel *Gymnothorax javanicus* and its likely precursor from the dinoflagellate *Gambierdiscus toxicus*. *J. Am. Chem. Soc.* **112**, 4380–6.

[111] Murata, M., Legrand, A.M., Ishibashi, Y. and Yasumoto, T. (1989) Structures of ciguatoxin and its congener. *J. Am. Chem. Soc.* **111**, 8929–31.

[112] Lewis, R.J., Sellin, M., Poli, M.A., Norton, R.S., MacLeod, J.K. and Sheil, M. (1991) Purification and characterization of ciguatoxins from moray eel (*Lycodontis javanicis*, Muraenidae). *Toxicon* **29**, 1115–27.

[113] Satake, M., Morohashi, A., Oguri, H., Oishi, T., Hirama, M., Harada, N. and Yasumoto, T. (1997c) The absolute configuration of ciguatoxin. *J. Am. Chem. Soc.* **119**, 11325–6.

[114] Lewis, R.J., Vernoux, J-P. and Brereton, I.M. (1998) Structure of Caribbean ciguatoxin isolated from *Caranx latus*. *J. Am. Chem. Soc.* **120**, 5914–20.

[115] Lewis, R.J. and Holmes, M.J. (1993) Origin and transfer of toxins involved in ciguatera. *Comp. Biochem. Physiol.* **106C**, 615–28.

[116] Lewis, R. and Jones, A. (1997) Characterisation of ciguatoxins and ciguatoxin congeners present in ciguateric fish by gradient reverse phase HPLC/mass spectrometry. *Toxicon* **35**, 159–68.

[117] Yasumoto, T., Igarashi, T., Legrand, A-M,, Cruchet, P., Chinain, M., Fujita, T. and Naoki, H. (2000) Structural elucidation of ciguatoxin congeners by fast-atom bombardment tandem mass spectrometry. *J. Am. Chem. Soc.* **122**, 4988–9.

[118] Satake, M., Murata, M. and Yasumoto, T. (1993c) Gambierol – a new toxic polyether compound isolated from the marine dinoflagellate *Gambierdiscus toxicus*. *J. Am Chem. Soc.* **115**, 361–2.

[119] Murata, M., Naoki, H., Iwashita, T., Matsunaga, S., Sasaki, M., Yokoyama, A. and Yasumoto, T. (1993) Structure of maitotoxin. *J. Am. Chem. Soc.* **115**, 2060–2.

[120] Yasumoto, T., Bagnis, R. and Vernoux, J.P. (1976) Toxicities of the surgeonfishes – II Properties of the principal water-soluble toxin. *Bull. Jap. Soc. Sci. Fish.* **42**, 359–65.
[121] Yasumoto, T., Hashimoto, Y., Bagnis, R., Randall, J.E. and Banner, A.H. (1971) Toxicity of the surgeonfishes. *Bull. Jap. Soc. Sci. Fish.* **37**, 724–34.
[122] Holmes, M.J., Lewis, R.J. and Gillespie, N.C. (1990) Toxicity of Australian and French Polynesian strains of *Gambierdiscus toxicus* (Dinophyceae) grown in culture: characterization of a new type of maitotoxin. *Toxicon* **28**, 1159–72.
[123] Nagai, H., Torigoe, K., Satake, M., Murata, M., Yasumoto, T. and Hirota, H. (1992) Gambieric acids – unprecedented potent antifungal substances isolated from cultures of a marine dinoflagellate *Gambierdiscus toxicus*. *J. Am. Chem. Soc.* **114**, 1102–3.
[124] Chinain, M., Faust, M.A. and Pauillac, S. (1999) Morphology and molecular analyses of three toxic species of *Gambierdiscus* (Dinophyceae): *G. pacificus*, sp. nov., *G. australes*, sp. nov., and *G. polynesiensis*, sp. nov. *J. Phycol.* **35**, 1282–96.
[125] Holmes, M.J. and Lewis, R.J. (1994) Purification and characterization of large and small maitotoxins from cultured *Gambierdiscus toxicus*. *Natural Toxins* **2**, 64–72.
[126] Lewis, R.J., Holmes, M.J., Alewood, P.F. and Jones, A. (1994) Ionspray mass spectrometry of ciguatoxin-1, maitotoxin-2 and -3, and related marine polyether toxins. *Natural Toxins* **2**, 56–63.
[127] Holmes, M.J. (1998) *Gambierdiscus yasumotoi* sp. nov. (Dinophyceae), a toxic benthic dinoflagellate from southeastern Asia. *J. Phycol.* **34**, 661–8.
[128] Norris, D.R., Bomber, J.W. and Balech, E. (1985) Benthic dinoflagellates associated with ciguatera from the Florida Keys. I. *Ostreopsis heptagona* sp. nov. In: *Toxic Dinoflagellates*. Anderson DM, White AW and Baden DG (eds), pp. 39–44, Elsevier, New York.
[129] Faust, M.A., Morton, S.L. and Quod, J.P. (1996) Further SEM study of marine dinoflagellates: the genus *Ostreopsis* (Dinophyceae). *J. Phycol.* **32**, 1053–65.
[130] Tosteson, T.R., Ballantine, D.L., Tosteson, C.G., Bardales, A.T., Durst, H.D. and Higerd, T.B. (1986) Comparative toxicity of *Gambierdiscus toxicus*, *Ostreopsis* cf. *lenticularis*, and associated microflora. *Mar. Fish. Rev.* **48**, 57–9.
[131] Tindall, D.R., Miller, D.M. and Tindall, P.M. (1990) Toxicity of *Ostreopsis lenticularis* from the British and United States Virgin Islands. In: *Toxic Marine Phytoplankton*. Granéli E, Sundström L, Edler L and Anderson DM (eds), pp. 424–9, Elsevier, New York.
[132] Mercado, J.A., Viera, M., Tosteson, T.R., González, I., Silva, W. and Escalona de Motta, G. (1995) Differences in the toxicity and biological activity of *Ostreopsis lenticularis* observed using different extraction procedures. In: *Harmful Marine Algal Blooms*. Lassus P, Arzul G, Erard E, Gentien P and Marcaillou C (eds), pp. 321–6, Lavoisier, New York.
[133] Meunier, F., Mercado, J.A., Molgo, J., Tosteson, T.R. and Escalona de Motta, G. (1997) Selective depolarization of the muscle membrane in frog-nerve muscle preparations by a chromatographically purified extract of the dinoflagellate *Ostreopsis lenticularis*. *Brit. J. Pharmacol.* **121**, 1224–30.
[134] Quod, J.P. (1994) *Ostreopsis mascarenensis* sp. nov. (Dinophyceae), dinoflagellé toxique associé á la ciguatera dans L'Océan Indien. *Cryptogamnie Algol.* **15**, 243–51.
[135] Holmes, M.J., Lee, F.C., Teo, S.L.M. and Khoo, H.W. (1998) A survey of benthic dinoflagellates on Singapore reefs. In: *Harmful Algae*. Reguera B, Blanco J, Fernández ML and Wyatt T (eds), pp. 50–1, Xunta de Galicia and Intergovernmental Oceanographic Commission of UNESCO.
[136] Usami, M., Satake, M., Ishida, S., Inoue, A., Kan, Y. and Yasumoto, T. (1995) Palytoxin analogs from the dinoflagellate *Ostreopsis siamensis*. *J. Am. Chem. Soc.* **117**, 5389–90.

[137] Onuma, Y., Satake, M., Ukena, T., Roux, J., Chanteau, S., Rasolofonirina, N., Ratsimaloto, M., Naoki, H. and Yasumoto, T. (1999) Identification of putative palytoxin as the cause of clupeotoxism. *Toxicon* **37**, 55–65.

[138] Lawrence, J.E., Grant, J., Quilliam, M.A., Bauder, A.G. and Cembella, A.D. (2000) Colonization and growth of the toxic dinoflagellate *Prorocentrum lima* and associated fouling macroalgae on mussels in suspended culture. *Mar. Ecol. Prog. Ser.* **201**, 147–54.

[139] Dickey, R.W., Bobzin, S.C., Faulkner, D.J., Bencsath, F.A. and Andrzejewski, D. (1990) Identification of okadaic acid from a Caribbean dinoflagellate *Prorocentrum concavum*. *Toxicon* **28**, 371–7.

[140] Faust, M.A. (1993) Three new benthic species of *Prorocentrum* (Dinophyceae) from Twin Cays, Belize: *P. maculosum* sp. nov., *P. foraminosum* sp. nov. and *P. formosum* sp. nov. *Phycologia* **32**, 410–18.

[141] Zhou, J. and Fritz, L. (1993) Ultrastructure of two toxic marine dinoflagellates, *Prorocentrum lima* and *Prorocentrum maculosum*. *Phycologia* **32**, 444–50.

[142] Gamboa, P.M., Park, D.L. and Fremy, J-M. (1992) Extraction and purification of toxic fractions from barracuda (*Sphyraena barracuda*) implicated in ciguatera poisoning. In: *Proceedings of the Third International Conference on Ciguatera Fish Poisoning*, Puerto Rico 1990. Tosteson TR (ed.), pp. 13–24, Polyscience, Quebec.

[143] Yasumoto, T., Oshima, Y. and Yamaguchi, M. (1978) Occurrence of a new type of shellfish poisoning in the Tohoku district. *Bull. Jap. Soc. Sci. Fish.* **44**, 1249–55.

[144] Yasumoto, T., Murata, M., Oshima, Y., Sano, M., Matsumoto, G.K. and Clardy, J. (1985) Diarrhetic shellfish toxins. *Tetrahedron* **41**, 1019–25.

[145] Lee, J.S., Yanagi, T., Kenma, R. and Yasumoto, T. (1987) Fluorometric determination of diarrhetic shellfish toxins by high-performance liquid chromatography. *Agric. Biol. Chem.* **51**, 877–81.

[146] Stabell, O.B., Steffenak, I. and Aune, T. (1991) An evaluation of the mouse bioassay applied to extracts of 'diarrhoetic' shellfish toxins. *Fd. Chem. Toxic.* **30**, 139–44.

[147] Morton, S.L., Petitpain, D.L., Busman, M. and Moeller, P.D.R. (2000) Production of okadaic acid and Dinophysis toxins by different species of *Prorocentrum*. In: *9th International Conference on Harmful Algal Blooms*, Hobart, Tasmania, 7–11 February 2000, conference abstracts p 183.

[148] Ten-Hage, L., Delaunay, N., Pichon, V., Couté, A., Piuseux-Dao, S. and Turquet, J. (2000a) Okadaic acid production from the marine dinoflagellate *Prorocentrum arenarium* Faust (Dinophyceae) isolated from Europa Island coral reef ecosystem. *Toxicon* **38**, 1043–54.

[149] Morton, S.L., Moeller, P.D.R., Young, K.A. and Lanoue, B. (1998) Okadaic acid production from the marine dinoflagellate *Prorocentrum belizeanum* Faust isolated from the Belizean coral reef ecosystem. *Toxicon* **36**, 201–6.

[150] Morton, S.L. (1998) Morphology and toxicity of *Prorocentrum faustiae* sp. nov., a toxic species of non-planktonic dinoflagellate from Heron Island, Australia. *Bot. Mar.* **41**, 565–9.

[151] Aikman, K.E., Tindall, D.R. and Morton, S.L. (1993) Physiology and potency of the dinoflagellate *Prorocentrum hoffmannianum* (Faust) during one complete growth cycle. In: *Toxic Phytoplankton in the Sea*. Smayda TJ and Shimizu Y (eds), pp. 463–8, Elsevier, Amsterdam.

[152] Hu, T.M., Curtis, J.M., Walter, J.A. and Wright, J.L.C. (1999) Hoffmanniolide: a novel macrolide from *Prorocentrum hoffmannianum. Tetrahedron Lett.* **40**, 3977–80.

[153] Murakami, Y., Oshima, Y. and Yasumoto, T. (1982) Identification of okadaic acid as a toxic component of a marine dinoflagellate *Prorocentrum lima*. Bull. *Jap. Soc. Sci. Fish.* **48**, 69–72.

[154] Hu, T.M., deFreitas, A.S.W., Doyle, J., Jackson, D., Marr, J., Nixon, E., Pleasance, S., Quilliam, M.A., Walter, J.A. and Wright, J.L.C. (1993) New DSP toxin derivatives isolated from toxic mussels and the dinoflagellates, *Prorocentrum lima* and *Prorocentrum concavum*. In: *Toxic Phytoplankton Blooms in the Sea*. Smayda TJ and Shimizu Y (eds), pp. 507–12, Elsevier, New York.

[155] Sechet, V., Quilliam, M.A. and Rocher, G. (1998) Diarrhetic shellfish poisoning (DSP) toxins in *Prorocentrum lima* in axenic and non-axenic batch culture: detection of new compounds and kinetics of production. In: *Harmful Algae*. Reguera B, Blanco J, Fernández ML and Wyatt T (eds), pp. 485–488, Xunta de Galacia and International Oceanographic Commission of UNESCO.

[156] Hu, T.M., Curtis, J.M., Walter, J.A. and Wright, J.L.C. (1995a) Identification of DTX-4, a new water-soluble phosphatase inhibitor from the toxic dinoflagellate. *J. Chem. Soc. Commun.* 597–9.

[157] Yasumoto, T., Seino, N., Murakami, Y. and Murata, M. (1987) Toxins produced by benthic dinoflagellates. *Biol. Bull.* **172**, 128–31.

[158] Schmitz, F.J. and Yasumoto, T. (1991) The 1990 United States–Japan seminar on bioorganic marine chemistry, meeting report. *J. Nat. Prod.* **54**, 1469–90.

[159] Holmes, M.J., Lee, F.C., Khoo, H.W. and Teo, S.L.M. (2001) Production of 7-deoxy okadaic acid by a New Caledonian strain of *Prorocentrum lima* (Dinophyceae). *J. Phycol.* **37**, 280–8.

[160] Torigoe, K., Murata, M., Yasumoto, T. and Iwashita, T. (1988) Prorocentrolide, a toxic nitrogenous macrocycle from a marine dinoflagellate, *Prorocentrum lima*. *J. Am. Chem. Soc.* **110**, 7876–7.

[161] Hu, T.M., Curtis, J.M., Walter, J.A., McLachlan, J. and Wright, J.L.C. (1995b) Two new water-soluble DSP toxin derivatives from the dinoflagellate *Prorocentrum maculosum*: possible storage and excretion products. *Tetrahedron Lett.* **36**, 9273–6.

[162] Hu, T.M., deFreitas, S.W., Curtis, J.M., Oshima, Y., Walter, J.A. and Wright, J.L.C. (1996) Isolation and structure of prorocentrolide B, a fast-acting toxin from *Prorocentrum maculosum*. *J. Nat. Prod.* **59**, 1010–14.

[163] Ten-Hage, L., Turquet, J., Quod, J-P., Puiseux-Dao, S. and Coute, A. (2000b) *Prorocentrum borbonicum* sp. nov. (Dinophyceae), a new toxic benthic dinoflagellate from the southwestern Indian Ocean. *Phycologia* **39**, 296–301.

[164] Tindall, D.R., Miller, D.M. and Bomber, J.W. (1989) Culture and toxicity of dinoflagellates from ciguatera endemic regions of the world. *Toxicon* **27**, 83.

[165] Quod, J.P., Turquet, J., Diogene, G. and Fessard, V. (1995) Screening of extracts of dinoflagellates from coral reefs (Reunion Island, SW Indian Ocean), and their biological activities. In: *Harmful Marine Algal Blooms*. Lassus P, Azrul G, Erard E, Gentien P and Marcaillou C (eds), pp. 815–20, Lavoisier, New York.

[166] Tindall, D.R., Dickey, R.W., Carlson, R.D. and Morey-Gaines, G. (1984) Ciguatoxigenic dinoflagellates from the Caribbean Sea. In: *Seafood Toxins*. Ragelis EP (ed.), pp. 225–40, ACS Symposium Series no. 262, American Chemical Society, Washington DC.

[167] Lu, C.K., Lee, G.H., Huang, R. and Chou, H.N. (2001) Spiro-prorocentrimine, a novel macrocyclic lactone from a benthic *Prorocentrum* sp. of Taiwan. *Tetrahedron Lett.* **42**, 1713–16.

[168] Satake, M., Ofuiji, K., Naoki, H., James, K.J., Furey, A., McMahon, T., Silke, J. and Yasumoto, T. (1998) Azaspiracid, a new marine toxin having unique spiro ring assemblies, isolated from Irish mussels, *Mytilus edulis*. *J. Am. Chem. Soc.* **120**, 9967–8.

4   Involvement of Na$^+$ in the Actions of Ciguatoxins and Brevetoxins that Stimulate Neurotransmitter Release and Affect Synaptic Transmission

JORDI MOLGO* and EVELYNE BENOIT

4.1   INTRODUCTION

Poisoning by toxins of marine organisms is a substantial worldwide hazard that may occur by the ingestion of contaminated seafood. The increased frequency and wide distribution of toxic algal blooms has become a major environmental and health problem around the world and the bloom-forming microalgae are, therefore, of great ecological, medical and economic interest. A number of dinoflagellate species are known to produce bioactive compounds, and some of these compounds are among the most potent nonproteinaceous toxins known [1]. Dinoflagellate toxins affect finfish, shellfish, birds, marine mammals and humans, most frequently by vectorial transport through the various food webs.

Neurotoxins have drawn scientists' attention because of the increasing number of human poisonings and the socio-economic impact of these incidents. An important by-product of these investigations has been the discovery of an array of highly specific toxins which have become essential tools for dissecting biological processes and for understanding the multitude of steps involved in synaptic transmission.

Among the large number of neurotoxins that promote neurotransmitter release from nerve terminals, the majority selectively target presynaptic ion channels and modulate their activity, thereby indirectly eliciting the exocytotic

---

* Corresponding author

Abbreviations: ACh, acetylcholine; BAPTA/AM 1,2-bis(2-aminophenoxy)ethane-N,N,N', N'-tetraacetic acid/acetoxy methyl ester; BoNT/A, botulinum type-A neurotoxin; PbTx, brevetoxin from *Gymnodinium breve* (previously designated *Ptychodiscus brevis*); C-CTX, Caribbean ciguatoxin; EGTA, ethyleneglycol-bis-(-amino-ethyl)-tetraacetic acid; EPPs, endplate potentials; FM1-43, (N-(3-triethyl ammoniumpropyl)-4-(p-dibutylaminostyryl) pyridinium dibromide); IP$_3$, inositol 1,4,5-trisphosphate; MEPPs miniature endplate potentials; NMDA, N-methyl-D-aspartate; P-CTX, Pacific ciguatoxin; SNAP-25, 25 kDa synaptosome-associated protein; SNARE, <u>S</u>oluble <u>N</u>-ethylmaleimide-sensitive fusion protein <u>A</u>ttachment Protein <u>R</u>eceptor; TTX, tetrodotoxin.

This chapter is dedicated to Professor Paul Lechat, member of the French Academy of Medicine, for his continuous support.

Perspectives in Molecular Toxinology
Edited by A Ménez © 2002 John Wiley & Sons, Ltd

release of synaptic vesicles. Less common are neurotoxins that stimulate the release machinery by acting upon cell surface signal-transducing receptors or specific neuroexocytosis pathways [2]. The elucidation of the mechanisms of action of these toxins has been a major driving force in advancing our understanding of how synapses work. Much of the functioning of the nervous system is based on three basic processes: the conduction of nerve impulses, the release of neurotransmitters by nerve terminals, and the detection and decoding of neurotransmitters by the postsynaptic cell. Synaptic transmission across most chemical synapses involves the almost synchronous release from the presynaptic nerve terminal of a number of multi-molecular packets of neurotransmitter, named quanta [3]. At the vertebrate neuromuscular junction, spontaneous and nerve-evoked quantal acetylcholine (ACh) release can be easily detected by conventional electrophysiological measurements of endplate potential changes or currents. The discovery that motor nerve terminals contain large numbers of synaptic vesicles filled with neurotransmitter led to the vesicular hypothesis of quantal secretion, which holds that each quantum of transmitter is confined within a single vesicle and is released by exocytosis (reviewed by [4]). Nerve impulse-evoked, synchronous quantal ACh release requires the influx of Ca^{2+} through voltage-sensitive Ca^{2+} channels located at specialised regions of the nerve terminal membrane known as 'active zones'. Ca^{2+} influx can be modulated by the pattern or frequency of nerve stimulation or by toxins that target specific ion channels.

In this article, we review studies of ciguatoxins and brevetoxins, potent marine neurotoxins purified from dinoflagellates and poisonous fish, which activate Na^+ channels in axons and motor nerve terminals. In addition, we analyse the consequences of these actions on synchronous and asynchronous neurotransmitter release, synaptic efficacy and synaptic vesicle recycling, an area of research in which we have been particularly interested.

4.2   ORIGINS AND CHEMICAL STRUCTURES OF CIGUATOXINS AND BREVETOXINS

The ciguatoxins are a large family of lipid-soluble, heat-stable, cyclic polyether molecules isolated from fish inhabiting tropical and sub-tropical areas of the Pacific Ocean [5–12] and the Caribbean Sea [13, 14]. The chemical structure of the principal ciguatoxins is shown in Figures 4.1 and 4.2.

Ciguatoxins are responsible for a widespread, complex poisoning of humans, known as ciguatera, which develops after the consumption of coral reef fish that have acquired the toxins through their food chain. This severe form of fish poisoning may present with either acute or chronic intoxication syndromes and constitutes a global health problem. Ciguatoxins pose a health risk at concentrations above 0.1 ppb, and ciguatera produces a polymorphic syndrome with neurological, gastrointestinal and cardiovascular symptoms in humans [15–29].

(a)

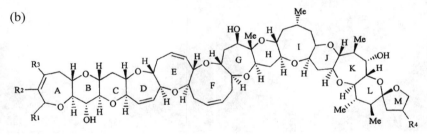

Toxins,	[M+H]$^+$	Origin	R$_1$	R$_4$
P-CTX-1B,	1111	moray-eel	-CH=CH-CHOH-CH$_2$OH	OH
P-CTX-2A2,	1095	moray-eel	-CH=CH-CHOH-CH$_2$OH	H
P-CTX-2B2,	1095	moray-eel	-CH=CH-CHOH-CH$_2$OH	H
P-CTX-4A,	1061	parrot fish G. toxicus	-CH=CH-CH=CH$_2$	H
P-CTX-4B,	1061	G. toxicus	-CH=CH-CH=CH$_2$	H

(b)

Toxins,	[M+H]$^+$	Origin	R$_1$	R$_2$	R$_3$	R$_4$
P-CTX-2A1,	1057	moray-eel parrot fish	H	OH	OH	H
P-CTX-3B,	1023	parrot fish	H	H	H	H
P-CTX-3C,	1023	moray-eel parrot fish G. toxicus	H	H	H	H

Figure 4.1 The chemical structure of the two types of Pacific-ciguatoxins (P-CTXs) extracted from fish and their dinoflagellate congeners which have a polyether backbone. The type 1 P-CTX (a) and type 2 P-CTX (b) possess 60 and 57 carbon atoms, respectively, in 13 fused ether rings. The protonated molecular masses ([M + H]$^+$) of the P-CTXs, and the toxins' origins, are indicated. Note, that P-CTXs obtained from fish flesh and viscera are more highly oxygenated than are the congeners isolated from the dinoflagellate G. toxicus, and that P-CTXs-type 2 lack the R1 (C1-C4) side-chain present in type 1, and have an eight-rather than a seven-membered ring E. Data were obtained from [9, 11, 12, 34–36]

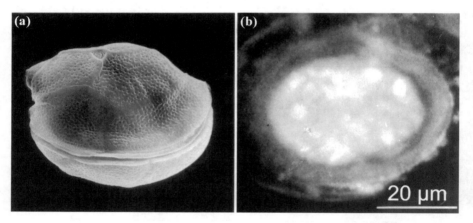

Figure 4.2 Structure of the first fully characterized Caribbean-ciguatoxin (C-CTX-1) which was purified from the fish called the 'horse-eye jack' (*Caranx latus*), according to [14]. Note that C-CTX-1 lacks the spiroketal at C52, compared to P-CTXs types 1 and 2 (Figure 4.1, panels a and b), which is replaced by an additional fused, six-membered cyclic ether

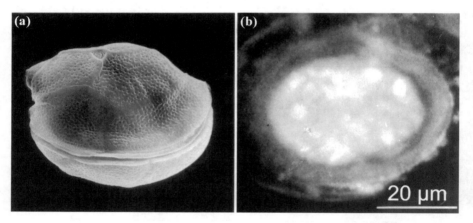

Figure 4.3 Images of the *Gambierdiscus toxicus* dinoflagellate obtained either by electron scanning microscopy of a fixed specimen (a) or by confocal laser scanning microscopy of a living specimen (b). Panel b contains a 'look through' projection of a series of optical sections. The calibrations in (a) and (b) are identical

In the Pacific, there is strong evidence that the dinoflagellate *Gambierdiscus toxicus* [30–32] (Figure 4.3) is the originator of at least some of the ciguatoxins, because they have been purified from wild or toxic cultured strains of *G. toxicus* dinoflagellates [8, 33–35]. In addition, Pacific CTX-1B (P-CTX-1B, also called P-CTX-1), the most potent toxin extracted from moray-eels (*Gymnothorax javanicus, Lycodontis javanicus*), is produced by acid-catalysed spiroisomerisation and oxidative modification of P-CTX-4A or P-CTX-4B produced by the dinoflagellate *G. toxicus* [1, 10, 12, 36].

The brevetoxins, isolated from harmful algal blooms known as 'red tides', are another family of potent, lipid-soluble, polyether compounds (see Figure 4.4),

(a)

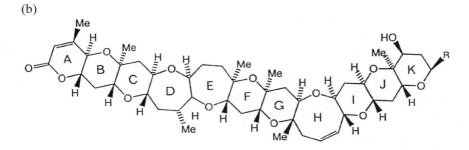

Toxins	R
PbTx-1	-CH₂-C(=CH₂)CHO
PbTx-7	-CH₂-C(=CH₂)CH₂OH
PbTx-10	-CH₂-CH(-CH₃)CH₂OH

(b)

Toxins	R	Others
PbTx-2	-CH₂-C(=CH₂)CHO	
PbTx-3	-CH₂-C(=CH₂)CH₂OH	
PbTx-5	-CH₂-C(=CH₂)CHO	37-COCH₃
PbTx-6	-CH₂-C(=CH₂)CHO	27,28-β-epoxide
PbTx-8	-CH₂-CO-CH₂Cl	
PbTx-9	-CH₂-CH(-CH₃)CH₂OH	

Figure 4.4 The two different backbone structures of the ten naturally occurring brevetoxins (PbTxs, for *Ptychodiscus brevis* toxins; the dinoflagellate has recently been restored to its earlier taxonomic designation of *Gymnodinium breve*). Note that brevetoxin A (a) and brevetoxin B (b) share a lactone in the A ring, and note the similarities in the H–I–J–K rings in the brevetoxin B backbone and the G–H–I–J rings in brevetoxin A. Data were obtained from [40, 46, 47]

and they have been purified to homogeneity from cultures of the unarmoured dinoflagellate *Gymnodinium breve* [37], formerly known as *Ptychodiscus brevis* [38–40]. Brevetoxins are implicated in a severe form of microalgae-derived poisoning known as neurotoxic shellfish poisoning [41–43].

Presently, ten brevetoxins (PbTx-1 to PbTx-10) are known [40, 44–47], and the complete synthesis of brevetoxin A has recently been achieved [48]. These compounds have been implicated in the massive fish kills that accompany blooms of *G. breve* [40, 49] and, more recently, in an unprecedented catastrophe resulting in the death of a large number of manatees (*Trichechus manatus latirostris*) along the south-west coast of Florida [50]. Death from brevetoxicosis is not necessarily acute and may occur after chronic inhalation [51] and/or ingestion (the accumulation of brevetoxins in filter-feeding shellfish has been found to poison human consumers).

4.3   VOLTAGE-DEPENDENT Na$^+$ CHANNELS AS TARGETS FOR CIGUATOXINS AND BREVETOXINS

Voltage-gated Na$^+$ channels are integral transmembrane proteins that respond to changes in membrane potential and are involved in the generation and propagation of action potentials along myelinated axons, nerve terminals and muscle fibres. The highest density of voltage-gated Na$^+$ channels in the nervous system occurs at the node of Ranvier, a specialisation of myelinated axons that enables high speed propagation of electrical signals. Therefore, activation of the channels in response to membrane depolarisation is essential for rapid conduction of electrical signals, and Na$^+$ channels control many aspects of signal transduction in both nerve and muscle tissues [52]. The channels are finely tuned, and either increased or decreased channel activity can adversely affect nerve and muscle functions. The voltage-gated Na$^+$ channel from various excitable animal tissues has a major α-subunit (M$_r$ of 240–280 kDa) composed of about 2000 amino acids organised in four repeating homologous domains (I to IV), each consisting of six putative α-helical transmembrane segments (S1–S6). The four domains fold together and create a central pore that determines the conductance and selectivity properties of the channel (reviewed by [53–56]). Neurotoxins from a variety of sources have been essential tools in characterising neurotoxin receptor-sites and the molecular properties of Na$^+$ channels.

Brevetoxins bind with high affinity to neurotoxin receptor-site 5 of the neuronal Na$^+$ channel protein, as revealed by direct binding studies using radio-labelled [^3H]PbTx-3, and by binding assays that show noncompetitive interactions between receptor-site 5 and a variety of natural toxin probes specific for receptor-sites 1–4 [44, 57]. Brevetoxins bind to the α-subunit of the Na$^+$ channel [58]. Data obtained from studies using a photo-labelled derivative of PbTx-3 and site-directed antibody mapping suggest that receptor-site 5 is specifically located in the region of interaction of transmembrane segments S6 and S5 of domains I and IV, respectively [59]. The ciguatoxins share with the brevetoxins a common locus of action at receptor-site 5 of the neuronal Na$^+$ channel protein [11, 60–63].

4.4 ALTERATIONS OF NEURONAL Na$^+$ CURRENT AND MEMBRANE POTENTIAL BY CIGUATOXINS AND BREVETOXINS

After exposure to nanomolar concentrations of P-CTX-1B, a fraction of the nodal Na$^+$ current of single myelinated axons fails to inactivate, giving rise to a late Na$^+$ current (Figure 4.5a) [64–66]. The peak Na$^+$ current amplitude and voltage-dependence are not modified by the toxin in the range of concentrations studied. In contrast, the late Na$^+$ current has a more negative activation (Figure 4.5b), and reversal potential than does the peak Na$^+$ current. For example, in the presence of P-CTX-1B, the late Na$^+$ current is activated at the resting membrane potential (-70 mV). These results indicate that P-CTX-1B modifies a fraction of Na$^+$ channels which remain permanently open, especially at the resting membrane potential of myelinated axons. Whether the Na$^+$ channels responsible for the late Na$^+$ current are a subset of channels distinct from those responsible for the peak Na$^+$ current remains to be elucidated. Although it is clear that P-CTX-1B affects the biophysical properties of a fraction of Na$^+$ channels, some of the pharmacological properties of the toxin-modified Na$^+$ channels remain unaffected. For example, they are sensitive to tetrodotoxin (TTX), to the local anaesthetic lidocaine, to increases in the external Ca^{2+} concentration, and to increases in external osmolality by D-mannitol, tetra-methylammonium or sucrose [66].

Nanomolar concentrations of P-CTX-1 have also been reported to modify voltage-gated Na$^+$ channels in peripheral sensory neurons, which provides an explanation for the sensory neurological disturbances associated with ciguatera fish poisoning. In rat dorsal root ganglion neurons, the main actions of P-CTX-1 are: (i) hyperpolarising shifts in the voltage dependence of TTX-sensitive Na$^+$ channel activation and inactivation, as well as a rapid TTX-sensitive rise in the membrane leakage current in cells expressing TTX-sensitive Na$^+$ channels; and (ii) a significant increase in the rate of recovery from TTX-resistant Na$^+$ channel inactivation. In addition, P-CTX-1 causes a reduction of peak currents flowing through both channel types but it has no effect on their activation and inactivation kinetics [67]. Finally, in cell-attached membrane patches from isolated para-sympathetic neurons, P-CTX-1 markedly increases the open probability of single TTX-sensitive Na$^+$ channels in response to depolarising voltage steps, without altering the unitary conductance. Interestingly, the spontaneously opened Na$^+$ channels do not close even at hyperpolarised membrane potentials [68].

The effects of brevetoxins on Na$^+$ channels are similar to those of ciguatox-ins, in that they shift the voltage-dependence of channel activation to more negative membrane potentials, inhibit their inactivation, and prolong the mean

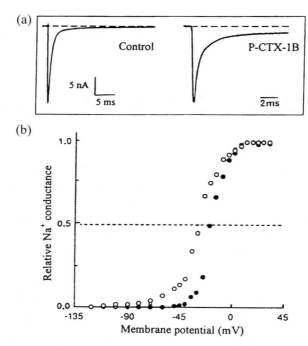

Figure 4.5   Effects of P-CTX-1B on the nodal Na$^+$ current recorded in a single frog myelinated axon (a) and on the Na$^+$ conductance–membrane potential relationship (b). The Na$^+$ current was recorded under voltage-clamp conditions during depolarising steps to 0 mV (a) or to various membrane potentials (b), preceded by 50 ms hyperpolarisations to −120 mV, applied at 0.5 Hz, from a holding potential of −70 mV. (a) contains the traces of the Na$^+$ current in the absence (left) and in the presence (right) of 10 nM P-CTX-1B. The dashed lines show the zero current level. The peak (filled circles) and late (open circles) Na$^+$ conductances shown in (b) were calculated as a function of the membrane potential, in the presence of 10 nM P-CTX-1B. Data are expressed with respect to their maximum values obtained at large depolarisations. Note the negative shift in the late Na$^+$ conductance–membrane potential curve compared to that of the peak Na$^+$ conductance (for further details, see [65])

open time of TTX-sensitive Na$^+$ channels in various excitable cells [69–73]. Also, PbTx-3 has been reported to modify Na$^+$ channel gating in neuronal cell lines B50 and B104 (derived from rat brain), in which single Na$^+$ channel currents were recorded from cell-attached membrane patches [74]. In the neuronal cells, PbTx-3 caused a significant increase in the frequency of channel reopening, which indicated a slowing of Na$^+$ channel inactivation. In addition, PbTx-3 shifted the threshold potential for Na$^+$ channel opening and promoted channel opening near the resting membrane potential of the cells (about −60 mV).

As a consequence of the persistent activation of Na$^+$ channels at the resting membrane potential, ciguatoxins and brevetoxins produce an entry of Na$^+$ ions into cells, which depolarises the membrane. In various nerve cells, myelinated axons and motor nerve terminals, this depolarisation, in turn, triggers elevated

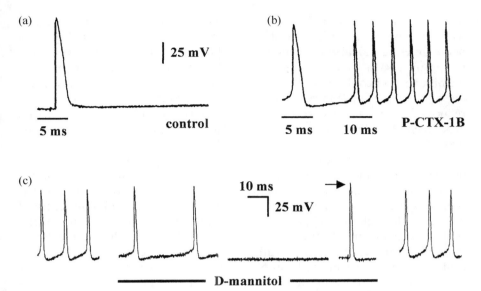

Figure 4.6 Spontaneous and repetitive nodal action potentials induced by P-CTX-1B in myelinated axons and their suppression by a D-mannitol-produced a 50% increase in the osmolality of the external medium. Data were obtained by the conventional current-clamp technique. (a) shows a control action potential evoked by a 0.5 ms depolarising stimulus. After addition of 10 nM P-CTX-1B to the standard physiological solution, spontaneous and repetitive action potentials are observed (b). (c) shows that the addition of 100 mM D-mannitol to the external solution containing P-CTX-1B (10 nM) suppresses spontaneous action potentials, but does not affect the action potential evoked by a depolarizing stimulus of 0.5 ms duration (arrow). (Modified from [65])

frequencies (60–100 Hz) of spontaneous and/or repetitive action potential discharges (Figure 4.6a and b) [60, 64–66, 69, 70 75–83]. It is worth noting that, in myelinated axons, the local anaesthetic lidocaine, increases in the external Ca^{2+} concentration, and increases in external osmolality by D-mannitol, tetramethylammonium or sucrose, first decrease the frequency of spontaneous and repetitive action potentials induced by ciguatoxins, and then progressively suppress the action potentials (Figure 4.6c) [66, 82, 83]. However, under these various conditions, action potentials evoked by depolarising stimuli can still be obtained, suggesting that the membrane excitability previously increased by ciguatoxins is reduced to a level similar to that under control conditions.

4.5   CIGUATOXIN- AND BREVETOXIN-MEDIATED SWELLING OF THE NODES OF RANVIER OF MYELINATED AXONS

An important consequence of the persistent activation of voltage-dependent Na$^+$ channels induced by the various ciguatoxins and brevetoxins, at the resting membrane potential, is nodal swelling of single myelinated axons (Figure 4.7), without apparent modification of the morphology of the internodal parts of nerve fibres, as revealed by using the styryl dye FM1-43 (N-(3-triethyl ammo-niumpropyl)-4-(p-dibutylaminostyryl) pyridinium dibromide) to stain the nerve membrane, and confocal laser scanning microscopy [65, 66, 82–85]. The swelling of the nodes of Ranvier increases the capacitance of the membrane and reduces the conduction velocity of myelinated axons, which is consistent with experimental findings [86, 87] and clinical reports of ciguatera fish poisoning [88, 89].

The nodal swelling is prevented by blocking voltage-dependent Na^+ channels with TTX, and it is reversed by increasing the osmolality of the external solution with D-mannitol, tetramethylammonium chloride or sucrose. It has been proposed that ciguatoxins and brevetoxins cause a continuous entry of Na^+ ions through Na^+ channels permanently activated at the resting membrane potential and through Na^+ channels which open during the toxin-induced spontaneous action potential discharges. This continuous Na^+ entry leads to an increase in intracellular Na^+ concentration that directly or indirectly disturbs the osmotic equilibrium between intra- and extra-cellular fluids. Therefore, an influx of water occurs in order to restore both the osmotic equilibrium and the intracellular Na^+ concentration to initial levels [90, 65, 66, 82–85].

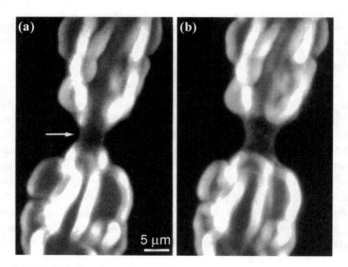

Figure 4.7 P-CTX-1B-induced swelling of the node of Ranvier (arrow) of a single frog myelinated axon. Images were obtained under control conditions (a) and after 50 minutes' exposure to 10 nM P-CTX-1B (b). Membrane structures were stained with the fluorescent dye FM1-43 for 15 minutes, and they were then washed free of dye before imaging with a confocal laser scanning microscope. Each collected series of optical sections scanned at 0.13 μm increments was used to calculate a 'look-through' projection, thus enabling the isolated nerve fibre to be digitally reconstructed. The scale bar in (a) also applies to (b)

4.6 CIGUATOXINS AND BREVETOXINS INCREASE
INTRACELLULAR Ca^{2+} LEVELS IN NEURONAL CELLS

Ca^{2+} plays a key role in regulating nerve membrane excitability and the release of neurotransmitters from nerve terminals [3, 4, 91, 92]. Thus, the regulation of intracellular Ca^{2+} levels is of paramount importance in controlling synaptic transmission. Indeed, transient Ca^{2+} entry into motor nerve terminals through voltage-gated Ca^{2+} channels is known to play a pivotal role in the regulation of synchronous quantal transmitter release, while asynchronous quantal release depends mainly on background Ca^{2+} levels. Increased excitability of axons and depolarisation of nerve terminals during the action of ciguatoxins and brevetoxins is expected to open Ca^{2+} channels, elevate intracellular Ca^{2+} levels, and enhance the efficiency of neurotransmitter release and synaptic transmission at various chemical synapses [93, 40, 75, 76, 79, 80].

The primary routes for ciguatoxin- and brevetoxin-elicited Ca^{2+} entry into nerve cells or nerve terminals include the (i) voltage-dependent Ca^{2+} channels, (ii) Na$^+$-Ca^{2+} exchanger (in the reversed mode), and (iii) neurotransmitter-receptor-operated ion channels. Nanomolar concentrations of P-CTX-1B have been shown to increase intracellular Ca^{2+} levels in differentiated neuroblastoma (NG108-15) cells, as assessed by fura-2-based microfluorometric recordings [94, 95]. The brevetoxin PbTx-1 has also been reported to produce rapid and concentration-dependent increases of intracellular Ca^{2+} concentration in fluo-3–loaded cerebellar granule neurons [96]. In those cells, N-methyl-D-aspartate (NMDA) receptor antagonists and tetanus neurotoxin (an inhibitor of Ca^{2+}-dependent exocytotic neurotransmitter release) significantly reduced the Ca^{2+} response and excitatory amino acid release, and they protected neurons against PbTx-1's neurotoxicity. Interestingly, although simultaneous blockade of L-type Ca^{2+} channels and the Na$^+$−Ca^{2+} exchanger reduced the intracellular Ca^{2+} response to a level below that produced by NMDA receptor blockade, it failed to attenuate PbTx-1's neurotoxicity completely. This observation suggests that, in addition to intracellular Ca^{2+} load, neuronal vulnerability is governed in those neurons by the NMDA receptor, Ca^{2+} influx pathway.

The increased levels of intracellular Ca^{2+} caused by P-CTX-1B occur even in NG108-15 cells bathed with a Ca^{2+}-free medium supplemented with the Ca^{2+} chelator EGTA (Figure 4.8). Also, TTX, which blocks voltage-gated Na$^+$ channels, completely prevents the P-CTX-1B-induced increase in Ca^{2+} levels [94, 95, 97]. These two observations suggest that Na$^+$ influx through voltage-gated Na$^+$ channels is responsible for Ca^{2+} mobilisation from intracellular Ca^{2+} stores.

The mechanism whereby P-CTX-1B induces Ca^{2+} mobilisation has not yet been elucidated. However, strong evidence indicates that an enhanced influx of Na$^+$ stimulates the production of inositol 1,4,5-trisphosphate (IP$_3$) in synaptosomes [98–100]. In addition, P-CTX-1B-induced Ca^{2+} mobilisation prevents the subsequent action of bradykinin [95], which suggests that the intracellular Ca^{2+}

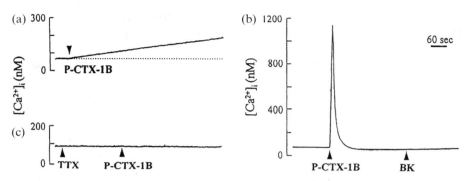

Figure 4.8 Mobilisation of intracellular Ca^{2+} concentration by P-CTX-1B (2.5 nM, (a); 25 nM, (b)) and its inhibition by 1 μM TTX (c) in differentiated neuroblastoma (NG108-15) cells. Ca^{2+} levels were measured by microspectrofluorometry with the Ca^{2+}-sensitive probe fura-2/AM. Note the sustained (a) and transient (b) increases in the intracellular Ca^{2+} concentration caused by 2.5 and 25 nM P-CTX-1B, respectively, and the P-CTX-1B (25 nM)-elicited inhibition of the subsequent action of 1 μM bradykinin (BK). Arrow heads indicate addition of toxins or drugs to the nominally Ca^{2+}-free medium supplemented with 1 mM EGTA. (Modified from [95])

store stimulated by the toxin is the same as the one activated by bradykinin [101, 102]; i.e. the IP_3-releasable Ca^{2+} store. Thus, an Na^+-dependent, Ca^{2+}-mobilisation mechanism involving IP_3-sensitive intracellular stores could explain the fact that P-CTX-1B raises intracellular Ca^{2+} in neuroblastoma cells bathed in Ca^{2+}-free medium containing EGTA.

Investigation of the effect of P-CTX-1B on Ca^{2+} levels in fluo-3-loaded bovine chromaffin cells in primary culture, with high-resolution confocal laser scanning microscopy, revealed a transient increase in the fluorescence of the cells' cytoplasm, with a particularly strong signal in a restricted area of the optical section corresponding to about 20 % of the cells' surface [103]. Such localised Ca^{2+} signals are typical responses of chromaffin cells to agonists that mobilise Ca^{2+} from internal stores [104, 105].

4.7 QUANTAL TRANSMITTER RELEASE CHANGES CAUSED BY CIGUATOXINS AND BREVETOXINS

Quantal ACh release from motor nerve terminals is one of the pivotal events in synaptic transmission, and ACh quanta are released from clear synaptic vesicles spontaneously or in response to neural activity triggered by presynaptic action potentials. Some of the molecular processes governing neurotransmitter release are now becoming better understood [91, 106, 107, 108]. After nerve terminal depolarisation and subsequent entry of Ca^{2+}, synaptic vesicles move toward the plasma membrane and dock at the active zone. Primed synaptic vesicles rapidly fuse with the plasma membrane after a marked elevation in intracellular Ca^{2+}, and they release their neurotransmitter content in the synaptic cleft. Synaptic

vesicle recycling is a critical feature of synaptic transmission because it ensures a
constant supply of releasable transmitter at the nerve terminal.

4.7.1 SYNCHRONOUS EVOKED QUANTAL TRANSMITTER RELEASE

At the vertebrate neuromuscular junction, as a consequence of the changes in
axonal excitability caused by P-CTX-1B or C-CTX-1, trains of spontaneous
action potentials may reach the nerve terminals spontaneously. Sometimes the
trains are set off after a single nerve stimulus. The presynaptic action potentials
trigger repetitive endplate potentials (EPPs) which, if they attain the threshold
potential, generate action potentials in the muscle fibre (Figure 4.9) that can
reach frequencies of up to 100 Hz [75, 81]. However, this synaptic activity is
usually transient, probably due to both depression of nerve-evoked neurotransmitter release and toxin-induced depolarisation of nerve endings.

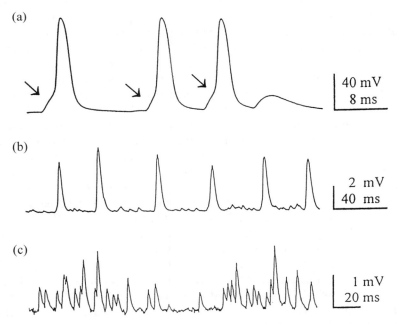

Figure 4.9 Spontaneous repetitive muscle action potentials, EPPs, and MEPPs recorded in
the junctional region of the same muscle fibre during the action of C-CTX-1 (120 nM). Note
the (i) action potentials triggered by spontaneous EPPs (arrows in (a)), (ii) repetitive EPPs
resulting from synchronous quantal transmitter release (b), and (iii) high frequency of MEPPs
appearing at a time when nerve stimulation does not evoke quantal transmitter release (c). All
data were obtained from the same neuromuscular junction of a frog *cutaneous pectoris* muscle
in which excitation-contraction was uncoupled. The resting membrane potential during
recordings was −81 mV

In junctions equilibrated with low Ca^{2+}-high Mg^{2+} medium, repetitive EPPs are scarce, probably because high Mg^{2+} changes the activation potential for Na^+ channels. Under those conditions, P-CTX-1B first increases the mean quantal content of EPPs, then reduces it before irreversibly blocking nerve-evoked transmitter release. The increased quantal ACh release elicited by P-CTX-1B probably is not due to enhanced Ca^{2+} influx through voltage-dependent Ca^{2+} channels, but to the reversed-mode operation of the Na^+– Ca^{2+} exchanger, which allows Ca^{2+} entry in exchange for intracellular Na^+, as has been reported for isolated cholinergic nerve endings [75, 109, 110]. At the time transmitter release evoked by nerve impulses is blocked by P-CTX-1B or C-CTX-1, spontaneous quantal ACh release, recorded as miniature endplate potentials (MEPPs) or currents, is greatly enhanced (Figure 4.9).

4.7.2 ASYNCHRONOUS QUANTAL TRANSMITTER RELEASE

P-CTX-1B, P-CTX-4B and C-CTX-1 dramatically increase spontaneous quantal ACh release from motor nerve terminals [75, 81, 97]. The increased MEPP frequency induced by the toxins occurs after a latency of about 20 minutes and is sustained for around 3 hours, and the increased spontaneous release of ACh quanta leads to exhaustion of transmitter release after 3 hours or so. Ciguatoxins have similar effects on isolated neuromuscular preparations bathed with a Ca^{2+}-free medium supplemented with the Ca^{2+} chelator EGTA [84, 90, 97]. Activation of Na^+ channels and the subsequent entry of Na^+ ions into nerve terminals appear to be responsible for ciguatoxins' enhancement of asynchronous quantal neurotransmitter release, because TTX blockade of Na^+ influx into the terminals prevents the neurotransmitter release. Nanomolar concentrations of PbTx-3 also increase MEPP frequency at the neuromuscular junction, presumably by activating Na^+ channels at motor nerve terminals, since blockade of Na^+ influx into the terminals by TTX reverses the effect [70, 111, 112]. In addition, PbTx-3 increases the frequency of MEPPs in junctions exposed to a Ca^{2+}-free medium supplemented with EGTA. Under those conditions, the sustained increase in asynchronous quantal ACh release leads to an exhaustion of transmitter release, and the effects are prevented by TTX. Furthermore, the increase in MEPP frequency still occurs after loading the nerve terminals with the Ca^{2+}-chelator dimethyl-BAPTA/AM, or after pre-treatment with various agents known to prevent Ca^{2+} release from intracellular pools. Notably, PbTx-3 does not increase MEPP frequency after removal of extracellular Na^+, even in the presence of isotonic Ca^{2+} Ringer's solution. Therefore, it appears that Na^+ entry through TTX-sensitive Na^+ channels, and an excess of Na^+ in the nerve terminal *per se*, are involved in the activation of quantal ACh release induced by PbTx-3 [113]. There is good evidence that conditions or neurotoxins eliciting intracellular accumulation of Na^+ augment transmitter release at nerve terminals [114]. However, the mechanism by which Na^+ triggers quantal transmitter release in nerve terminals is not yet clear.

Elevation of intracellular Na^+ concentration can increase intracellular Ca^{2+} concentration by releasing Ca^{2+} from nerve terminal stores or by Ca^{2+} entering from the external medium. After long periods of equilibration in Ca^{2+}-free-EGTA-containing medium, it is unlikely that ciguatoxins or PbTx-3 cause Ca^{2+} entry either through voltage-sensitive Ca^{2+} channels or through the reversed operation of the Na^+-Ca^{2+} exchanger. Therefore, it is likely that Na^+-dependent Ca^{2+} mobilisation from nerve terminal Ca^{2+}-stores accounts for the increase in asynchronous quantal ACh release observed in Ca^{2+}-free medium (see section 4.6). However, as reported with PbTx-3, neither the various pharmacological agents known to inhibit Ca^{2+} release from intracellular pools, nor the Ca^{2+}-chelator dimethyl-BAPTA prevented the increase of quantal transmitter release caused by the toxin. Therefore, it has been hypothesised that Na^+ *per se* promotes the increase of quantal transmitter release, in the absence of a rise in intracellular Ca^{2+} [113].

Essential clues for the mechanisms of the neurotransmitter release process have been obtained during the last 10 years, and the proteins involved in the mechanism of docking and fusion of synaptic vesicles at the plasma membrane during exocytosis have been identified (reviewed by [2, 108]). Ca^{2+}-dependent exocytosis at the nerve terminal involves the synaptic core (Soluble N-ethylmaleimide-sensitive fusion protein Attachment Protein Receptor, called SNARE) complex which is composed of the t-SNARE's syntaxin and 25 kDa, synaptosome-associated protein (SNAP-25), and the v-SNARE's vesicle-associated membrane protein (VAMP/synaptobrevin), and is a stable heterotrimer which can associate with the putative Ca^{2+}-sensor protein, synaptotagmin. Several lines of evidence indicate that synaptotagmin, a major Ca^{2+}-binding protein of synaptic vesicles, is the Ca^{2+} sensor involved in exocytosis, and is required for Ca^{2+}-dependent, synchronous neurotransmitter release in nerve terminals, possibly by binding to phospholipids of the plasma membrane, binding to syntaxin, or both (reviewed by [115–117]). However, it is not known whether synaptotagmin acts as a Na^+-sensor in Na^+-dependent–Ca^{2+}-independent neurotransmitter release caused by P-CTX-1B and PbTx-3. Interestingly, an unexpected and direct interaction between neuronal Na^+ channels and synaptotagmin was recently discovered by means of co-immunoprecipitation, Western blotting analysis, and *in vitro* assays using recombinant proteins [118]. Thus, a Na^+ channel–synaptotagmin complex could be involved in Na^+-dependent exocytosis. However, the majority of the synaptotagmin complexed with Na^+ channels was shown to be distinct from the synaptotagmin-SNARE protein complex [118], that selectively interacts with presynaptic Ca^{2+} channels [119, 120]. Further studies are required to determine the physiological significance of the synaptotagmin interaction with Na^+ channels, and the basis for the Na^+-dependent–Ca^{2+}-independent neurotransmitter release detected in motor nerve terminals treated with P-CTX-1B and PbTx-3.

4.8   SWELLING OF MOTOR NERVE TERMINALS *IN SITU*

Nerve membrane structures and perisynaptic Schwann cell somata of living isolated neuromuscular preparations can be stained passively with the styryl dye FM1-43 [121] and imaged at high resolution with confocal laser scanning microscopy [90]. The inability of the FM1-43 dye to penetrate nerve membranes and the persistence of the staining, due to its partition only into the outer membrane leaflet, renders the dye particularly useful in detection by confocal laser scanning microscopy of changes in the area projected from three-dimensional image series of nerve terminals.

P-CTX-1B, P-CTX-4B and PbTx-3 cause a time-dependent swelling of motor nerve terminals in isolated frog neuromuscular preparations bathed with standard physiological medium or a nominally Ca^{2+}-free medium containing EGTA [84, 90, 113]. Swelling of nerve terminals is also produced by C-CTX-1 (Figure 4.10). All of the ciguatoxins studied produce an increase in the projected area

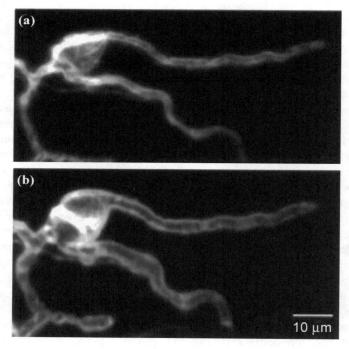

Figure 4.10   C-CTX-1-induced swelling of the motor nerve terminal and perisynaptic Schwann cell soma of a single frog neuromuscular junction. Images are reconstitutions of 'look through' projections before (a), and after 2 hours' exposure to 120 nM C-CTX-1 (b). The living nerve structures were stained with the dye FM1-43 for 15 minutes and then washed free of dye for 10 minutes before imaging with the confocal laser scanning microscope. Scale bar in (b) applies to (a)

per unit length of nerve terminals (which gives an index of the volume changes) which is larger during the first hour of exposure to toxin than during the second or third hours of exposure.

Na$^+$ channel blockade with TTX completely prevents the swelling of the nerve terminals caused by P-CTX-1B, P-CTX-4B, C-CTX-1 and PbTx-3 [84, 90, 113]. Thus, it appears that the swelling is related to both Na$^+$ entry into the terminals and sustained quantal transmitter release induced by the toxins. These results suggest an important contribution of the neurotransmitter release process to the changes in nerve terminal projected area induced by the toxins. The increase in the projected surface area of nerve terminals could be explained by the fusion of synaptic vesicles with the nerve terminal axolemma during the enhanced quantal transmitter release. Interestingly, when transmitter release is markedly reduced by botulinum type-A neurotoxin (BoNT/A), which selectively cleaves the nerve terminal's SNAP-25 protein [122], P-CTX-1B and PbTx-3 still induce an increase in the projected area of the terminal. However, the increase is only about one-third of that reported at normal junctions. These results suggest that the changes in the projected area of BoNT/A-poisoned terminals, in which transmitter release is drastically reduced, can not be attributed to the incorporation of vesicular membranes into the axolemma, but might be due to an osmotic action, whereby the increased Na$^+$ load induced by the toxins causes water influx into the terminals.

Although the fine processes of the perisynaptic, non-myelinating Schwann cells covering the nerve terminal branches can not be volumetrically resolved, Schwann cell somata also appear to be affected by the swelling of the terminal. However, prolonged exposure to the toxins does not cause detectable changes in the total length of nerve terminal branches at individual neuromuscular junctions.

4.9   IMPAIRMENT OF SYNAPTIC VESICLE RECYCLING BY CIGUATOXINS AND BREVETOXINS

The process of exocytosis implies that there is a close apposition of the synaptic vesicle membrane to the nerve terminal axolemma, fusion of the lipid matrix of the two membranes, and subsequent pore formation between the interior of the vesicle and the extracellular space. After fusion with the nerve terminal axolemma, synaptic vesicles reform by an endocytic recycling mechanism [123, 124], and the process of recycling is crucial for the maintenance of a functional population of synaptic vesicles during prolonged transmitter release. Two hypotheses, both involving endocytosis of the vesicular membrane, have been proposed to explain the recycling process [125]. The first of the hypotheses, the coated vesicle model [123], proposes that vesicular membrane components merge with the plasma membrane and are subsequently recovered and possibly sorted in coated pits. These pinch off as coated vesicles that either fuse with a sorting endosome from which new vesicles emerge or uncoat to become synaptic vesicles directly. The alternative 'kiss-and-run' model [126, 127] proposes that 'empty' vesicles are retrieved intact from the plasma membrane after secretion occurs via a fusion pore [128, 129] they are then immediately refilled with transmitter and re-enter the secretion-competent pool. It is still unclear under what conditions and in what proportion the empty synaptic vesicles are either retrieved directly without losing their identity (by the 'kiss-and-run' mechanism), or whether the vesicular membrane merges with the plasma membrane and is recycled by endocytosis.

A powerful optical assay allowing the direct study of synaptic vesicle recycling in living motor nerve terminals has been developed using the fluorescent styryl dye FM1-43 [90, 121, 130–132]. If quantal transmitter release from motor nerve terminals is stimulated in the presence of FM1-43, the dye is trapped in recycled synaptic vesicles and an alignment of brightly stained spots is observed. These fluorescent spots are consistent with the distribution of synaptic vesicle clusters inside nerve terminals. The staining is stable under resting conditions, but further stimulation of the terminals causes reversible destaining of the fluorescent spots due to exocytosis of synaptic vesicles loaded with the dye. The activity-dependent destaining elevates the concentration of dye in the extracellular medium during exocytosis [133]. This unique property of the FM1-43 dye has been exploited to gain novel insights into vesicular recycling and to measure optically synaptic vesicle turnover during endo-exocytosis in living nerve terminals [121, 132].

Confocal laser scanning microscopy used to monitor synaptic vesicle recycling in FM1-43-stained motor nerve terminals from frog neuromuscular preparations depolarised for 5 minutes with an isotonic, high K^+ (60 mM) medium containing Ca^{2+} (1.8 mM), and then washed with dye-free normal medium, revealed numerous fluorescent spots inside the nerve terminals. However, labelling did not occur in terminals in which quantal transmitter release was stimulated with P-CTX-1B for 60–180 minutes, either in Ca^{2+}-free or in Ca^{2+}-containing standard physiological medium [90]. Similar results were obtained with C-CTX-1 (M. Marquais, C. Mattei, E. Benoit, J-P. Vernoux, R. Lewis and J. Molgó, unpublished results). These results strongly suggest that P-CTX-1B and C-CTX-1 block synaptic vesicle recycling in motor nerve terminals.

Ultrastructural examination of motor nerve terminals from frog neuromuscular junctions in which transmitter release was exhausted, by exposure to P-CTX-1B and PbTx-3 for three hours, revealed a marked depletion of small clear synaptic vesicles and large dense core vesicles, and the presence of numerous cisternae and infoldings of the axolemma (Figure 4.11). These results strongly confirm that small clear synaptic vesicle recycling is altered in the neurotoxin-treated nerve terminals, and that the terminals release transmitter quanta, independently of the presence of extracellular Ca^{2+}, until they are depleted of synaptic vesicles. Most nerve terminals which had a marked reduction in the

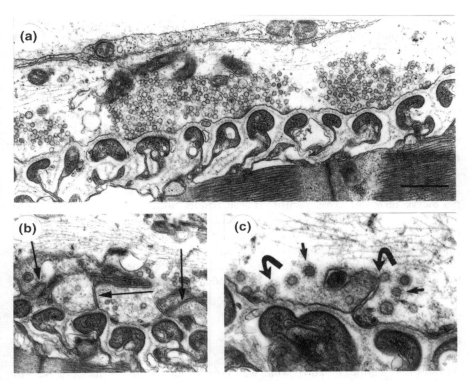

Figure 4.11 Electron micrographs of longitudinal sections from frog neuromuscular junctions exposed for three hours, in a Ca^{2+}-free medium supplemented with the Ca^{2+} chelator EGTA, under control conditions (a) and after treatment with 50 nM PbTx-3 (b and c). Note the normal appearance of the control nerve terminal (a), the large reduction in the number of synaptic vesicles and the presence of axolemma infoldings emerging between active zones (b), and at higher magnification, several coated vesicles (small arrows) and coated pits (curved arrows) associated with an axolemma infolding (c). The scale bar in panel a corresponds to 0.5 μm in panels a and b, and to 0.25 μm in (c). (Modified from [113])

number of synaptic vesicles were swollen and exhibited a few coated pits sometimes associated with axolemma infoldings and clathrin-coated vesicles at the outer margin of the active-zone regions [76, 97, 113]. Electron microscopic studies have shown that the membranes of synaptic vesicles can be retrieved via clathrin-coated pits and vesicles [123], and genetic, biochemical and morphological evidence has corroborated these findings [125, 134–136]. Clathrin-mediated endocytosis of synaptic vesicles at motor nerve terminals of the neuromuscular junction represents a specialised form of endocytosis allowing the reuptake of synaptic vesicle constituents, and it appears to be quantitatively the most important mechanism for synaptic vesicle recycling [137]. Therefore, it is likely that during the three hours' exposure to P-CTX-1B and PbTx-3 almost all synaptic vesicles contained in the nerve terminal were emptied by fusion with the plasma membrane, and they were not replaced by others that had been recycled by clathrin-mediated endocytosis. The few synaptic vesicles that arrived at the clathrin-coated stage during the action of P-CTX-1B and PbTx-3 probably were unable to complete the recycling process.

No obvious alterations in the number of synaptic vesicles have been detected in frog motor nerve terminals treated with BoNT/A (to completely block neurotransmitter release), prior to exposure to P-CTX-1B and PbTx-3 for three hours (J. Molgó and J.X. Comella, unpublished results; [113]). Similarly, neuromuscular preparations treated with TTX, before exposure to P-CTX-1B or PbTx-3 for three hours, exhibited nerve terminals having a normal appearance and unchanged numbers of synaptic vesicles and large dense core vesicles per nerve terminal cross section. These results indicate that Na^+ load into motor nerve terminals by P-CTX-1B and PbTx-3 strongly inhibits synaptic vesicle recycling. At the present time, it is unclear whether P-CTX-1B and PbTx-3 affect the recycling process directly; i.e., by interacting with some of the various proteins required for endocytosis [124], or indirectly through the entry of Na^+. Further studies are needed to answer this interesting question.

4.10   PERSPECTIVES

At the present time, much of what we know about the involvement of Na^+ in the ability of ciguatoxins and brevetoxins to stimulate neurotransmitter release and affect synaptic transmission mechanisms comes from electrophysiological, morphological and microspectrofluorimetric measurements in living tissues and cells, and from ultrastructural studies of fixed tissues. In addition, the above-mentioned, detailed studies of the effects of ciguatoxins and brevetoxins have markedly improved our understanding of how they affect Na^+ channels, nerve conduction, and neurotransmitter release processes. Several synaptic mechanisms are now being examined with the help of these neurotoxins, and these studies should yield new insights into synaptic transmission mechanisms, particularly in the central nervous system. Although elucidating the three-dimensional structure of ciguatoxins and brevetoxins remains a challenge for future research, the synthesis of ciguatoxin and brevetoxin derivatives possessing refined structural features should help to determine the molecular mechanisms underlying the toxins' selective action on the Na^+ channel protein[1].

While this manuscript was in press, the total synthesis of the complex P-CTX-3C was reported by Hirama and co-workers [138], which is an impressive achievement in organic chemistry [139]. It is expected that the strategy used for the synthesis of P-CTX-3C will help to create new synthetic probes for elucidating ciguatoxin's essential chemical structural factors interacting with voltage-sensitive Na^+ channels, and expedite the preparation of anti-ciguatoxin antibodies that may be useful not only for detecting ciguateric fish, but also for basic research.

The molecular mediators of Na$^+$-induced synaptic changes and the mechanisms involved still remain intriguingly elusive, and their identification remains an important challenge. Further studies of the proteins involved with the various actions of the ciguatoxins and brevetoxins should provide detailed knowledge about the proteins' functional properties and, it is hoped, pave the way for a detailed molecular understanding of how the toxins affect cell function.

ACKNOWLEDGEMENTS

We thank Dr Anne-Marie Legrand and Dr Mireille Chinain (Institut Territorial de Recherches Médicales Louis Malardé, Papeete, Tahiti, French Polynesia) for providing the ciguatoxins and the samples of G. *toxicus* dinoflagellates used in our studies. We are grateful to Mrs Lucette Faille for technical assistance and Dr Brigitte Berland (Centre d'Océanologie de Marseille, Station Marine d'Endoume) for the scanning electron micrograph image contained in Figure 4.3. We thank Dr Arnold Kreger, (University of Maryland School of Medicine, USA) for his assistance in copyediting the manuscript. The ciguatoxin and brevetoxin studies in the authors' laboratory were supported by the CNRS and by the Direction des Systèmes de Forces et de la Prospective (grant 01 34 029).

REFERENCES

[1] Yasumoto, T. and Murata, M. (1993). Marine toxins. *Chem. Rev.* **93**, 1897–909.
[2] Schiavo, G., Matteoli, M. and Montecucco, C. (2000). Neurotoxins affecting neuroexocytosis. *Physiol. Rev.* **80**, 717–66.
[3] Katz, B. (1969). The release of neurotransmitter substances. Liverpool Univ. Press, Liverpool, U.K.
[4] Van Der Kloot, W. and Molgó, J. (1994). Quantal acetylcholine release at the vertebrate neuromuscular junction. *Physiol. Rev.* **74**, 899–991.
[5] Scheuer, P.J., Takahashi, W., Tsutsumi, J. and Yoshida, T. (1967). Ciguatoxin isolation and chemical nature. *Science* **155**, 1267–68.
[6] Tachibana, K., Nukina, M., Joh, Y. and Scheuer, P. (1987). Recent developments in the molecular structure of ciguatoxin. *Biol. Bull.* **172**, 122–7.
[7] Legrand, A.M., Litaudon, M., Genthon, J.N., Bagnis R. and Yasumoto, T. (1989). Isolation and some properties of ciguatoxin. *J. Appl. Phycol.* **1**, 183–8.
[8] Legrand, A.M., Fukui, M., Cruchet, P., Ishibashi, Y. and Yasumoto, T. (1992). Characterization of ciguatoxins from different fish species and wild *Gambierdiscus toxicus*. In: *Proceedings of the Third International Conference on Ciguatera Fish Poisoning* (T.R. Tosteson, ed.), Polyscience Publications, Quebec. pp. 25–32.
[9] Murata, M., Legrand, A.M., Ishibashi, Y. and Yasumoto, T. (1989). Structures and configurations of ciguatoxin and its congener. *J. Am. Chem. Soc.* **111**, 8929–31.
[10] Murata, M., Legrand, A.M., Ishibashi, Y., Fukui, M. and Yasumoto, T. (1990). Structures and configurations of ciguatoxin from the Moray eel *Gymnothorax javanicus* and its likely precursor from the dinoflagellate *Gambierdiscus-toxicus*. *J. Am. Chem. Soc.* **112**, 4380–6.

[11] Lewis, R.J., Sellin, M., Poli, M.A., Norton, R.S., Macleod, J.K. and Sheil, M.M. (1991). Purification and characterization of ciguatoxins from Moray eel (*Lycodontis javanicus*, muraenidae). *Toxicon* **29**, 1115–27.

[12] Satake, M., Fukui, M., Legrand, A.M., Cruchet, P. and Yasumoto, T. (1998). Isolation and structures of new ciguatoxin analogs, 2,3-dihydroxyCTX3C and 51-hydroxyCTX3C, accumulated in tropical reef fish. *Tetrahedron Lett.* **39**, 1197–8.

[13] Vernoux, J.P. and Lewis, R.J. (1997). Isolation and characterisation of Caribbean ciguatoxins from the horse-eye jack (*Caranx latus*). *Toxicon* **35**, 889–900.

[14] Lewis, R.J., Vernoux, J.P. and Brereton, M. (1998). Structure of Caribbean ciguatoxin isolated from *Caranx latus*. *J. Am. Chem. Soc.* **120**, 5914–20.

[15] Bagnis, R., Kuberski, T. and Laugier, S. (1979). Clinical observations on 3,009 cases of ciguatera (fish poisoning) in the South Pacific. *Am. J. Trop. Med. Hyg.* **28**, 1067–73.

[16] Lawrence, D.N., Enriquez, M.B., Lumish, R.M. and Maceo, A. (1980). Ciguatera fish poisoning in Miami. *J.A.M.A.* **244**, 254–8.

[17] Whiters, N.W. (1992). Ciguatera fish poisoning. *Ann. Rev. Med.* **33**, 97–111.

[18] Gillespie, N.C., Lewis, R.J., Pearn, J., Burke, A.T.C., Holmes, M.J., Bourke, J.B. and Shields, W.J. (1986). Ciguatera in Australia: Occurrence, clinical features, pathophysiology and management. *Med. J. Aust.* **145**, 584–90.

[19] Gollop, J.H. and Pon, E.W. (1992). Ciguatera: a review. *Hawaii Med. J.* **51**, 91–9.

[20] Swift, A.E.B. and Swift, T.R. (1993). Ciguatera. *J. Toxicol.-Clin. Toxicol.* **31**, 1–29.

[21] Laurent, D., Bourdy, G., Amade, P., Cabalion, P. and Bourret, D. (1993). La gratte ou ciguatera. Ses remèdes traditionnels dans le Pacifique Sud. Ed. ORSTOM, Paris.

[22] Lange, W.R. (1994). Ciguatera fish poisoning. *Am. Fam. Phys.* **50**, 579–84.

[23] Glaziou, P. and Legrand, A.M. (1994). The epidemiology of ciguatera fish poisoning. *Toxicon* **32**, 863–73.

[24] Quod, J.P. and Turquet, J. (1996). Ciguatera in Reunion Island (SW Indian Ocean): epidemiology and clinical patterns. *Toxicon* **34**, 779–85.

[25] Molgó, J., Benoit, E., Legrand, A.M. and Kreger, A.S. (1999). Bioactive agents involved in fish poisoning: an overview. In: *Proceedings of the 5th Indo-Pacific Fish Conference* (B. Séret and J.Y. Sire, eds), Soc. Fr. Ichtyol., Paris. pp. 721–38.

[26] Lehane, L. and Lewis, R.J. (2000). Ciguatera: recent advances but the risk remains. *Int. J. Food Microbiol.* **61**, 91–125.

[27] Lewis, R.J. (2001). The changing face of ciguatera. *Toxicon* **39**, 97–106.

[28] Pottier, I., Vernoux, J-P. and Lewis, R.J. (2001). Ciguatera fish poisoning in the Caribbean islands and Western Atlantic. *Rev. Environ. Contam. Toxicol.* **168**, 99–141.

[29] Pearn, J. (2001). Neurology of ciguatera. *J. Neurol. Neurosurg. Psychiatry* **70**, 4–8.

[30] Bagnis, R., Chanteau, S. and Yasumoto, T. (1977). Découverte d'un agent étiologique vraisemblable de la ciguatera. *C.R. Acad. Sci. Ser. B, Paris* **28**, 105–8.

[31] Yasumoto, T., Nakajima, I., Bagnis, R. and Adachi, R. (1977). Finding a dinoflagellate as a likely culprit of ciguatera. *Bull. Jpn. Soc. Sci. Fish.* **43**, 1021–6.

[32] Adachi, R. and Fukuyo, Y. (1979). The thecal structure of a marine toxic dinoflagellate *Gambierdiscus toxicus* gen. et sp. nov. collected in a ciguatera-endemic area. *Bull. Jpn. Soc. Sci. Fish.* **45**, 67–71.

[33] Holmes, M.J., Lewis, R.J., Poli, M.A. and Gillespie, N.C. (1991). Strain dependent production of ciguatoxin precursors (gambiertoxins) by *Gambierdiscus toxicus* (*Dinophyceae*) in culture. *Toxicon* **29**, 761–75.

[34] Satake, M., Murata, M. and Yasumoto, T. (1993). The structure of CTX3c, a ciguatoxin congener isolated from cultured *Gambierdiscus toxicus*. *Tetrahedron Lett.* **34**, 1975–8.

[35] Satake, M., Ishibashi, Y., Legrand A.M. and Yasumoto, T. (1997). Isolation and structure of ciguatoxin-4A, a new ciguatoxin precursor, from cultures of dinoflagellate

Gambierdiscus toxicus and parrotfish *Scarus gibbus*. *Biosci. Biotech. Biochem.* **60**, 2103–5.

[36] Lewis, R.J. and Holmes, M.J. (1993). Origin and transfer of toxins involved in ciguatera. *Comp. Biochem. Physiol.* **106C**, 615–28.

[37] Davis, C.C. (1948). *Gymnodinium brevis* sp. nov., a cause of discolored water and animal mortality in the Gulf of Mexico. *Bot. Gaz.* **109**, 358–60.

[38] Lin, Y.Y., Risk, M., Ray, S.M., Engen, D.V., Clardy, J., Golik, J., James, J.C. and Nakanishi, K. (1981). Isolation and structure of brevetoxin B from the 'red tide' dinoflagellate *Ptychodiscus brevis* (*Gymnodinium breve*). *J. Am. Chem. Soc.* **103**, 6773–5.

[39] Shimizu, Y., Chou, H.N., Bando, H., Van Duyne, G., and Clardy, J.C. (1986). Structure of brevetoxin A (GB-1 toxin), the most potent toxin in the Florida red tide organism *Gymonidinium breve* (*Ptychodiscus brevis*). *J. Am. Chem. Soc.* **108**, 505–14.

[40] Baden, D.G. (1989). Brevetoxins: unique polyether dinoflagellate toxins. *FASEB J.* **3**, 1807–17.

[41] Morris, P.D., Campbell, D.S., Taylor, T.J. and Freeman, J.I. (1991). Clinical and epidemiological features of neurotoxic shellfish poisoning in North Caroline. *Am. J. Public Health* **81**, 471–4.

[42] Dickey, R., Jester, E., Granade, R., Mowdy, D., Moncreiff, C., Rebarchik, D., Robl, M., Musser, S. and Poli, M. (1999). Monitoring brevetoxins during a *Gymnodinium breve* red tide: comparison of sodium channel specific cytotoxicity assay and mouse bioassay for determination of neurotoxic shellfish toxins in shellfish extracts. *Nat. Toxins* **7**, 157–65.

[43] Poli, M.A., Musser, S.M., Dickey, R.W., Eilers, P.P. and Hall, S. (2000). Neurotoxic shellfish poisoning and brevetoxin metabolites: a case study from Florida. *Toxicon* **38**, 981–93.

[44] Poli, M.A., Mende, T.J. and Baden, D.G. (1986). Brevetoxins, unique activators of voltage-sensitive sodium channels, bind to specific sites in rat brain synaptosomes. *Mol. Pharmacol.* **30**, 129–35.

[45] Baden, D.G. and Tomas, C.R. (1988). Variations in major toxin composition for six clones of *Ptychodiscus brevis*. *Toxicon* **26**, 961–3.

[46] Gawley, R.E., Rein, K.S., Kinoshita, M. and Baden D.G. (1992). Binding of brevetoxins and ciguatoxin to the voltage-sensitive sodium channel and conformational analysis of brevetoxin B. *Toxicon* **30**, 780–85.

[47] Rein, K.S., Baden, D.G. and Gawley, R.E. (1994). Conformational analysis of the sodium channel modulator, brevetoxin A, comparison with brevetoxin B conformations, and a hypothesis about the common pharmacophore of the 'site 5' toxins. *J. Org. Chem.* **59**, 2101–6.

[48] Nicolaou, K.C., Yang, Z., Shi, G., Gunzner, J.L., Agrios, K.A. and Gartner, P. (1998). Total synthesis of brevetoxin A. *Nature* **392**, 264–9.

[49] Culotta, A. (1992). Red menace in the world's oceans. *Science* **257**, 1476–7.

[50] Bossart, G.D., Baden, D.G., Ewing, R.Y., Roberts, B. and Wright, S.D. (1998). Brevetoxicosis in manatees (*Trichechus manatus latirostris*) from the 1996 epizootic: gross, histologic, and immunohistochemical features. *Toxicol. Pathol.* **26**, 276–82.

[51] Pierce, R.H. (1986). Red tide (*Ptychodiscus brevis*) toxin aerosols: a review. *Toxicon* **24**, 955–66.

[52] Barchi, R.L. (1998). Ion channel mutations affecting muscle and brain. *Curr. Opin. Neurol.* **11**, 461–8.

[53] Catterall, W.A. (1995). Structure and function of voltage-gated ion channels. *Annu. Rev. Biochem.* **64**, 493–531.

[54] Catterall, W.A. (2000). From ionic currents to molecular mechanisms: the structure and function of voltage-gated sodium channels. *Neuron* **26**, 13–25.

[55] Gordon, D. (1997). Sodium channels as targets for neurotoxins: mode of action and interaction of neurotoxins with receptor sites on sodium channels. In: *Toxins and signal transduction, cellular and molecular mechanisms of toxins action series* (Y. Gutman and P. Lazarowici, eds), Harwood Acad. Pub., The Netherlands. pp. 119–49.

[56] Marban, E., Yamagishi, T. and Tomaselli, G.F. (1998). Structure and function of voltage-gated sodium channels. *J. Physiol. (London)* **508**, 647–57.

[57] Sharkey, R.G., Jover, E., Couraud, F., Baden, D.G. and Catterall, W.A. (1987). Allosteric modulation of neurotoxin binding to voltage-sensitive sodium channels by *Ptychodiscus brevis* toxin 2. *Mol. Pharmacol.* **31**, 273–8.

[58] Trainer, V.L., Thomsen, W.J., Catterall, W.A. and Baden, D.G. (1991). Photoaffinity labeling of the brevetoxin receptor on sodium channels in rat brain synaptosomes. *Mol. Pharmacol.* **40**, 988–94.

[59] Trainer, V.L., Baden, D.G. and Catterall, W.A. (1994). Identification of peptide components of the brevetoxin receptor site of rat brain sodium channels. *J. Biol. Chem.* **269**, 19904–9.

[60] Bidard, J.N., Vijverberg, H.P.M., Frelin, C., Chungue, E., Legrand, A.M., Bagnis, R., and Lazdunski, M. (1984). Ciguatoxin is a novel type of Na^+ channel toxin. *J. Biol. Chem.* **259**, 8353–7.

[61] Lombet, A., Bidard, J.N. and Lazdunski, M. (1987). Ciguatoxin and brevetoxins share a common receptor site on the neuronal voltage-dependent Na^+ channel. *FEBS Lett.* **219**, 355–9.

[62] Pauillac, S., Blehaut, J., Cruchet, P., Lotte, C. and Legrand, A.M. (1995). Recent advances in detection of ciguatoxins in French Polynesia. In: *Harmful marine algal blooms* (P. Lassus, G. Arzul, E. Erard, P. Gentien and C. Marcaillou, eds), Lavoisier Intersept Ltd. pp. 801–8.

[63] Dechraoui, M.Y., Naar, J., Pauillac, S. and Legrand, A.M. (1999). Ciguatoxins and brevetoxins, neurotoxic polyether compounds active on sodium channels. *Toxicon* **37**, 125–43.

[64] Benoit, E., Legrand, A.M. and Dubois, J.M. (1986). Effects of ciguatoxin on current and voltage clamped frog myelinated nerve fibre. *Toxicon* **24**, 357–64.

[65] Benoit, E., Juzans, P., Legrand, A.M. and Molgó, J. (1996). Nodal swelling produced by ciguatoxin-induced selective activation of sodium channels in myelinated nerve fibers. *Neuroscience* **71**, 1121–31.

[66] Benoit, E., Mattei, C., Legrand, A.M. and Molgó, J. (1999). Ionic basis of the neurocellular actions of ciguatoxins implicated in ciguatera fish poisoning. In: *Proceedings of the 5th Indo-Pacific Fish Conference* (B. Séret and J.Y. Sire, eds), Soc. Fr. Ichtyol., Paris. pp.745–58.

[67] Strachan, L.C., Lewis, R.J. and Nicholson, G.M. (1999). Differential actions of Pacific ciguatoxin-1 on sodium channel subtypes in mammalian sensory neurons. *J. Pharmacol. Exp. Ther.* **288**, 379–88.

[68] Hogg, R.C., Lewis, R.J. and Adams, D.J. (1998). Ciguatoxin (CTX-1) modulates single tetrodotoxin-sensitive sodium channels in rat parasympathetic neurones. *Neurosci. Lett.* **252**, 103–6.

[69] Huang, J.M.C., Wu, C.H. and Baden, D.G. (1984). Depolarizing action of a red-tide dinoflagellate brevetoxin on axonal membranes. *J. Pharmacol. Exp. Ther.* **229**, 615–21.

[70] Atchison, W.D., Scrugs Luke, V., Narahashi, T. and Vogel, S.M. (1986). Nerve membrane sodium channels as the target site of brevetoxins at neuromuscular junctions. *Br. J. Pharmacol.* **89**, 731–38.

[71] Sheridan, R.E. and Adler, M. (1989). The actions of a red tide toxin from *Ptychodiscus brevis* on single sodium channels in mammalian neuroblastoma cells. *FEBS Lett.* **247**, 448–52.

[72] Gawley, R.E., Rein, K.S., Jeglitsch, G., Adams, D.J., Theodorakis, E.A., Tiebes, J., Nicolaou, K.C. and Baden, D.G. (1995). The relationship of brevetoxin 'length' and A-ring functionality to binding and activity in neuronal sodium channels. *Chem. & Biol.* **2**, 533–41.

[73] Jeglitsch, G., Rein, K., Baden, D.G. and Adams, D.J. (1998). Brevetoxin-3 (PbTx-3) and its derivatives modulate single tetrodotoxin-sensitive sodium channels in rat sensory neurons. *J. Pharmacol. Exp. Ther.* **284**, 516–25.

[74] Purkerson, S.L., Baden, D.G. and Fieber, L.A. (1999). Brevetoxin modulates neuronal sodium channels in two cell lines derived from rat brain. *Neurotoxicology* **20**, 909–20.

[75] Molgó, J., Comella, J.X. and Legrand, A.M. (1990). Ciguatoxin enhances quantal transmitter release from frog motor nerve terminals. *Br. J. Pharmacol.* **99**, 695–700.

[76] Molgó, J., Benoit, E., Comella, J.X. and Legrand, A.M. (1992). Ciguatoxin: a tool for research on sodium-dependent mechanisms. In: *Methods in Neuroscience, Neurotoxins* (P.M. Conn, ed.), Academic Press, New York. vol. **8**, pp. 149–64.

[77] Molgó, J., Benoit, E., Mattei, C. and Legrand, A.M. (1999). Neuropathology of ciguatera fish poisoning: involvement of voltage-dependent sodium channels. *Neuropathol. Appl. Neurobiol.* **25** (Suppl.1), 4–5.

[78] Benoit, E. and Legrand, A.M. (1994). Gambiertoxin-induced modifications of the membrane potential of myelinated nerve fibres. *Memoirs Qld Mus.* **34**, 461–4.

[79] Hamblin, P.A., McLachlan, E.M. and Lewis, R.J. (1995). Sub-nanomolar concentrations of ciguatoxin-1 excite preganglionic terminals in guinea pig sympathetic ganglia. *Naunyn Schmiedeberg's Arch. Pharmacol.* **352**, 236–46.

[80] Brock, J.A, McLachlan, E.M., Jobling, P. and Lewis, R.J. (1995). Electrical activity in rat tail artery during asynchronous activation of postganglionic nerve terminals by ciguatoxin-1. *Br. J. Pharmacol.* **116**, 2213–20.

[81] Marquais, M., Vernoux, J.P., Molgó, J., Sauviat, M.P. and Lewis, R.J. (1998). Isolation and electrophysiological characterization of a new ciguatoxin extracted from Caribbean fish. In: *Harmful Algae.* (B. Reguera, J. Blanco, M.L. Fernandez and T. Wyatt, eds), Xunta de Galicia-Intergovernmental Oceanographic Commission of UNESCO. pp. 476–7.

[82] Mattei, C., Molgó, J., Marquais, M., Vernoux, J.P. and Benoit, E. (1999). Hyperosmolar D-mannitol reverses the increased membrane excitability and the nodal swelling caused by Caribbean ciguatoxin-1 in single frog myelinated axons. *Brain Res.* **847**, 50–8.

[83] Mattei, C., Molgó, J., Legrand, A.M. and Benoit, E. (1999). Ciguatoxines et brévétoxines: dissection de leurs actions neurobiologiques. *J. Soc. Biol.* **193**, 329–344.

[84] Mattei, C., Benoit, E., Juzans, P., Legrand, A.M. and Molgó, J. (1997). Gambiertoxin (CTX-4B), purified from wild *Gambierdiscus toxicus* dinoflagellates, induces Na$^+$-dependent swelling of single frog myelinated axons and motor nerve terminals *in situ.* *Neurosci. Lett.* **234**, 75–8.

[85] Mattei, C., Dechraoui, M.Y., Molgó, J., Meunier, F.A., Legrand, A.M. and Benoit, E. (1999). Neurotoxins targeting receptor site 5 of voltage-dependent sodium channels increase the nodal volume of myelinated axons. *J. Neurosci. Res.* **55**, 666–73.

[86] Flowers, A.E., Capra, M.F. and Cameron, J. (1988). The effects of ciguatoxin on nerve conduction parameters in teleost fish. In: *Progress in Venom and Toxin research, Proceedings of the first Asia-Pacific Congress on Animal, Plant and Microbial Toxins,* (P. Gopalakrishnakone and C.K. Tan, eds), Natl. Univ. Singapore. pp. 411–17.

[87] Cameron, J., Flowers, A.E. and Capra, M.F. (1991). Effects of ciguatoxins on nerve excitability in rats (Part I). *J. Neurol. Sci.* **101**, 87–92.

[88] Allsop J.L., Martini L., Lebris H., Pollard J., Walsh J. and Hodgkinson, S. (1986).
      Neurologic manifestations of ciguatera. 3 cases with a neurophysiologic study and
      examination of one nerve biopsy. *Rev. Neurol. (Paris)* **142**, 590–7.
[89] Cameron, J., Flowers, A.E. and Capra, M.F. (1991). Electrophysiological studies
      on ciguatera poisoning in man (Part II). *J. Neurol. Sci.* **101**, 93–7.
[90] Molgó, J., Juzans, P. and Legrand, A.M. (1994). Confocal laser scanning micro-
      scopy: a new tool for studying the effects of ciguatoxin (CTX-1b) and mannitol at
      motor nerve terminals of the neuromuscular junction *in situ. Memoirs Qld Mus.* **34**,
      577–85.
[91] Zucker, R.S. (1996). Exocytosis: a molecular and physiological perspective.
      Neuron **17**, 1049–55.
[92] Augustine, G.J. (2001). How does calcium trigger neurotransmitter release ? *Curr.
      Opin. Neurobiol.* **11**, 320–6.
[93] Wu, C.H. and Narahashi, T. (1988). Mechanism of action of novel marine neuro-
      toxins on ion channels. *Ann. Rev. Pharmacol. Toxicol.* **28**, 141–61.
[94] Molgó, J., Shimahara, T., Comella, J.X., Morot Gaudry-Talarmain Y. and
      Legrand, A.M. (1992). Ciguatoxin-induced changes in acetylcholine release and
      in cytosolic calcium levels. *Bull. Soc. Pathol. Ex.* **85**, 486–8.
[95] Molgó, J., Shimahara, T. and Legrand, A.M. (1993). Ciguatoxin, extracted from
      poisonous morays eels, causes sodium-dependent calcium mobilization in NG108-
      15 neuroblastoma × glioma hybrid cells. *Neurosci. Lett.* **158**, 147–50.
[96] Berman, F.W. and Murray, T.F. (2000). Brevetoxin-induced autocrine excitotoxi-
      city is associated with manifold routes of Ca^{2+} influx. *J. Neurochem.* **74**, 1443–51.
[97] Molgó, J., Comella J.X., Shimahara, T. and Legrand, A.M. (1991). Tetrodotoxin-
      sensitive ciguatoxin effects on quantal release, synaptic vesicle depletion, and
      calcium mobilization. *Ann. N. Y. Acad. Sci.* **635**, 485–9.
[98] Gusovsky, F., Hollingsworth, E.B. and Daly, J.W. (1986). Regulation of phos-
      phatidylinositol turnover in brain synaptoneurosomes: stimulatory effects
      of agents that enhance influx of sodium ions. *Proc. Natl. Acad. Sci. USA* **83**,
      3003–7.
[99] Gusovsky, F., McNeal, E.T. and Daly, J.W. (1987). Stimulation of phosphoinosi-
      tide breakdown in brain synaptoneurosomes by agents that activate sodium influx:
      antagonism by tetrodotoxin, saxitoxin, and cadmium. *Mol. Pharmacol.* **32**, 479–87.
[100] Carrasco, M.A., Morot-Gaudry, Y. and Molgó, J. (1996). Ca^{2+}-dependent
      changes of acetylcholine release and IP_3 mass in *Torpedo* cholinergic synapto-
      somes. *Neurochem. Int.* **29**, 637–43.
[101] Yano, K., Higashida, H., Inoue, R. and Nozawa, Y. (1984). Bradykinin-induced
      rapid breakdown of phosphatidylinositol 4,5-bisphosphate in neuroblastoma ×
      glioma hybrid NG108-15 cells. *J. Biol. Chem.* **259**, 10201–7.
[102] Higashida, H. and Ogura, A.. (1991). Inositol trisphosphate/calcium-dependent
      acetylcholine release evoked by bradykinin in NG108-15 rodent hybrid cells. *Ann.
      N. Y. Acad. Sci.* **635**, 153–66.
[103] Mattei, C., Benoit, E., Darchen, F., Legrand, A.M. and Molgó, J. (1999). Acti-
      vation of Na^+ channels by ciguatoxin-1B produces a localized increase of intracel-
      lular Ca^{2+} in chromaffin cells. *Biochimie* **81** (Suppl. 6), S191.
[104] Burgoyne, R.D. (1991). Control of exocytosis in adrenal chromaffin cells. *Bio-
      chem. Biophys. Acta* **1071**, 174–202.
[105] Meunier, F.A., Mattei, C., Chameau, P., Lawrence, G., Colasante, C., Kreger,
      A.S., Dolly, J.O. and Molgó, J. (2000). Trachynilysin mediates SNARE-dependent
      release of catecholamines from chromaffin cells *via* external and stored Ca^{2+}. *J.
      Cell Sci.* **113**, 1119–25.
[106] Rahamimoff, R. and Fernandez, J.M. (1997). Pre- and postfusion regulation of
      transmitter release. *Neuron* **18**, 17–27.

[107] Neher, E. (1998). Vesicle pools and Ca^{2+} microdomains: new tools for understanding their roles in neurotransmitter release. *Neuron* **20**, 389–99.
[108] Jahn, R. and Südhof, T.C. (1999). Membrane fusion and exocytosis. *Annu. Rev. Biochem.* **68**, 863–911.
[109] Molgó, J., Morot-Gaudry-Talarmain Y., Legrand A.M. and Moulian, N. (1993). Ciguatoxin extracted from poisonous moray eels (*Gymnothorax javanicus*) triggers acetylcholine release from *Torpedo* cholinergic synaptosomes via reversed Na^+-Ca^{2+} exchange. *Neurosci. Lett.* **160**, 65–8.
[110] Morot Gaudry, Y., Molgó, J., Meunier, F.A., Moulian, N. and Legrand, A.M. (1996). Reversed mode Na^+-Ca^{2+} exchange activated by ciguatoxin (CTX-1b) enhances acetylcholine from Torpedo cholinergic synaptosomes. *Ann. N. Y. Acad. Sci.* **779**, 404–6.
[111] Wu, C.H., Huang, J.M.C., Vogel, S.M., Luke, V.S., Atchison, W.D. and Narahashi, T. (1985). Actions of *Ptychodiscus brevis* toxins on nerve and muscle membranes. *Toxicon* **23**, 481–8.
[112] Meunier, F.A., Colasante, C. and Molgó, J. (1994). Sodium-dependent increase of quantal acetylcholine release from motor endings by brevetoxin (PbTx-3). *J. Physiol. (Paris)* **88**, 387.
[113] Meunier, F.A., Colasante, C. and Molgó, J. (1997). Sodium-dependent increase in quantal secretion induced by brevetoxin-3 in Ca^{2+}-free medium is associated with depletion of synaptic vesicles and swelling of motor nerve terminals *in situ*. *Neuroscience* **78**, 883–93.
[114] Meir, A., Ginsburg, S., Butkevich, A., Kachalsky, S.G., Kaiserman, I., Ahdut, R., Demirgoren, S. and Rahamimoff, R. (1999). Ion channels in presynaptic nerve terminals and control of transmitter release. *Physiol. Rev.* **79**, 1019–88.
[115] Littleton, J.T. and Bellen, H.J. (1995). Synaptotagmin controls and modulates synaptic-vesicle fusion in a Ca^{2+}-dependent manner. *Trends Neurosci.* **18**, 177–83.
[116] Calakos, N. and Scheller, R.H. (1996). Synaptic vesicle biogenesis, docking, and fusion. A molecular description. *Physiol. Rev.* **76**, 1–29.
[117] Schiavo, G., Osborne, S.L. and Sgouros, J.G. (1998). Synaptotagmins: more isoforms than functions? *Biochem. Biophys. Res. Commun.* **248**, 1–8.
[118] Sampo, B., Tricaud, N., Levêque, C., Seagar, M., Couraud, F. and Dargent B. (2000). Direct interaction between synaptotagmin and the intracellular loop I-II of neuronal voltage-sensitive sodium channels. *Proc. Natl. Acad. Sci. USA* **97**, 3666–71.
[119] Levêque, C., el Far, O., Martin-Moutot, N., Sato, K., Kato, R., Takahashi, M. and Seagar, M.J. (1994). Purification of the N-type calcium channel associated with syntaxin and synaptotagmin. A complex implicated in synaptic vesicle exocytosis. *J. Biol. Chem.* **269**, 6306–12.
[120] Martin-Moutot, N., Charvin, N., Levêque, C., Sato, K., Nishiki, T., Kozaki, S., Takahashi, M. and Seagar, M. (1996). Interaction of SNARE complexes with P/Q-type calcium channels in rat cerebellar synaptosomes. *J. Biol. Chem.* **271**, 6567–70.
[121] Betz, W.J., Mao, F. and Bewick, G.S. (1992). Activity-dependent fluorescent staining and destaining of living vertebrate motor nerve terminals. *J. Neurosci.* **12**, 363–75.
[122] Blasi, J., Chapman, E.R., Link, E., Binz, T., Yamasaki, S., De Camilli, P., Südhof, T.C., Niemann, H. and Jahn, R. (1993). Botulinum neurotoxin A selectively cleaves the synaptic protein SNAP-25. *Nature* **365**, 160–3.
[123] Heuser, J.E. and Reese, T.S. (1973). Evidence for recycling of synaptic vesicle membrane during transmitter release at the frog neuromuscular junction. *J. Cell Biol.* **57**, 315–44.
[124] De Camilli, P. and Takei, K. (1996). Molecular mechanisms in synaptic vesicle endocytosis and recycling. *Neuron* **16**, 481–6.

[125] Koenig, J.H. and Ikeda, K. (1989). Disappearance and reformation of synaptic vesicle membrane upon transmitter release observed under reversible blockage of membrane retrieval. *J. Neurosci.* **9**, 3844–60.

[126] Ceccarelli, B. and Hurlbut, W.P. (1980). Vesicle hypothesis of the release of quanta of acetylcholine. *Physiol. Rev.* **60**, 396– 441.

[127] Fesce, R., Grohovaz, F., Valtorta, F. and Meldolesi, J. (1994). Neurotransmitter release: fusion or 'kiss-and-run'? *Trends Cell Biol.* **4**, 1– 4.

[128] Almers, W. and Tse, F.W. (1990). Transmitter release from synapses: does a preassembled fusion pore initiate exocytosis ? *Neuron* **4**, 813–18.

[129] Neher, E. (1993). Secretion without full fusion. *Nature* **363**, 497–8.

[130] Betz, W.J. and Bewick, G.S. (1992). Optical analysis of synaptic vesicle recycling at the frog neuromuscular junction. *Science* **255**, 200–3.

[131] Betz, W.J. and Wu, L.G. (1995). Synaptic transmission. Kinetics of synaptic-vesicle recycling. *Curr. Biol.* **5**, 1098–101.

[132] Cochilla, A.J., Angleson, J.K. and Betz, W.J. (1999). Monitoring secretory membrane with FM1-43 fluorescence. *Ann. Rev. Neurosci.* **22**, 1–10.

[133] Henkel, A.W., Lubke, J. and Betz, W.J. (1996). FM1-43 dye ultrastructural localization in and release from frog motor nerve terminals. *Proc. Natl. Acad. Sci. USA* **93**, 1918–23.

[134] Koenig, J.H. and Ikeda, K. (1996). Synaptic vesicles have two distinct recycling pathways. *J. Cell Biol.* **135**, 797–808.

[135] Zhang, B., Koh, Y.H., Beckstead, R.B., Budnik, V., Ganetzky, B. and Bellen, H.J. (1998). Synaptic vesicle size and number are regulated by a clathrin adaptor protein required for endocytosis. *Neuron* **21**, 1465–75.

[136] Ringstad, N., Gad, H., Low, P., Di Paolo, G., Brodin, L., Shupliakov, O. and De Camilli, P. (1999). Endophilin/SH3p4 is required for the transition from early to late stages in clathrin-mediated synaptic vesicle endocytosis. *Neuron* **24**, 143–54.

[137] Teng, H., Cole, J.C., Roberts, R.L., and Wilkinson, R.S. (1999). Endocytic active zones: hot spots for endocytosis in vertebrate neuromuscular terminals. *J. Neurosci.* **19**, 4855–66.

[138] Hirama, M., Oishi, T., Uehara, H., Inoue, M., Maruyama, M., Oguri, H. and Satake, M. (2001). Total synthesis of ciguatoxin CTX3C. *Science* **294**, 1904–07.

[139] Markó, I.E. (2001) The art of total synthesis. *Science* **294**, 1842–43.

Part II   Animal Toxins and
New Methodologies

5 Role of Discovery Science in Toxinology: Examples in Venom Proteomics

JAY W. FOX*, JOHN D. SHANNON, BJARKI STEFANSSON,
AURA S. KAMIGUTI, R. DAVID G. THEAKSTON, SOLANGE
M.T. SERRANO, ANTONIO C.M. CAMARGO and NICHOLAS
SHERMAN

5.1 INTRODUCTION

Much of our current understanding of biology is a result of the overarching paradigm that structure defines function. Over the past thirty years there have been tremendous advances in biological and biomedical sciences. Directly related to these advances is the development of sophisticated, sensitive instrumentation for the analyses of biomolecular structures. Examples of such instrumentation that were critical to structural determinations during the late 1970s and throughout the 1980s were the high-sensitivity gas phase automated protein sequencer [1] and the automated DNA sequencers [2]. These two instruments allowed for the analysis of protein and DNA structure at levels of sensitivity and high throughput that gave rise to exponential growth in the protein and DNA sequence databases. The field of venom toxinology certainly took advantage of these technologies with many cDNA and protein sequences of toxins published during this time and continuing through to the present [3–5]. Understanding of the structures of these venom proteins has certainly advanced the field of toxinology and most importantly has shed light on nontoxic structural homologues identified in other biological systems ([6]).

The most recent technological advances stimulating biological and biomedical research are protein identification and sequencing by mass spectrometry together with gene expression analysis using high-density 'gene chips' [7–9]. These technologies have led to the very recent explosion of knowledge in the field of proteomics, and functional genomics promises to continue to expand our understanding of biological systems and the interactivity of these structural 'networks' at the cellular, tissue, organ and organ system levels.

---

* Corresponding author

Perspectives in Molecular Toxinology
Edited by A Ménez © 2002 John Wiley & Sons, Ltd

The sheer power of these techniques has suggested to some investigators that an additional approach to scientific analysis of biological systems should be considered. As opposed to the traditional 'hypothesis-driven' approach to scientific exploration, perhaps a different approach, one that exploits these new technologies, should be followed. This concept has best been stated by Dr Leroy Hood, of the Institute for Systems Biology [10]. Essentially, Dr Hood's thesis supports a complementary approach to the hypothesis-driven system, which he has termed 'discovery-driven' science. In the latter case, of discovery-driven science, all the current tools of high-throughput protein identification and gene expression profiling are brought to bear on a particular biological system in order to elucidate the biomolecules active in that system, and more importantly to clarify their functional relationships during various biological processes [11, 12].

In this chapter we will explore the use of some of these technologies in increasing our understanding of the very complex protein systems represented by snake venoms with an appraisal of this discovery-based scientific approach to venom toxinology.

5.2   2-D GEL ELECTROPHORESIS/PROTEOME ANALYSIS OF SNAKE VENOMS

The use of two-dimensional gel electrophoresis as an analytical tool to explore complex protein systems has been available for over 25 years [13]. However, until recently full exploitation of the informational content in the 2-D gel was limited to a comparison of the patterns of proteins observed in the gels. Although techniques for the electrophoretic elution or blotting to a matrix have been available for approximately 15 years [14, 15], in practical terms this is a very low throughput approach requiring Edman chemistry protein sequence analysis of the proteins. Although feasible, there are significant restrictions associated with this approach dictated by the amounts of protein required for analysis and the strict requirement for a free amino terminal. With the advent of high-sensitivity protein identification/sequence analysis using mass spectrometry [9], in conjunction with new techniques and instrumentation for 2-D gel electrophoresis [16], it is now possible to readily identify proteins resolved by 2-D gel electrophoresis at femtomole to attomole levels with a reasonable throughput, in spite of the fact that the technology is still rather complex and expensive.

One group of investigators has examined venom complexity and diversity using 2-D gel electrophoresis [17] with very interesting results. In this study, the venoms from the yellow-lipped sea snake, *Laticauda colubrina* (Hydrophiidae) and from the land snake, Russell's viper, *Vipera russelli* (Viperidae) were analysed by 2-D gel electrophoresis and some of the proteins in the gel were identified by amino acid analysis and/or N-terminal Edman sequencing. This work clearly underscored the potential for complex venom analysis by 2-D gel electrophoresis, but also showed the limitations of protein identification associated with amino acid analysis and N-terminal sequencing in association with 2-D gel electrophoresis.

As an extension of their efforts we have examined the proteome of the venom from *Dispholidus typus* (Boomslang) using both the mass mapping approach and *de novo* mass spectrometric sequencing approaches to determine the potential of these technologies in high-throughput analysis of venom proteins. We next examined *Crotalus atrox* (Western diamondback rattlesnake) and *Bothrops jararaca* (Jararaca) venoms; these two viperid snakes produce similar pathologies associated with human envenoming [18]. The proteomes of these snake venoms were compared and some interesting observations follow from these comparisons. In the case of *C. atrox* venom, we have further examined the use of narrower pI ranges, in an attempt to determine how many proteins might *not* be observed with a broad range pI focusing due to crowding often observed with broad range pI gels. In another example of the use of this technology, we have examined intraspecies venom proteomes. Here, instead of using pooled venoms, we have compared the proteomes of the venom extracted from individual snake specimens. The results of these studies are described below.

5.2.1 *DISPHOLIDUS TYPUS* (BOOMSLANG) PROTEOME

As genomic analyses of various organisms are completed, an approach can be taken to circumvent intense mass spectrometric sequence analysis of tryptic peptides using the simpler and faster method of 'mass mapping' [19]. The mass mapping approach is based on the measurement of the masses of tryptic peptides generated *in situ* of Coomassie- or silver-stained spots/bands in acrylamide gels followed by comparison of those peptide masses with computer-generated tryptic peptide masses for all sequences in the databases [20]. The major limitation of this approach is the strict need for a known database for the proteome/genome of the organism being studied. Slight variations in the tryptic peptide masses due to simple amino acid substitutions of homologous proteins from organisms lacking a complete genome/proteome database will not be identified using this approach.

In Figure 5.1 is shown a comparison of a 1-D gel of Boomslang venom approximately aligned with the 2-D gel of the venom. The five major protein bands from the 1-D gel were analysed by liquid chromatography/mass spectrometry/mass spectrometry (LC/MS/MS) and the most abundant proteins in each band identified. Approximately 100 proteins can be seen in the corresponding 2-D gel, clearly indicating the power of the technique to resolve complex protein systems. At least five proteins observed in the 1-D gel were identified from the major bands using LC/MS/MS sequencing methods. The spots in the 2-D gel were analysed using the mass mapping approach. With this technique, no spots were identified, exemplifying the need for a 'known' genome for successful application of this approach. For example, we have performed similar mass mapping analyses of 2-D gels from human cell lines

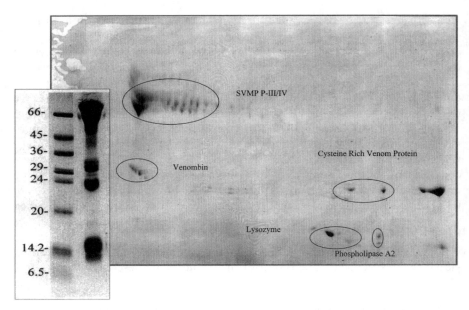

Figure 5.1   Comparison of 1-D and 2-D gel electrophoresis of *Dispholidus typus* (Boomslang) venom. Acrylamide (12%) was used for the 1-D gel and 20 μg of pooled venom from four specimens of Boomslang was electrophoresed under reducing conditions. The five major bands were sequenced by LC/MS/MS with a Finnigan LCQ mass spectrometer. For the 2-D gel, 3–10 IPG strips were used for the first dimension followed by electrophoresis in a 8–18% acrylamide gel of 500 μg of pooled venom. The identification of the major proteins in the five major bands from the 1-D gel is shown next to the isoforms of those proteins resolved in the 2-D gel

with approximately 55% of the proteins identified. Nevertheless, all the protein spots in the 2-D gel of the venom could in principle be identified using the LC/MS/MS approach. Therefore, our conclusion from this study is that the currently available mass mapping techniques with samples of unknown or incomplete genomes will likely not be productive for complete protein identification. However, using LC/MS/MS, although certainly a more laborious system, would result in most if not all spots in the gel being identified.

5.3   PROTEOMIC COMPARISON OF THE VENOMS FROM TWO DIFFERENT SNAKE SPECIES PRODUCING SIMILAR PATHOLOGIES OF ENVENOMING

The pathologies associated with envenoming by *C. atrox* and *B. jararaca* are noted to be similar [21] thus suggesting that the toxic components of the venoms are similar. We therefore compared the proteomes of these venoms by 2-D PAGE. The 2-D gels for pooled venoms from *C. atrox* and *B. jararaca* are shown in Figure 5.2. Using a computer software analysis of the gels (Compugen),

Crotalus atrox Venom   Bothrops jararaca Venom

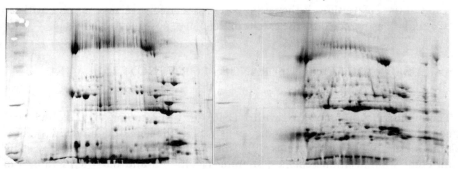

Figure 5.2 Comparison of 2-D gel electrophoresis of pooled *Crotalus atrox* venom and pooled *Bothrops jararaca* venom. 1.8 mg of pooled venoms from each snake was applied to a 3–10 pH IPG strip followed by electrophoresis into a 8–18 % acrylamide gel

Table 5.1 Results from computerised comparison of digitised images of 2D gels of *Crotalus atrox* and *Bothrops jararaca* pooled venoms.

	Total Spots Identified	Unmatched Spots	Up regulated Spots	Down regulated Spots
C. atrox (reference gel)	206	163	7	9
B. jararaca	213	147	—	—

approximately 206 spots were detected in the *C. atrox* gel compared to 213 in the *B. jararaca* gel. Again, using the gel analysis software, the two gels were electronically superimposed to provide a visual sense of the similarities between the two venoms. Only a few spots are identified by the software to be identical as represented by the black images in the electronically reconstructed gel image (see Plate I). The computer analysis of the gel images show that far more spots on the gels were denoted as non-identical (Table 5.1). It is well established from the literature that many, if not most, of the components in these venoms are homologous, but not identical, due to amino acid substitutions and post-translational modifications differentiating them and thus contributing to their non-identical migration profiles in the gels. In this sense, the computer-generated comparative analysis of these venoms does in fact reflect the exact nature of the venoms, i.e. similar compositions sharing many homologous proteins, yet certainly most of the homologous proteins in the two venoms are not identical. The 2-D gels also demonstrate the various levels of different protein classes in the gels. Although proteins from both venoms populate many regions of the two gels, some venoms clearly have a greater number and diversity of proteins in a particular region compared to the other venom. This suggests that although the two venoms do have many homologous proteins in common, the relative amounts of the homologues may vary greatly. It is therefore likely that the end result of these variations is the pathologies associated with envenoming by these two snakes, as reflected by the extent of diversity shown in the 2-D gels.

5.4   USE OF DIFFERENT pI RANGES FOR VENOM 2-D PAGE PROTEOME ANALYSIS

The main problem associated with 2-D PAGE of proteomics is the absence of complete entry of proteins into the gel as well as the lack of complete visualisation of electrophoresed proteins [22]. This problem has been addressed by using different solubilisation schemes, staining techniques and narrow pI ranges in the first dimension [22]. Nevertheless, the problem is still 'what you cannot see, you cannot analyse'. In the case of snake venoms, most proteins in the complex mixture are relatively soluble and have molecular weights of less than 100 kDa, therefore most, if not all proteins would be expected to migrate into the gel. However, visualisation of any individual protein in the gel would still depend on the amount of that protein, its ability to accept effectively a particular staining protocol and the relative congestion caused by other migrating proteins with properties similar to those of the protein of interest. To address the latter question, we performed 2-D electrophoresis of C. atrox venom using successively narrower pI ranges. It can be seen in Figure 5.3 that as the pI range narrows there is an increasing resolution of the proteins within that range, resulting in the resolution of additional spots that were not visualised in the broader pI range gels. Thus, as one would expect, if a complete visualisation of the proteome of a venom is required, additional steps must be taken, including loading different amounts of venom, use of different staining techniques and the use of a variety of pI ranges to decrease areas of protein congestion in the gel. The end result will probably show that the venom contains many more proteins than originally thought and is hence much more complex than a simple broad range pI 2-D gel would indicate.

5.5   PROTEOME COMPARISON OF VENOMS FROM IDENTICAL SNAKE SPECIES

Questions surrounding the similarity of venom from individual specimens of the same species of snakes have long been considered in the field of toxinology [23]. In addition to the scientific importance of these questions regarding the effects of environment, diet, age, etc. on the composition of snake venoms, the practical consideration of the nature of venom used for anti-venom production must also be addressed. For example, is it better to use pooled venoms rather than venom from a single snake? Normally the former option is considered to be preferable because it is assumed that a pool of venom from a large number of

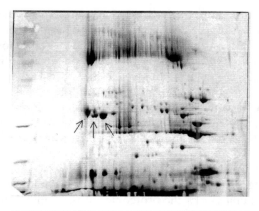

A. pH 3–10 IPG Strip

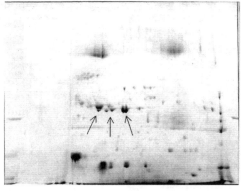

B. pH 4–7 IPG Strip

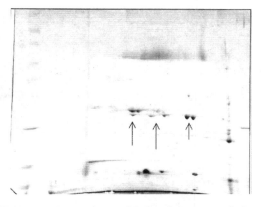

C. pH 4.5–5.5 IPG Strip

Figure 5.3 Comparison of 2-D gel patterns of *Crotalus atrox* venom using different 1st dimension conditions. In gel **A**, the broad pI range 3–10 pI IPG strip was used; in gel **B**, a 4–7 pI range IPG strip was used; and in gel **C**, the very narrow 4.5–5.5 pI range IPG strip was used. In all cases 1 mg of pooled *C. atrox* venom was applied to the IPG strip followed by electrophoresis into 8–18% acrylamide gels. Arrows in each gel point to regions of spot congestion in the broad range gel (**A**) that are resolved in the narrower pI ranges (**B** and **C**)

snakes of different ages and sexes is more likely to produce antibodies which cover the maximum number of toxins present in the venom of that species. However, although unlikely, it may be that a single snake possesses venom containing more numerous and abundant important toxic components than the venoms of other snakes or pooled venoms from many snakes. Likewise, which region or environmental conditions should be considered in terms of venom collection to yield a venom pool with the best antigenic composition for development of anti-venom? The potential for proteomic analysis of venoms, both pooled and from individual specimens, allows for a rational consideration of these questions.

To address these questions we analysed venom from six individual specimens of *B. jararaca* collected from geographically distinct regions in São Paulo State, Brazil. The results of the 2-D gel analysis of the proteomes of these individual snakes are shown in Figure 5.4. It can easily be seen that there are notable differences between the proteomes of these individual snakes. The question as to the source of these differences (assuming identical genomes) remains to be answered. A computerised comparison of the 2-D gel images from two of the specimens is presented in Plate II. These two particular images were chosen for comparison since they show major differences. From the colourised comparison image the distinctive nature of the two venoms are highlighted. From the computerised analysis, 119 spots were identified in the image of gel 1 and 224 spots in the image of gel 2 (Table 5.2). The analysis also identified 40 spots as identical in the two images.

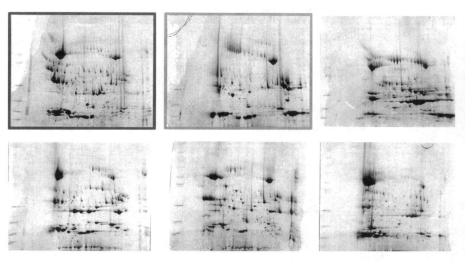

Figure 5.4 Comparison of 2-D gel patterns of venom from six individual specimens of *Bothrops jararaca*. Venom (1.8 mg) from each snake was applied to 3–10 IPG strips followed by electrophoresis into 8–18 % acrylamide gels

Table 5.2 Computerised comparison of digitised images of 2D gels of venom from two individual specimens of *Bothrops jararaca*.

	Total Spots Identified	Unmatched Spots	Up regulated Spots	Down regulated Spots
B. jararaca sample #4 (reference gel)	119	79	8	10
B. jararaca sample #3	224	156	—	—

REFERENCES

[1] Hewick, R.M., Hunkapiller, M.W., Hood, L.E. and Dreyer, W.J. (1981) A gas–liquid solid phase peptide and protein sequenator. *Journal of Biological Chemistry.* **256(15)**, 7990–7.
[2] Horvath, S.J., Firca, J.R., Hunkapiller, T., Hunkapiller, M.W. and Hood L. (1987) 'An automated DNA synthesizer employing deoxynucleoside 3'-phosphoramidites' in *Methods in Enzymology.* Vol. **154**, 314–26.
[3] Bjarnason, J.B., and Fox, J.W. (1995) 'Snake Venom Metallo-Endopeptidases' in *Methods in Enzymology – Proteolytic Enzymes*, Vol. **248**, 345–68.
[4] Hite, L.A., Jia, L-G., Bjarnason, J.B., and Fox, J.W. (1994) cDNA sequences for four snake venom metalloproteinases: structure, classification and their relationship to mammalian reproductive proteins. *Archives of Biochemistry and Biophysics.*, **308**, 182–91.
[5] Fox, J.W. and Long, C. (1998) 'The ADAMs/MDC family of proteins and their relationships to the snake venom metalloproteinases' in *Snake Venom Enzymes* (Bailey, G., ed.), pp.151–78. Alaken Press, Ft. Collins, CO.
[6] Bjarnason, J.B. and Fox, J.W. (1998) 'Introduction to the Reprolysins' in *Handbook of Proteolytic Enzymes* (Barrett, A.J., Rawlings, N.D., and Woessner, J.F., eds), pp.1247–54, Academic Press. San Diego, CA.
[7] Ekstrom, S., Onnerfjord, P., Nilsson, J., Bengtsson, M., Laurell, T. and Marko-Varga, G. (2000) Integrated microanalytical technology enabling rapid and automated protein identification. *Analytical Chemistry.* **72(2)**, 286–93.
[8] Califano, A. Stolovitzky, G. and Tu, Y. (2000) Analysis of gene expression microarrays for phenotype classification. *Proceedings of the International Conference on Intelligent Systems for Molecular Biology*; ISMB. **8**: 75–85.
[9] Hunt, D.F., Zhu, N.Z. and Shabanowitz, J. (1989) Oligopeptide sequence analysis by collision-activated dissociation of multiply charged ions. *Rapid Communications in Mass Spectrometry.* **3(4)**, 122–4.
[10] Hood. L. (2000) 'Discovery Science' a lecture presented at the 13th Meeting of Methods in Protein Structure Analysis of the International Association for Protein Structure Analysis and Proteomics, Charlottesville, VA, August 2000.
[11] Su, Y.A., Bittner, M.L., Chen, Y., Tao, L., Jiang, Y., Zhang, Y., Stephan, D.A. and Trent, J.M. (2000) Identification of tumor-suppressor genes using human melanoma cell lines UACC903, UACC903(+6), and SRS3 by comparison of expression profiles. *Molecular Carcinogenesis.* **28(2)**, 119–27.
[12] Chambers, G., Lawrie, L., Cash, P. and Murray, G.I. (2000) Proteomics: a new approach to the study of disease. *Journal of Pathology.* **192(3)**, 280–8.
[13] O'Farrell, P.H. (1975) High resolution two-dimensional electrophoresis of proteins. *Journal of Biological Chemistry.* **250**, 4007–21.

[14] Futcher, B., Latter, G.I., Monardo, P., McLaughlin, C.S. and Garrels, J.I. (1999) A sampling of the yeast proteome. *Molecular and Cellular Biology*. **19(11)**, 7357–68.

[15] Hong, H.Y., Yoo, G.S. and Choi, J.K. (2000) Direct Blue 71 staining of proteins bound to blotting membranes. *Electrophoresis* **21(5)**, 841–5.

[16] Blomberg, L. Blomberg, J. Norbeck, S.J. Fey, P.M. Larsen, M. Larsen, P. Roepstorff, H. Degand, M. Boutry, A. Posch and A. Gorg. (1995) Interlaboratory reproducibility of yeast protein patterns analyzed by immobilized pH gradient two-dimensional gel electrophoresis. *Electrophoresis*. **16**, 1935–45.

[17] Rioux, V., Gerbod, M.-C., Bouet, F., Menez, A. and Galat, A. (1998) Divergent and common groups of proteins in glands of venomous snakes. *Electrophoresis* **19**, 788–96.

[18] Braud, S., Bon, C. and Wisner, A. (2000) Snake venom proteins acting on hemostasis. *Biochimie* **82(9–10)**, 851–9.

[19] Yates, J.R. III, Speicher, S., Griffin, P.R. and Hunkapiller, T. (1993) Peptide mass maps: a highly informative approach to protein identification. *Analytical Biochemistry*. **214**, 397–408.

[19] Pook, C.E., Wuster, W., Thorpe, R. (2000) Historical biogeography of the Western Rattlesnake (Serpentes: viperidae: *Crotalus viridis*), inferred from mitochondrial DNA sequence information. *Molecular Phylogenetics & Evolution* **15(2)**, 269–82.

[20] Clauser, K.R., Baker, P.R. and Burlingame, A.L. (1999) Role of accurate mass measurement $(+/- 10\,\text{ppm})$ in protein identification strategies employing MS or MS/MS and database searching. *Analytical Chemistry* **71(14)**, 2871–6.

[21] Warrell, D.A. (1996). Injuries, envenoming, poisoning and allergic reactions caused by animals, Section 8.4.1. In: *Oxford Textbook of Medicine*, 3rd Edition (Weatherall, D.J., Ledingham, G.G., Warrell, D.A.). Oxford Medical Publications, OUP, pp.1124–51.

[22] Gygi, S.P., Corthals, G.L., Zhang, Y., Rochon, Y., and Aebersold, R. (2000) Evaluation of two-dimensional gel electrophoresis-based proteome analysis technology. *Proceedings of the National Academy of Sciences* **97(17)**, 9390–5.

[23] Serrano, S.M.T., Mentele, R., Sampaio, C.A. and Fink, E. (1995). Purification, characterization, and amino acid sequence of a serine proteinase, PA-BJ, with platelet-aggregating activity from the venom of *Bothrops jararaca*. *Biochemistry* **34**, 7186–93.

[24] Paine, M.J., Desmond, H.P., Theakston, R.D. and Crampton, J.M. (1992). Purification, cloning, and molecular characterization of a high molecular weight hemorrhagic metalloprotease, jararhagin, from *Bothrops jararaca* venom. Insights into the disintegrin gene family. *J. Biol. Chem.* **267**, 22869–76.

[25] Ownby, C.L., Selistre de Araujo, H.S., White, S.P., Fletcher, J.E. (1999). Lysine 49 phospholipase A2 proteins. *Toxicon* **37**, 411–45.

[26] Nishida, S., Fujimura, Y., Miura, S., Ozaki, Y., Usami, Y., Suzuki, M., Titani, K., Yoshida, E., Sugimoto, M., Yoshioka, A. and Fukui, H. (1994). Purification and characterization of bothrombin, a fibrinogen-clotting serine protease from the venom of *Bothrops jararaca*. *Biochemistry* **33**, 1843–9.

6 Proteomics of Venom Peptides

RETO STÖCKLIN* and PHILIPPE FAVREAU

While the discovery of a new bio-active compound from a natural extract is usually initiated by the observation of a biological activity, we propose a reversed approach going backwards from structure to function. This new strategy is based on a combination of genomics through sequencing of cDNA from venom glands, proteomics through systematic identification and characterisation of the venom components, and bio-computing through the development of a unique bibliographical and sequence database on venomous animals and their venoms.

To achieve our goal, we intend to obtain massive quantities of proteomic data by analysis of chosen crude venoms and to enhance the value of the information obtained through bio-informatics with a view to diagnostic and drug development. Although strategies designed for larger proteins will be discussed, this chapter focuses on the proteomics of venom peptides, which is illustrated by several examples that follow a more general introduction to the field.

6.1 PROTEOMICS

The term 'proteome' refers to the total protein profile of a cell or a tissue at a given time and can vary during the development of an organism, maturation of cell types and tissues and progression of disease. Proteomics is most widely known as a post-genomic science, in which the protein profile of an organism, a tissue or any biological sample is compared to the accumulated nucleic acid-based knowledge from the genome projects. The DNA information accumulated from the genome projects might enable us to predict the proteins that can potentially be generated, but not when, in which tissue, in what form or at what level. For example, the examination of the genome is not sufficient to determine post-translational modifications of proteins or to assess the effects of environmental factors, ageing or diseases [1, 2]. In this respect, mass spectrometry (MS) remains the work-horse of protein identification and structural elucidation, although this technique has limitations when it comes to complete

---

* Corresponding author

Perspectives in Molecular Toxinology
Edited by A Ménez © 2002 John Wiley & Sons, Ltd

characterisation of unknown biomolecules e.g. by MS/MS *de novo* sequencing. In a typical experiment, the biological sample is analysed by two-dimensional polyacrylamide gel electrophoresis (2D-PAGE), and the stained protein spots are digested in-gel by a specific enzyme such as trypsin. It is usually possible to extract a substantial portion of the proteolytic fragments from the gel and to submit them to a mass spectrometry peptide mapping. This is often completed with additional MS/MS fragmentation of the major signals to generate additional partial sequence information. The peptide mass maps thus obtained allow fast automated matching against protein databases [3, 4, 5]. These techniques have paved the way for high throughput protein identification and given birth to several novel approaches not necessarily based on 2D-PAGE sample preparation, which is not well adapted to peptides, proteins below 10 kDa or basic compounds. There is growing use of micro- and nano-systems for sample preparation and chromatographic purification specifically adapted to MS analyses of peptides, polypeptides and larger proteins. Finally, the most challenging and promising area is probably the MS/MS-based automated *de novo* sequencing, which can be extremely fast and efficient in some cases. This technique is an excellent complement to the Edman degradation, and remains a method of choice for the determination of post-translational modifications. However, some problems still need to be overcome: MS/MS sequencing is not yet able to distinguish between leucine and isoleucine in most cases, the fragmentation is not always complete and the dedicated data analysis software often gives rise to ambiguous results, particularly in automated mode.

It is thus essential to distinguish between protein identification through genomic databases matching of partial peptide mass fingerprints, and protein characterisation through complete *de novo* MS/MS and/or Edman sequencing of unknown compounds.

6.2   MASS SPECTROMETRY

Mass spectrometry is an analytical technique by which the molecular mass of molecules can be measured. Its application to large biomolecules has expanded in the last twenty years with the development of new instruments that allow precise and sensitive measurements over a broad mass range using soft ionisation methods. A mass spectrometer is typically made of four distinct parts: (1) the ion source, in which the molecules are introduced to be ionised (usually protonated to carry positive charges) and accelerated, is followed by (2) an analyser that allows a separation of the different ions according to their mass-to-charge ratio (m/z) before they reach (3) the detector from where the acquired information is transmitted to the (4) data system that is crucial for both data acquisition and interpretation. The raw mass spectra thus obtained represent m/z values on the horizontal axis and relative intensities for the vertical axis. The results are not directly quantitative, as the number of ions reaching the detector is also dependent on the analytical conditions and ionisation efficiency, which can differ from one compound to another. Through software deconvolution, the mass spectra are often transformed to display the molecular mass in daltons (Da) on the horizontal scale.

6.2.1 ION SOURCES

The most widely used ion sources for peptide and protein analyses are electrospray (ES) and matrix-assisted laser desorption (MALDI). ES is an atmospheric pressure soft ionisation technique in which the sample in solution is introduced into the source at continuous flow and is electrostatically sprayed with help of a nitrogen flow that facilitates the desorption of the ions into the mass spectrometer. ES allows for analyses directly in physiological fluids at atmospheric pressure and is also ideal for on-line coupling of liquid chromatography to the mass spectrometer (LC-MS). Each compound is ionised in a heterogeneous manner leading to multiple signals corresponding to different charge states of the same molecule, the so-called molecular envelope: if protonated, the different ions generated by each compound will correspond to $[M + nH^+/n]$ signals on the m/z mass scale, where M is the molecular mass, H^+ the proton and n the number of protons carried by each molecular species. In-source fragmentation is possible in some cases (by increasing the cone voltage), but requires pure samples as no ion selection is possible. MALDI requires preparation of the samples that are spotted on a target together with a matrix that co-crystallises and absorbs the energy of the laser pulses leading to the desorption under high vacuum of the molecules that are accelerated towards the mass analyser. MALDI allows for very fast analysis, as hundreds of samples can be loaded simultaneously on the same target and spectra can be acquired within seconds. The mass spectra thus obtained are relatively simple, as MALDI usually generates singly charged ions such as $M + H^+$, which is comfortable when analysing complex mixtures. Some MALDI instruments are equipped for MS/MS in-source fragmentation of selected ions by post-source decay (PSD) for partial peptide sequencing.

6.2.2 MASS ANALYSERS

The three common mass analysers include quadrupoles (Q), ion-traps (TRAP) and time-of-flights (TOF), each with specific characteristics. Quadrupole analysers were developed some 40 years ago and are extremely reliable and flexible, but are limited in resolution and mass range (up to a maximum m/z of 2000–4000 generally, which is well adapted to electrospray). For a given voltage, the quadrupolar electric field that is generated by two distinct voltages allows only a single m/z value to stay in a stable oscillatory state and reach the detector. During data acquisition, the instrument is thus usually scanned from the lower m/z value to a higher one over a few seconds, and the process is repeated several times for better statistics. TOF analysers are the simplest, as they basically measure the relative time taken by each of the compounds that have received the same energy to fly through a tube of a given length and to reach the detector: the heavier the molecule and the fewer charges it carries, the longer its molecular ion will take to reach the detector. TOF require pulsed inlet of the sample and accurate measurement of flying times. However, these analysers are not restricted in mass range and they do not suffer from the major drawbacks of scanning systems, as all ions from each laser pulse that fly through the tube are acquired, offering significant increase in sensitivity. This is ideal for the measurement of high molecular masses, but usually with poor resolution and accuracy. However, most of these analysers are nowadays fitted with a reflectron option, made of electrostatic lenses that improve the resolution and thus the accuracy, with the drawback of mass range restriction and sensitivity losses. Ion traps allow for controlled accumulation of a broad range of m/z ions in a stable circular pathway inside the trap and their successive liberation towards the detector with very fast scan speed facilities. For structural investigations, specific m/z ions can also be selected (discarding all other ions) and fragmented by collision-induced dissociation (CID) inside the trap. Generated fragment ions can further be selected inside the trap for an additional MS/MS fragmentation step (MS^n). However, MS/MS data can be acquired only down to $\frac{1}{3}$ of the parent ion m/z value, which is restrictive, especially for *de novo* sequencing. Although less accurate and apparently less user-friendly, ion traps tend to be very promising as they are extremely sensitive and of high resolution, they cover a wide mass range and they permit several levels of MS/MS fragmentation. Further developments are possible, especially in terms of data acquisition and data analysis software.

6.2.3   MASS SPECTROMETERS

Most of the instruments are a combination of the two different ionisation sources and the three types of analysers described above, each with advantages and drawbacks, depending on the application. The main features required from a mass spectrometer are accuracy, sensitivity, mass range, resolution, dynamic range, throughput and ease of use; each instrument being a compromise and financial consideration often influencing the choice. ES-MS exists with all three types of analysers, although quadrupoles are by far the most commonly used. MALDI is usually coupled to a TOF (MALDI-TOF-MS) as it is the only analyser offering an unlimited mass range. Tandem mass spectrometers are made of two successive mass analysers (MS1 and MS2) with a collision cell in between, which allows for MS/MS analysis with CID. This is widely used for example to select one ion in MS1, to have it fragmented in the collision cell, and to measure the molecular masses of the generated fragments in MS2 for structural investigations (such as peptide sequencing or elucidation of post-translational modifications) up to 3000 daltons approximately. This permits for example detailed investigation of complex mixtures during LC-MS analyses when efficient switching to the LC-MS/MS mode with data acquisition is possible within seconds. Depending on the cleavage site on a peptide chain and on the fragment that retains the charge, the fragments observed can allow for simultaneous C- and N-terminal sequencing, together with the observation of internal fragments. The most common of these instruments are triple quadrupoles in which both MS1 and MS2 are quadrupoles. Ion traps do basically not need any tandem architecture as they allow for several successive MS/MS steps in a single experiment inside the same analyser (MS^n). Although expensive and requiring advanced expertise, hybrid MS/MS electrospray instruments with a quadrupole followed by a TOF in reflectron mode (Q-TOF) belong to the best available compromises, some models even offering additional MALDI ionisation facilities for high-throughput peptide mapping. Emerging instruments dedicated to proteomics include TOF-TOF with both ES and MALDI ionisation sources, Lift-TOF (closely related) and TRAP-TOF. However, whatever their architecture, the recent developments of these mass spectrometers reveal that the limiting factor is the software. The data system is the crucial part of an instrument and both acquisition and manual or automatic data analysis systems suffer from this fast evolution and are presently the focus of much attention.

6.3   ANIMAL VENOM PROTEOMICS

Venoms are highly complex and concentrated liquid mixtures, usually made up of more than one hundred different components, each with a specific biological activity. This includes small organic and mineral components, histamine and other allergens, polyamines, alkaloids, small peptides, polypeptide toxins acting on the neuromuscular junction, ion channel modulators, cytotoxins, antibacterial peptides, phospholipases, enzymes, etc. Some 1500 different peptide toxins and enzymes of venomous origin have been described so far from a molecular point of view. Although this number is continuously increasing, it is representative only of a very small portion of the existing venom components. The latest developments in mass spectrometry include high-throughput protein identification and characterisation facilities directly from complex mixtures. The systematic screening of the complete protein profile in a crude venom is performed through an appropriate combination of sample extraction and preparation, electrophoresis, chromatography and mass spectrometry. For the full characterisation of novel compounds, complementary techniques are generally used to complete or confirm the results. These can include amino acid analysis, automated gas phase Edman sequencing, molecular biology techniques or comparative MS/MS analysis of the natural and synthetic compounds. Such venom proteomics will be illustrated through several distinct examples.

6.3.1   PROTEOMIC STRATEGIES

For proteomic applications, venoms differ from human tissues by two main features. On the one hand, while the human genome was finalised during the year 2000, few genomic data are available from the venom glands, and this situation will continue for years, probably decades. The reason for this is the biodiversity of the thousands of living venomous animal species, each with its own specific venom made of dozens or hundreds of constituents. This means that the isolated proteins need to be fully characterised, and partial mapping strategies through database mass matching are not well adapted. On the other hand, venoms present the advantage of being highly concentrated fluids, active components are relatively easy to isolate and material is usually available in sufficient amounts to allow sample preparation at the analytical or preparative level without necessarily depending on micro- or nano-technologies.

The 2D-PAGE approach to venom proteomics is not the best strategy in our opinion. First of all because it is not well suited to peptides or small proteins which are the main components of many venoms. The resolving power will also in many cases not be sufficient to differentiate amongst a family of protein isoforms as often observed in venoms. Furthermore, although extremely powerful for protein identification, the analysis of protein spots isolated by 2D-PAGE is limited to low sample amounts, requires in-gel reduction, alkylation and enzymatic treatment (generally with trypsin) to generate fragments small enough to be extracted from the gel with a view to peptide mass mapping. These operations are time-consuming and will provide only partial sequence information: only a portion of the fragments generated by the enzymatic treatment will be recovered from the spot and only part of these fragments will provide MS/MS fragmentation spectra of good quality. To date, the results that can be obtained from 2D-gel spots will in no case allow a full characterisation of unknown compounds. Partial mapping strategies through database mass matching of 2D-PAGE gel spots can, however, be appropriate in several cases: they may help to confirm the presence or structure of a known compound in a sample; partial sequence information can be used to design degenerated oligonucleotides that can be used for hybridisation screening; peptides can also be synthesised from the same information and used to raise specific antibodies to be employed in immuno-screening of cDNA libraries; and they may be used to look for homologies with known compounds from other genomes. However, we consider this latter approach to be extremely speculative and useful only if additional work is undertaken to demonstrate the structural and functional homologies.

Liquid chromatography processes such as reversed-phase, ion exchange, size exclusion, affinity extraction or related electrophoretic techniques will thus be preferred as they allow large-scale purification and easier recovery of the enzymatically generated fragments. While small linear peptides up to 3000 Da can be analysed by MS/MS straight from relatively complex mixtures, those with disulphide bonds will nevertheless require a preliminary treatment to generate reduced cysteine residues with free S–H groups. In the case of larger polypeptides, proteins or glycoproteins, the sample will first have to be purified and submitted to partial degradation through reduction, alkylation and enzymatic or chemical treatments to generate fragments of appropriate size for MS/MS fragmentation and/or automated Edman sequencing.

6.3.2 DIRECT ANALYSIS OF CRUDE VENOMS

Mass spectrometry is like a method of choice for the analysis of complex biological extracts, and for the identification and the characterisation of their content in biomolecules. Thus, the survey of known compounds or the search for novel bioactive compounds in a raw sample is illustrated by MS investigation of crude animal venoms using distinct strategies in order to identify the total protein profile. On the one hand, a reversed-phase liquid chromatography system can be coupled on-line to the mass spectrometer, which in a single step allows measurement of the molecular masses of the different eluting compounds every few seconds, and simultaneously collection of the fractions (on-line LC-ES-MS). On the other hand, the direct analysis of crude venom or fractions thereof can also be achieved by MALDI-TOF-MS, with simultaneous measurement of the different molecular masses. These two approaches yield a specific mass map of the venom analysed, which can be used like a 'molecular fingerprint' representative of the species to which the venomous animal belongs, known as toxin mass maps or venom mass fingerprints. These sets of data can further be compared to molecular masses calculated from described toxins whose sequences are listed in specialised databases such as Swiss-Prot [6] or Venoms [7], allowing easy targeting of potentially novel bioactive compounds with a view to drug development. On-line LC-ES-MS was the first chemotaxonomic technique used to classify unambiguously venomous animals (cobra snakes of the genus *Naja*) by virtue of their venom mass fingerprints [8, 9, 10]. This was followed by similar experiments on venoms of tarantulas in the genus *Brachypelma* by HPLC and MALDI-TOF-MS [11]. We compared both methods for the analysis of crude tarantula venoms (from the *Pterinochilus* group) and used the MS profiles as a chemical taxonomy pattern to separate species of tarantulas based on the mass fingerprinting differences of their venom [12]. Although more complex and time-consuming, the additional chromatographic dimension of on-line LC-ES-MS significantly increased sensitivity and accuracy when compared to MALDI-TOF-MS, which suffers from suppressive effects and poor resolution in the linear mode used. However, recent investigation of *Tityus serrulatus* scorpion venom revealed that the off-line MALDI-TOF-MS analysis of HPLC fractions can be a more powerful method for searching for poorly represented molecules in a highly complex mixture. Using two different MS analytical methods, 380 different molecular masses were found to constitute the two main toxic fractions from this scorpion venom [13]. In the frame of an evolutionary study, the venom

composition of 70 bamboo vipers (*Trimeresurus stejnegeri*) from different populations throughout Taiwan, two off-shore islands and one population from the
adjacent mainland, was investigated by mass spectrometry. The results revealed
that the main components detected correspond to a suite of PLA_2 enzymes, with
a number of possible isoforms present in each sample. Interestingly, the distribution of these isoforms did not correspond to evolutionary lineage as determined
by phylogenetic analysis of mitochondrial genes, leading to interesting hypotheses [14].

6.3.3   IDENTIFICATION AND CHARACTERISATION OF ISOLATED TOXINS

Mass spectrometry also plays a key role in structural investigations of purified
compounds. The use of chemical and enzymatic methods in combination with
mass spectrometry (ES-MS, MALDI-TOF-MS, LC-MS and/or MS/MS) and
automated sequencing using Edman's chemistry, permits determination of the
structure of a complex toxin, and characterisation of its post-translational
modifications. In this respect, mass spectrometry has been used on several
occasions for the analysis of individual toxins [15–19]. It is the method of choice
for the localisation of the disulphide bonds, for the study of uncommon or
modified amino acids, for the characterisation of glycoproteins or that of
complex biopolymers with blocked N-terminal residues that can not be sequenced with conventional Edman chemistry. Mass spectrometry also allows
the study of noncovalent interactions in solution, or monitoring of the *in vitro*
or *in vivo* behaviour and metabolism of proteins.

6.3.4   ON-LINE LC-ES-MS OF CRUDE *APIS MELLIFERA* (HONEY BEE) VENOM

The analysis of 10 μg of crude honey-bee venom (*Apis mellifera*) by on-line LC-
ES-MS using a microbore HPLC system allowed us to detect more than 85
different molecular masses (data not shown). All major compounds known
from this widely studied venom, such as histamine, melittin, apamin, the
mast-cell degranulating peptide or secapin, could be detected. Several different
glycoforms of the phospholipase A_2 were also identified, surprisingly together
with the nonglycosylated form of the enzyme. Interestingly, we detected melittin's inactive precursor promelittin and the complete series of ten conversion
intermediates to the mature peptide (Table 6.1). This allowed us to demonstrate
in vivo that promelittin is progressively transformed into active melittin through
dipeptidase processing, thus confirming previous *in vitro* results describing the
maturation with dipeptidylpeptidase IV [20]. These results further suggest the
presence of the maturation enzyme in the venom [21].

Table 6.1 Promelittin and the complete series of conversion intermediates to the mature melittin as detected by on-line LC-ES-MS. The result confirm *in vivo* the maturation process from the inactive promelittin precursor to the active melittin through processing by an N-terminal dipeptidylpeptidase.

Compound	Calculated mass (Da)	Measured mass (Da)
Propeptide [22–43] – Melittin [44–69]		
APEPEPAPEPEAEADAEADPEA – [44–69]	5060.7	—
EPEPAPEPEAEADAEADPEA – [44–69]	4892.5	4891.8
EPAPEPEAEADAEADPEA – [44–69]	4666.3	4665.9
APEPEAEADAEADPEA – [44–69]	4440.1	4439.4
EPEAEADAEADPEA – [44–69]	4271.9	4271.4
EAEADAEADPEA – [44–69]	4045.6	4044.8
EADAEADPEA – [44–69]	3845.4	3844.9
DAEADPEA – [44–69]	3645.2	3644.6
EADPEA – [44–69]	3459.1	3458.5
DPEA – [44–69]	3258.9	3258.5
EA – [44–69]	3046.7	3046.2
Melittin major, mature [44–69]	2846.5	2846.1

6.3.5 MALDI-TOF-MS ANALYSIS OF CRUDE VENOMS

Direct MALDI-TOF-MS analysis of crude venoms is a fast and efficient preliminary screening method. In order to complete our database with venom mass fingerprints, we have analysed hundreds of venom samples of different species or distinct origins by MALDI-TOF-MS using different analytical conditions. We thus obtained a minimum of two mass spectra for each sample, one focusing on the low molecular mass range, the other on larger components. This is incremented with a list of all detected molecular masses and a comparison with described sequences from the venom of each species. MALDI-TOF-MS mass spectra of the high molecular mass range obtained with the venom of three snake species are illustrated in Figure 6.1. A relatively complex picture was obtained with *Bothrops jararaca* (Figure 6.1A) venom (single specimen, Butantan Institute, São Paolo, Brazil) revealing the presence of several high molecular mass components that probably correspond to enzymes, several of which having been characterised from this venom, as for example the phospholipases A_2 in the 14 kDa mass range [22]. A large number of components were also detected in the low mass range (data not shown), several of them obviously corresponding to BPPs: bradykinin-potentiating peptides [23]. In collaboration with Prof. Antonio Camargo and colleagues from the Center for Applied Toxinology (CAT/CEPID, Butantan Institute in São Paolo and University of Rio Claro, Brazil), we are presently screening the full peptide profile of this venom, focusing on BPPs and their biological activity. Almost nothing is known about the venom of *Atheris nitschei* (Great Lakes bush viper, batch

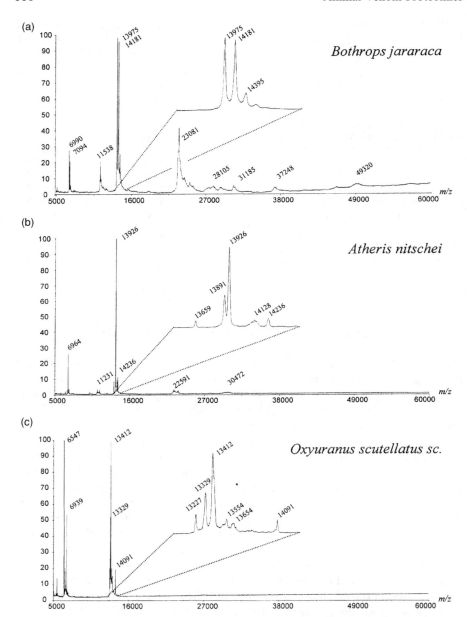

Figure 6.1 MALDI-TOF mass spectra of crude venoms from (a) *Bothrops jararaca*, (b) *Atheris nitschei* and (c) *Oxyuranus scutellatus scutellatus*. Analyses were carried out on a PerSeptive Biosystems Voyager-Elite instrument in positive ionisation mode. Samples (40 ng) were co-crystallised with 3,5-dimethoxy-4-hydroxycinnamic acid (sinapinic acid) in 30 % acetonitrile containing 0.1 % TFA. Spectra were obtained in linear mode and 256 scans were averaged. Insets show zoomed areas corresponding to phospholipases A_2

PA954, lot L1101 from Latoxan, Valence, France), and its analysis revealed the dominant presence of phospholipase A_2 isoforms and other proteins in the 22–23 kDa and 30–31 kDa mass ranges (Figure 6.1B). As expected from the literature and illustrated in Figure 6.1C, the venom of the Australian taipan (*Oxyuranus scutellatus scutellatus*, batch PA1001, Lot L1330 from Latoxan, Valence, France) contains several phospholipases A_2, but also several polypeptides in the 6500–7000 Da mass range that could correspond to short-chain post-synaptic alpha-neurotoxins [24].

6.3.6   PROTEOMICS OF *CONUS* VENOM

Conus venoms represent ideal candidates for the straightforward characterisation of peptides by mass spectrometry. The venoms of these molluscs are indeed composed of numerous peptides exhibiting extensive post-translational modifications, making them valuable tools for the development of proteomics. A small portion of a batch of *Conus textile* crude venom gland extract (10 specimens collected in New Caledonia) was thus subjected to several experiments with the aim of characterising most of its constituents. A preliminary high-resolution on-line LC-ES-MS analysis (Figure 6.2A) of the crude venom allowed us to detect more than 150 peptides, most of them ranging from 1000–3000 Da. This molecular mass fingerprint can be used as a chemotaxonomic specific marker for this particular *Conus* species as is the case for snake or spider venoms. Moreover, a closer examination of the data also enabled the detection of some post-translational modifications such as bromination through the particular isotopic pattern of the resolved spectra. The LC-MS procedure could then be repeated with the crude venom that was previously subjected to reduction of the disulphide bonds and the resulting fractions were collected for further analyses. At this point, a preliminary overview of the peptide content of the venom was obtained including accurate molecular mass, retention time for the native and reduced forms, as well as the number of disulphide bridges of each component. The subsequent characterisation of compounds could be performed by MS/MS as demonstrated here with the characterisation of the scratcher peptide [25]. The amino acid sequence as well as post-translational modifications such as C-terminal amidation and hydroxyproline could be unambiguously assigned (Figure 6.2B). Numerous novel conopeptides could be fully sequenced in the same manner from this venom and a large number of partial sequences were also obtained. However, one must bear in mind that most of the work requires time and expertise as automated data processing for *de novo* MS/MS sequencing still remains more or less reliable. To date, our experience tends to favour manual or semi-automated modes for complete *de novo* sequencing of peptides. To this effect, large efforts have already been made and software developments still remain the stumbling block for large-scale MS/MS *de novo* sequencing.

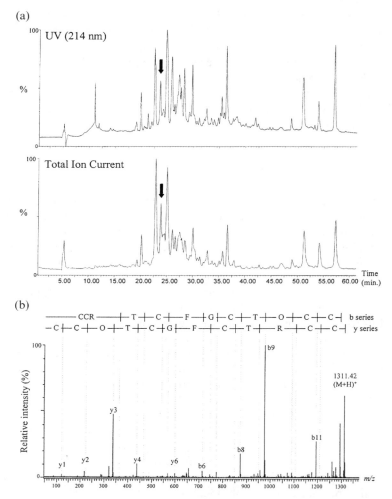

Figure 6.2   UV and Total Ion Current (TIC) chromatograms obtained by on-line LC-ES-MS of the crude venom gland extract from *Conus textile* (a). The chromatographic separation was performed with a Hewlett-Packard model 1100 HPLC system using a C$_{18}$ reversed-phase column (218TP54, Vydac, Mojave, CA, USA). The chromatography was operated at a flow rate of 600 μl/min. A linear gradient was initiated to 9 % ACN in 5 min, and then to 72 % ACN over 70 min. Mass spectrometric analyses were carried out on a Q-Tof 2 ES-MS/MS orthogonal tandem instrument (Micromass, Altrincham, UK) in positive-ionisation mode. High-resolution mass spectra were acquired every four seconds from *m/z* 400 to 1800 during the whole chromatographic process. An example of an MS/MS mass spectrum is shown in (b) for a doubly charged species at 656.2 *m/z*. Complete fragmentation of the parent ion was obtained by adjustment of the collision energy (15–40 eV) using constant collision gas flow. The multiply-charged spectrum was transformed into a singly-charged axis using the Max-Ent3 routine from the MassLynx3.5 software (Micromass) to allow complete sequence analysis of the scratcher peptide. Amino acids are listed with one-letter codes, O is for hydroxyproline and the C-terminal sequence was found to be amidated

6.3.7 DISCOVERY OF NOVEL SARAFOTOXINS IN *ATRACTASPIS* SNAKE VENOMS

Sarafotoxins (SRTX) are 21-amino-acid peptides containing two disulphide bridges that were first isolated from the snake venom of the Israeli burrowing asp, *Atractaspis engaddensis* [26]. They are markedly similar to endothelins [27], which is remarkable since endothelins are natural compounds of the mammalian vascular system, while sarafotoxins are highly toxic components of snake venom. These toxins exhibit strong vasoconstrictor activity and cause cardiac arrest, probably as a result of coronary vasoplasm. Sarafotoxins largely contribute to the understanding of the structure–function relationship of endothelins, as they bind to and act through the same receptors in vascular and brain tissues. We have analysed crude venom from a related snake, *Atractaspis microlepidota microlepidota*, looking for potentially novel endothelin-like toxins. Although more than 150 molecular masses were detected by on-line LC-ES-MS analysis of the crude venom, we were surprised not to find any mass in the expected SRTX range between 2400–2600 Da. However, a series of some 20 masses around 2800–3000 Da caught our attention, and was thus submitted to *de novo* MS/MS amino acid sequencing (Figure 6.3). This analysis allowed C-terminal sequencing straight from LC-MS fractions, revealing a new family of 24-amino-acid sarafotoxins with an additional C-terminal 'Asp-Glu-Pro' pattern that follows the typical 21-amino-acid SRTX motif. These results confirmed those obtained in parallel by molecular biology through cDNA cloning of the venom gland of the same snake specimen, and allowed us to characterise five novel endothelin-type peptides [28]. One of these, SRTX-m, was synthesised, and its NMR structure revealed interesting features [29]. Biological activity and specificity are presently under investigation.

6.4 CONCLUSION

The different themes presented in this article illustrate the diversity of proteomic approaches to the field of venom peptides and proteins. The flourishing of techniques in the field of mass spectrometry has yielded a panel of methods facilitating protein identification and characterisation, which can be successfully applied to the study of complex protein mixtures. Information such as the venom's molecular mass fingerprint, *in vivo* biochemical processing and the partial or complete sequence that can thus be obtained generates new perspectives in molecular toxinology. We aim to use these new tools and to develop innovative strategies for high-throughput targeting of novel bioactive compounds directly from natural extracts with a view to drug discovery.

(a)

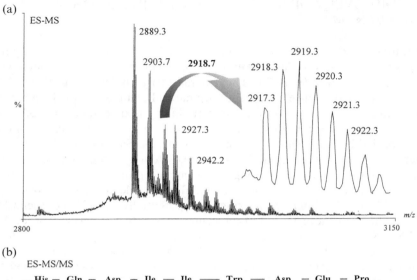

(b)

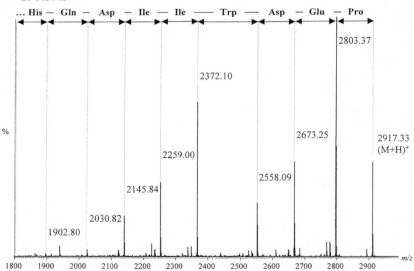

Figure 6.3 High-resolution mass spectrum of an HPLC fraction of crude *Atractaspis micro-lepidota microlepidota* snake venom (a). The deconvoluted mass spectrum (singly-charged axis, M + H$^+$ signals) was obtained from the raw data using the MaxEnt1 routine from MassLynx3.5. The inset presents an enlarged view of the isotopic envelope around 2920 *m/z*. Part of the MS/MS spectrum obtained from the triply-charged species at 973.11 m/z corresponding to a monoisotopic molecular mass of 2916.3 Da (b). Complete fragmentation of the parent ion was obtained after reduction of the disulphide bridges by adjustment of the collision energy (15– 40 eV). The multiply-charged spectrum was deconvoluted in a singly-charged axis using the MaxEnt3 routine (Micromass MassLynx 3.5 software) to allow *de novo* sequence analysis. The C-terminal sequence presented here relies on the b series fragment ions

ACKNOWLEDGEMENTS

We are grateful to Frédéric Ducancel, Frédéric Le Gall, Bernard Maillère, Robert Thai and André Ménez (DSV/DIEP, CEA-Saclay, France); Badia Amekraz, Christophe Moulin and Philippe Pierrard (DEN/DPC/SPCA, CEA-Saclay, France); Danielle Ianzer, Mirian Hayashi, Fernanda Portaro, Mario Palma and Antonio Camargo (Center for Applied Toxinology, CAT/CEPID, Butantan Institute, São Paolo, Brazil); Amos Bairoch (Swiss Institute for Bioinformatics), Denis Hochstrasser (University Hospital), Robin Offord (University Medical Center) and Keith Rose (GeneProt Inc.) in Geneva; as well as Guillaume Cretton, Sophie Michalet and Sylvie Stöcklin from our laboratories.

REFERENCES

[1] Banks, R.E., Dunn, M.J., Hochstrasser, D.F., Sanchez, J.C., Blackstock, W., Pappin, D.J. and Selby, P.J. (2000) Proteomics: new perspectives, new biomedical opportunities. *Lancet* **356**, 1749–56.

[2] Fischer, E.H. (1997) In: *New frontiers in functional genomics*. Wilkins MR, Williams KL, Appel RD, Hochstrasser DF (eds). Berlin: Springer Verlag.

[3] Binz, P.A., Muller, M., Walther, D., Bienvenut, W.V., Gras, R., Hoogland, C., Bouchet, G., Gasteiger, E., Fabbretti, R., Gay, S., Palagi, P., Wilkins, M.R., Rouge, V., Tonella, L., Paesano, S., Rossellat, G., Karmime, A., Bairoch, A., Sanchez, J.C., Appel, R.D. and Hochstrasser, D.F. (1999) A molecular scanner to automate proteomic research and to display proteome images. *Anal. Chem.* **71**, 4981–8.

[4] Gevaert, K. and Vandekerckhove, J. (2000) Protein identification methods in proteomics. *Electrophoresis* **21**, 1145–54.

[5] Wilkins, M.R., Gasteiger, E., Bairoch, A., Sanchez, J.C., Williams, K.L., Appel, R.D. and Hochstrasser, D.F. (1999) Protein identification and analysis tools in the ExPASy server. *Methods Mol. Biol.* **112**, 531–52.

[6] Bairoch, A. and Apweiler, R. (2000) The SWISS-PROT protein sequence database and its supplement TrEMBL in 2000. *Nucleic Acids Res.* **28**, 45–8.

[7] Stöcklin, R. and Cretton, G. *VENOMS:* The ultimate database on venomous animals – *Module 1:* 'Snakes' – Venomous snakes of the world (CD-Rom). 2nd edition, 2000. Atheris Laboratories, Geneva, Switzerland.

[8] Gillard, C., Virelizier, H., Arpino, P. and Stöcklin, R. (1997). Classification of the white *Naja* by on-line LC-ES-MS. In: *Eighteenth International Symposium on Capillary Chromatography* (ISSC Ed.), *ISSC, Riva del Garda, Italy*, **Vol. III**, 2192–7.

[9] Stöcklin, R. and Mebs, D. (1995) Analysis and identification of snake venoms by mass spectrometry. In: Institut Pasteur, ed. *1st International Congress on Envenomations and their Treatments*. Paris: *Institut Pasteur*, **189** (abstract).

[10] Stöcklin, R., Mebs, D., Boulain, J.C., Panchaud, P.A., Virelizier, H. and Gillard-Factor, C. (2000) Identification of snake species by toxin mass fingerprinting of their venoms. *Methods Mol. Biol.* **146**, 317–35.

[11] Escoubas, P., Celerier, M.L. and Nakajima, T. (1997) High-performance liquid chromatography matrix-assisted laser desorption/ionization time-of-flight mass spectrometry peptide fingerprinting of tarantula venoms in the genus *Brachypelma*: chemotaxonomic and biochemical applications. *Rapid Commun. Mass Spectrom.* **11**, 1891–9.

[12] Escoubas, P., Chamot-Rooke, J., Stöcklin, R., Whiteley, B.J., Corzo, G., Genet, R. and Nakajima, T. (1999) A comparison of matrix-assisted laser desorption/ionization time-of-flight and liquid chromatography electrospray ionization mass spectrometry methods for the analysis of crude tarantula venoms in the *Pterinochilus* group. *Rapid Commun. Mass Spectrom.* **13**, 1861–8.

[13] Pimenta, A.M.C., Stöcklin, R., Favreau, P., Bougis, P.E. and Martin-Eauclaire, M.-F. (2001) Moving pieces in a proteomic puzzle: mass fingerprinting of toxic fractions from the venom of *Tityus serrulatus* (SCORPIONES, Buthidae). *Rapid Commun. Mass Spectrom.* **15**, 1562–72.

[14] Malhotra, A., Creer, S., Stöcklin, R., Favreau, P., Thorpe, R.S. and Chou, W.-H. (2000) Intraspecific variation in venom composition: taxonomists meet toxinologists. *IST Meeting Abstracts Book*, p.169.

[15] Castañeda, O., Sotolongo, V., Amor, A.M., Stöcklin, R., Harvey, A.L., Engström, A., Wernstedt, C. and Karlsson, E. (1994) Characterization of a potassium channel toxin from the sea anemone *Stichodactyla helianthus*. *Toxicon* **33**, 603–13.

[16] Kalume, D.E., Stenflo, J., Czerwiec, E., Hambe, B., Furie, B.C., Furie, B. and Roepstorff, P. (2000) Structure determination of two conotoxins from *Conus textile* by a combination of matrix-assisted laser desorption/ionization time-of-flight and electrospray ionization mass spectrometry and biochemical methods. *J. Mass Spectrom.* **35**, 145–56.

[17] Kolarich, D. and Altmann, F. (2000) N-Glycan analysis by matrix-assisted laser desorption/ionization mass spectrometry of electrophoretically separated nonmammalian proteins: application to peanut allergen Ara h 1 and olive pollen allergen Ole e 1. *Anal. Biochem.* **285**, 64–75.

[18] Okada, K., Uyehara, T., Hiramoto, M., Kato, H. and Suzuki, T. (1973) Application of mass spectrometry to sequence analysis of pyroglutamyl peptides from snake venoms: contribution to the confirmation of the amino acid sequence of bradykinin-potentiating peptides B, C and E isolated from the venom of *Agkistrodon halys blomhoffii*. *Chem. Pharm. Bull. (Tokyo)* **21**, 2217–23.

[19] Tyler, M.I., Retson-Yip, K.V., Gibson, M.K., Barnett, D., Howe, E., Stöcklin, R., Turnbull, R.K., Kuchel, T. and Mirtschin, P. (1997) Isolation and amino acid sequence of a new long-chain neurotoxin with two chromatographic isoforms from the venom of the Australian death adder (*Acanthophis antarcticus*). *Toxicon* **33**, 603–13.

[20] Kreil, G., Haiml, L. and Suchanek, G. (1980) Stepwise cleavage of the pro part of promelittin by dipeptidylpeptidase IV. Evidence for a new type of precursor–product conversion. *Eur. J. Biochem.* **111**, 49–58.

[21] Stöcklin, R., Thai, R., Gos, P., Amekraz, B., Ducancel, F. and Ménez, A. (2000b) Analyses of honeybee and burrowing asp venoms by mass spectrometry. *Toxicon* **38**, 1649–50 (abstract).

[22] Serrano, S.M., Reichl, A.P., Mentele, R., Auerswald, E.A., Santoro, M.L., Sampaio, C.A., Camargo, A.C. and Assakura, M.T. (1999) A novel phospholipase A2, BJ-PLA2, from the venom of the snake *Bothrops jararaca*: purification, primary structure analysis, and its characterization as a platelet-aggregation-inhibiting factor. *Arch. Biochem. Biophys.* **367**, 26–32.

[23] Murayama, N., Hayashi, M.A., Ohi, H., Ferreira, L.A., Hermann, V.V., Saito, H., Fujita, Y., Higuchi, S., Fernandes, B.L., Yamane, T. and de Camargo, A.C. (1997) Cloning and sequence analysis of a *Bothrops jararaca* cDNA encoding a precursor of seven bradykinin-potentiating peptides and a C-type natriuretic peptide. *Proc. Natl. Acad. Sci. U S A* **94**, 1189–93.

[24] Zamudio, F., Wolf, K.M., Martin, B.M., Possani, L.D. and Chiappinelli, V.A. (1996) Two novel alpha-neurotoxins isolated from the taipan snake, *Oxyuranus scutellatus*, exhibit reduced affinity for nicotinic acetylcholine receptors in brain and skeletal muscle. *Biochem.* **35**, 7910–16.

[25] Olivera, B.M., Rivier, J., Clark, C., Ramilo, C.A., Corpuz, G.P., Abogadie, F.C., Mena, E.E., Woodward, S.R., Hillyard, D.R. and Cruz, L.J. (1990) Diversity of *Conus* neuropeptides. *Science* **249**, 257–63.
[26] Kloog, Y., Ambar, I., Sokolovsky, M., Kochva, E., Wollberg, Z. and Bdolah, A. (1988) Sarafotoxin, a novel vasoconstrictor peptide: phosphoinositide hydrolysis in rat heart and brain. *Science* **242**, 268–70.
[27] Yanagisawa, M., Kurihara, H., Kimura, S., Tomobe, Y., Kobayashi, M., Mitsui, Y., Yazuki, Y., Goto, K. and Masaki, T. (1988) A novel potent vasoconstrictor peptide produced by vascular endothelial cells. *Nature* **332**, 411.
[28] Ducancel, F., Wery, M., Hayashi, M.A.F., Muller, B.H., Stöcklin, R. and Ménez, A. (1999) Les sarafotoxines de venins de serpent. In: Georges Cohen, ed. *Annales de l'Institut Pasteur / Actualités:* 'Les venins' *Institut Pasteur, Paris.* **10**(2), 183–94.
[29] Lamthanh, H., Volpon, L., Ducancel, F., Bdolah, A., Wollberg, Z., Stöcklin, R., Ménez, A. and Lancelin, J.M. (2000) NMR structure study of a new sarafotoxin detected in the venom of *Atractaspis microlepidota microlepidota*: SRTX-m. *IST Meeting Abstracts Book*, p. 54.

7 High-resolution NMR of Venom Toxins in Nanomolar Amounts

MURIEL DELEPIERRE

7.1 INTRODUCTION

Venoms are complex mixtures. Proteins are the principal component but venoms contain a large number of other constituents. Venoms are either produced by a specialised gland and injected by a specialised apparatus or are present throughout the tissues of poisonous animals or plants. They differ widely in composition and toxicity, depending primarily on the species of origin. Whereas snake venoms contain mostly toxins and enzymes, scorpion venoms, with few exceptions, have little or no enzymatic activity [1]. Scorpion toxins contain mostly polypeptide toxins together with nucleotides, lipids, mucopeptides, biogenic amines and other unknown substances [2, 3]. Venomous arthropods also store various complex neuroactive compounds including not only proteins and polypeptides but also low-molecular-weight organic compounds. The venom glands of these animals also contain acylpolyamines consisting of a hydrophobic moiety linked to a side-chain mainly composed of aminopropyl and aminobutyl units [4, 5]. A large variety of nonpeptide molecules are also found in aquatic poisoning species such as dinoflagellates, certain species of cyanobacteria that produce the paralytic shellfish poisoning (PSP) toxins [6] and certain bivalves, producing the neurotoxic shellfish poisoning (NSP) toxins of the brevetoxin family [7]. Sea anemones and marine snails also contain toxic compounds.

Envenomation represents a considerable public health problem. Nevertheless, these toxins do constitute excellent tools for biological research, particularly for neurobiology and ion channel characterisation. Indeed, the high specificity and exquisite biological activity of these toxins make them useful tools for investigating the function of voltage-sensitive and ligand-gated ion channels [8].

The relative concentrations of the various components of venom differ greatly [9] and many venom components have yet to be identified. For example, the venom of each species of scorpion contains around 70 distinct peptides. Given that there are about 1500 known species of scorpion in the world we can estimate that there are 100 000 different polypeptides in scorpion venoms, of

Perspectives in Molecular Toxinology
Edited by A Ménez © 2002 John Wiley & Sons, Ltd

which less than 1 % are known [10]. In most cases, a single gland contains too little material for chemical analysis and structure elucidation. It is therefore not surprising that the first molecules to be identified and characterised were those present in venom in large amounts. For example, the toxins in scorpion venoms that act on sodium channels were the first to be identified and characterised not only because they are very toxic to humans but also because they are the most abundant in the venom. Toxins active on potassium channels are present as a minor component of scorpion venom and were identified soon afterwards. However, the increasing interest in venom constituents has led to efforts to improve the separation and identification of all venom compounds.

7.2 STRUCTURAL CHARACTERISATION BY NMR USING SMALL AMOUNTS OF MATERIAL

Knowledge of three-dimensional structure is the key to understanding the interactions of toxins with their receptors. NMR and X-ray crystallography are the two most powerful tools available for structure determination and have been extensively used to obtain structural information about various molecules. X-ray crystallography is limited by the necessity of obtaining crystals whereas NMR is limited by the size of the molecule and the low sensitivity of the method. These two techniques are, therefore, complementary. NMR is a powerful analytical tool for structure determination, especially for small molecules, for obtaining information on the dynamics of macromolecules over a large time scale, and for studying molecular interactions. It is therefore not surprising that this technique is widely used to study toxins. However, NMR is inherently less sensitive than almost all other analytical methods. This is due to the small energy gap between ground and excited states. Of all the spectroscopic methods used to characterise molecules, NMR is by far the least sensitive. Modern instrumentation in infrared spectroscopy has made it possible to obtain spectra with only a few picograms of compounds [11] and mass spectrometry can be used for solutions with concentrations less than a few hundred picomoles per microliter and even for attomolar concentrations [12]. In contrast, NMR is rarely the method of choice for analysis of compounds present in trace amounts despite its great potential for structural characterisation and its nondestructive nature. NMR usually requires a concentration in the millimolar range at least and this low sensitivity of NMR has precluded its routine use in trace analysis.

A number of approaches have been used to increase the intrinsic sensitivity of NMR experiments. The first involves performing the experiment in a higher static magnetic field (B0), but cost increases exponentially with field strength and this approach remains limited as theoretically the signal-to-noise ratio varies with Larmor frequency to the three power halves [13], but other factors may also have an effect.

A second approach complementary and parallel to the development of high field strength, is increasing the performance of probes. The probe, an interface between the magnet, the transmitter and the receiver, is one of the main elements linked to the sensitivity of the spectrometer. To improve the performance of a probe, the signal-to-noise ratio must be increased, by increasing the signal and/or decreasing the noise.

7.2.1   THE SIGNAL-TO-NOISE RATIO

The noise in NMR experiments is mostly thermal, originating from the coil. The sample itself can produce noise if it is a conductor. The voltage associated with noise over a certain bandwidth at the resonant frequency depends on the resistance and temperature of the coil. Reducing either the resistance or the temperature of the coil, decreases the noise. Thus sensitivity increases (i) if temperature decreases, leading to the development of super-conducting probes, or (ii) if the resistance of the coil decreases, leading to different coil configurations.

It is possible to limit the receiver coil noise and to increase the coil quality factor by using high-temperature super-conducting materials or cryogenically cooled coils. Radio-frequency coils and associated electronic apparatus are cooled to 25 K or below, resulting in a 12- to 16-fold reduction in data acquisition time for a similar signal-to-noise ratio. Improvements in sensitivity by a factor of four to five over a conventional 5 mm solution probe have been reported [14, 15].

Coil resistance depends on various factors including the perimeter of the conductor and its length. For identical sample volumes, while unit currents flowing through saddle-shaped coils and solenoidal coils create similar B_1 fields and the coils receive similar signals from the sample, the resistance of a saddle-shaped coil is much higher than that of a solenoidal coil because it is longer. This higher resistance results in a lower signal-to-noise ratio for saddle-shaped coils. Thus solenoidal coils have higher intrinsic sensitivity than saddle-shaped coils. However, the choice of coil is governed by the appropriate orientation of the radio-frequency field for the detection of nuclear magnetisation. Although solenoidal coils are more efficient than saddle-shaped coils they require the sample to be placed perpendicular or at an angle to the magnet axis [16–19].

7.2.2   SMALL-VOLUME PROBES

Gains in sensitivity can also be obtained by optimising the sample volume for the specific availability or solubility of a given sample. Considerable efforts have been made in recent years in NMR probe development, driven by the need to characterise small samples. The use of small-diameter tubes was made popular twenty years ago by Shoolery, who recognised that 'real-life chemical and biological problems will always create ever-smaller samples that need to be analysed with ever-increasing sensitivity'[20].

A few decades ago, when sample size was limited, microsample cells with a spherical cavity were used to obtain high-resolution spectra. However, it was very difficult first to eliminate lineshape distortions due to magnetic susceptibility discontinuities, a consequence of sample geometry, and second to make perfectly spherical microcells. Therefore small-sample volume probes were developed of which three types are currently available:

3 mm microprobes

These microprobes use small-diameter (3 mm) vertical sample tubes and conventional geometry probes with volumes of 120 μl to 150 μl. These probes, designed to couple HPLC with NMR, are built with saddle-type RF coils. They are widely used in the pharmaceutical industry and are increasingly used in academic laboratories for the characterisation of small samples of natural products and other rare samples.

These microprobes, introduced in the early 1990s [21, 22], were used to elucidate the structures of extremely complex natural products, such as brevetoxins, aguatoxins and maitotoxins. Total assignment of the ^{1}H and ^{13}C NMR spectra of brevetoxin-3 (PbTX-3), a polycyclic ether found in dinoflagellates, was achieved at submicromolar levels, using 0.95 μmoles of sample dissolved in 130 μl d_6-benzene [23] with 3 mm microdetection probes (Nalorac Z.SPEC^R MID 500-3 and MC-500-3). The ^{13}C, ^{1}H heteronuclear multiple quantum correlated spectrum (HMQC) was acquired in less than 10 h with assignment of all carbon-bearing protons.

The structure of ciguatoxin was also resolved using this technology. Ciguatera biotoxins are present at concentrations of a few parts per billion in subtropical and tropical finfish species. Ciguatoxin is a potent activator of voltage-dependent sodium channels which causes severe gastrointestinal and neurological dysfunction in humans. ^{13}C and ^{1}H signal assignments for cuigatoxin were obtained with as little as 0.1 μmol dissolved in 130 μl of d_5-pyridine [24]. However, the time required for a simple 2-D ^{13}C and ^{1}H correlated experiment was prohibitively long (231 h).

Sample volume can be further reduced from the 130 μl or 150 μl nominal volumes used in 3 mm micro NMR tubes to 70–80 μl using Shigemi tubes. This doubles the effective working concentration, thereby decreasing the acquisition time for a similar signal-to-noise ratio. In Shigemi NMR microcells, the composition of the glass is matched to the dielectric properties of the solvent used. Therefore, the bottom of the tube that is not in the receiver coil is filled with glass [25]. The same amount of ciguatoxin used in the studies described above may be dissolved in only 67 μl of solvent if a Shigemi 3 mm NMR microcell is used. This halves the experimental time required [24].

The amount of material necessary and/or experimental time can be further decreased using 3 mm cryogenic probes. Assignments of the ^{13}C and ^{1}H signals of strychnine were obtained in one hour using 40 μg (120 nmol) of compound [26].

Microcoil NMR probes

The use of microcoil NMR probes was first reported in 1994 [27–29]. These probes contain capillaries with nanolitre to microlitre volumes and are built with a solenoidal coil of about 1 mm in length. They were developed to make it possible to couple capillary HPLC or electrophoresis with NMR [30–33]. Solenoidal NMR coils less than 1 mm long and with diameters as small as 150 μm are highly effective because the detection sensitivity is inversely proportional to coil diameter. For a limited amount of sample, the signal received is maximal if the sample is concentrated in a minimal solvent volume, and the smallest possible radio frequency coil is constructed to enclose this sample volume. Empirically, the total sample volume should be at least 10 times larger than the coil volume, that is the volume observed (V_{obs}) by the transmitter coil, to obtain high-resolution spectra comparable with those from an infinite sample (standard probes). The mass sensitivity of the NMR detection coil increases as coil diameter decreases [34]. However, the use of coils less than 1 mm long requires some form of susceptibility matching to obtain narrow line width. By immersing the coil in Fluorinert FC-43 a nonconducting liquid composed of a mixture of perfluorinated C12-branched tributylamines with a mean molecular mass of 670 g and a volume magnetic susceptibility very close to that of copper, Olson et al. [29] obtained a line width of 0.6 Hz for pure ethylbenzene. The test was conducted using a solenoidal coil of 357 μm in diameter with an observed volume of 5 nl. They reported the proton spectrum, recorded in 10 minutes, of 3 μg (3.3 nmol) of a heptapeptide with a high signal-to-noise ratio.

Nevertheless, the concentrations used in these experiments are not realistic in all cases (several hundreds of mM) and may not be compatible with studies of biological molecules. Improvements have therefore been made to give a 1–20 mM working range with an observed volume of 200 nl [35]. The ^{1}H spectrum of chloroquine was obtained in one hour with an observed volume of 131 nl, corresponding to 1.2 μg (2.4 nmol) of sample in the detection cell [35].

Further volume reduction was achieved in 1998 with the introduction of 1.7 mm submicro inverse-detection gradient NMR (SMIDG-NMR techniques) [36]. Using this technology, the authors were able to characterise fully an 8% impurity, cryptolepinone, corresponding to 0.04 μmol cryptolepinone contained in a 0.55 μmol sample of cryptolepine with a working sample volume of 20–30 μl (Nalorac SMIDG-600–1.7 submicro NMR probe). The authors later characterised 0.25 μmol (75 μg) of a related compound, a novel alkaloid, using the same technique [37].

To extend structure determination further, ^{13}C direct detection probes were constructed with microcoils as long as 2.5 mm, giving a V_{obs} of 100 nl. Inverse detection probes with a 1.4 mm long microcoil corresponding to a 550 nl V_{obs} were also constructed. These probes, developed for natural abundance ^{13}C NMR, make it possible to work with a few tens of micrograms of sample rather than the milligram quantities required by conventional ^{13}C NMR [38]. Of course, ^{13}C enrichment, when possible, can further reduce these quantities. Thus with a micro inverse detection probe with a 1.9 mm long solenoidal coil and a Vobs of 745 μl, ^{1}H and ^{13}C signal assignments of chloroquine, a first-line anti-malarial drug, were obtained at natural abundance using 40 nanomoles (13 μg) of chloroquine, corresponding to a concentration of 53.5 mM [39]. Complete ^{13}C decoupled HSQC (heteronuclear single quantum coherence) took less than 4 h.

Although the results presented here are impressive, most of these probes built with microcoils are still prototypes. Furthermore, they are selective for protons or have only dual selectivity for ^{13}C and ^{1}H. The probes most commonly available commercially are 3 mm probes and the Nano.nmr probe.

The Nano.nmr probe

The Nano.nmr probe was first conceived as part of a programme to develop analytical methods for combinatorial chemistry, to characterise molecules covalently attached to solid-phase synthesis beads, thereby eliminating the requirement for sample cleavage [40–42]. The first version of the Nano.nmr probe could be used for proton detection only, but an inverse detection/gradient version is currently available for standard heteronuclear experiments on any nuclei from ^{15}N frequency to ^{31}P frequency.

The observed volume (V_{obs}) for the Nano.nmr probe is 40 μl. The coil volume need not be filled completely without compromising line widths. Sample concentration is the same as for standard probes. The cell has a 40 μl internal cavity within a 28 mm long, 4 mm diameter glass tube, accessible via a thick-walled filling tube. A specially designed Teflon plug is used with the sample cell to eliminate losses due to evaporation. The attainable spinning rate is 0.2–5 KHz.

The coil is placed immediately around the sample inside the dewar thus increasing the filling factor and giving a higher signal-to-noise ratio, with shorter pulses per tip angle per power available. The probe uses a solenoidal coil to maximise sensitivity [17, 18]. The solenoidal coil is 4 mm in diameter, is oriented at the magic angle and is made with zero susceptibility wire [43–45]. In addition, a high detection efficiency is achieved by placing 100 % of the sample in the receiver coil [46]. The disadvantage of this probe is the use of nonspherical samples resulting in the perturbation of the main magnetic field due to bulk magnetic susceptibility being uncorrectable with the current shim system. Distortions of the static field, induced by the sample, originate from any region in which the susceptibility of the sample changes. The three most important factors affecting magnetic susceptibility are: (i) molecular magnetic susceptibility anisotropy, which occurs mainly if molecules have restricted motion; (ii) susceptibility discontinuities resulting in the poor line shapes typically found in the spectra of very small liquid NMR samples and due to the interface between the tube and the liquid; and (iii) sample geometry.

Susceptibility variations due to sample geometry are absent only with the following three shapes: infinite cylinder, sphere and toroid. If sample size is limited, a micro-sample cell with a spherical cavity can be used to obtain high-resolution spectra but, as stated above, perfectly spherical microcells are very difficult to construct. High-resolution probes minimise magnetic susceptibility discontinuities around a liquid sample by using long cylindrical samples, which resemble infinite cylinders for the receiver coil. With regard to shifts, a cylinder can be considered infinitely long if its length-to-diameter ratio is greater than five [47]. Under these conditions, the V_{obs} corresponds at best to only one third of the total volume (V_{tot}) but all non-symmetric susceptibility interfaces are moved away from the active volume of the receiver coil. Thus, sample observation efficiency is low, whereas ideally observed volume should equal total volume. The Nano.nmr probe increases detection efficiency by placing 100% of the sample ($V_{obs} = V_{tot}$) in the receiver coil and by eliminating the resulting lineshape distortions by spinning the sample at the magic angle, $\theta = 54°7$. By spinning the sample at the magic angle relative to the z axis (B axis), the $1 - 3\cos^2\theta$ dependence of the susceptibility-broadening term in the NMR Hamiltonian reduces to zero and the effect disappears. Usually magic-angle spinning is conventionally used in solid-state NMR only to remove chemical shift anisotropy effects or to average dipolar couplings, whereas magnetic susceptibility effects are typically disregarded. However, it can also be applied successfully to the removal of the bulk magnetic susceptibility that causes the poor line shapes typically found in the spectra of very small liquid NMR samples [48]. A thorough analysis of the demagnetisation fields for cylindrical samples of finite length, such as those used for the Nano.nmr probe, was reported recently [43].

In conclusion:

- The Nano.nmr probe spins samples very rapidly (1–4 kHz) at the magic angle to eliminate the contribution of magnetic susceptibility to line width either around or within the sample. The probe requires a special sample cell and a turbine. The spinning rate is controlled by an optical fibre and a tachymeter.
- The Nano.nmr probe achieves high detection efficiency by placing 100% of the sample in the receiver coil, giving a very high sensitivity per nucleus.

- The lineshape and resolution of the Nano.nmr probe are similar to those of conventional liquid probes. Figure 7.1 shows a one-dimensional spectrum obtained for a small toxin, with a molecular mass of 4 kDa. The sample concentration is at most 1 mM and the pH is unknown, but is probably low (acidic conditions). Good lineshape, sensitivity and resolution can be achieved and small NH–Hα coupling constants can be measured directly from the spectrum.

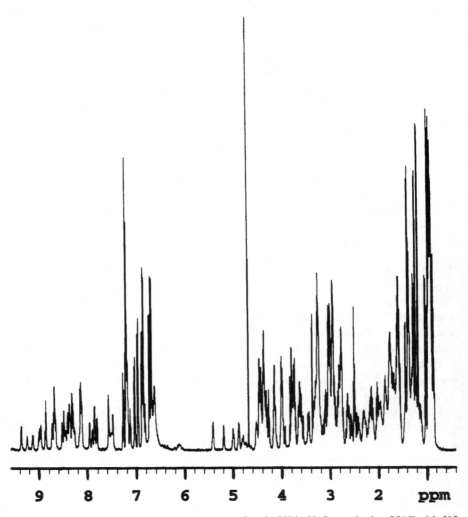

Figure 7.1 One-dimensional proton spectrum of toxin Pi7 in H_2O acquired at 35 °C with 512 scans. The spinning rate was 2585 Hz

- As it uses magic-angle spinning to eliminate susceptibility-mismatch line-broadening effects, shimming is equally good for both 40 μl and 4 μl samples [49]. However, the shimming procedure is very different from that typically used for liquid samples as it is strongly influenced by the radial position of the probe in the magnet. Adjustments concentrate on changes in X and Y followed by ZX and ZY and all other radial gradients.

- The Nano.nmr probe produces a very narrow water lineshape, even at the base of the water signal (Figure 7.1).

- For conventional solution NMR, the major potential disadvantage of the Nano.nmr probe is sideband aliasing in 2-D applications, because the sample must be spun to obtain a narrow line width [46]. This inconvenience can be easily overcome by changing the spinning rate or temperature. Rotor synchronisation, with a pulse sequence, similar to that used for solid-state experiments, is useful but much evaluation is still required.

- Another disadvantage lies in the use of sample spinning in the presence of both radio-frequency and magnetic-field inhomogeneities which results in modulation of the effective field. The performances of isotropic mixing in TOCSY experiments may therefore be dramatically reduced when using conventional composite-pulse mixing sequences [50]. Adiabatic mixing sequences are less susceptible to such modulations [51, 52] and have been shown to perform considerably better in TOCSY experiments at the magic angle [53].

7.3 NANO.NMR PROBE APPLICATIONS

Although the Nano.nmr probe was initially designed for the characterisation of compounds bound to their solid support [40, 44, 54], we rapidly showed that it was possible to use the Nano.nmr probe to study macromolecules in solution, in the same way as standard high-resolution NMR [49, 55]. Thus, the technique was successfully applied to solve the three-dimensional structure of two small scorpion toxins present in venom in small amounts.

Several scorpion venom toxins have been extensively studied in terms of structure, mode of action, and location of the active site. The most widely known are those specific for sodium and potassium channels. The principal molecular targets of these toxins are the voltage-dependent Na^+ and K^+ channels found in many excitable tissues that are responsible for action potentials in nerves and muscles. However, peptides that recognise calcium and chloride channels have also been described [56]. Scorpion toxins provide scientists with unique tools for investigation of such diverse areas as the ionic channels of excitable membranes, the phylogeny of proteins and the structure/function relationships of proteins.

The relative concentrations of scorpion toxin active on ionic channels differs greatly between venoms. Whereas sodium channel toxins are present in reasonable amounts in scorpion venom, the potassium channel toxins may account for less than 0.1 % of total venom. It is not surprising that the first toxin structure to be solved was that of a sodium channel toxin, namely the *Centruroides sculpturatus* variant-3 toxin [57, 58]. In contrast, although several potassium channel toxins were characterised in the early 1980s, their three-dimensional structures were not solved for several years. In addition, toxins affecting potassium channels make only a minor contribution to the effects of envenomation on humans.

The first toxin acting on potassium channels to be identified was noxiustoxin in 1982 [9]. Its three-dimensional structure was solved in 1995 [59]. The first structure to be solved was that of charybdotoxin [60], which was discovered in 1985 [61]. These toxins are very powerful tools and have been successfully used to characterise potassium channels [62] including the *Streptomyces lividans* K^+, the three-dimensional structure of which was recently solved [63].

Although the molecular basis of toxin specificity has been widely studied, there is still much that we need to know if we are to understand fully the structural and functional characteristics of toxins. Small differences, such as single amino acid substitutions within or near the critical binding region of the toxin, are important in receptor recognition and modifications of channel function. Thus, comparative studies of structurally similar toxins with different affinities for, or effectiveness against, their target receptors (ion-channels) would help to identify the crucial residues involved in these molecular interactions. Since the discovery of noxiustoxin, a large number of homologous peptides isolated from scorpions have been shown to block the two large families of potassium channels: high-conductance calcium-activated potassium channels (BK channels) and voltage-dependent channels (Kv channels, *shaker* and Kv1.3.). The structures of several of these toxins have been elucidated in the last ten years. This was made possible by recent developments including: (i) chemical synthesis associated with efficient refolding; (ii) proper folding of recombinant toxin; (iii) progress in extraction and purification; and (iv) recent developments in NMR making it possible to determine structure with only a few micrograms of compound.

A few years ago, the first toxin acting on potassium channels and with four disulphide bridges, Pi1, was isolated from the venom of the scorpion *Pandinus imperator* [64]. Its three-dimensional structure was determined by two-dimensional NMR methods, with nanomolar amounts of compound and the Nano.nmr probe. The structure of Pi1 is organised around a short α-helix spanning residues Ser8 to Thr18 and a β-sheet. These two elements of secondary structure are stabilised by the two disulphide bridges that are very well conserved in all short toxins, Cys2-Cys6 and Cys3-Cys7. The additional disulphide bridge, Cys4-Cys8, fixes the C-terminus to the turn following the α-helix [65].

Since the discovery of Pi1, several toxins acting on potassium channels and cross-linked by four disulphide bridges have been described. These toxins form a new family of toxins, subfamily 6, which currently has six different members. All the members of this family have an additional disulphide bridge which, in most cases, occupies the same position as the additional disulphide bridge in Pi1

[56]. The structures of maurotoxin (MTX) [66] and HstX1 [67] were resolved by NMR using synthetic toxins. Pi7, purified from the same venom as Pi1, is of particular interest in this subfamily as its target has not yet been identified although its predicted overall folding pattern is similar to that of all potassium channel toxins [68]. The three-dimensional structure of this toxin was thus determined by two-dimensional NMR methods with 50 nanomoles of compound using the Nano.nmr probe [50].

The overall folding pattern of the Pi7 peptide is identical to that of all other K^+ channel toxins isolated from scorpion venoms for which three-dimensional structure is known. Therefore it has been suggested that its apparent lack of biological activity on K^+-channels is due to differences in primary structure. It was found that the most significant difference between this toxin and other K^+ channel toxins is the net positive charge on the peptide. Pi7 was found to have an overall charge of $+3$, making it the least basic peptide of all those in subfamily 6. Unless there is an unknown subtype of K^+-channels for which Pi7 has evolved and is maintained in *Pandinus imperator* venom [69], the electrostatic interactions of the entire molecule with the channels may be responsible for its low affinity for the ion-channels tested to date.

Through these two studies, we have shown that it is possible to collect an adequate data set with nanomolar amounts of toxins and to determine the structure of the molecules using Nano.nmr probe technology. There are advantages of using a Nano.nmr probe for studies of such small molecules. First, the presence of numerous disulphide bridges may make it difficult to synthesise and properly fold these small toxins. Second, even if synthesis is successful, it is still important to determine whether the compound synthesised is identical to the native compound. Although this information can be obtained by other spectroscopic techniques, the use of small sample volume probes in NMR is unique in providing a wealth of information from very small quantities of material.

7.4 CONCLUSION

At a time when the trend is towards ultra-high field strength spectrometers, the reduction of sample size may be seen as a complementary alternative to these developments.

It is now possible to collect high-quality spectra in terms of resolution, sensitivity and lineshape with small amounts of product using small-volume probes. This is a very promising technique for situations in which sample quantity is a real limitation. The combined use of small-volume probes, 3 mm and below, together with cryogenic technology and Shigemi tubes will make it possible to investigate chemical structures at unprecedented low levels in a reasonable period of time, opening up new possibilities for the discovery of novel potent compounds.

ACKNOWLEDGMENTS

We would like to thank Dr Lourival Possani and his collaborators for providing the toxins and for having driven us to use the NMR small-sample volume technology.

REFERENCES

[1] Simard, J.M. and Watt, D.P. (1990) Venoms and toxins. In *The Biology of scorpion* (GA Polis, ed.) Stanford University Press 414–44. Stanford, California.

[2] Zlotkin, E., Miranda, F. and Rochat, H. (1978) Chemistry and pharmacology of buthinae scorpion venom. In *Handbook of Experimental Physiology* (Bettini S, ed.) **48**, 317–69. Springer-Verlag, Heidelberg.

[3] Possani, L.D. (1984) Structure of scorpion toxins. In *Handbook of Natural Toxins* (TU AT Ed) **2**, 513–50. Marcel Dekker Inc., New York.

[4] McCormick, K.C. and Meinwald, J. (1993) Neurotoxic acylpolyamines from spider venoms. *J. Chem. Ecol.* **19**, 2411–51.

[5] Escoubas, P., Diochot, S. and Gorzo, G. (2000) Structure and pharmacology of spider venom neurotoxins. *Biochimie* **82**, 893–907.

[6] Sato, S., Ogata, T., Borja, V., Gonzalzes, C., Fukuyo, Y., Kodoma, M. (2000) Frequent occurrence of paralytic shellfish poisoning as dominant toxins in marine puffer from tropical water. *Toxicon* **38**, 1101–9.

[7] Poli, M.A., Musser, S.M., Dickey, R.W., Eilers, P.P. and Hall, S. (2000) Neurotoxic shellfish poisoning and brevetoxin metabolites: a case study from Florida. *Toxicon* **38**, 981–93.

[8] Catterall, W.A. (1976) Purification of a toxic protein from scorpion venom which activates the action potential Na^+ ionophore. *J. Biol. Chem.* **251**, 5528–36.

[9] Possani, L.D., Martin, B.M. and Svendsen, I. (1982) The primary structure of noxiustoxin: a potassium channel-blocking peptide, purified from the venom of the scorpion *Centruroides noxius* Hoffmann. *Carlsberg Res. Commun.* **47**, 285–9.

[10] Possani, L.D., Becerril, B., Delepierre, M. and Tytgat, J. (1999) Scorpion toxins specific for Na+ channels. *Eur. J. Biochem.* **264**, 287–300.

[11] Putzig, C.L., Leugers, M.A., McKelvy, M.L., Mitchell, G.E., Nyquist, R.A., Papenfuss, R.R. and Yurga, L. (1994) Infra-red spectroscopy. *Anal. Chem.* **66**, 26R-66R.

[12] Worm, O., Roespstorff, P. and Mann, M. (1994) Improved resolution and very high sensitivity in Maldi Tof of matrix surfaces made by fast evaporation. *Anal. Chem.* **66**, 3281–7.

[13] Abragam, A. (1961) The Principle of Nuclear Magnetism. Oxford University Press, UK.

[14] Styles, P., Soffe, N.F., Scott, C.A., Cragg, D.A., Row, F., White, D.J. and White, P.C.J. (1984) A high resolution NMR probe in which the coil and preampifier are cooled with liquid helium. *J. Magn. Reson.* **60**, 397–404.

[15] Styles, P., Soffe, N.F. and Scott, C.A. (1989) An improved cryogenically cooled probe for high resolution NMR. *J. Magn. Reson.* **84**, 376–8.

[16] Hoult, D. (1996) Encyclopedia of Nuclear Magnetic Resonance. Eds: Grant, D. M, and Harris R.R. '*Sensitivity of the NMR experiment*' **7**, 4256–66.

[17] Hoult, D. and Richards, R.E. (1976) The signal-to-noise ratio of the nuclear magnetic resonance experiment. *J. Magn. Reson.* **24**, 71–85.

[18] Hoult, D.I. (1978) The NMR receiver: a description and analysis of design. *Prog. Nucl. Magn. Reson. Spect.* **12**, 41–77.

[19] Hill, H.D.W. (1995) Encyclopedia of Nuclear Magnetic Resonance. Eds: Grant, D. M, and Harris R.R. '*Probes for high resolution NMR*' **6**, 3762–8.

[20] Shoolery, J.N. (1979) Small coils for NMR microsamples in Topics in Carbon-13 NMR Spectroscopy **3**, 28–38.

[21] Crouch, R.C. and Martin, G.E. (1992) Micro inverse detection: a powerful technique for natural product structure elucidation. *J. Nat. Prod.* **55**, 1343–7.

[22] Crouch, R.C. and Martin, G.E. (1992) Comparative evaluation of conventional 5 mm inverse and micro-inverse detection probes at 500 MHz. *Magn. Reson. Chem.* **30**, S66–S70.

[23] Crouch, R.C., Martin, G.E., Dickey, R.W., Baden, D.G., Gawley, R.E. and Kein, K.S. (1995) Brevetoxin 3: total assignment of the ^{1}H and ^{13}C NMR spectra at the submicromole level. *Tetrahedron* **51**, 8409–22.

[24] Crouch, R.C., Martin, G.E., Musser, S.M., Grenade, H.R. and Dickey, R.W. (1995) Improvements in the sensitivity of inverse detected heteronuclear correlation spectra using micro inverse probes and microcells: HMQC and HMBC spectra of Caribbean ciguatoxin-preliminary structural inferences. *Tetrahedron Letters* **36**, 6827–30.

[25] Shigemi Inc, Allison Park Pennsylvania.

[26] Russel, D.J., Hadden, C.E., Martin, G.E., Gibson, A.A., Zens, A.P. and Cavolan, J.L. (2000) A comparison of inverse detected heteronuclear NMR performance: conventional versus cryogenic microprobe performance. *J. Nat. Prod.* **63**, 1047–9.

[27] Wu, N., Peck, T.L., Webb, A.G., Magin, R.L. and Sweedler, J.V. (1994) ^{1}H NMR spectroscopy on the nanoliter scale for static and on-line measurements. *Anal. Chem.* **66**, 3849–57.

[28] Wu, N., Peck, T.L., Webb, A.G., Magin, R.L. and Sweedler, J.V. (1994) Nanoliter volume sample cells for ^{1}H NMR. Application to on-line detection in capillary electrophoresis. *J. Am. Chem. Soc.* **116**, 7929–31.

[29] Olson, D.L., Peck, T.L., Webb, A.G., Magin, R.L. and Sweedler, J.V. (1995) High-resolution microcoil ^{1}H NMR for mass-limited, nanoliter-volume samples. *Sciences* **270**, 167–70.

[30] Wu, N., Peck, T.L., Webb, A.G. and Sweedler, J.V. (1995) On-line NMR detection of amino acids and peptides on microbore LC. *Anal. Chem.* **67**, 3101–7.

[31] Behnke, B., Schlotterbeck, G., Tallareck, U., Strohschein, S., Tseng, L-H., Keller, T., Albert, T. and Bayer, E. (1996) Capillary HPLC-NMR coupling: high resolution ^{1}H NMR spectroscopy on the nanoliter scale. *Anal. Chem.* **68**, 1110–15.

[32] Olson, D.L., Lacey, M.E. and Sweedler, J.V. (1998b) The nanoliter niche. *Anal. Chem. News and Features* 257A–264A.

[33] Olson, D.L., Lacey, M.E. and Sweedler, J.V. (1998a) High resolution microcoil NMR for analysis of mass-limited, nanoliter samples. *Anal. Chem.* **70**, 645–50.

[34] Peck, T.L., Magin, R.L. and Lauterbur, P.C. (1995) Design and analysis of micro-coils for NMR microscopy. *J. Magn. Reson. Ser. B* **108**, 114–24.

[35] Subramanian, R., Lam, M.M. and Webb, A.G. (1998) RF microcoil design for practical NMR of mass-limited samples. *J. Magn. Reson.* **133**, 227–31.

[36] Martin, G.E., Guido, J.E., Robins, R.H., Sharaf, M.H.M., Tackie, A.N. and Schiff, P.L. Jr. (1998) Submicro inverse-detection gradient NMR: a powerful new way of conducting structure elucidation studies with $< 0.05\,\mu$ mol samples. *J. Nat. Prod.* **61**, 555–9.

[37] Hadden, C.E., Sharaf, M.H.M., Guido, J.E., Robins, R.H., Tackie, A.N., Phoebe, C.H., Schiff, P.L. and Martin, G.E. (1999) 11-Isopropylcryptolepine: a novel alkaloid isolated from *Cryptolepis sanguinolenta* characterized using submicro NMR techniques. *J. Nat. Prod.* **62**, 238–40.

[38] Subramanian, R. and Webb, A.G. (1998) Design of soleonidal microcoils for high-resolution ^{13}C NMR spectroscopy. *Anal. Chem.* **70**, 2454–8.

[39] Subramanian, R., Sweedler, J.V. and Webb, A.G. (1999) Rapid two-dimensional inverse detected hetero nuclear correlation experiments with < 100 nmoles samples with solenoid microcoil NMR probes. *J. Am. Chem. Soc.* **121**, 2333–4.

[40] Fitch, W.L., Detre, G., Holmes, C.P., Shoolery, J.N. and Kiefer, P.A. (1994) High resolution ^{1}H NMR in solid-phase organic synthesis. *J. Org. Chem.* **59**, 7955–6.

[41] Keifer, P.A. (1997) High-resolution NMR techniques for solid-phase synthesis and combinational chemistry. *Drug Discovery Today* **2**, 468–78.

[42] Keifer, P.A. (1998) New methods for obtaining high-resolution NMR spectra of solid-phase synthesis resins, natural products and solution-state combinatorial chemistry libraries. *Drugs of the Future* **23**, 301–17.

[43] Barbara, T.M. (1994) Cylindrical demagnetization fields and microprobe design in high-resolution NMR. *J. Magn. Reson. Ser. A* **109**, 265–9.

[44] Manzi, A., Salimath, P.V., Spiro, R.C., Keifer, P.A. and Freeze, H.H. (1995) Identification of a novel glycosaminoglycan core-like molecule I: 500 MHz ^{1}H NMR analysis using nano-NMR probe indicates the presence of a terminal α-GalNac residue capping 4-methylumbelliferyl-β-D-xyloside. *J. Biol. Chem.* **270**, 9154–63.

[45] Fuks, L.F., Huang, F.C.S., Carter, C.M., Edelsein, W.A. and Roemer, P.B. (1992) Susceptibility, lineshape and shimming in high-resolution NMR. *J. Magn. Reson.* **100**, 229–42.

[46] Keifer, P.A., Baltusis, L., Rice, D.M., Tymiak, A.A. and Shoolery, J.N. (1996) A comparison of NMR spectra obtained for solid-phase synthesis resins using conventional high-resolution magic angle spinning, and high-resolution magic-angle-spinning probes. *J. Magn. Reson. Ser. A.* **119**, 65–75.

[47] Mozurkewich, G., Ringermacher, H.I. and Bolef, D.I. (1979) Effect of demagnetization on magnetic resonance line shapes in bulk samples: application to tungsten. *Phys. Rev. B.* **20**, 33–8.

[48] Garroway, A.N. (1982) Magic-angle sample spinning of liquid. *J. Magn. Reson.* **49**, 168–71.

[49] Delepierre, M., Roux, P., Chaffotte, A-F. and Goldberg, M.E. (1998) ^{1}H NMR characterization of renatured lysozyme obtained from fully reduced lysozyme under folding/oxidation conditions: a high-resolution liquid NMR study at magic-angle spinning. *Magn. Reson. Chem.* **36**, 645–50.

[50] Delepierre, M., Prochnicka-Chalufour, A., Boisbouvier, J. and Possani, L.D. (1999) Pi7, an orphan peptide from the scorpion *Pandinus imperator*: a ^{1}H NMR analysis using a Nano-nmr probe. *Biochemistry* **38**, 16756–65.

[51] Kupce, E., Schmidt, P., Rance, M. and Wagner, G. (1998) Adiabatic mixing in the liquid state. *J. Magn. Reson.* **135**, 361–7.

[52] Kupce, E. and Freeman, R. (1995) Adiabatic pulses for wideband inversion and broadband decoupling. *J. Magn. Reson. Ser A* **115**, 273–6.

[53] Kupce, E., Keifer, P.A. and Delepierre, M. (2001) Adiabatic TOCSY MAS in liquids. *J. Magn. Reson.* **148**, 115–20.

[54] Sarkar, S.K., Garigipati, R.S., Adams, J.L. and Keifer, P.A. (1996) An NMR method to identify nondestructively chemical compounds bound to a single solid-phase-synthesis bead for combinatorial chemistry application. *J. Am. Chem. Soc.* **118**, 2305–6.

[55] Delepierre, M. (1998) High resolution liquid NMR and magic-angle spinning. *J. Chim. Phys.* **95**, 235–40.

[56] Selisko, B., Garcia, C., Becerril, B., Gomez-Lagunas, F., Garay, C. and Possani, L.D. (1998) Cobatoxins 1 and 2 from *Centruroides noxius* Hoffmann constitute a subfamily of potassium-channel-blocking scorpions toxins. *Eur. J. Biochem.* **254**, 468–79.

[57] Fontecilla-Camps, J-C., Almassy, R.J., Suddath, F.L., Watt, D.D. and Bugg, C.E. (1980) Three-dimensional structure of a protein from scorpion venom: a new structural class of neurotoxins. *Proc. Natl. Acad. Sci. USA* **77**, 6496–500.

[58] Almassy, R.J., Fontecilla-Camps, J-C., Suddath, F.L. and Bugg, C.E. (1983) Structure of variant-3 scorpion neurotoxin from *Centruroides sculpturatus* Ewing, refined at 1.8 Å resolution. *J. Mol. Biol.* **170**, 497–527.

[59] Dauplais, M., Gilquin, B., Possani, L.D., Gurrola-Briones, G., Roumestand, C. and Ménez, A. (1995) Determination of the three-dimensional solution structure of noxiustoxin: analysis of structural differences with related short-chain scorpion toxins. *Biochemistry* **34**, 16563–73.

[60] Bontems, F., Roumestand, C., Gilquin, B., Ménez, A. and Toma, F. (1991) Refined structure of charybdotoxin: common motifs in scorpion toxins and insect defensins. *Science* **254**, 1521–3.

[61] Miller, C., Moczydlowski, E., Latorre, R. and Philipps, M. (1985) Charybdotoxin, a protein inhibitor of single Ca^{2+}-activated K^+ channels from mammalian skeletal muscle. *Nature* **313**, 316–18.

[62] MacKinnon, R., Heginbothan, L. and Abramson, T. (1990) Mapping the receptor site for charybdotoxin, a pore-blocking potassium channel inhibitor. *Neuron* **5**, 767–71.

[63] Doyle, D.A., Morais Cabral, J.M., Pfuetzner, R.A., Kuo, A., Gulbis, J.M., Cohen, S.L., Chait, B.T. and MacKinnon, R. (1998) The structure of the potassium channel: molecular basis of K^+ conduction and selectivity. *Science* **280**, 69–77.

[64] Olamendi-Portugal, T., Gomez-Lagunas, F., Gurrola, G.B. and Possani, L.D. (1996) A novel structural class of $K+$ channel-blocking toxin from the scorpion *Pandinus imperator*. *Biochem. J.* **315**, 977–81.

[65] Delepierre, M., Prochnika-Chalufour, A. and Possani, L.D. (1997) A novel potassium channel-blocking toxin from the scorpion *Pandinus imperator*: a ¹H NMR analysis using a Nano-nmr probe. *Biochemistry* **36**, 2649–58.

[66] Blanc, E., Sabatier, J.M., Kharat, R., Meunier, S., El Ayeb, M., Van Rietschoten, J. and Darbon, H. (1997) Solution structure of maurotoxin, a scorpion toxin from *Scorpio maurus*, with high affinity for voltage-gated potassium channels. *Proteins: Struct. Funct. Gen.* **29**, 321–33.

[67] Savarin, P., Romi-Lebrun, R., Zinn-Justin, S., Lebrun, B., Nakajima, J., Gilquin, B. and Ménez, A. (1999) Structural and functional consequences of the presence of a fourth disulfide bridge in the scorpion short toxins: solution structure of the potassium channel inhibitor HsTX1. *Protein Sci.* **8**, 2675–85.

[68] Olamendi-Portugal, T., Gomez-Lagunas, F., Gurrola, G.B. and Possani, L.D. (1998) Two similar peptides from the venom of the scorpion *Pandinus imperator*, one highly effective blocker and the other inactive on K^+ channels. *Toxicon* **36**, 759–70.

[69] Possani, L.D., Selisko, B. and Gurrola, G.B. (1999) Structure and function of scorpion toxins affecting K^+-channels. *Persp.Drug.Discov.Design.* **15/16**, 15–40.

Part III   Animal Toxins: From Fundamental Studies to Drugs?

8 Cone Snails and Conotoxins Evolving Sophisticated Neuropharmacology

BALDOMERO M. OLIVERA*, JULITA S. IMPERIAL and GRZEGORZ BULAJ

This article will provide a general perspective of work on *Conus* venoms. The article comprises four main sections: the molecules in *Conus* venoms; functional aspects; conopeptides in medicine and neuroscience; and future directions. The goal is to provide an overview framework for the *Conus* peptide system, rather than review the research literature in detail. Thus, the literature cited is not meant to be comprehensive, and in many cases, previous review articles rather than primary research articles are quoted.

8.1 *CONUS* VENOMS: THE MOLECULES

8.1.1 OVERVIEW

The unprecedented diversity of small peptides found in the venom of a single *Conus* species is the most striking feature of *Conus* venoms [1]. Although there is also clear evidence for the presence of both smaller organic molecules as well as larger proteins, relatively little work has been done on the nonpeptide venom components to date. Some *Conus* species have a high content of arachidonic acid [2], and of serotonin [3]; evidence for other yet uncharacterised organic molecules has been obtained. Of the larger polypeptides, the best characterised is conodipine, a phospholipase A_2 first isolated from *Conus magus* venom ducts, but widely distributed among *Conus* species [4].

The majority of biologically active components of *Conus* venoms are relatively small peptides, divided into two general classes: those with multiple disulphide bonds (between 2–5 disulphide crosslinks) and those that either have a single or no disulphide crosslink. In general, the different families of multiply disulphide-crosslinked peptides are more diverse and more widely

---

* Corresponding author

Perspectives in Molecular Toxinology
Edited by A Ménez © 2002 John Wiley & Sons, Ltd

distributed throughout the genus. The term 'conotoxin' has generally been applied to these Cys-rich peptides. In this article, we use the more generic 'conopeptides' for both classes, but continue to refer to individual peptides and peptide families with $\geq$ 2 disulphides as conotoxins or conotoxin families.

Although *Conus* peptides may be small, they assume specific conformations (like larger polypeptides, but unlike many smaller peptides of comparable size). For most *Conus* peptides, conformational stability is enhanced by multiple disulphide crosslinks, which restrict conformations accessible to the peptide. Thus, although many multiply disulphide-crosslinked *Conus* peptides are in the 10–20 amino acid (AA) size range, their three-dimensional conformations in solution can readily be determined by multidimensional NMR techniques or by X-ray crystallography (in a few cases).

8.1.2   POST-TRANSLATIONAL MODIFICATION OF CONOPEPTIDES

A striking feature of many *Conus* peptides is the unusually high frequency of post-translational modification observed [5]. Many of the modified amino acids are unusual, and some unprecedented (for example, the post-translational modification of tryptophan to 6-bromotryptophan). A list of presently known post-translational modifications in *Conus* peptides is given in Table 8.1.

Conus peptide families that are not multiply disulphide-crosslinked are often extensively post-translationally modified. In some cases, the post-translational modification is essential for achieving the biologically-active conformation – a good example is the role of γ-carboxylation of Glu residues (to γ-carboxyglutamate) in stabilising the helical conformation of some conopeptides, notably the conantokins (see Table 8.1).

Conus peptides are at an unusual biological intersection between conventional natural products and polypeptides translated by ribosomes from mRNA. Being at this intersection allows, on the one hand, the easy generation of diversity by direct mutation of open reading frames (which in turn leads to considerable evolutionary flexibility). On the other hand, access to functional groups other than those present in the standard 20 amino acids is provided by overlaying ribosomal translation with post-translational modification. Thus, in their venom peptides, the cone snails have successfully combined features of conventional gene products with the more diverse chemistry characteristic of natural products.

8.1.3   BIOSYNTHESIS OF CONOPEPTIDES

Each individual *Conus* peptide is translated from a specific mRNA transcript into a precursor polypeptide with a canonical prepropeptide organisation [6]. The conopeptide precursor comprises three regions: the 'pre' region or signal sequence at the N-terminus (typically 20–25 amino acids in length); the mature

Table 8.1 Post-translationally Modified Amino Acids and the *Conus* Peptides In Which They Were First Identified.

Non-Standard Amino Acid	Modification	Peptide	Sequence	Enzyme	Ref.
4-OH-Proline	Hydroxylation of proline	μ-GIIIA	RDCCTOOKKCCKDRQC KOQRCCA*	Proline hydroxylase	[23]
γ-Carboxyglutamate	Carboxylation of glutamic acid	Conantokin-G	GEγγLQγNQγLIRγKSN*	γ-Glutamate carboxylase	[24]
6-Br-Tryptophan	Bromination of tryptophan	Bromocontryphan	GCOwEPW†C*	Bromo peroxidase	[25]
D-Tryptophan	Isomerisation of tryptophan	Contryphan	GCOwEPWC*	Tryptophan epimerase	[26]
Pyroglutamate	Cyclisation of N-terminal Gln	Bromoheptapeptide	ZCGQAW†C*	Glutaminyl cyclase	[27]
Sulphated tyrosine	Sulphation of tyrosine	α-Epl	GCCSDPRCNMNNPY‡C*	Tyrosyl sulfotransferase	[28]
Glycosylated serine (or threonine)	O-glycosylation	κA-SIVA	ZKSLVPS§VITTCCG YDOGTMCOOCRCTNSC*	Polypeptide HexNAc transferase	[29]

γ = γcarboxyglutamate; Z = pyroglutamic acid; Y ‡ = tyrosine sulphate; w = D-tryptophan; W † = 6-L-bromotryptophan; 0 = 4-trans hydroxyproline; S § = Hex₃HexNAc₂-Ser. In addition to the modified AA above, many conopeptides have disulphide linkages (which may involve disulphide isomers) and C-terminal amidation (that requires the protein amidating oxygenases)

conopeptide at the C-terminus, always present in a single copy (from 8 to > 40 amino acids in length, with the majority between 12–30 AA); and between the signal sequence and the mature conopeptide regions is an intervening region (the 'pro' region) which for most conopeptides is 30–60 AA in length (in conotoxin families with a long N-terminal region before the first disulphide bond, an abbreviated pro region is found).

Thus, for the average 20 AA conopeptide, the open reading frame in the corresponding mRNA encodes an ~ 80 AA precursor. Between translation of the precursor on ribosomes and the final proteolytic cleavage that releases the biologically-active conopeptide, folding to form the correct disulphide cross-links must occur, and, in addition, the post-translational modifications found in the mature gene product are effected.

8.1.4 FUNCTIONAL DOMAINS OF CONOPEPTIDE PRECURSORS: THE GENERATION OF MOLECULAR DIVERSITY

Although a conopeptide precursor is encoded by a relatively small open reading frame, the three regions described above ('pre-', 'pro-' and 'mature toxin' regions) can be considered as separate functional domains. The N-terminal signal sequence targets the conopeptide to a secretory pathway. The C-terminal mature region functions as the final biologically-active *Conus* venom component after it is processed. Data suggest that the propeptide region has recognition signals for post-translational modification enzymes [7, 8]. Recognition signals embedded in propeptide sequence domains apparently direct modification enzymes to catalyse modification reactions on specific amino acids downstream, in the mature peptide region.

Although the amount of work done to date on conopeptide genes is limited, for at least some *Conus* peptides, the three functional domains of the precursor are separately encoded in three widely-separated, small exons (R. Schoenfeld, unpublished results; [9]). In the genes analysed so far, large introns separate the small exons corresponding to the open reading frame encoding a conopeptide precursor. An interspecific comparison (as well as intraspecific data) comparing sequences of homologous and orthologous peptides reveals a sharp contrast between mutation rates in the three functional domains (~ exons). Particularly striking is the difference in apparent mutation rates between the signal sequence and the mature conopeptide domains [10]. The former are extraordinarily conserved (when compared to signal sequences of other secreted gene products), while the latter exhibit an unprecedented level of interspecific hypermutation. The contrast between these two regions in the same gene product appears to be unique; in no other gene family has the extreme dichotomy between signal sequences and C-terminal regions of conopeptides been noted. The rate of mutation within the 'pro' region appears to be in the normal range, although the 'pre' and 'toxin' domains deviate dramatically from the rates expected for 'normal' genes.

8.1.5 PERSPECTIVES

The overview that has emerged is that *Conus* venom peptides are small but conformationally stable gene products with a high frequency of post-translationally modified amino acids. The mature conopeptide is ultimately the biologically relevant entity. However, in order to generate these highly potent venom components, appropriate modification enzymes that effectively increase the repertoire of functional groups beyond the side chains of the 20 standard amino acids act on the unmodified gene product, most likely by binding initially to recognition signals in the 'pro' region.

The remarkable conservation of signal sequences observed within conopeptide gene families suggests that in addition to targeting the precursor to a secretory pathway, these signal sequences may have other important roles. One possibility is that separate secretory pathways have evolved in the venom duct epithelium for different conopeptide families; this could require that the signal sequence targets a conopeptide precursor to a specific locus on the endoplasmic reticulum. Once transported across the endoplasmic reticulum, different conopeptide families may be sorted with different sets of maturation proteins such as specific chaperones (to facilitate proper folding of members of that conopeptide family into the correct disulphide-bonded framework), as well as particular modification enzymes (the probability of whether an amino acid will or will not be modified appears to be dependent on which family the conopeptide belongs to). This hypothesis is supported by the fact that the secretory granules found in the venom duct are not uniform in size, and fall into discrete size classes.

There are many parallels between the evolution of the array of conopeptides in the venom of a *Conus* species and drug development by a large pharmaceutical company. The hypermutation that occurs as species diverge from each other is equivalent to a combinatorial peptide library strategy for developing lead compounds. In a big drug company, a lead compound is further refined by applying sophisticated medicinal chemistry; the *Conus* equivalent is the post-translational modification enzymes which may enhance specificity and efficacy of an unmodified gene product.

8.2 CONOPEPTIDES: FUNCTIONAL ASPECTS

8.2.1 OVERVIEW

As outlined in the section above, most biologically-active components of *Conus* venom components are small, mostly disulphide-rich peptides. There may be 100–200 such peptides that can be expressed in the venom of a single species – why are these venoms so complex? In the discussion below, we develop a working conceptual framework: venom components can be grouped together based on the different broad physiological effects elicited by the venom as a whole. In many cases, groups of peptides act together to effect a particular physiological end-point. However, each individual peptide within such a group likely has a single molecular target; binding of the conopeptide to the target causes a drastic change in the function of the targeted molecule. Many of the targets of conopeptides (if not the majority) are either ion channels or receptors found in the peripheral nervous system and the neuromuscular circuitry [1].

8.2.2 BIOLOGY OF CONE SNAILS: A SYNOPSIS

The genus *Conus* is unusually species-rich – indeed, it is arguably the largest single genus of marine invertebrates presently living today (ca. 500 different species). The genus first radiated in the Eocene period, approximately 55 million years ago, and has generally been expanding in number of species with time [11]. The genus *Conus* can be subdivided on the basis of major prey (see Plate III). The largest number of species probably feed on polychaete worms (vermivorous). However, a significant number of *Conus* (50 or more species) eat other gastropods (molluscivorous) and an equal or perhaps even larger number envenomate and devour fish (piscivorous). In addition, there are *Conus* species that will also prey on hemicordates, echiuroids and bivalves either as major prey, or as a substantial fraction of total prey consumed [12].

It appears that some *Conus* species are highly specialised; some vermivorous species are known to only eat one type of polychaete worm. In contrast, there are some generalist species known to consume prey in three different phyla. Cone snails use their venom not only for prey capture, but also for defence against predators, and even for competitive interactions. Human stinging cases, which have long been known to be occasionally fatal [13], are almost certainly defensive in nature. Thus, even highly specialised *Conus* that may be eating only a phylogenetically narrow prey spectrum would be expected to have venom components targeted to predators or competitors in other phyla – this makes it difficult to predict *a priori* the phylogenetic range of activity for a given *Conus* peptide, even though the major prey of the species has been identified. Additionally, while some individual conopeptides are active only over a narrow range phylogenetically, other *Conus* peptides inhibit homologous targets over a wide range of phyla. Since most cone snails immobilise their prey, it seems generally true that some major peptides in a *Conus* venom are likely to interfere with neuromuscular transmission of the major prey.

8.2.3 MOLECULAR MECHANISMS

It is important to note that at the present time, the *Conus* peptides for which there is some mechanistic information comprise only a small fraction of those that are biochemically characterised. For most conopeptides, the underlying mechanisms that account for their biological activity have not yet been elucidated.

The two largest classes of conopeptide targets known are voltage-gated ion channels and ionotropic receptors (ligand-gated ion channels), which are key signalling components of nervous systems [10]. For the voltage-gated ion channels, inhibitory conopeptides that are channel blockers or which inhibit the transition of the channel from a closed to an open state are the largest class known. However, there are clearly other mechanisms; one that is well characterised is to delay or completely inhibit inactivation of the targeted voltage-gated ion channel, in effect causing a much longer open state for the ion channel (and much greater flow of ions across the plasma membrane each time the channel opens). Similarly, several different mechanisms for affecting the function of ligand-gated ion channels have been elucidated. Most conopeptides that target ionotropic receptors are competitive inhibitors that compete with the neurotransmitter for binding to the ligand site. However, non-competitive inhibitors have also been described. A third significant group of conopeptides are those which affect G-protein-coupled receptors.

We should note that while a wide variety of mechanisms are possible, the outstanding common characteristic of *Conus* peptides is their exceptional selectivity: an individual *Conus* peptide can discriminate between closely related members of the same ion channel or receptor family. The underlying basis for the high affinity and selectivity of *Conus* peptides has not been definitively explained. One hypothesis for such unusual selectivity is shown in Figure 8.1; as has previously been suggested, at least some *Conus* peptides may not just have one pharmacophore, but may instead have two distinct recognition/interaction faces. The proposed mechanism illustrated, which could explain high selectivity and specificity for some conopeptides, has been called the 'Janus ligand hypothesis' [10].

In many venoms, several conopeptides belonging to the same family are found; invariably, these target different molecular subtypes of the same ion channel or receptor family. Thus, conopeptides provide an exceptional opportunity to understand how to design ligands that discriminate between various isoforms of the large receptor and ion channel families that are so characteristic of the nervous system.

8.2.4   TOXIN CABALS

One useful concept to link the molecular mechanisms of individual conopeptides to the broad biological effects of *Conus* venom is the toxin *cabal* [10]. An intensive study of a few venoms has revealed a synergy in the activity of groups of different individual conopeptides, with the aggregate activity of the group leading to a powerful physiological effect on the injected animal [14]. The best understood in terms of underlying mechanisms of individual peptides is the inhibitory effect of fish-hunting *Conus* venoms on neuromuscular transmission

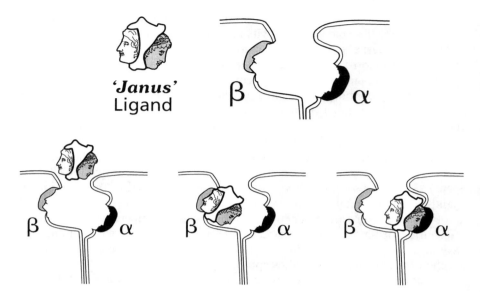

Figure 8.1 Proposed model for conotoxin target selectivity: the Janus Ligand Hypothesis. A cartoon representation of the 'Janus Ligand Hypothesis' [10], which postulates that there are two distinct recognition faces in some *Conus* peptides. The variation shown in the lower panel of the figure is one specific model based on the structure and kinetic properties of α-conotoxin MII [21, 22], known as 'dock-and-lock', in which one face of the peptide interacts with the target nicotinic receptor by first binding to a site on the β subunit docking step which orients the peptide and concentrates it in the vicinity of the second site on the α subunit, known as the locking site. The docking interaction provides a fast on-time, while the locking interaction results in a slow off-time. Since the peptide has sites on two different subunits (the β and α subunits of the nAChR in this specific example), such a mechanism leads to the high affinity and specificity observed for many *Conus* peptides.

The specific mechanism shown in the lower panel is not a unique explanation for the available data. Another variation of the Janus ligand hypothesis would be if a conformational change after initial binding to the first site led to an interaction with the second site. Such conformational changes could occur in the *Conus* peptide, in the receptor target, or in both

of their prey. The group of peptides that block normal neuromuscular transmission is known as the 'motor cabal' of conopeptides (the word 'cabal' normally refers to secretive groups plotting to overturn existing authority). A typical motor cabal might consist of peptides that block the presynaptic terminus (by inhibiting the voltage-gated calcium channels which allow the entry of extracellular Ca^{2+} required for neurotransmitter exocytosis), peptides that block the receptor at the postsynaptic terminus, both competitive and non-competitive, as well as peptides which abolish skeletal muscle action potentials, primarily voltage-gated sodium channel blockers.

Another cabal of toxins present in many fish-hunting cone venoms is the 'lightning-strike cabal', a group of conopeptides which effects instant immobilisation of prey [10, 14]. In this case, the conopeptides of the cabal appear to target axonal voltage-gated ion channels in the immediate vicinity of the venom injection site. The underlying strategy is to cause massive depolarisation of axons close to the injection site, leading to uncontrolled firing from these axons – the effect is equivalent to a focal electric shock, resulting in the prey being stunned (in many ways, a strategy analogous to that used by electric eels in the Amazon River). The underlying molecular mechanisms appear to involve conopeptides which delay inactivation of voltage-gated sodium channels, leading to much larger sodium influx, combined with peptides which inhibit voltage-gated potassium channels – the combination causes uncontrolled firing of affected axons. As a consequence, the venom injection immediately causes the fist to enter an immobilised, tetanic state.

8.2.5   INTERSPECIFIC DIVERGENCE: AN OVERVIEW

The diverging biology of cone snail species is presumably reflected in interspecific differences of venom components. Most easily rationalised is conopeptide divergence between species that have prey in different phyla: optimising conopeptides involved in prey capture for the major prey would be expected to be strongly selected for. Thus, major conopeptides from fish-hunting cone snail venoms are highly potent on vertebrate ion channels or receptors, and less potent when tested on molluscan neurons, while conopeptides from molluscivorous *Conus* venoms show the converse potency.

Clearly, the biological determinants that lead to most conopeptide divergence are much more subtle. Each species occupies its own ecological niche, and has a behavioural repertoire consistent with that niche. Thus, two fish-hunting *Conus* species may use entirely different strategies for approaching and capturing prey; the striking contrast between a harpoon-and-line strategy versus a net strategy was described previously [10]. In the former case, the fish is initially tethered rather precariously, and the lightning-strike cabal is essential for successful prey capture. However, for a *Conus* species using a net strategy, the lightning-strike cabal of conotoxins is unnecessary and generally found to be absent from these venoms. It was suggested that net-hunting piscivorous cone snails first cause a sedative-like effect on the prey even before injection, possibly by releasing small amounts of venom into their 'net'. Thus, fish-hunting cone snails that use a harpoon-and-line strategy have major toxins in their venoms that cause potent excitatory effects (components of the lightning-strike cabal), whereas there are absent in fish-hunting cone snails that use a net strategy. Instead, major conopeptides that quiet down neuronal circuitry are found (the 'nirvana cabal', see [15]).

With respect to prey capture, selective pressures leading to interspecific divergence may be so subtle that they could be difficult to identify. Some harpoon-and-line fish-hunting *Conus* species are extremely aggressive: when a fish is introduced, such species adopt a foraging behaviour that leads to aggressively tracking down the fish. Other *Conus* are more cautious and approach fish only when they are relatively immobile. Thus, precisely where on the body of the fish different *Conus* species strike may depend on how the snail approaches a fish, and how the potential fish prey behave – these could potentially select for different conopeptides (for example, axonal ion channels may be different in distinct sections of the periphery, and this could select for different conopeptides of the lightning-strike cabals of distinct fish-hunting *Conus*). In addition, the effective predators on different *Conus* species may be divergent as well. Thus, what might appear to be relatively subtle biological considerations may play a major role in interspecific *Conus* peptide divergence.

8.3   CONOPEPTIDES IN MEDICINE AND NEUROSCIENCE

8.3.1   INTRODUCTION

Important applications for the conopeptides have been developed; they are now widely used reagents in experimental neuroscience, and a few conopeptides have been developed for diagnostic and therapeutic application. The brief overview below of this facet of conopeptide work demonstrates the great potential for the future. The basis for both the medical (i.e., therapeutic and diagnostic) uses of conotoxins, as well as applications in basic neuroscience research, is the high specificity and potency of conopeptides. For applications in neuroscience, a highly specific conopeptide is equivalent to, and in some cases can be used even more flexibly than, a gene knockout mouse. For therapeutic applications, conopeptides provide a paradigm for developing neuroactive drugs with far fewer side effects, because of their exquisite target selectivity.

8.3.2   MEDICAL APPLICATIONS

The most advanced therapeutic application of conopeptides is the development of ω-conotoxin MVIIA, which specifically blocks one molecular form of voltage-gated calcium channels, as a potential analgesic drug [16, 17]. This 27-amino-acid peptide, originally purified by J.M. McIntosh, an undergraduate student at the University of Utah in the early 1980s, has been through Phase III clinical trials, and the company developing it, Elan Corporation, has received an 'approvable' letter from the United States Food and Drug Administration (which is just prior to final approval as a commercial drug). The commercial product, to be given the generic of 'ziconotide' is chemically identical to the natural peptide isolated from the venom of the piscivorous species *Conus magus* (the magician's cone). The main therapeutic application of ziconotide is to alleviate malignant cancer pain, particularly in patients that have become tolerant to morphine.

The peptide has promise as a therapeutic for two reasons. First, unlike opioid drugs, it is not an agonist for a G-protein-coupled receptor, and therefore the problem of downregulation of such receptors (which is the basis for tolerance by patients) does not apply. The second property of ω-conotoxin MVIIA is its almost absolute selectivity for those calcium channel complexes that have a α_{1B} subunit. In the mammalian spinal cord, it appears that this class of voltage-gated calcium channels is restricted to the dorsal horn, and may be found as the major Ca channel subtype only in those synapses between C-fibres and the spinal cord neurons that carry pain signals to the higher centres in the brain. Thus, the underlying mechanism for analgesia is the unprecedented selectivity of this conopeptide for a single subtype of voltage-gated calcium channels.

Several other conopeptides are being developed for therapeutic applications; furthest along is the NMDA receptor antagonist, conantokin-G, which has just entered clinical trials for intractable epilepsy. Several reviews of peptides in various stages of clinical development (most in early stages) have recently been published [18, 17].

8.3.3   *CONUS* VENOMS: A TOOLBOX FOR THE NEUROSCIENTIST

Conopeptides are now in widespread use in neuroscience research; over 2000 publications to date describe experiments that use conopeptides. The utility of these peptides in neuroscience falls into several major categories: the most widespread use of conotoxins is as tools for blocking synaptic transmission. Several peptides, notably ω-conotoxin GVIA and ω-conotoxin MVIIC, have become standard reagents for blocking the voltage-gated calcium channel sub-types present in most mammalian presynaptic termini, thereby inhibiting neuro-transmitter release and synaptic transmission. In addition, a number of different conopeptides are being used to discriminate *in situ* between various molecular forms of an ion channel or receptor family. Among the peptides that are widely used for these purposes are the ω-conotoxins (to identify different calcium channel subtypes), the α-conotoxins (to functionally define subunits of nicotinic acetylcholine receptor complexes) and the μ-conotoxins (to discrimin-ate between various types of voltage-gated sodium channels).

Given the accelerated pace at which mechanisms of the ca. 50 000 different peptides in *Conus* venoms are being elucidated, it appears almost certain that the use of these pharmacological agents as tools for the neuroscientist will continue to increase in the near future.

8.4   FUTURE DIRECTIONS

8.4.1   INTRODUCTION

The data available to date on *Conus* peptides provide the overview outlined above, and also lead to new questions being framed. Much of the future work on *Conus* peptides is readily predictable. Undoubtedly, many more peptide sequences will be elucidated, either by directly isolating peptides from venom, or through molecular cloning. The mechanisms and targets of more *Conus* peptides will be identified. How different peptides in venom act together in new cabals should also begin to emerge. The correlation between the biochemistry and physiology of the individual components of the venom, and the biology of the organism, is likely to proceed more slowly – understanding differences between the biology of different species is a rate-limiting step. Since these future research directions are predictable, the sections that follow focus instead on issues that are less obvious and/or those which contribute to a conceptual scientific framework.

8.4.2   STRUCTURE AND FOLDING

One major question is how these small peptides are properly folded into their biologically-active three-dimensional conformations. The problem is a subset of one of the most important remaining questions in biology, how one-dimensional amino acid sequences give rise to the three-dimensional shapes of proteins. Although the final correctly folded peptides are unusually small, large superfamilies of *Conus* peptides that diverge greatly in amino acid sequence are all routinely folded in the snail venom ducts to form the same disulphide-crosslinked scaffold. Understanding the specific folding pathway for each of these peptide superfamilies may lead not only to important information about the *Conus* system *per se*, but could shed some light on the general problem of protein folding.

When it was first discovered that conotoxins were initially translated as prepropeptide precursors, since most smaller peptides equilibrate rapidly between many different conformations, it seemed reasonable to postulate that the larger size of the precursor provided enough weak interactions to stabilise a specific conformation in the mature conotoxin region at the C-terminus, which could then be locked in by the formation of covalent crosslinks [6]. However, this proposition was examined experimentally by Price-Carter and Goldenberg for one specific conotoxin, ω-conotoxin MVIIA [19], and they found that the presence of the propeptide region had basically no effect on folding of the mature conotoxin region. As an alternative, it is suggested above that different conotoxin gene families may have specialised secretory pathways, and that proper folding may be facilitated by interacting with other proteins. For those *Conus* peptides with six cysteine residues, there are 15 different isomeric conformations, with the biologically active conformation essentially the only one found in venom. Elucidating how the snails have extended the rules for protein folding to much smaller peptides with disulphide linkages may provide insights into 'normal' protein folding.

A second related issue requiring systematic investigation is the link between amino acid sequence and 'small secondary structures'. The conopeptide conformations deduced so far, primarily from multidimensional NMR spectroscopy, have suggested that small secondary structural features occur at high frequency in these peptides. In addition to more global motifs (for example, the cysteine knot found in the O-superfamily of *Conus* peptides), small α-helical regions and β hairpin loops are regularly found in *Conus* peptide structures. As AA sequences and NMR structures of more and more peptides that belong to a particular superfamily accumulate, a correlation between sequence and micro secondary structure in the peptides should become increasingly apparent. Such information could well be important for learning how to predict three-dimensional conformations of new conopeptides. Of more general significance, it has been postulated that such small secondary structural elements may seed folding in larger proteins. A systematic analysis of *Conus* peptides may provide one link between amino acid sequences and micro secondary structures, and allow an experimental system for evaluating the role of such secondary structures in seeding proper folding.

8.4.3   DELIVERY DESIGN AND POST-TRANSLATIONAL MODIFICATION

One problem that has not been incisively addressed in understanding *Conus* peptides is how the physical properties of the peptide might be optimised for efficient delivery to the site that is physiologically relevant. The concept of cabals helps to define the scope of this problem. Conopeptides of the motor cabal in fish-hunting cone snails, if they are to be efficacious, must spread as rapidly as possible throughout the body of the prey. In contrast, it would appear that the 'lightning-strike cabal' of peptides, in order to rapidly elicit the tetanic immobilisation of fish prey, must be concentrated on the axons in the neighbourhood of the injection site; random peptide dispersal would be counterproductive. Thus, some peptides in the venom appear to be selected for efficient dispersal from the site of injection, while others are meant to target the closest relevant neuronal membranes in a concentrated bolus. Some differences between the *Conus* peptides in the lightning-strike cabal vs. the motor cabals of fish-hunting snails are suggestive; thus, larger peptides (> 30 AA) with some post-translational modifications may be favoured for cabals where localisation near the injection site is desirable. In the next few years, much more information should be available to assess more subtle features of these post-injection delivery aspects of *Conus* peptides.

The functional significance of a post-translational modification can potentially occur at several levels. It is tempting to assume that a post-translational modification directly enhances the interaction between a *Conus* peptide and its receptor target, but other considerations may be even more important. Are some post-translational modifications important for the delivery aspects described above? Alternatively, do some post-translational modifications facilitate folding a *Conus* peptide into its relevant three-dimensional conformation?

8.4.4 *IN VITRO* EXTENSION OF PEPTIDE EVOLUTION: DRUG DEVELOPMENT

Although the development of *Conus* peptides as therapeutics is being actively explored, there are undeveloped areas that ultimately may have even greater potential with regard to drug development. Given that *Conus* peptides are highly selective ligands which can discriminate between closely homologous members of ion channel and receptor superfamilies, the possibility of extending the hypermutation practised by the cone snails over tens of millions of years to *in vitro* systems is a promising technology not yet realised (although the desirability of developing this was suggested over a decade ago [20]). In principle, by judicious use of combinatorial peptide library techniques, it should be possible to screen for a set of conotoxin structures that specifically target one member of a complex ion channel or receptor superfamily. Since *Conus* peptides are presumably already preselected to have the appropriate framework for interacting with homologous target sites present in these families of ion channel or receptor targets, screening for peptides with the appropriate specificities seems feasible, at least theoretically.

As more mechanistic insights are gained into the molecular basis for the high selectivity of *Conus* peptides for their specific molecular targets (mechanisms such as the Janus ligand hypothesis in Figure 8.1), it should be increasingly possible to take a set of peptides specific for a particular target and use these as leads for designing a peptidomimetic or nonpeptidic ligand specific for only one molecular isoform of an ion channel or receptor family, but which might have more desirable bioavailability properties (i.e. the ability to cross the gut/bloodstream or blood/brain barrier). Thus, a combination of mechanistic insights into the molecular mechanisms that cone snails have evolved through their peptides and the use of the peptides or derivatives thereof as starting leads for drug development could very well lead to new levels of sophistication in neuropharmacology.

ACKNOWLEDGEMENTS

The work of the authors on which these perspectives are based was entirely funded by the National Institute of General Medical Sciences (GM48677).

REFERENCES

[1] Olivera, B.M., Rivier, J., Clark, C., Ramilo, C.A., Corpuz, G.P., Abogadie, F.C., Mena, E.E., Woodward, S.R., Hillyard, D.R. and Cruz, L.J. (1990). Diversity of *Conus* neuropeptides. *Science* **249**, 257–63.
[2] Nakamura, H., Kobayashi, J., Ohizumi, Y. and Hirata, Y. (1982). The occurrence of arachidonic acid in the venom duct of the marine snail *Conus textile*. *Experientia* **39**, 897.

[3] McIntosh, J.M., Foderaro, T.A., Li, W., Ireland, C.M. and Olivera, B.M. (1993). Presence of serotonin in the venom of *Conus imperialis*. *Toxicon* **32**, 1561–6.

[4] McIntosh, J.M., Ghomashchi, F., Gelb, M.H., Dooley, D.J., Stoehr, S.J., Giordani, A.B., Naisbitt, S.R. and Olivera, B.M. (1995). Conodipine-M, a novel phospholipase A2 isolated from the venom of the marine snail *Conus magus*. *J. Biol. Chem.* **270**, 3518–26.

[5] Craig, A.G., Bandyopadhyay, P. and Olivera, B.M. (1999). Post-translationally modified peptides from *Conus* venoms. *Eur. J. Biochem.* **264**, 271–5.

[6] Woodward, S.R., Cruz, L.J., Olivera, B.M. and Hillyard, D.R. (1990). Constant and hypervariable regions in conotoxin propeptides. *EMBO J.* **1**, 1015–20.

[7] Bandyopadhyay, P.K., Colledge, C.J., Walker, C.S., Zhou, L.-M., Hillyard, D.R. and Olivera, B.M. (1998). Conantokin-G precursor and its role in γ-carboxylation by a vitamin K-dependent carboxylase from a *Conus* snail. *J. Biol. Chem.* **273**, 5447–50.

[8] Hooper, D., Lirazan, M.B., Schoenfeld, R., Cook, B., Cruz, L.J., Olivera, B.M. and Bandyopadhyay, P. (2000) Post-translational modification: a two-dimensional strategy for molecular diversity of *Conus* peptides. In *Peptides for the New Millennium: Proceedings of the Sixteenth American Peptide Symposium* (Fields, G.B., Tam, J.P. and Barany, G., eds), Kluwer Academic Publishers, Dordrecht, The Netherlands, pp. 727–9.

[9] Olivera, B.M., Walker, C., Cartier, G.E., Hooper, D., Santos, A.D., Schoenfeld, R., Shetty, R., Watkins, M., Bandyopadhyay, P. and Hillyard, D.R. (1999). Speciation of cone snails and interspecific hyperdivergence of their venom peptides. Potential evolutionary significance of introns. *Ann. N.Y. Acad. Sci.* **870**, 223–37.

[10] Olivera, B.M. (1997). E.E. Just Lecture (1996). *Conus* venom peptides, receptor and ion channel targets and drug design: 50 million years of neuropharmacology. *Mol. Biol. Cell.* **8**, 2101–9.

[11] Kohn, A.J. (1990). Tempo and mode of evolution in Conidae. *Malacologia* **32**, 55–67.

[12] Kohn, A.J. (1959). The ecology of *Conus* in Hawaii. *Ecology Monograph* **29**, 47–90.

[13] Rumphius, G.E. (1705) D'Amboinsche Rariteikamer. Fr. Halma, Amsterdam.

[14] Terlau, H., Shon, K., Grilley, M., Stocker, M., Stühmer, W. and Olivera, B.M. (1996). Strategy for rapid immobilization of prey by a fish-hunting cone snail. *Nature* **381**, 148–51.

[15] Olivera, B.M. and Cruz, L.J. (2001). Conotoxins, in retrospect. *Toxicon* **39**, 7–14.

[16] Miljanich, G.P. (1997). Venom peptides as human pharmaceuticals. *Science & Medicine* Sept/Oct., 6–15.

[17] Olivera, B.M. (2000) ω-Conotoxin MVIIA: from marine snail venom to analgesic drug. In *Drugs from the Sea* (Fusetani, N., ed.), Karger, Basel, Switzerland, pp. 75–85.

[18] Jones, R.M. and Bulaj, G. (2000). Conotoxins – new vistas for peptide therapeutics. *Current Pharmaceutical Design* **6**, 1249–55.

[19] Price-Carter, M., Gray, W.M. and Goldenberg, D.P. (1996). Folding of ω-conotoxins: 2. Influence of precursor sequences and protein disulfide isomerase. *Biochemistry* **35**, 15547–57.

[20] Olivera, B.M., Rivier, J., Scott, J.K., Hillyard, D.R. and Cruz, L.J. (1991). Conotoxins. *J. Biol. Chem.* **266**, 22067–70.

[21] Cartier, G.E., Yoshikami, D., Gray, W.R., Luo, S., Olivera, B.M. and McIntosh, J.M. (1996). A new α-conotoxin which targets α3β2 nicotinic acetylcholine receptors. *J. Biol. Chem.* **271**, 7522–8.

[22] Shon, K., Koerber, S.C., Rivier, J.E., Olivera, B.M. and McIntosh, J.M. (1997). Three-dimensional solution structure of α-conotoxin MII, and α3β2 neuronal nicotinic acetylcholine receptor-targeted ligand. *Biochemistry* **36**, 15693–700.

[23] Stone, B.L. and Gray, W.R. (1982) Occurrence of hydroxyproline in a toxin from the marine snail *Conus geographus*. *Arch. Biochem. Biophys.* **216**, 756–67.

[24] McIntosh, J.M., Olivera, B.M., Cruz, L.J. and Gray, W.R. (1984). H-carboxy-glutamate in a neuroactive toxin. *J. Biol. Chem.* **259**, 14343–6.

[25] Jimenez, E.C., Craig, A.G., Watkins, M., Hillyard, D.R., Gray, W.R., Gulyas, J., Rivier, J.E., Cruz, L.J. and Olivera, B.M. (1997). Bromocontryphan : post-translational bromination of tryptophan. *Biochemistry* **36**, 989–94.

[26] Jimenez, E.C., Olivera, B.M., Gray, W.R. and Cruz, L.J. (1996). Contryphan is a D-tryptophan-containing *Conus* peptide. *J. Biol. Chem.* **281**, 28002–5.

[27] Craig, A.G., Jimenez, E.C., Dykert, J., Nielsen, D.B., Gulyas, J., Abogadie, F.C., Porter, J., Rivier, J.E., Cruz, L.J., Olivera, B.M. and McIntosh, J.M. (1997). A novel post-translational modification involving bromination of tryptophan : identification of the residue, L-6-bromotryptophan, in peptides from *Conus imperialis* and *Conus radiatus* venom. *J. Biol. Chem.* **272**, 4689–98.

[28] Loughnan, M., Bond, T., Atkins, A., Cuevas, J., Adams, D.J., Broxton, N.M., Livett, B.G., Down, J.G., Jones, A., Alewood, P.F. and Lewis, R.J. (1998). α-conotoxin EpI, a novel sulfated peptide from *Conus episcopatus* that selectively targets neuronal nicotinic acetylcholine receptors. *J. Biol. Chem.* **273**, 15667–74.

[29] Craig, A.G., Zafaralla, G., Cruz, L.J., Santos, A.D., Hillyard, D.R., Dykert, J., Rivier, J.E., Gray, W.R., Imperial, J., DelaCruz, R.G., Sporning, A., Terlau, H., West, P.J., Yoshikami, D. and Olivera, B.M. (1998). An O-glycosylated neuroexcitatory *Conus* peptide. *Biochemistry* **37**, 16019–25.

9 Toxin Structure and Function: What Does Structural Genomics Have To Offer?

RAYMOND S. NORTON

More than 13 000 structures determined by X-ray crystallography and/or nuclear magnetic resonance (NMR) spectroscopy are currently deposited in the Protein Data Bank ([1]; http://www.rcsb.org/pdb/). These include several hundred polypeptide and protein toxin structures, and there is no doubt that these structures have made an enormous contribution to our understanding of the mechanism of action of these toxins [2–4].

This chapter addresses the questions 'How much does the structure of a polypeptide or protein toxin contribute to an understanding of its toxicity?' and 'What impact will the rapidly developing field of structural genomics have in this area?'. These questions assume greater significance as we move towards an era of complete genome sequences for a range of organisms, including venomous species. In parallel with these genome sequencing efforts are large-scale structural genomics initiatives which aim to determine experimental structures for all proteins from a given species, or at least all proteins detected in a proteome analysis of that species. How will the plethora of structures that emerges from such initiatives, some with new protein folds but many more with known folds, enhance our understanding of how these proteins function?

9.1 MANY FUNCTIONS BUT FEW FOLDS

The 20 amino acid residues commonly found in proteins combine to generate an enormous variety of amino acid sequences, varying in length from just a few residues to many thousands. The peptides, polypeptides and proteins corresponding to these sequences adopt, in most instances, well-defined, folded structures which are, in turn, a prerequisite for biological activity. These peptide chains have a limited number of ordered secondary structures available to them [5], but the combination of these ordered secondary structure elements with unordered regions generates a diverse, although by no means unlimited,

Perspectives in Molecular Toxinology
Edited by A Ménez © 2002 John Wiley & Sons, Ltd

repertoire of tertiary structures. It has been proposed that the number of protein families (that is, groups of proteins with recognisable sequence similarity) in Nature is around 1000, but that the number of unique protein folds is lower [6]. Among the polypeptide and protein structures currently deposited in the Protein Data Bank, the number of distinct protein folds is 500–700 depending on the classification system employed [7–9].

A striking feature of this plethora of protein structures is a skewed distribution in favour of a few commonly recurring folds. One fold that is strongly represented among polypeptide and protein toxins is the Inhibitor Cystine Knot (ICK). In fact, this structural motif occurs in globular polypeptides and proteins from a range of phylogenetically diverse organisms, including fungi, plants, cone-shells, scorpions and spiders [10–12]. The motif is essentially a small, well-defined scaffold that has been used by Nature for a variety of functions, but most commonly the inhibition of ion channels and enzymes. It also shows promise as a compact scaffold for mimicking larger proteins in biotechnological applications. The core of the motif is a three-disulphide bridge 'cystine knot' and a small, triple-stranded, anti-parallel β-sheet. A cystine knot is also found in the Growth Factor Cystine Knot (GFCK) family of proteins, which includes platelet-derived growth factor and transforming growth factor β2 [13, 14] but the secondary structure patterns of the protein backbone participating in the cystine knots of the ICK and GFCK motifs are not superimposable. The backbone cyclic version of the ICK motif is known as the Cyclic Cystine-knot (CCK) motif [15], and a two-disulphide bridge subset of the ICK motif is referred to as the cystine-stabilised β (CSB) motif [16].

The presence of the ICK motif is predictable with reasonable success from the pattern of half-cystine gaps in the sequence, with the nature of the intervening residues providing little or no guide to its presence or absence [12]. It is characterised by the presence of a cystine knot consisting of a ring formed by two of the disulphide bridges (C_1-C_4 and C_2-C_5) and the interconnecting backbone (C_1-C_2 and C_4-C_5) through which the third disulphide bridge passes (C_3-C_6). Although referred to as a knot, this description is only accurate where the backbone is cyclic and the knot is unable to be untied without breaking the chain; strictly speaking, non-cyclic cystine-knots should be referred to as 'pseudo-knots'.

The ICK fold represents one of the smallest globular polypeptide folds known, having a triple-stranded β-sheet comprising as few as 25 residues. Its uniformity across diverse phyla, including the fixed positions of the half-cystine residues relative to the β-sheet residue packing, implies that it is a particularly favourable topology energetically, as with other common structural motifs found in proteins, such as the β-barrel and immunoglobulin folds. The loops between the half-cystines are hyper-variable even in closely-related species of cone shells, generating multiple functions while employing the same underlying scaffold. Thus, the ICK motif essentially represents a tight scaffold for the

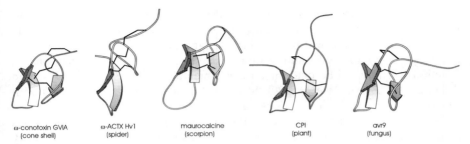

Figure 9.1 3-D structures of examples of ICK-motif polypeptides from different classes of organism. The PDB accession numbers are 2CCO, 1AXH and 4CPA for ω-conotoxin GVIA, ω-ACTX-Hv1 and CPI, respectively (http://www.rcsb.org/pdb/). These examples all have only three disulphide bridges. The cone-shell polypeptides are the most compact while the spider and scorpion polypeptides have large loops between the fifth and sixth half-cystine residues and the plant and fungal polypeptides have loops of non-zero length between the third and fourth half-cystine residues. This diagram was generated using MOLSCRIPT [67]

presentation of side-chains for miscellaneous, usually inhibitory or modulatory, functions. Being cystine stabilised, the remaining residues are available for the display of functional groups, rather than being required for structural purposes.

The structures of several polypeptides based on the ICK motif are shown in Figure 9.1. These structures have been found mainly in fungi, plants, cone-shells, scorpions and spiders; as a broad generalisation, it appears that animals use ICK polypeptides to target ion channels whereas plants use them to target proteases. The recent isolation of a phenoloxidase inhibitor from the housefly *Musca domestica* that matches the sequence motif and has been modelled in 3-D to the motif structure [17] suggests that the ICK motif has an even wider distribution. For our current discussion, however, it represents an excellent example of Nature's ability to decorate a stable structural motif with a variety of functional groups to generate a variety of functions, many of them toxic.

9.2 STRUCTURAL GENOMICS

In addition to the example of the ICK motif described in the preceding section, there are many other examples in Nature of common folds serving diverse biological purposes. This highlights the absolute requirement for functional studies in characterising any new protein, toxin or otherwise. In the field of structural genomics, target selection is a key issue [18]. Where there is high sequence similarity to a class of proteins with known structure and biological activity, for example venom phospholipases, it will usually be assumed that the newly identified protein is yet another member of that class and therefore not a target for structural genomics, even though there may be good reasons for determining its structure experimentally as a basis for future detailed studies of its structure–function relationships. By contrast, the proteins targetted initially by structural genomics will have little or no sequence similarity to known classes of protein. Once their folds are determined either experimentally or by modelling [19], it will be possible to suggest functions, but, as illustrated quite clearly by the ICK family of polypeptides and proteins, their precise function will have to be determined experimentally. This issue has also been discussed by Thornton *et al.* [20]. Structural genomics must be coupled with functional characterisation.

What can we expect, therefore, as the first genome sequences for venomous species become available? A number of isotoxins with high sequence similarity to known toxic polypeptides and proteins will be identified, expanding the picture already emerging from analysis of cDNA libraries (e.g. [21]). There will be little incentive for experimental structure determination of these isotoxins unless a particular molecule shows unexpected activity compared with its close cousins, in which case structural analysis may be called for to explain the difference. As the degree of sequence similarity with toxins of known structure diminishes to around 50%, the case for experimental structure determination grows, although it must be noted here that the accuracy of molecular modelling continues to improve apace [19, 22], and this may represent a satisfactory alternative in many instances. One factor in this decision will be the likely functional importance of loops in the molecule, as it is in these parts of the molecule where modelling is likely to suffer from the greatest uncertainty.

As the degree of sequence similarity drops to 20% or below, which is the case for many members of the ICK family, homology modelling can no longer be relied upon and approaches such as threading and *ab initio* structure prediction have to be adopted [23, 24]. Here the case for experimental structure determination becomes compelling. Nevertheless, even when the structure is available, functional characterisation will still be necessary. This is true regardless of whether or not the structure matches a known fold because, as we saw with the ICK motif, structure may give some clues as to what general class of activity a given toxin may have, for example an ion channel blocker or a protease inhibitor, but the precise details of its target (or range of targets) can only be established experimentally. The nexus between structure and function is as tight for toxins as it is for any other class of polypeptide and protein.

A major goal of structural genomics is to develop a comprehensive picture of the 'protein structure universe'. As structural genomics initiatives get underway based on genomes from bacterial, plant and mammalian species, we can expect many new folds to be revealed. Gaps in our knowledge of the protein structure universe will persist, however, for reasons that are well illustrated by the work of a Canadian consortium on the thermophilic archaeon *Methanobacterium thermoautotrophicum* [25]. Of the 1871 open reading frames in this genome, about 30% correspond to membrane-associated proteins and were not analysed in this study (the issue of membrane protein structures is discussed below). Proteins with clear homologues in the PDB (about 27%) were also excluded. Of the approximately 900 proteins remaining, 424 were chosen for cloning, expression and structural analysis by NMR and/or X-ray crystallography. Under a standard set of growth conditions, most proteins (~80%) could be expressed but many were not soluble (< 0.5 mg/l). Proteins of mass > 20 kDa were submitted to crystallisation trials, while those < 20 kDa were ^{15}N-labelled for NMR analysis. Of the latter group, 100 soluble proteins were tested, with 33 having excellent two-dimensional ^{15}N–^{1}H correlation spectra indicative of folded structures, ten having spectra consistent with a folded structure but with some peaks missing, and 57 having poor spectra attributed to a lack of structure and/or non-specific aggregation. While problems of low expression and poor solubility of expressed protein can probably be overcome in most cases by optimisation of the expression system, including possibly the use of cell-free expression [26], and problems of aggregation and a lack of folded structure can be addressed (although not necessarily solved) by manipulation of solution conditions or modification of the construct, the experience of this consortium suggests that many proteins of interest will not be amenable to high-throughput structural analysis of the sort advocated by proponents of structural genomics.

Focusing on how structural genomics might be applied in toxinology, its initial applications will probably be more appropriately described as structural proteomics. One can imagine a thorough analysis of the proteome of a venom gland or other venom apparatus identifying the range of proteins expressed at any time in a venomous species (the proteomics approach is described elsewhere in this volume in Chapters 5 by Fox et al. and 6 by Stöcklin and Favreau). The suite of polypeptides and proteins thus identified would constitute the initial targets for structural analysis (with the smaller polypeptides being produced by either solid-phase peptide synthesis or bacterial expression). Subsequent analysis of the complete genome may well identify close relatives of toxins already identified by proteomics or even unrelated proteins that prove to act as toxins. Nevertheless, in establishing a profile of the toxins present in a particular venom, including post-translationally modified homologues, which are a common feature of venoms [27], proteomics is likely to be the more informative approach.

Can we expect structural genomics or structural proteomics analyses of venomous species to yield new polypeptide or protein folds? The answer is almost certainly yes, although, as exemplified above for the ICK motif, toxins have many scaffolds in common with polypeptides and proteins from non-venomous species. The structure of the sea anemone potassium channel blocker ShK toxin is a good example of a small protein toxin with a unique fold first identified in a venom-producing species [28]. Given that many toxins have to act extra-cellularly in a foreign and potentially harsh environment, their structural repertoire may be skewed relative to the proteome of a mammalian cell in favour of more stable and soluble structures. For the same reason, they may also be more amenable to structural analysis, with the proportion of toxins having low solubility or poorly ordered structures being lower than in the example outlined above.

The structure of ShK toxin also provides a good example of another aspect of protein structure and function that is relevant here, namely that quite different protein structures can evolve to have the same function [29]. Sea anemone potassium channel blockers, typified by ShK, are helical, whereas scorpion potassium channel blockers such as charybdotoxin have a three-stranded anti-parallel β-sheet with an α-helix lying across strands 2 and 3. Despite their quite different overall structures, the channel-binding surfaces of the scorpion and sea anemone polypeptides have roughly the same dimensions and use similar residues to interact with the channel. Thus, having identified either the sea anemone or scorpion toxin fold as one associated with potassium channel blockade, it would be a mistake to ignore this potential activity in unrelated folds from other venomous species. Indeed, the more recently characterised κ-conotoxin PVIIA, which has yet another fold, also blocks voltage-gated K^+ channels and even appears to use the same residues as the sea anemone and scorpion toxins to achieve blockade [30].

Finally, we need to bear in mind the important role of a host of other factors in determining protein structure, such as pH, temperature, metal ions, cofactors and the like. Not only might these change the structure, they may in fact be critical to whether a biologically relevant structure is determined. NMR is well suited to a rapid assessment of the importance of these parameters. And of course there are many examples in the literature of proteins that do not adopt a stable, ordered structure until they interact with other proteins in the cell, so a knowledge of what these partners are will become important as the less well behaved members of the proteome are tackled.

9.3 MEMBRANE PROTEINS: A MAJOR CHALLENGE

Membrane proteins present a significant challenge to structural biology. Although there have been some spectacular successes in solving the structures of integral membrane proteins by X-ray crystallography, the potassium channel structure of MacKinnon and co-workers [31] being an example of particular relevance to ion channel toxins, obtaining suitable crystals remains a major hurdle. Electron crystallography [32] and cryo-electron microscopy [33, 34] offer alternative, albeit lower resolution, approaches to membrane protein structure determination that are not dependent on the generation of three-dimensional crystals. Solid-state NMR continues to advance, with structures determined for several membrane-spanning polypeptides in lipid bilayers [35, 36], but at its current state of development its strength lies more in an ability to provide precise information on selected regions of membrane-bound proteins (e.g. [37]) than in determining their complete structures. It will be some time before membrane protein structure determination becomes routine, although there are strong incentives to development in this area, not the least of which is the importance of integral membrane proteins such as ion channels and G-protein coupled receptors as drug targets.

As noted above, some 30 % of the proteins encoded by most genomes are anticipated to be membrane proteins. Among polypeptide and protein toxins, membrane-active toxins that act by disrupting cell membranes represent a major category, ranging from polypeptides such as melittin and the magainins [38] to large proteins such as perfringolysin O, which forms massive oligomeric pores in cholesterol-containing membranes [39–41]. However, an important distinction needs to be drawn between these cytolytic toxins and integral membrane proteins in that most of the toxins are water soluble. Thus, although the functionally important state of these toxins is membrane-bound, valuable structural information can be obtained at least on the water-soluble, 'pre-lytic' form. As candidates for structural genomics analyses, therefore, these types of toxin are tractable, although a complete understanding of their action will require the application of techniques for studying integral membrane protein structures.

A good example of this is provided elsewhere in this volume in the chapter on cholesterol-dependent cytolysins by Hotze and Tweten. Another is provided by our own work on equinatoxin II (EqtII), a potent pore-forming protein of 179 residues, isolated from the sea anemone *Actinia equina* [42, 43]. It is essentially identical with tenebrosin-C, isolated from the Australian anemone, *Actinia tenebrosa*, and the first of this class of toxins to be sequenced in full [44]. EqtII is representative of a class of toxins found in all sea anemones and characterised by a high pI, a molecular mass of $\sim 20\,\mathrm{kDa}$, affinity for sphingomyelin, and permeabilising activity in model lipid and cell membranes [45, 46]. The nature of the interaction of EqtII with lipids in bilayer membranes and the specific role of sphingomyelin in pore formation is not understood at the molecular level.

Recently, we described the NMR assignments and secondary structure for EqtII [47] and the 3-D structure has since been solved [48]. The structure consists of a β-sandwich with an amphiphilic α-helix at the N-terminus, as shown in Figure 9.2. Spectroscopic and calorimetric analyses detect conformational changes when the molecule binds to lipid membranes [43, 49, 50], with an apparent increase in α-helical content. The N-terminal amphiphilic helix may be involved in these changes, and it is possible that this helix stabilises the toxin on the surface of the lipid membrane and assists the insertion of other parts of the protein into the lipid phase.

In principle, at least two steps are involved in the formation of an EqtII pore: initial binding of monomers to the membranes; and subsequent oligomerisation into a functional pore. Oligomerisation does not occur in aqueous solution but as the molecule approaches a sphingomyelin-containing membrane we do not know what changes ensue. The pore is estimated to consist of 3–4 monomers [51]. However, it has been shown that the most abundant form of the toxin on the membrane is the monomer and that the fraction of toxin associated in pores is actually small [45, 51]. At lipid/toxin ratios $> 400{:}1$, more than 80 % of the

Figure 9.2 Ribbon representation of the solution structure of equinatoxin II, determined from NMR data [48]. This protein is highly soluble and monomeric in aqueous solution, and yet its active form is an oligomeric pore in membranes containing sphingomyelin. This diagram was generated using MOLMOL [68]

protein is bound to large unilamellar vesicles [52], with 100–200 EqtII monomers bound per vesicle. However, from permeabilisation experiments, the number of pores is never more than 2–3 per vesicle at a lipid/toxin molar ratio of 10:1, and is certainly less at a lipid/toxin ratio of 400:1 [51–53]. This clearly indicates that most of the toxin is not in the oligomeric form and that NMR studies of the interaction of EqtII with lipid bilayers would report on lipid–protein interactions rather than pore formation. However, it is possible to cross-link EqtII in the membrane [45, 46] and this approach can be used to study the oligomeric pore form. EqtII thus exemplifies many of the issues faced in determining the mechanism of action of membrane-active toxins. Not only are our commonly-used structural tools less well suited to membrane-bound proteins, but capturing the protein in the desired form once in contact with the membrane can also present a challenge.

The other category of membrane proteins that is of particular interest to toxinologists is of course the toxin targets themselves, many of which are integral membrane proteins such as ion channels. Although many polypeptide and protein toxins are small and soluble enough to enable their structures to be determined readily by NMR or X-ray crystallography, a complete picture of their mechanism of action requires structural information on their target-bound states, as discussed further in the following section. Thus, limitations on our ability to study the structures of membrane-bound proteins are as much a concern in the toxins field as in others.

9.4 DOES PROTEIN STRUCTURE EXPLAIN FUNCTION?

The primary goal of structural genomics/proteomics initiatives currently under-way is to develop a complete picture of the protein structure universe. Francis Crick [54] remarked 'If you want to understand function, study structure', albeit while acknowledging the importance of a multi-faceted approach. But while the structure of a protein is necessary for an understanding of how that protein functions, it is generally not sufficient. This is no way a criticism of structural genomics/proteomics *per se*, but it is an important consideration in placing this approach into context in the genomics era. We saw in the example of the inhibitor cystine knot family of polypeptides that structure did not dictate function. However, even when we know the structure *and* the function, further work is usually required to understand *how* a protein functions.

To illustrate this point we briefly consider the sea-anemone-derived toxin ShK, which is a potent potassium channel blocker. ShK is a 35-residue poly-peptide which proved to have a novel fold [28]. The K^+-channel binding surface of the molecule has been mapped [55–57], as shown in Figure 9.3, and the contributions to its structure and activity of the disulphide bridges [58] and a helical N-capping motif [59] have been investigated. It would appear, therefore, that we have an excellent basis for understanding its interaction with the

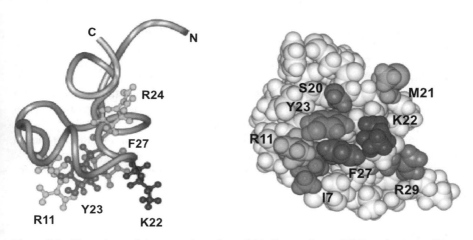

Figure 9.3 Two views of the potassium channel binding surface of ShK toxin. **Left:** ribbon diagram of the closest-to-average structure of ShK [28] showing the side chains of key binding residues of ShK in ball-and-stick representation. **Right:** CPK representation with the side chains of Arg11, Lys22, Tyr23, Ile7, Ser20, Met21, Phe27 and Arg29 shaded differently. Arg24 is obscured by Met21 in this view of the structure

potassium channel at the molecular level. Nonetheless, when we have constructed models of its docking with the voltage-gated channel Kv1.3 ([60];
Lanigan *et al.*, to be published), a number of questions remain concerning
the roles of several residues identified as being important in the interaction. In
some cases, for example Lys22, which plugs the pore of the channel, it is easy to
account for their importance. In others, such as Tyr23, the molecular basis for
their significance is less clear, and there are several side chains that appear to
make close contact with the channel, for example Leu25 and Ser26, that can be
replaced by Ala with little effect on binding affinity [57]. The finding that not all
side chains in the binding surface show up as important in an Ala scan is
consistent with observations in other protein–protein interactions. For
example, in the growth hormone–receptor complex, eight of the 31 residues in
the interface contribute 85 % of the binding energy and more than half make no
significant contribution to the affinity [61]. A similar result was found for a
growth hormone–monoclonal antibody complex, where only five residues were
critical for binding [62]. But these observations highlight the fact that even a
high-resolution protein structure will not explain all aspects of its function, and
that further analysis at the molecular level will always be called for.

Another interesting permutation of the ShK-potassium channel interaction
has been observed in the case of the analogue ShK-Dap22 [60]. In this analogue
the critical Lys22 in ShK has been replaced with the positively-charged, non-
natural amino acid, diaminopropionic acid in order to target the unique ring of
His404 residues located at the outer entrance to the Kv1.3 pore. The Dap22 side
chain does indeed interact with His404 whereas Lys22 in the native toxin
interacts with residues deeper in the channel pore in the vicinity of the K^+
selectivity filter, and ShK-Dap22 blocks Kv1.3 with a high degree of specificity
and picomolar potency [60]. Nevertheless, modelling of the ShK-Dap22
docking with Kv1.3, guided by extensive complementary mutational data,
suggests that ShK-Dap22 does not dock in the same configuration as the native
toxin, implying that the altered interactions of this one side chain are sufficient
to re-orient the toxin in the pore–vestibule region of the channel. This is an
unexpected result that further underscores the need for careful analysis of
protein–protein interactions at the molecular level.

The importance of mapping protein–protein interactions in understanding
how cells function is reflected in recent initiatives in this area. The example
described above illustrates the importance of extending this effort to the molecular level.

9.5   FUTURE PROSPECTS

We have seen that structural genomics (or proteomics) initiatives are likely to
leave large gaps in our knowledge in the area of toxin structure and function.
Proteins that do not express well or are uncooperative once expressed because

of instability, poor solubility or aggregation are likely to be put to one side in the first instance regardless of their biological importance. Even once the fold of a protein is established, it will not always be straightforward to guess its function. And for the 30% or so of proteins that are membrane-bound, progress will be much slower than for their soluble cousins. Notwithstanding these limitations, a number of structural genomics efforts are already underway throughout the world [26, 63, 64]. What will their impact be?

For academic structural biologists in every field, not just toxinology, they create threats and opportunities. Within five years or thereabouts, the prospects for finding a new protein fold will have shrunk dramatically unless one is prepared to tackle the tough problems left untouched. On the other hand, a knowledge of the likely fold for a new protein identified by proteomics or genomics will enable experiments designed to probe the activity of the molecule and define its structure–function relationships to proceed in parallel with detailed structural studies, and for the one to inform the other more effectively and efficiently than at present. There will also be major advances in the tools of the trade, with more facilities for structure determination and more efficient methods of data analysis for both NMR [65] and crystallography [66]. These developments will advantage all structural biologists, not just those who are part of large-scale structural genomics initiatives, and they will enhance our capacity to study the tougher candidates alluded to above. Thus, toxinologists, like biologists generally, can look forward to an era where determining the structure of their favourite toxin ceases to be a bottleneck in the process of understanding how it functions. Indeed, trying to do so *without* a high-resolution structure will become a rare pastime. To paraphrase Winston Churchill, structural genomics may come to be seen not as the end for individual structural biologists, or even the beginning of the end, but rather the end of the beginning.

ACKNOWLEDGEMENTS

I thank Paul Pallaghy, Mark Hinds and Mark Lanigan for assistance with the figures and a number of colleagues for their comments on the text.

REFERENCES

[1] Bernstein, F.C., Koetzle, T.F., Williams, G.J.B., Meyer, E.F., Brice, M.D., Rodgers, J.R., Kennard, O., Shimanouchi, T. and Tasumi, M. (1977) The Protein Data Bank: a computer-based archival file for macromolecular structures. *J. Mol. Biol.* **112**, 535–42.

[2] Gouaux, E. (1997) The long and short of colicin action: the molecular basis for the biological activity of channel-forming colicins. *Structure* **5**, 313–17.

[3] Norton, R.S. (1998) Structure and function of peptide and protein toxins from marine organisms. *Journal of Toxicology-Toxin Reviews* **17**, 99–130.

[4] Souza, D.H.F., Selistre-de-Araujo, H.S. and Garratt, R.C. (2000) Determination of the three-dimensional structure of toxins by protein crystallography. *Toxicon* **38**, 1307–53.

[5] Richardson, J.S. (1981) The anatomy and taxonomy of protein structure. *Adv. Protein Chem.* **34**, 167–339.

[6] Brenner, S.E., Chothia, C. and Hubbard, T.J.P. (1997) Population statistics of protein structures: lessons from structural classifications. *Curr. Opin. Struct. Biol.* **7**, 369–76.

[7] Orengo, C.A., Michie, A.D., Jones, S., Jones, D.T., Swindells, M.B. and Thornton, J.M. (1997) CATH – a hierarchic classification of protein domain structures. *Structure* **5**, 1093–108.

[8] Thornton, J.M., Orengo, C.A., Todd, A.E. and Pearl, F.M. (1999) Protein folds, functions and evolution. *J. Mol. Biol.* **293**, 333–42.

[9] Zhang, C. and De Lisi, C. (2001) Protein folds: molecular systematics in three dimensions. *Cell Mol. Life Sci.* **58**, 72–9.

[10] Pallaghy, P.K., Nielsen, K.J., Craik, D.J. and Norton, R.S. (1994) A common structural motif incorporating a cystine knot and triple-stranded β-sheet in toxic and inhibitory polypeptides. *Protein Sci.* **3**, 1833–9.

[11] Narasimhan, L., Singh, J., Humblet, C., Guruprasad, K. and Blundell, T. (1994) Snail and spider toxins share a similar tertiary structure and 'cystine motif'. *Nature Struct. Biol.* **1**, 850–2.

[12] Norton, R.S. and Pallaghy, P.K. (1998) The cystine knot structure of ion channel toxins and related polypeptides. *Toxicon* **36**, 1573–83.

[13] McDonald, N.Q. and Hendrickson, W.A. (1993) A structural superfamily of growth factors containing a cystine-knot motif. *Cell* **73**, 421–4.

[14] Isaacs, N.W. (1995) Cystine knots. *Curr. Opin. Struct. Biol.* **5**, 391–5.

[15] Craik, D.J., Daly, N.L. and Waine, C. (2001) The cystine knot motif in toxins and implications for drug design. *Toxicon* **39**, 43–60.

[16] Kobayashi, Y., Takashima, H., Tamaoki, H., Kyogoku, Y., Lambert, P., Kuroda, H., Chino, N., Watanabe, T.X., Kimura, T., Sakakibara, S. and Moroder, L. (1991) The cystine-stabilized α-helix: a common structural motif of ion-channel blocking neurotoxic peptides. *Biopolymers* **31**, 1213–20.

[17] Daquinag, A.C., Sato, T., Koda, H., Takao, T. Fukuda, M., Shimonishi, Y. and Tsukamoto, T. (1999) A novel endogenous inhibitor of phenoloxidase from *Musca domestica* has a cystine motif commonly found in snail and spider toxins. *Biochemistry* **38**, 2179–88.

[18] Brenner, S.E. (2000) Target selection for structural genomics. *Nature Struct. Biol.* **7** Suppl 967–9.

[19] Sanchez, R., Pieper, U., Melo, F., Eswar, N., Marti-Renom, M.A., Madhusadan, M.S., Mirkovic, N. and Sali, A. (2000) Protein structure modeling for structural genomics. *Nature Struct. Biol.* **7** Suppl. 986–90.

[20] Thornton, J.M., Todd, A.E., Milburn, D., Borkakoti, N. and Orengo, C.A. (2000) From structure to function: approaches and limitations. *Nature Struct. Biol.* **7** Suppl 991–4.

[21] Lirazan, M.B., Hooper, D., Corpuz, G.P., Ramilo, C.A., Bandyopadhyay, P., Cruz, L.J. and Olivera, B.M. (2000) The spasmodic peptide defines a new conotoxin superfamily. *Biochemistry* **39**, 1583–8.

[22] Murzin, A.G. (2001) Progress in protein structure prediction. *Nature Struct. Biol.* **8**, 110–12.

[23] Baker, D. (2000) A surprising simplicity to protein folding. *Nature* **405**, 39–42.

[24] Jones, D.T. (2000) Protein structure prediction in the postgenomic era. *Curr. Opin. Struct. Biol.* **10**, 371–9.

[25] Christendat, D., Yee, A., Dharamsi, A., Kluger, Y., Savchenko, A., Cort, J.R., Booth, V., Mackereth, C.D., Sarikidis. V., Ekiel, I., Kozlov, G., Maxwell, K.L.,

Wu, N., McIntosh, L.P., Gehring, K., Kennedy, M.A., Davidson, A.R., Pai, E.F., Gerstein, M., Edwards, A.M. and Arrowsmith, C.H. (2000) Structural proteomics of an archaeon. *Nature Struct. Biol.* **7**, 903–9.

[26] Yokoyama, S., Hirota, H., Kigawa, T., Yabuki, T., Shirouzu, M., Terada, T., Ito, Y., Matsuo, Y., Kuroda, Y., Nishimura, Y., Kyogoku, Y., Miki, K., Masui, R. and Kuramitsu, S. (2000) Structural genomics projects in Japan. *Nature Struct. Biol.* **7** Suppl 943–5.

[27] Craig, A.G., Bandyopadhyay, P. and Olivera, B.M. (1999) Post-translationally modified neuropeptides from *Conus* venoms. *Eur. J. Biochem.* **264**, 271–5.

[28] Tudor, J.E., Pallaghy, P.K., Pennington, M.W. and Norton, R.S. (1996) Solution structure of ShK toxin, a novel potassium channel inhibitor from a sea anemone. *Nature Struct. Biol.* **3**, 317–20.

[29] Dauplais, M., Lecoq, A., Song, J., Cotton, J., Jamin, N., Gilquin, B., Roumestand, C., Vita, C., de Medeiros, C.L.C., Rowan, E.G., Harvey, A.L. and Menez, A. (1997) On the convergent evolution of animal toxins. Conservation of a diad of functional residues in potassium channel-blocking toxins with unrelated structures. *J. Biol. Chem.* **272**, 4302–9.

[30] Jacobsen, R.B., Koch, E.D., Lange-Malecki, B., Stocker, M., Verhey, J., Van Wagoner, R.M., Vyazovkina, A., Olivera, B.M. and Terlau, H. (2000) Single amino acid substitutions in kappa-conotoxin PVIIA disrupt interaction with the Shaker K$^+$ channel. *J. Biol. Chem.* **275**, 24639–44.

[31] Doyle, D.A., Morais Cabral, J., Pfuetzner, R.A., Kuo, A., Gulbis, J.M., Cohen, S.L., Chait, B.T. and MacKinnon, R. (1998) The structure of the potassium channel: molecular basis of K$^+$ conduction and selectivity. *Science* **280**, 69–77.

[32] Murata, K., Mitsuoka, K., Hirai, T., Walz, T., Agre, P., Heymann, J.B., Engel, A. and Fujiyoshi, Y. (2000) Structural determinants of water permeation through aquaporin-1. *Nature* **407**, 599–605.

[33] Kuhlbrandt, W. and Williams, K.A. (1999) Analysis of macromolecular structure and dynamics by electron cryo-microscopy. *Curr. Opin. Chem. Biol.* **3**, 537–43.

[34] Sato, C., Ueno, Y., Asai, K., Takahashi, K., Sato, M., Engel, A. and Fujiyoshi, Y. (2001) The voltage-sensitive sodium channel is a bell-shaped molecule with several cavities. *Nature* **409**, 1047–51.

[35] Marassi, F.M., Ma, C., Gratkowski, H., Straus, S.K., Strebel, K., Oblatt-Montal, M., Montal, M. and Opella, S.J. (1999) Correlation of the structural and functional domains in the membrane protein Vpu from HIV-1. *Proc Natl Acad Sci USA* **96**, 14336–41.

[36] Fu, R., Cotton, M. and Cross, T.A. (2000) Inter- and intramolecular distance measurements by solid-state MAS NMR: determination of gramicidin A channel dimer structure in hydrated phospholipid bilayers. *J. Biomol. NMR* **16**, 261–8.

[37] Gröbner, G., Burnett, I.J., Glaubitz, C., Choi, G., Mason, A.J. and Watts, A. (2000) Observations of light-induced structural changes of retinal within rhodopsin. *Nature* **405**, 810–13.

[38] Bechinger, B. (1999) The structure, dynamics and orientation of antimicrobial peptides in membranes by multidimensional solid-state NMR spectroscopy. *Biochim. Biophys. Acta* **1462**, 157–83.

[39] Shatursky, O., Heuck, A.P., Shepard, L.A., Rossjohn, J., Parker, M.W., Johnson, A.E. and Tweten, R.K. (1999) The mechanism of membrane insertion for a cholesterol-dependent cytolysin: a novel paradigm for pore-forming toxins. *Cell* **99**, 293–9.

[40] Heuck, A.P., Hotze, E.M., Tweten, R.K. and Johnson, A.E. (2000) Mechanism of membrane insertion of a multimeric beta-barrel protein. Perfringolysin O creates a pore using ordered and coupled conformational changes. *Mol. Cell.* **6**, 1233–42.

[41] Billington, S.J., Jost, B.H. and Songer, J.G. (2000) Thiol-activated cytolysins: structure, function and role in pathogenesis. *FEMS Microbiol. Lett.* **182**, 197–205.

[42] Maček, P. and Lebez, D. (1988) Isolation and characterization of three lethal and hemolytic toxins from the sea anemone *Actinia equina* L. *Toxicon* **26**, 441–51.

[43] Belmonte, G., Menestrina, G., Pederzolli, C., Krizaj, I., Gubenšek, F., Turk, T. and Maček, P. (1994) Primary and secondary structure of a pore-forming toxin from the sea anemone, *Actinia equina* L., and its association with lipid vesicles. *Biochim. Biophys. Acta* **1192**, 197–204.

[44] Simpson, R.J., Reid, G.E., Moritz, R.L., Morton, C. and Norton, R.S. (1990) Complete amino acid sequence of tenebrosin-C, a cardiac stimulatory and haemolytic protein from the sea anemone *Actinia tenebrosa*. *Eur. J. Biochem.*, **190**, 319–28.

[45] Belmonte, G., Pederzolli, C., Maček, P. and Menestrina, G. (1993) Pore formation by the sea anemone cytolysin equinatoxin II in red blood cells and model lipid membranes. *J. Memb. Biol.* **131**, 11–22.

[46] Maček, P., Belmonte, G., Pederzolli, C. and Menestrina, G. (1994) Mechanism of action of equinatoxin II, a cytolysin from the sea anemone *Actinia equina* L. belonging to the family of actinoporins. *Toxicology* **87**, 205–27.

[47] Zhang, W., Hinds, M.G., Anderluh, G., Hansen, P.E. and Norton, R.S. (2000) Sequence-specific resonance assignments of the potent cytolysin equinatoxin II. *J. Biomol. NMR* **18**, 281–2.

[48] Hinds, M.G., Zhang, W., Anderluh, G., Hansen, P.E. and Norton, R.S. (2002) Solution structure of the eukaryotic pore-forming cytolysin equinatoxin II: implications for pore-formation *J. Mol. Biol.* **315**, 1219–29.

[49] Menestrina, G., Cabiaux, V. and Tejuca, M. (1999) Secondary structure of sea anemone cytolysins in soluble and membrane bound form by infrared spectroscopy. *Biochem. Biophys. Res. Commun.* **8**, 174–80.

[50] Poklar, N., Fritz, J., Maček, P., Vesnaver, G. and Chalikian, T.V. (1999) Interaction of the pore-forming protein equinatoxin II with model lipid membranes: A calorimetric and spectroscopic study. *Biochemistry* **38**, 14999–15008.

[51] Tejuca, M., Dalla Serra, M., Ferreras, M., Lanio, M.E. and Menestrina, G. (1996) Mechanism of membrane permeabilization by sticholysin I, a cytolysin isolated from the venom of the sea anemone *Stichodactyla helianthus*. *Biochemistry* **35**, 14947–57.

[52] Anderluh, G., Barlic, A., Podlesek, Z., Maček, P., Pungerčar, J., Gubenšek, F., Zecchini, M.L., Dalla Serra, M. and Menestrina, G. (1999) Cysteine-scanning mutagenesis of an eukaryotic pore-forming toxin from sea anemone: topology in lipid membranes. *Eur. J. Biochem.* **263**, 128–36.

[53] Maček, P., Zecchini, M., Pederzolli, C., Dalla Serra, M. and Menestrina, G. (1995) Intrinsic tryptophan fluorescence of equinatoxin II, a pore-forming polypeptide from the sea anemone *Actinia equina* L, monitors its interaction with lipid membranes. *Eur. J. Biochem.* **234**, 329–35.

[54] Crick, F. (1988) *What Mad Pursuit*. p150. Basic Books Inc., New York, NY.

[55] Pennington, M.W., Mahnir, V.M., Khaytin, I., Zaydenberg, I., Byrnes, M.E. and Kem, W.R. (1996) An essential binding surface for ShK toxin interaction with rat brain potassium channels. *Biochemistry* **35**, 16407–11.

[56] Kem, W.R., Pennington, M.W. and Norton, R.S. (1999) Sea anemone toxins as templates for the design of immunosuppressant drugs. *Persp. Drug Discov. Design* **16**, 111–29.

[57] Rauer, H., Pennington, M., Cahalan, M. and Chandy, K.G. (1999) Structural conservation of the pores of calcium-activated and voltage-gated potassium channels determined by a sea anemone toxin. *J. Biol. Chem.* **274**, 21885–92.

[58] Pennington, M.W., Lanigan, M.D., Kalman, K., Nguyen, A., Mahnir, V., Rauer, H., McVaugh, C.T., Behm, D., Donaldson, D., Chandy, K.G., Kem, W.R. and

Norton, R.S. (1999) Role of disulfide bonds in the structure and potassium channel blocking activity of ShK toxin. *Biochemistry* **38**, 14549–58.

[59] Lanigan, M.D., Tudor, J.E., Pennington, M.W. and Norton, R.S. (2001) The N-cap motif in ShK toxin and its role in helix stabilization. *Biopolymers* **58**, 422–36.

[60] Kalman, K., Pennington, M.W., Lanigan, M.D., Nguyen, A., Rauer, H., Mahnir, V., Gutman, G.A., Paschetto, K., Kem, W.R., Grissmer, S., Christian, E. P., Cahalan, M.D., Norton, R.S. and Chandy, K.G. (1998) ShK-Dap[22]: A potent Kv1.3–specific immunosuppressive polypeptide. *J. Biol. Chem.* **273**, 32697–707.

[61] Clackson, T. and Wells, J.A. (1995) A hot spot of binding energy in a hormone–receptor interface. *Science* **267**, 383–386.

[62] Jin, L. and Wells, J.A. (1994) Dissecting the energetics of an antibody-antigen interface by alanine shaving and molecular grafting. *Protein Sci.* **3**, 2351–7.

[63] Heinemann, U. (2000) Structural genomics in Europe: slow start, strong finish? *Nature Struct. Biol.* **7** Suppl 940–2.

[64] Terwilliger, T.C. (2000) Structural genomics in North America. *Nature Struct. Biol.* **7** Suppl 935–9.

[65] Montelione, G.T., Zheng, D., Huang, Y.J., Gunsalus, K.C. and Szyperski, T. (2000) Protein NMR spectroscopy in structural genomics. *Nature Struct. Biol.* **7** Suppl, 982–5.

[66] Lamzin, V.S. and Perrakis, A. (2000) Current state of automated crystallographic data analysis. *Nature Struct. Biol.* **7** Suppl, 978–81.

[67] Kraulis P. (1991) MOLSCRIPT: a program to produce both detailed and schematic plots of protein structures. *J. Appl. Crystallogr.* **24**, 946–50.

[68] Koradi, R., Billeter, M. and Wüthrich, K. (1996) MOLMOL: a program for display and analysis of macromolecular structures. *J. Mol. Graph.* **14**, 51–5.

10 The Binding Sites of Animal Toxins Involve Two Components: A Clue for Selectivity, Evolution and Design of Proteins?

ANDRÉ MÉNEZ*, DENIS SERVENT and SYLVAINE GASPARINI

10.1 INTRODUCTION

The presence of discrete poisonous principles in venom was postulated more than two centuries ago [1]. However, due to a lack of tools, animal toxins remained undiscovered until new chromatographic approaches, instrumental for protein purification, emerged (reviewed in [2, 3]). Then, in a period of no more than three decades, a multitude of toxins were extracted from venom and subjected to various molecular studies. Physiological, pharmacological and biochemical analyses have rapidly revealed the diversity of their biological activities [4–10]. Structural analyses have shown that these activities are mostly exerted by miniproteins, which often possess less than 120 amino acids, a high density of disulphide bonds that stabilise the toxins and a high content of secondary structures like any typical globular protein [5, 8].

Animal toxins are currently classified in structural families [5, 8] with less than sixty toxin folds that have been identified (unpublished). The adherence of a toxin to one structural family is defined by the length of its amino acid sequence and (perhaps principally) by the number and location of its half-cystines which define a signature of the common file. However, amino acid sequence deviations commonly occur within a structural family. A number of subfamilies may therefore emerge and may be characterised by slightly different or even entirely distinct biological activities [for example, reviewed in 2, 10–12]. Within each toxin subfamily, a number of specifically conserved residues reflect the common biological activity shared by its toxin members. In other words, animal toxins are made of a complex and subtle variety of conserved and

* Corresponding author

Perspectives in Molecular Toxinology
Edited by A Ménez © 2002 John Wiley & Sons, Ltd

variable elements from which a diversity of biological properties, including fine specificities, have emerged. How? This review aims to shed some light on this intriguing question.

For most, if not all toxins, binding to a specific target, usually accessible from the extracellular side, constitutes a critical step for a toxic activity to be exerted. Binding may then be followed by a blockage, an activation, a modulation of the target's function or, if the toxin possesses an enzymatic activity, by a catalytic transformation of the target itself or a nearby substrate, as observed with phospholipase A_2-folded toxins [13]. Therefore, elucidation of binding determinants and clarification of their properties should help understanding of the intimate modes of actions of animal toxins. It would be beyond the scope of this paper to review all the body of data that is available on the binding determinants displayed by animal toxins. Therefore, we will only focus on some of those obtained with animal toxins that act on potassium Kv1 channels or nicotinic acetylcholine receptors (AChR), for which models can now be derived from crystallography-based structures of related proteins [14, 15]. We shall see how these data show that binding determinants of these toxins, and perhaps of all toxic proteins from animals, involve two components that will be presented in this chapter and which may contribute to understanding both the pleiotropic properties and the molecular evolution of animal toxins and to engineering new activities on toxin scaffolds.

10.2   ON THE MOLECULAR TARGETS

10.2.1   THE VOLTAGE-GATED POTASSIUM CHANNELS

Potassium channels form a diverse superfamily of integral membrane proteins in excitable and non-excitable cells, which carry out many biological tasks such as electrical signalling, signal transduction and osmotic balance [16, 17]. These tasks are based on the ability of the channels to catalyse the rapid and highly specific permeation of K^+ ions. Potassium channels are remarkably ubiquitous since their sequences are found in genomes of eukaryotics, eubacterials and archeals. Voltage-gated potassium channels (Kv) constitute one of the two major classes of potassium channels. In this review we will mostly focus on the Kv1 subfamily, for which several subtypes (Kv1.1 to Kv1.7) of their pore-forming subunits, likely to be structurally highly similar, have been identified [18, 19]. Kv1 channels are tetramers of identical or similar subunits organised in a fourfold symmetry around the water-filled ion-conduction pathway. The transmembrane topologies of each subunit, derived from sequence analyses of their genes, possess six hydrophobic and α-helical transmembrane segments termed S1–S6. Direct structural information on Kv channels comes from the structurally homologous bacterial protein KcsA [14]. The P-region of the channels is a re-entrant pore-loop, comprised between the transmembrane-spanning segments S5 and S6, which constitutes the extracellular entryway of the pore and the selectivity filter (see Plate IV). The P-region has 72% identity between the different Kv1 subtypes and is located in a sort of vestibule where several animal toxins bind and hence block ion channel function. The 'selectivity filter' is the narrowest part of the pore. It is a sort of tube of 3 Å diameter which extends normal to the membrane plane for about 10–15 Å. The pore is occluded on the intracellular side of the membrane by the crossing of four transmembrane helices, organised in a 'teepee'-like structure, approximately 34 Å long. This is probably the region harbouring the conformational switch, which opens and closes the pore.

10.2.2   THE NICOTINIC ACETYLCHOLINE RECEPTORS (AChRs)

AChRs are localised in postynaptic membranes of various synapses from the peripheral and central systems [20]. Binding of two molecules of neurotransmitter, acetylcholine, to one molecule of AChR causes a specific increase in conductance for Na^+, K^+ and Ca^{2+} ions and hence generates postsynaptic depolarisation. AChRs belong to the ligand-gated ion channel family (LGIC) as 5HT3, $GABA_A$, GABAc and glycine receptors, which all contain five homologous transmembrane subunits arranged as pentamers with a central channel [21]. Each subunit includes a large extracellular N-terminal domain of about 210 residues, linked to four transmembrane regions (M1 to M4), with a cytoplasmic part between the M3 and M4 domains and a short extracellular C-terminal tail [20, 21, 22]. Snake α-neurotoxins block specifically and with high affinity the function of muscular AChR [23].

The subunit composition and the stoichiometry of AChRs vary with tissue localisation, causing distinct pharmacological and electrophysiological properties [22]. Muscular-type AChRs are found at neuromuscular junctions and in the electric organs of *Torpedo* and electric eels. They possess four different subunits arranged in the pentameric order αγαδβ and αεαδβ in the foetal and adult muscle, respectively. In the peripheral and central nervous systems, some receptors are $\alpha_n\beta_m$ with various combinations of α (from α2 to α6) and β (from β2 to β4) subunits. Other AChRs may be homopentameric with five identical subunits α7, α8, α9 [24]. Recently, a novel subunit, α10, was cloned [25, 26]. It forms a functional receptor upon association with α9 subunit. Electron microscopy studies reveal that AChR is a transmembrane protein 115 Å long, with approximately half of its structure exposed extracellularly [27]. The architecture of the extracellular domain of AChR may be inferred from the recently solved crystal structure of the functionally and structurally related molluscan acetylcholine-binding protein [15]. Each AChR subunit may adopt an immunoglobulin-like topology and the ligands, including animal toxins, may bind at the interfaces between two subunits (see Plate V). This domain is linked allosterically to the receptor channel, the affinities of ligands depending on the different receptor states (resting or desensitised). The external diameter of the receptor is about 80 Å, whereas the diameter of the pore is around 30 Å in the synaptic part and only 7 Å in the channel [28].

10.3   ON THE BINDING SITES DISPLAYED BY ANIMAL TOXINS

Our knowledge about the sites by which animal toxins block AChRs or Kv1 channels has been derived from various studies. Comparison of amino acid sequences of natural toxin variants yielded a first line of information [reviewed in 2, 29]. Then came the use of (i) chemically modified toxin derivatives [2, 29]; (ii) chemically synthesised toxins and their variants [6, 30, 31] and (iii) recombinant toxins and their mutants [31–35]. Also, the site by which fasciculin binds to acetylcholinesterase was identified from crystallographic studies of the toxin-target complexes [36, 37]. For a few toxins, the binding surfaces were deduced from docked toxin-target complexes [38–41]. From all these studies, it appears that a binding determinant displayed by an animal toxin involves around 5–15 amino acids and covers a homogeneous surface of about 1000 Å^2 [42–49]. This size is comparable to that of the determinants displayed by non toxic proteins [50].

10.4   WHAT CAN WE LEARN FROM COMPARISONS OF BINDING SITES DISPLAYED BY ANIMAL TOXINS?

10.4.1   AN UNDERSTANDING OF THE STRUCTURAL BASIS FOR THE BIOLOGICAL DIVERSITY OF ANIMAL TOXINS: ONE MOULD FOR MULTIPLE MISSIONS

A comparison of the partially identified 'toxic sites' of two three-fingered toxins, a cardiotoxin [51] and a neurotoxin [52] isolated from *Naja nigricollis* venom, has revealed that though the two determinants slightly overlap each other, they cover topographically different regions of the three-fingered fold. Hence, it was concluded that the sites by which two structurally similar toxins bind to unrelated targets may be displayed in different regions of the fold [5, 51, 52]. This conclusion has now been extended to other examples. Thus, α-dendrotoxin and BPTI respectively bind to Kv1 channels and trypsin, through residues that are in opposite regions of their common fold, [53]. Similarly, short-chain scorpion toxins block Kv1 channels [43, 44] or apamin-sensitive potassium channels [54], using residues that are respectively located on the topographically distinct β-sheet and helix of their common fold. Finally, the two three-fingered toxins α-neurotoxin and dendroaspin block respectively AChR [45, 55] and the platelet integrin GP IIb-IIIa [56–58], through determinants that are essentially located on the second and third loops, respectively. Therefore, binding determinants can be spread in virtually any exposed region of a toxin fold, in contrast to paratopes that are confined in the same region of an immunoglobulin [59]. This does not imply, however, that all determinants displayed by a toxin fold will necessarily be topographically distinct. For example, the site by which charybdotoxin blocks potassium channels [43, 44] is made of residues that are located on the exposed side of the β-sheet, just like those by which the recently engineered scorpion toxin mimics the CD4 and hence blocks gp120 from HIV [60]. Therefore, we now better understand how a diversity of toxic [5, 61, 8, 62] and non-toxic [63, 64, 60] activities can be expressed using a limited number of small protein scaffolds.

10.4.2   A MOLECULAR RATIONALE FOR THE PLEIOTROPIC PROPERTIES OF ANIMAL TOXINS

To the diversity of animal toxins corresponds a variety of targets (ion channels, receptors, enzymes, etc.), that are offered by preys and predators. Possibly reflecting a co-evolution of these two complementary worlds a remarkable pleiotropism is observed: an ion channel can be bound by different toxins of similar or different folds and conversely a toxin can bind to several members of a receptor family. How can the binding determinants of toxins and receptors account for this complex and subtle situation? A bird's eye view of the data accumulated during the past decade reveals that three major characteristics of binding sites might explain it.

First, the residues of a toxin-binding site contribute unequally in terms of binding energy. For example, an extensive mutational study of a short-chain toxin [42], erabutoxin a (Ea) from the sea snake *Laticauda semifasciata* [65], revealed in 1993 that 'a hierarchy of individual contributions seems to prevail within the functional ionogenic surface, and (...) among these, Arg-33 appears to be the most important' [42]. The same phenomenon was observed with other toxins, including scorpion toxins [43, 44]. These observations agreed well with those derived from the pioneering alanine scanning experiments carried out on hormones [66]. Nowadays, residues that play a predominant energetic role in binding sites are called 'hot spots of binding energy' [67–72]. Therefore, the binding surfaces displayed by animal toxins involve few key residues that contribute predominantly to binding to their targets.

As a second major characteristic, a binding site displayed by a toxin is composed of at least two groups of complementary residues, defined with respect to the toxin subfamily: 'the conserved residues and the variable residues'. Thus, the three-fingered neurotoxins from snakes possess more than 100 toxin members, including both short-chain toxins and long-chain toxins, which all block AChRs [29]. The site by which the short-chain toxin Ea binds to muscular-type (*Torpedo*) AChR involves 10 residues, spread on the three toxin loops [45], and among these some are virtually conserved throughout the whole neurotoxin family [42]. It was proposed that these invariant residues 'might constitute a common core through which curaremimetic toxins establish conservative contacts with nAChRs' [45]. The second group of functional residues is more variable. For example, the functional residues located at the tip of the first loop of Ea vary greatly among toxins and are even absent in the long-chain toxins. It was suggested that these short-chain toxin-specific residues may 'contribute to the selectivity of Ea for some nAChRs' [45]. The short-chain toxins, for example, have a lower affinity to human receptors as compared to long-chain toxins [73].

The third major characteristic of binding sites displayed by toxin folds is their remarkable flexibility, which allows them to accommodate substantial structural deviations. This feature has now received much experimental support showing, in particular, that toxins can exert virtually the same function through non-identical binding solutions. More recent additional evidence which supports the three major characteristics defining the binding site of an animal toxin, will now be presented.

How does a toxin bind to different receptor subtypes?

Animal toxins are not always highly selective. Often, a particular toxin is capable of binding with high affinities to several members of a receptor family. This situation is understandable from a deterministic viewpoint: with pleiotropic tools, venomous animals have a larger range of possibilities to subdue prey. But what is the detailed molecular basis that is associated with such multiple binding properties? This will be illustrated now with two animal toxins: BgK from the sea anemone *Bunodosoma granulifera* [74], a 37-amino-acid miniprotein with 3 disulphides, which blocks the homomeric Kv1.1, Kv1.2 and Kv1.3 channels [75] and α-cobratoxin (α-Cbtx) from *Naja kaouthia* [76], a 71-amino-acid toxin with 5 disulphides which can block the muscular-type (Kd in pM range) or the neuronal α7 (Kd in nM range) AChRs [11].

How does a sea anemone toxin bind to three Kv1 channels?

The sea anemone toxin BgK binds to Kv1.1, Kv1.2 and Kv1.3 channels using a differential number of residues, namely 5, 7 and 11 residues, respectively [75]. Lys25, Tyr26 and Ser23 are involved in binding to the three channels, with Lys25 and Tyr26 that bind with high energy to the three channels and Ser23 that plays a more important binding role for Kv1.3 than for the other two channels. An alignment of all known amino acid sequences of analogous sea anemone toxins that block Kv1 channels [41], reveals that these three residues are strictly conserved. As it will be seen later, they may be involved in a primary binding core for all sea anemone toxins to block Kv1 channels. Additional residues are implicated in binding to two out of three subtypes. These are Asn19 and Gln24 for binding to Kv1.1 and Kv1.3 and, His13 and Phe6 for binding to Kv1.2 and Kv1.3. Finally, few residues are important for binding to a single-channel subtype. These are Arg12 for Kv1.2 and Lys7, Leu17, Thr22 and Arg27 for Kv1.3. These channel-specific residues are all variable in the family of sea anemone toxins.

How does a long-chain neurotoxin bind to two AChRs?

α-Cbtx is a long-chain neurotoxin which, like other snake neurotoxins [77–83], adopts a three-fingered fold, with three adjacent fingers rich in β-sheet and emerging from a small globular core [80]. As shown with 36 toxin mutants, for α-Cbtx to bind to both the muscular and α7 neuronal AChRs [47, 48], the toxin 'uses' the same core of six residues composed of Trp25, Asp27, Phe29, Arg33, Arg36 and Phe65, that tend to be highly conserved among long-chain neurotoxins [29]. Among these, Arg33 exerts the highest binding contribution toward both receptors. Then, Trp25 and Asp27 play comparable, though weaker binding roles with both receptors, whereas Phe29, Arg36 and Phe65 contribute with very different binding energies to the two subtypes. Other residues are implicated in the specific recognition of a single receptor subtype. Thus, Ala28, Lys35 and Cys26–Cys30 are uniquely involved in binding to the neuronal AChR. In contrast, Lys23 and Lys49 are involved in binding to the muscular-type receptor only. They are both highly (almost strictly) conserved in both long-chain and short-chain neurotoxins, playing a major role in terms of energy of binding to muscular-type receptors.

Conclusion

Therefore, although BgK and α-Cbtx have different phylogenetic origins and bind to unrelated targets, they adopt the same strategy to bind with high affinities to different receptor subtypes. In agreement with our initial proposal [45], they use a common core of residues involved in binding to multiple (all?) receptor subtypes, together with a number of additional residues exclusively involved in binding to some receptor subtypes. The common binding core tends to be composed of highly conserved residues through the toxin family and to play an important role in terms of binding energy. The target-specific residues tend to be more variable, though this is not a strictly respected situation. We may wonder, of course, whether or not a toxin recognises homologous regions in the different receptor subtypes. Although we have no definitive proof for this, calculated models of toxin-channel complexes support this view (see below). Possibly, the common binding core establishes contacts with some invariant receptor residues, whereas the other residues interact with subtype-specific receptor residues. This, however, remains to be demonstrated.

Functional convergence: How do structurally different toxins bind to similar targets?

To cause paralysis or death, toxins must efficiently affect crucial physiological processes of the prey. Most of these processes are homologous throughout the Animal Kingdom and hence are controlled by similar molecular elements. This is the case, for example, for nerve conduction whose functioning is regulated by similar ion channels. It is no surprise, therefore, that phylogenetically different venomous animals have developed toxins that act on homologous molecular targets of distinct prey. In this respect, it is well established that snakes, sea anemones, scorpions and cone snails produce structurally unrelated toxins, which block Kv1 channels with varied affinities [84, 85] and in a mutually competitive manner [53]. Some of these toxins also block calcium-activated potassium channels (KCa) of intermediate (IKCa) or big (BK) conductance. Snakes and cone snails produce toxins that block AChRs, also with variable efficacy and in a competitive manner [77]. This situation, therefore, poses the question as to how, at a molecular level, animal toxins have developed similar blocking functions using unrelated moulds.

A similar binding dyad in structurally unrelated toxins that block potassium channels similarly

Various toxins from scorpions, sea anemones and snakes are structurally unrelated (see Plate VI) though they commonly block Kv1 and/or calcium-activated potassium channels. Three lines of evidence suggest that blockage results from binding to the same region of the channel. First, toxins from snakes, scorpions, sea anemones and cone snails are mutually competitive inhibitors for Kv1 channels, present in rat brain [53]. Second, mutagenesis of channels and studies with chimeric channels formed by toxin-sensitive and toxin-insensitive pore-forming subunits indicate that the S5–S6 linker constitutes the binding site of toxins from scorpions [86, 87], snakes [88] and conus [89, 90]. Third, scorpion toxins [91, and reviewed in 92] and κ-conotoxin PVIIA from cone snail [93] act as pore blockers.

The sites by which these toxins bind to Kv1 and calcium-activated potassium channels involve five to eleven residues, among which a pair of similar residues is present in all toxins and highly conserved throughout each toxin family. The first and most critical residue of this dyad is a conserved lysine. Thus, in scorpion toxins, the conserved K27 is important for the binding of (i) CTX (Charybdotoxin) to *slo* calcium-activated potassium channels [44], (ii) CTX and AgTx2 (Agitoxin 2) to the *Shaker* channel [43, 94], (iii) MgTx (Margatoxin) to Kv1.3 [95] and (iv) KTX (Kaliotoxin) and CTX to Kv1.3 [96, 39]. In sea anemone toxins the conserved K25 is involved in the binding of (i) ShK and BgK to rat brain Kv1 channels [46, 97], (ii) ShK to Kv1.3 and to IKCa1 calcium-activated potassium channel [98] and (iii) BgK to Kv1.1, Kv1.2 and Kv1.3 [75]. Finally, in snake toxins, the conserved lysine K5 is critical for α-DTX to bind to rat brain channels [53] and to Kv1.1 and Kv1.2 [41]. The residue equivalent to K5 (K3) is also important for κ-DTX to bind to a chimeric *Shaker*-Kv1.1 channel [99]. The second residue of the dyad is a hydrophobic residue, often with an aromatic side chain. It is Y36 in CTX [44], F25 in AgTx2 [94, 43] and KTX [96, 39], Y26(23) in BgK [46] and ShK 98], L9(7) in α-DTX [53] and κ-DTX [99]. In all toxins, the two residues of the dyad superimpose well, the critical lysine being constantly separated from the hydrophobic residue by 7 ± 1 Å [46, 53]. The two dyad residues clearly assure an important binding energy in all toxin-channel complexes, however they cannot be considered as the unique predominant binding hot spots in all cases [67–72].

Is this binding dyad a coincidence? Two lines of evidence suggest that this is not the case. First, it can be correctly predicted. Thus, knowing only the NMR structure of PVIIA [100, 101], a conotoxin that blocks Kv1 channels and inhibits competitively binding of other toxins [102, 103, 53], Savarin *et al.*, [101] proposed in 1998 that among the different possible lysine and aromatic residues displayed by the toxin, only K7 with F9 or F23 should play the role of the binding dyad (Plate VI). In 2000, Jacobsen *et al.* [102] made an alanine walk along the toxin sequence and demonstrated the validity of the proposal. Second, the dyad of various toxins seems to bind to the same region in Kv1 channels, as inferred from both complementary mutagenesis and assisted docking of toxin-Kv1 complexes [39, 41, 49, 96–98, 102, 104–111]. Inspection of these data suggest that toxins from scorpions, sea anemones and cone snails tend to interact with residues from the vestibule of the channel, the dyad lysine being often at the centre of the groove formed by the turrets, just over or plugging into the pore. The dyad hydrophobic residue also seems to interact with similar residues in the different Kv1 channels. These residues (447–449 in the *Shaker*) can be either variable or conserved, and are located in a loop that connects the pore helix and the transmembrane helix, adjacent to the selectivity filter. This may not be the unique possible binding scenario for toxins to block competitively Kv1 channels. Thus, the toxin α-DTX from snake may be off-centred by binding with a turret and two adjacent subunits [99]. Therefore, though being conserved in a toxin, the dyad may be involved in a larger range of binding possibilities toward homologous targets.

The binding dyad of Kv1-specific blocking toxins is assisted by variable residues

Residues other than those forming the critical dyad are present at the sites by which animal toxins bind to Kv1 channels. If we superimpose the dyads in the structures of the different toxins, assuming they all play homologous binding role in the different Kv1 channels, the other functional residues vary in number, nature and hardly superimpose in the different toxins. Some of these residues are likely to provide toxins with particular selectivity, like F6 in BgK that favours binding to Kv1.2 and Kv1.3 but not to Kv1.1 [75].

A binding dyad in structurally different AChR-blocking toxins?

Long-chain α-toxins from snakes [77] and the α-conotoxins ImI from cone snail [112, 113] are structurally different but they both block the neuronal α7 AChR. The site by which the snake toxin α-cobratoxin blocks this receptor involves at least nine residues: Trp25, Asp27, (Cys26–Cys30), Ala28, Phe29, Arg33, Lys35, Arg36 and Phe65, mostly located on the toxin central loop and in particular at its tip [48]. Among these, Arg33, Phe29 and Asp27 play a predominant binding role. Complementary mutagenesis [48] indicates that Arg33 is located in proximity to Tyr187 and Pro193 in the functional loop C of the receptor, and does not seem to be in contact with other receptor loops (loops 2, 3 and 4).

α-ImI isolated from the venom of *Conus imperialis* is a 12-residues-containing α-conotoxin with two disulphide bonds [112], which interacts selectively with the α7 neuronal AChR subtype [114]. Mutation analyses have revealed that Asp7, Pro8, Arg9 in the first conotoxin loop constitute a major binding determinant and that Trp10 in the second loop plays a complementary role [113, 115, 116]. It was also observed that the helical-type scaffold of the α-ImI functional site can superimpose with the tip of the central loop of α-Cbtx [115, 117], with the Cα of Arg7 and Trp10 in the conotoxin that nicely fit with those of Arg33 and Phe29 in α-Cbtx, the side chains of the corresponding residues being, however, oriented differently. Moreover, mutant cycle analyses have shown that Arg7 in ImI is coupled to Tyr195 in the α7 receptor [118] and Arg33 in α-Cbtx is coupled to Tyr187 in the α7 receptor [48]. Therefore, Arg7 in the conotoxin and Arg33 in α-Cbtx may bind to the same 'aromatic pocket' of the α7 receptor, a region that is critical for the binding of various ligands via cation-π interactions [119]. In addition, it has been suggested that an aromatic residue (Trp10 in ImI and Phe29 in α-Cbtx) may play a homologous binding function, and together with the arginine, it may form a dyad of binding residues [117]. However, if we impose a superimposition of the two arginine residues, the side chains of the aromatic residues show marked spatial deviations. In summary, these data suggest that to interact with the same α7 neuronal AChR, structurally unrelated toxins from cones or snakes 'use' a key arginine residue, and possibly an aromatic residue, as a common binding anchor, together with many different residues.

On the flexibility of binding sites: multiple possible solutions to block a target

As mentioned before, the members of toxin subfamilies share not only highly similar structures, as inferred from conservation of their half-cystine framework, but they also exert similar overall functions. Nevertheless, as a result of individual deviations, which include differences in size, sequences in the loops that define disulphide connections and sometimes the number of disulphides, the members of toxin families, sometimes called (perhaps incorrectly) isoforms, may be classified differentially [12, 29]. Do the differences that characterise isotoxins simply reflect innocent mutations or instead, do they intervene in the functional properties of the toxins? To shed light on this question, we will examine the binding sites of two types of 'isotoxins' produced by closely related species.

A first example is given by BgK and ShK, two structurally similar sea anemone toxins respectively produced by *Bunodosoma granulifera* and *Stichodactyla helianthus*, which bind competitively and with high affinity to Kv1.3 [75, 98]. A comparison of the sites by which the two toxins bind to Kv1.3 reveals two groups of residues. First, the two toxins use a group of four identical residues, located at homologous positions, and which display high energy of binding. These are Lys25, Tyr26, Ser23 and Arg27, the latter being energetically more important in ShK than in BgK. These residues are strictly or highly (Arg27) conserved within the family of five known sea anemone toxins [41]. Second, the two binding sites involve a group of additional residues that are not only different in the two toxins but that are also variable within the five known sea anemone toxins [41]. Within this group, two residues occupy homologous positions. These are Thr22 and Gln24 in BgK that are replaced by His22 and Met24 in ShK (using BgK's numbering). The other additional residues are important in one toxin only. Thus, Leu25 is exclusively important in ShK whereas Phe6, Lys7, His13, Leu17 and Thr22 are important in BgK only. Therefore, two structurally similar BgK and ShK adopt nonidentical binding solutions to block Kv1.3. Interestingly, some of the additional binding residues are located in regions that display structural deviations in the two toxins. Thus, Leu17 and Asn19 are in a loop of BgK that does not exist in ShK, where an helical structure is observed.

Another example is given by Ea and α-Cbtx, two snake α-neurotoxins, which bind competitively to *Torpedo* AChR with high affinities [42, 45, 47, 48]. Though these toxins possess a similar three-fingered structure [80, 81], they can be discriminated by a few local deviations, which illustrate their respective adherence to the groups of short-chain and long-chain neurotoxins. Thus, Ea possesses a longer first loop, whereas α-Cbtx has both an additional disulphide bond at the tip of the central loop and a longer C-terminal tail. The site by which Ea binds to a muscular-type AChR (from *Torpedo*) involves residues at the tip of loop I (Gln7, Ser8, Gln10), on the central loop (Lys27, Trp29, Asp31, Arg33, Ile36, Glu38) and in loop III (Lys47) [42, 45]. These results globally agree with mutational investigations made with the other short toxin NmmI from the African cobra *N. m. mossambica* [120–122]. To bind to the same muscular-type AChR, α-Cbtx 'uses' six identical residues that occupy homologous positions in the superimposed structures of the two toxins. These are residues in loop II [Lys23 (27), Trp25 (29), Asp27 (31) Phe29(32)and Arg33] and loop III [Lys49 (47)] [47]. These residues are highly conserved within the large family of snake neurotoxins [29] and they tend to be energetically important in both cases. The two toxins also 'use' differential residues that tend to be more variable throughout the family of neurotoxins [41]. First, the loop I (especially its tip) involves residues that are functionally important in Ea, whereas the whole loop I is not functionally important in α-Cbtx. Possibly, the loop I in Ea reaches a binding pocket that is inaccessible to the shorter loop I of α-Cbtx. We may note that the residue of Ea whose mutation causes the highest change in binding energy is Ser8, which is absent in long-chain toxins but strictly conserved among short-chain toxins. Second, Glu38, at the base of the central loop, is important in Ea whereas the corresponding Asp38 in α-Cbtx is not. Third, the critical position Arg36 in α-Cbtx is replaced in Ea by an isoleucine whose mutation into an arginine causes a substantial affinity increase for AChR. Fourth, the functional Phe65 is uniquely found at the C-terminal of the long toxin. Therefore, though the two toxins block the same AChR competitively and with comparably high affinities [47, 48], they achieve this function through non-identical binding solutions. They use a core of conserved residues which tend to have a strong energy of binding, assisted by a number of residues variable in nature, location and binding energy.

Interestingly, the local structural deviations that discriminate the two types of toxins occur in the binding site of one toxin, i.e. the tip of loop I in the short-chain toxin, and the tip of loop II and the C-terminal tail in the long-chain toxin. Has this correlation any significance? The question is all the more intriguing, since short-and long-chain neurotoxins can be found simultaneously in the same venom. Though we have no definitive explanation yet, it is possible that the observed deviations reflect a specific adaptation of the long toxins, which are thus able to bind not only to muscular AChR but also to a neuronal AChR (see below). Therefore, structurally and functionally similar toxins can accommodate substantial deviations, even in their functional site, without a major effect on their primary binding ability. Binding sites displayed by animal toxins, therefore, seem highly flexible.

Conclusions: the pleiotropic properties of animal toxins can be rationalised by the presence of two complementary components in their binding determinants

We have reviewed data showing how binding sites could explain how (i) a toxin binds to several members of a receptor family; (ii) structurally different toxins bind competitively to the same receptor or to receptors of the same family; and (iii) structurally related toxins bind competitively to the same receptor or to receptors of the same family. Strikingly, these pleiotropic properties can be rationalised by a simple principle: when binding to the same region of structurally related targets, the binding determinants of animal toxins involve two components. These are a common anchor, composed of strictly or highly conserved residues, and a group of target-specific residues that vary in number, location and nature from one toxin to another.

In each of the three situations above, the size of the common anchor may be different. For example, BgK uses a anchor of three residues for binding to Kv1.1, Kv1.2 and Kv1.3 whereas the BgK and ShK use an anchor of four identical residues to bind to Kv1.3. However, and this is perhaps the most interesting finding, a smallest common binding denominator can be detected through all situations. Thus, the toxins that bind competitively to Kv1 channels all possess a binding dyad composed of a lysine and a hydrophobic residue (frequently an aromatic residue), separated from each other by approximately 7 Å, whereas AChR-specific neurotoxins use an arginine and perhaps an aromatic residue. These minimal binding anchors are highly conserved in each toxin family and they tend to play an important energetic binding role although they are not necessarily the predominant 'hot spot' of binding in the interacting surfaces [66–70]. We suggest a close cooperation between a minimal common anchor that may provide a generic recognition of a family of receptors and the assisting residues that may assure high affinity and probably specificity for some members of the receptor family.

This principle of using two components in a binding determinant is clearly advantageous for toxins in defining multiple fine-tuned receptor specificities through a small number of mutations and to evolve readily toward new binding directions. This principle might develop independently in various discrete binding regions of a receptor, which would explain how a diversity of toxins can bind, sometimes competitively and sometimes noncompetitively, to different members of a receptor family, as observed with sodium channels [125].

Does this 'two binding components' principle apply to other cases? The presence of two similar components in binding determinants has been observed in various protein–protein interactions. This is the case for the elegant studies that describe the colicin-immunity protein complexes [71, 72], where the conserved core of an immunity protein (Im9) forms a clear binding-energy hot spot [71]. This was also observed in determinants of self-peptide-bound major histocompatibility complexes [123] and possibly in determinants of complexes involving SH2 and SH3 domains [124]. This situation has been called 'dual recognition' [71–72]. However, we feel that such a name could be confused with the term 'dual specificity' that has been given to phosphatases that dephosphorylate both phosphothreonine and phosphotyrosine residues [126]. We propose therefore to call the above principle, the 'two binding components' of protein interacting determinants. Finally, we should note that other strategies might apply to other animal toxins. Thus, the 'dock and lock' principle, also called the 'Janus Ligand Hypothesis' has been proposed to account for the specificity properties of cone snail toxins [127].

10.5   HOW DO BINDING SITE PROPERTIES HELP TO UNDERSTAND THE EVOLUTION OF ANIMAL TOXINS?

At least two pieces of evidence suggest that toxin folds may undergo divergent evolution in a phylum-dependent manner. First, toxin folds used by venomous animals from the same phylum can display binding sites almost anywhere on their surface. This is particularly well illustrated by the three-fingered and BPTI folds produced by snakes and by the α/β scorpion folds (see 10.4.1). Second, the gene domains encoding mature toxins undergo an accelerated rate of mutation, as suggested by studies of genes encoding snake PLA2s [128–130], conotoxins [131–134] and three-fingered toxins [128]. A mutator mechanism, which remains to be identified, might be responsible for this rapid process [134]. However, mutations seem not to occur homogeneously along toxin folds. The loops and especially their tips undergo many mutations whereas few elements, like disulphides, are strictly conserved, the half-cysteine organisation appearing even as a kind of signature of a conserved fold [5, 63, 64, 128]. Therefore, divergent evolution of a fold may be restricted to regions that are not structurally essential. We still do not know how such a discriminative process (if any) might occur. Recent genetic studies on cone snail toxins [133, 134] have suggested that a protein with 'stencil-like' properties might protect selectively the cysteine codons from mutation. Therefore, a fold produced in venom glands of an homogeneous group of animals (snakes, sea anemones, cone snails or scorpions, etc.) might undergo an accelerated rate of evolution through unclear mechanisms (if any) that retain the fold and favour mutations in exposed loops.

Two lines of evidence suggest that convergence also intervenes in the evolutionary process of animal toxins. First, toxins with entirely different folds and produced by phylogenetically unrelated venomous animals can bind competitively to the same region of a target. Second, the sites by which these toxins bind to the targets share similar binding anchors. Therefore, if we accept the general view that divergence is the primary driving force of toxin evolution, ion channels, receptors and other targets are anticipated to display binding domains that might act as molecular sieves to select complementary binding surfaces in toxins. A similar conclusion emerged from a recent very elegant study, which demonstrated that the hinge region of Fc immunoglobulin is a consensus domain that can be selectively bound by natural and artificial protein ligands of unrelated structures [135].

Many questions, however, still remain to be answered. First, what are the characteristics, if any, of the binding domains which, in Kv1 channels, AChRs and other targets, select complementary binding surfaces in toxins? Flexibility is possibly an important parameter [70, 135], but there might be others. Second, how is a target-specific binding site 'born' at the surface of a toxin during evolution? Are binding anchors selected first? Third, is there a preferential evolutionary pathway for the interconversion of two target-specific binding surfaces on a fold, as recently observed *in vitro* with peptides [136]?

These are questions that can be approached experimentally using *in vitro* evolution procedures [137–139].

10.6 DESIGN OF NOVEL TOXINS

The molecular principles that govern the binding properties of animal toxins being better understood, several groups, including ours, are currently trying to exploit this knowledge with a view to engineering toxins with new selective binding properties. At least two types of structure-based designs have been explored. One approach aims to modulate the original binding specificity profile of a natural toxin, whereas the other aims to provide a toxin scaffold with an entirely new binding property.

10.6.1 MODULATING THE ORIGINAL SPECIFICITY PROFILES OF NATURAL TOXINS

A first example is offered by the natural sea anemone toxin ShK which blocks Kv1.1, Kv1.3, Kv1.4 and Kv1.6 with affinities ranging from 11–312 pM [109, 140]. Replacement of the critical Lys22 of the toxin by the shorter, positively charged and non-natural amino acid diaminopropionic acid (Dap) suffices for the toxin to bind more weakly to Kv1.1, Kv1.4 and Kv1.6, with affinities ranging between 1.8 and 37 nM, whilst its affinity for Kv1.3 remained almost unchanged (23 pM instead of 11 pM for the natural toxin). This change was guided by a study of the docking of the toxin in a model of the channel, inspired by the structure of the bacterial KcsA channel [14]. The ShK-Dap22 analogue appears to be a selective blocker of Kv1.3 channels, which is the unique Kv channel expressed in T lymphocytes [141]. Having a low toxicity in a rodent model, ShK-Dap22 was proposed to constitute a potential immunosuppressant for prevention of graft rejection [109]. Though the other sea anemone toxin BgK shares a similar structural fold with ShK and the same binding anchor to Kv1 channels, including a key lysine [142, 97, 41, 46, 75, 97, 98], substitution of the Lys25 into Dap has virtually no effect on BgK's specificity for Kv1.1, Kv1.2 and Kv1.3 [75]. The reasons for these differences are not yet understood.

The binding specificity of BgK could nevertheless be modulated by exploiting the 'two binding components' principle described above. The sole substitution of the variable residue F6 into an alanine sufficed to cause a specific affinity decrease for Kv1.2, Kv1.3 but not for Kv1.1 [75] or Kv1.6 [41]. The resulting mutant is over 500-fold more potent for these two latter channels than for the first two.

Chandy's group has also engineered a scorpion toxin analogue that blocks a Ca^{2+}-activated potassium channel (IKCa1) with a 30-fold higher affinity than Kv1.3 [39, 140], the natural toxin blocking the two channels with approximately similar potencies. This structure-based design resulted from two docking models of the toxin-channel complexes, experimentally supported by double cycle mutant analyses. The models revealed a cluster of negatively charged residues in Kv1.3, which is in close proximity to Lys32 and not present in IKCa1. Introduction of a negatively charged glutamate at position 32 of the toxin caused a 20-fold affinity decrease for Kv1.3 and only six-fold for IKCa1. This result further shows that a structure-based approach can allow a redefinition of the selectivity profile of a toxin by substituting few of its key residues.

Obviously, the same strategy can be applied to other animal toxins. For example, the selectivity of the 16 residue-containing α-conotoxin PnIA toward AChR subtypes is controlled by a single substitution [143, 144]. Substitution A10L suffices to convert the parent toxin PnIA from an α3β2-preferring ligand to a high-affinity α7-preferring ligand.

Clearly, toxins with fine-tuned receptor specificity can be designed through a small number of mutations.

10.6.2   DESIGN OF NOVEL FUNCTIONS ON TOXIN FOLDS

How can we introduce a novel binding function on a toxin scaffold [145–147]? One possible way is to 'copy' or 'transfer' a binding site present in another protein on a host toxin scaffold. The most logical approach consists in searching for secondary structures that nicely superimpose in both the host toxin and the donor protein and to introduce the residues known to be important for the binding function of the donor protein at homologous positions in the host toxin scaffold. This approach proved to be successful in various instances.

One example concerns fasciculin [148], a blocker of acetylcholinesterase, whose functional residues have been transferred at homologous positions of the structurally similar AChR α-neurotoxin blocker [149]. Thus, loop I and half of loop II and the C-terminal of fasciculin 2 were transferred into a short-chain α-neurotoxin. In this operation, the chimera has lost its original AChR binding ability, but acquired an acetylcholinesterase binding potency that was only 15-fold lower than that of the parent fasciculin 2. Interestingly, the specific geometry of the region that binds to acetylcholinesterase was concomitantly transferred into the host protein, as judged from a comparison of the X-ray structures of the host, donor and chimeric proteins [150].

Also, in 1995, the His3 metal-binding site along with two further side chains present on two β-sheet strands of human carbonic anhydrase B, was grafted onto two neighbouring strands of the β-sheet of the scorpion charybdotoxin [151]. Four additional mutations were introduced to prevent unfavourable contacts. The engineered miniprotein was synthesised chemically. The original Kv1 channel blockage activity was lost, but the secondary and tertiary structures were maintained, as indicated by CD and NMR measurements, respectively. Affinity for Cu(II) ions was determined by fluorescence titration of the unique Trp in the chimeric miniprotein, yielding a Kd of 42 nM. From competition experiments, the affinity for Zn(II) ions was lower, with a Kd of 5.3 μM.

Following a similar strategy, several residues of the CDR2-like β-sheet loop of human CD4 have been transferred onto the similarly organised β-sheet of a scorpion toxin [152]. The chemically synthesised chimera, in which the N-terminal was truncated by four residues, adopts the same fold as the parent scorpion toxin. Moreover, as shown by an ELISA, the chimera inhibited the binding of soluble gp120 from HIV to a HSA-CD4 hybrid protein with an apparent IC50 of 40 μM. More recently, the affinity was further increased 100-fold (IC50 = 400 nM) and this construct was able to prevent infection of primary cells by HIV-1LAI and HIV-1BaL [60]. Since then, the binding affinity of the miniCD4 for gp120 has been improved dramatically (to be published).

Therefore, transfer of binding sites from one fold to another one is feasible, at least when the transferred residues are displayed by similar secondary structures in both the host and donor proteins.

10.7 CONCLUSIONS

As correctly predicted by Fontana in 1781 [1], animal venom contain many potent principles. These are increasingly well known and, perhaps more importantly, the molecular features that provide them with their activities are being understood to such an extent that it is now possible to tackle major unsolved questions such as the molecular evolution of toxic activities and to design novel 'toxin'-like substances. It should even be possible now to envisage more ambitious strategies for generating rationally novel activities on toxin scaffolds, *ab initio*. We are now facing this most exciting perspective. The acquired knowledge on the nature and properties of the binding sites displayed by animal toxins will probably be of great help in overcoming the obvious difficulties that will arise during this challenge. If we succeed, the doors might be opened to the design, at will, of miniproteins with novel potential drug-like activities.

ACKNOWLEDGEMENTS

We wish to thank warmly F. Izabelle, F. Lefèvre and D. Patron for their constant help and Drs J.C. Boulain, P. Fromageot, R. Genet, B. Gilquin, D. Gordon and R. Ménez for fruitful discussions.

REFERENCES

[1] Fontana, F. (1781) Traité sur le venin de la vipère, sur les poisons américains sur le laurier-cerise et sur quelques autres poisons végétaux.

[2] Karlsson, E. (1979) Chemistry of proteins toxins in snake venoms. In *Snake Venoms, Handb. Exp. Pharm.* Vol. 52, (Chen-Yuan Lee, ed.), pp. 159–212, Springer-Verlag Berlin.

[3] Karlsson, E., Rydén, L. and Brewer, J. (1989) Ion exchange chromatography in *Protein Purification. Principles, High Resolution Methods and Applications,* (Janson, J-C and Rydén, L, eds). VCH Publishers Inc., New York.

[4] Harvey, A.L. (1991), Snake toxins in *Int. Encyclo. Pharmac. and Ther.* Section 134. Pergamon Press Inc., New York.

[5] Ménez, A., Bontems, F., Roumestand, C., Gilquin, B. and Toma, F. (1992) Structural basis for functional diversity of animal toxins. *Proc. of the Royal Soc. of Edinburgh* **99B**, 83–103.

[6] Rappuoli, R. and Montecucco, C. (1997) *Guidebook to Protein toxins and their use in Cell Biology.* A Sambrook and Tooze Publication, Oxford University Press.

[7] Harvey, A.L. (1997) Recent studies on dendrotoxins and potassium ion channels. *Gen. Pharmacol.* **28**, 7–12.

[8] Ménez, A. (1998) Functional architectures of animal toxins: a clue to drug design? *Toxicon* **36**, 1557–72.

[9] Rochat, H. and Martin-Eauclaire, M.-F. (2000), *Animal Toxins: facts and protocols, Methods and Tools in Biosciences and Medicine,* Birkhäuser Verlag, Basel.

[10] Servent, D. and Ménez, A. (2001) Snake neurotoxins that interact with nicotinic acetylcholine receptors. *Neurotoxicology Handbook.* Chap. 20. Humana Press, **1**, 385–425.

[11] Servent, D., Winckler-Dietrich, V., Hu, H.Y., Kessler, P., Drevet, P., Bertrand, D. and Ménez, A. (1997) Only snake curaremimetic toxins with a fifth disulfide bond have high affinity for the neuronal alpha 7 nicotinic receptor. *J. Biol. Chem.* **272**, 24279–86.

[12] Tytgat, J., Chandy, K.G., Garcia, M.L., Gutman, G.A., Martin-Eauclaire, M.F., van der Walt, J.J. and Possani, L.D. (1999) A unified nomenclature for short-chain peptides isolated from scorpion venoms: alpha-KTx molecular subfamilies. *Trends Pharmacol. Sci.* **20**, 444–7.

[13] Kini, R.M. (1997) *Venom Phospholipases A₂ enzymes: Structure, function and mechanism* (Kini; R.M, ed.) John Wiley and Sons, Chichester.

[14] Doyle, A.D., Morais Cabral, J., Pfuetzner, R.A., Kuo, A., Gulbis, J.M., Cohen, S.L., Chait, B.T., and MacKinnon, R. (1998) The structure of the potassium channel: molecular basis of K^+ conduction and selectivity. *Science* **280**, 69–77.

[15] Brejc, K., van Dijk, W.J., Klaassen, R.V., Schuurmans, M., van Der Oost, J., Smit, A.B. and Sixma, T.K. (2001) Crystal structure of an ACh-binding protein reveals the ligand-binding domain of nicotinic receptors. *Nature* **411**, 269–76.

[16] Hille, B. (1992) *Ionic channels of excitable membranes.* 2nd edition, Sinauer Associates, Sunderland, MA.

[17] Miller, C. (2000) An overview of the potassium channel family. *Genome Biology.* **1**. 1–5.

[18] Gutman, G.A. and Chandy, K.G. (1993) Nomenclature for vertebrate voltage-gated K^+ channels. *Sem. Neurosci.* **5**, 101–6.

[19] Pongs, O. (1999) Voltage-gated potassium channels: from hyperexcitability to excitement. *FEBS Lett.* **452**, 31–5.

[20] Changeux, J.-P. (1990) Functional architecture and dynamics of the nicotinic acetylcholine receptors: An allosteric Ligand-gated channel. *Fidia Research Foundation Neuroscience Award Lectures* **4**, 21–168.

[21] Galzi, J.L. and Changeux, J.-P. (1992) The nicotinic acetylcholine receptor, a model of ligand-gated ion channels, in *Membrane Proteins: Structures, Interactions and Models* (A. Pullman, ed.). Kluwer Academic, The Netherlands, 127–46.

[22] Devillers-Thiéry, A., Galzi, J.L., Eiselé, J.L., Bertrand, S., Bertrand, D. and Changeux, J.-P. (1993) Functional Architecture of the Nicotinic Acetylcholine Receptor: A Prototype of Ligand-gated Ion Channels. *J. Membrane Biol.* **136**, 97–112.

[23] Changeux, J.-P., Kasai, M. and Lee, C.Y. (1970) Use of a snake venom toxin to characterize the cholinergic receptor protein. *Proc. Natl. Acad. Sci. USA* **67**, 1241–7.

[24] Lindstrom, J. (1999) *Purification and cloning of nicotinic acetylcholine receptors.* In *Neuronal Nicotinic Receptors, Pharmacology and Therapeutic Opportunities* (Arneric, S.P. and Brioni, J.D., eds.), pp. 3–23, Wiley-Liss, New York.

[25] Lustig, L.R., Peng, H., Hiel, H., Yamamoto, T. and Fuchs, P.A. (2001) Molecular Cloning and Mapping of the Human Nicotinic Acetylcholine Receptor alpha10 (CHRNA10). *Genomics* **73**, 272–83.

[26] Elgoyhen, A.B., Vetter, D.E., Katz, E., Rothlin, C.V., Heinemann, S.F. and Boulter, J. (2001) Alpha 10: A determinant of nicotinic cholinergic receptor function in mammalian vestibular and cochlear mechanosensory hair cells. *Proc. Natl. Acad. Sci. USA* **98**, 3501–6.

[27] Unwin, N. (1993) Nicotinic acetylcholine receptor at 9 Å resolution. *J. Mol. Biol.* **229**, 1101–24.

[28] Miyazawa, A., Fujiyoshi, Y., Stowell, M. and Unwin, N. (1999) Nicotinic acetylcholine receptor at 4.6 Å resolution: Transverse tunnels in the channel wall. *J. Mol. Biol.* **288**, 765–86.

[29] Endo, T. and Tamiya, N. (1991) Structure–function relationships of postsynaptic neurotoxins from snake venoms in *Snake Toxins* (Harvey, A.L., ed.), pp. 165–222, Pergamon Press, Inc., New York.

[30] Sabatier, J.M. (2000) Chemical synthesis and characterization of small proteins: example of scorpion toxins. In *Animal Toxins : facts and protocols. Methods and Tools in Biosciences and Medicine,* (Rochat, H. and Martin-Eauclaire, M.-F. eds). pp. 197–219, Birkhäuser Verlag, Basel.

[31] Boulain, J.-C., Ducancel, F. Mourier, G., Drevet, P. and Ménez; A. (2000) 'Three-fingered' toxins from hydrophid and elapid snakes: artificial procedures to overproduce wild-type and mutated curaremimetic toxins. In *Animal Toxins: facts and protocols. Methods and Tools in Biosciences and Medicine,* (Rochat, H. and Martin-Eauclaire, M.-F. eds). pp. 229–45, Birkhäuser Verlag, Basel.

[32] Ducancel, F., Boulain, J.-C., Trémeau, O.and Ménez, A. (1989) Direct expression in *E. coli* of a functionally active protein A–snake toxin fusion protein. *Protein Eng.* **3**, 139–43.

[33] Ducancel, F., Guignery-Frelat, G., Tamiya, T., Boulain, J-C. and Ménez, A. (1989b) Postsynaptically toxins and proteins with phospholipase structure from snake venoms: complete amino acid sequences deduced from cDNAs and production of a toxin with staphylococcal protein A gene fusion vector. In *Natural toxins, Characterization, Phamacology and Therapeutics* (Ownby C.L. and Odelle, G.V., eds) Pergamon Press, Oxford.

[34] Ducancel, F., Bouchier, C., Tamiya, T., Boulain, J.-C. and Ménez, A. (1991) Cloning and expression of cDNAs encoding snake toxins. In *Snake Toxins* (Harvey, A.L., ed.), pp. 385–414, Pergamon Press, Inc., New York.

[35] Park, C.S., Hausdorff, S.F. and Miller, C. (1991) Design, synthesis, and functional expression of a gene for charybdotoxin, a peptide blocker of K+ channels. *Proc. Natl. Acad. Sci. USA* **88**, 2046–50.

[36] Bourne, Y., Taylor, P. and Marchot, P. (1995) Acetylcholinesterase inhibition by fasciculin: crystal structure of the complex. *Cell* **83**, 503–12.

[37] Harel, M., Kleywegt, G.J., Ravelli, R.B., Silman, I. and Sussman, J.L. (1995) Crystal structure of an acetylcholinesterase–fasciculin complex: interaction of a three-fingered toxin from snake venom with its target. *Structure* **3**, 1355–66.

[38] Germain, N., Mérienne, K., Zinn-Justin, S., Boulain, J.-C., Ducancel, F. and Ménez, A. (2000) Molecular and structural basis of the specificity of a neutralizing acetylcholine receptor-mimicking antibody, using combined mutational and molecular modeling analyses. *J. Biol. Chem.* **275**, 21578–86.

[39] Rauer, H., Lanigan, M.D., Pennington, M.W., Aiyar, J., Ghanshani, S., Cahalan, M.D., Norton, R.S. and Chandy, K.G. (2000) Structure-guided transformation of charybdotoxin yields an analog that selectively targets Ca(2+)-activated over voltage-gated K(+) channels. *J. Biol. Chem.* **275**, 1201–8.

[40] Kalman, K., Pennington, M.W., Lanigan, M.D., Nguyen, A., Rauer, H., Mahnir, V., Paschetto, K., Kem, W.R., Grissmer, S., Gutman, G.A., Christian, E.P., Cahalan, M.D., Norton, S. and Chandy, K.G. (1998) ShK-Dap22, a potent Kv1.3-specific immunosuppressive polypeptide. *J. Biol. Chem.* **273**, 32697–707.

[41] Racapé, J. (2001) PhD thesis. Recherche et analyse des éléments moléculaires par lesquels les toxines animales lient les canaux Kv1. Université Paris-Sud (Paris XI).

[42] Pillet, L., Trémeau, O., Ducancel, F., Drevet, P., Zinn-Justin, S., Pinkasfeld, S., Boulain, J.-C. and Ménez, A. (1993) Genetic engineering of snake toxins. Role of invariant residues in the structural and functional properties of a curaremimetic toxin, as probed by site-directed mutagenesis. *J. Biol. Chem.* **268**, 909–16.

[43] Goldstein, S.A.N., Pheasant, D.J. and Miller, C. (1994) The charybdotoxin receptor of a *Shaker* K+ channel: peptide and channel residues mediating molecular recognition. *Neuron* **12**, 1377–88.

[44] Stampe, P., Kolmakova-Partensky, L. and Miller, C. (1994) Intimations of K+ channel structure from a complete functional map of the molecular surface of charybdotoxin. *Biochemistry* **33**, 443–50.

[45] Trémeau, O., Lemaire, C., Drevet, P., Pinkasfeld, S., Ducancel, F., Boulain, J.-C. and Ménez, A. (1995) Genetic engineering of snake toxins. The functional site of Erabutoxin a, as delineated by site-directed mutagenesis, includes variant residues. *J. Biol. Chem.* **270**, 9362–9.

[46] Dauplais, M., Lecoq, A., Song, J., Cotton, J., Jamin, N., Gilquin, B., Roumestand, C., Vita, C., de Medeiros, C.L.C., Rowan, E.G., Harvey, A.L. and Ménez, A. (1997) On the convergent evolution of animal toxins. Conservation of a dyad of functional residues in potassium channel-blocking toxins with unrelated structures. *J. Biol. Chem.* **272**, 4302–9.

[47] Antil, S., Servent, D. and Ménez, A. (1999) Variability among the sites by which curaremimetic toxins bind to *Torpedo* acetylcholine receptor, as revealed by identification of the functional residues of α-Cobratoxin. *J. Biol. Chem.* **274**, 34851–8.

[48] Antil-Delbeke, S., Gaillard, C., Tamiya, T., Corringer, P.J., Changeux, J.-P., Servent, D. and Ménez, A. (2000) Molecular determinants by which a long chain toxin from snake venom interacts with the neuronal alpha 7–nicotinic acetylcholine receptor. *J. Biol. Chem.* **275**, 29594–601.

[49] MacKinnon, R., Cohen, S.L., Kuo, A., Lee, A. and Chait, B.T. (1998) Structural conservation in prokaryotic and eukaryotic potassium channels. *Science* **280**, 106–9.

[50] Lo Conte, L.L., Chothia, C. and Janin, J. (1999) The atomic structure of protein-protein recognition sites. *J. Mol. Biol.* **285**, 2177–98.

[51] Gatineau, E., Takechi, M., Bouet, F., Mansuelle, P., Rochat, H., Harvey, A.L., Montenay-Garestier, T. and Ménez, A. (1990) Delineation of the functional site of a snake venom cardiotoxin: preparation, structure, and function of monoacetylated derivatives. *Biochemistry* **29**, 6480–9.

[52] Ménez, A., Gatineau, E., Roumestand, C., Harvey, A.L., Mouawad, L., Gilquin, B. and Toma, F. (1990) Do cardiotoxins possess a functional site? Structural and chemical modification studies reveal the functional site of the cardiotoxin from *Naja nigricollis*. *Biochimie* **72**, 575–88.

[53] Gasparini, S., Danse, J.M., Lecoq, A., Pinkasfeld, S., Zinn-Justin, S., Young, L.C., de Medeiros, C.L.C., Rowan, E.G., Harvey, A.L. and Ménez, A. (1998) Delineation of the functional site of alpha-dendrotoxin. The functional topographies of dendrotoxins are different but share a conserved core with those of other Kv1 potassium channel-blocking toxins. *J. Biol. Chem.* **273**, 25393–403.

[54] Sabatier, J.M., Zerrouk, H., Darbon, H., Mabrouk, K., Benslimane, A., Rochat, H., Martin-Eauclaire, M.F. and Van Rietschoten, J. (1993) P05, a new leiurotoxin I-like scorpion toxin: synthesis and structure–activity relationships of the alpha-amidated analog, a ligand of Ca(2+)-activated K+ channels with increased affinity. *Biochemistry* **32**, 2763–70.

[55] Weber, M. and Changeux, J.-P. (1974) Binding of *Naja nigricollis* [3H] α-toxin to membrane fragments from *Electrophorus* and *Torpedo* electric organs. *Mol. Pharmacol.* **10**, 15–34.

[56] Jaseja, M., Lu, X., Williams, J.A., Sutcliffe, M.J., Kakkar, V.V., Parslow, R.A. and Hyde, E.I. (1994) ¹H-NMR assignments and secondary structure of dendroaspin, an RGD-containing glycoprotein IIb-IIIa (alpha IIb-beta 3) antagonist with a neurotoxin fold. *Eur. J. Biochem.* **226**, 861–8.

[57] Sutcliffe, M.J., Jaseja, M., Hyde, E.I., Lu, X. and Williams, J.A. (1994) Three-dimensional structure of the RGD-containing neurotoxin homologue dendroaspin. *Nat. Struct. Biol.* **1**, 802–7.

[58] Williams, J.A., Lu, X., Rahman, S., Keating, C. and Kakkar, V. (1993) Dendroaspin: a potent integrin receptor inhibitor from the venoms of *Dendroaspis viridis* and *D. jamesonii. Biochem. Soc. Trans.* **21**, 73S.

[59] Chothia, C., Lesk, A.M., Tramontano, A., Levitt, M., Smith-Gill, S.J., Air, G., Sheriff, S., Padlan, E.A., Davies, D. and Tulip, W.R. (1989) Conformations of immunoglobulin hypervariable regions. *Nature* **342**, 877–83.

[60] Vita, C., Drakopoulou, E., Vizzavona, J., Rochette, S., Martin, L., Ménez, A., Roumestand, C., Yang, Y.S., Ylisastigui, L., Benjouad, A. and Gluckman, J.C. (1999) Rational engineering of a miniprotein that reproduces the core of the CD4 site interacting with HIV-1 envelope glycoprotein. *Proc. Natl. Acad. Sci. USA* **96**, 13091–6.

[61] Ménez, A. and Dauplais, M. (1997) As deadly as a scorpion sting. *Science Spectra* **8**, 44–50.

[62] Gurevitz, M., Zilberberg, N., Froy, O., Turkov, M., Wilunsky, R., Karbat, I., Anglister, J., Shaanan, B., Pelhate, M., Adams, M.E., Gilles, N. and Gordon, D. Diversification of toxic sites on a conserved protein scaffold. A scorpion recipe for survival. Chapter 13 in Section III of this book.

[63] Drenth, J., Low, B., Richardson, J.S. and Wright, J.S. (1980) The toxin–agglutinin fold. A new group of small protein strucutres organized around a four disulfide core. *J. Biol. Chem.* **255**, 2652–5.

[64] Bontems, F., Roumestand, C., Gilquin, B., Ménez, A. and Toma, F. (1991) Refined structure of charybdotoxin: common motifs in scorpion toxins and insect defensins. *Science* **254**, 1521–3.

[65] Tamiya, N. (1975) Sea Snake Venoms and Toxins, in *The biology of sea snakes*, (ed. W. A. Dunson) University Park Press, pp. 385–415.

[66] Cunningham, B.C. and Wells, J.A. (1989) High-resolution epitope mapping of hGH-receptor interactions by alanine-scanning mutagenesis. *Science* **244**, 1081–5.

[67] Clackson, T. and Wells, J.A (1995) A hot spot of binding energy in a hormone – receptor interface. *Science* **267**, 383–6.

[68] Bogan, A.A. and Thorn, K.S. (1998), The anatomy of hot spots in protein interfaces. *J. Mol. Biol.* **280**, 1–9.

[69] Hu, Z., Ma, B., Wolfson, H. and Nussinov, R. (2000) Conservation of polar residues as hot spots at protein interfaces. *Proteins* **39**, 331–42.

[70] Ma, B., Wolfson, H.J. and Nussinov, R. (2001) Protein functional epitopes: hot spots, dynamics and combinatorial libraries. *Curr. Opin. Struct. Biol.* **11**, 364–9.

[71] Kühlmann, U.C., Pommer, A.J., Moore, G.R., James, R. and Kleanthous, C. (2000) Specificity in protein–protein interactions: the structural basis for dual recognition in endonuclease colicin–immunity protein complexes. *J. Mol. Biol.* **301**, 1163–78.

[72] Li, W., Hamill, S.J., Hemmings, A.M., Moore, G.R., James, R. and Kleanthous, C. (1998) Dual recognition and the role of specificity-determining residues in colicin E9 DNase-immunity protein interactions. *Biochemistry.* **37** 11771–9.

[73] Ishikawa, Y., Kano, M., Tamiya, N. and Shimada, Y. (1985) Acetylcholine receptors of human skeletal muscles: A species difference detected by snake neurotoxins. *Brain. Res.* **346**, 82–8.

[74] Aneiros, A., Garcia, I., Martinez, J.R., Harvey, A.L., Anderson, A.J., Marshall, D.L., Engstrom, A., Hellman, U. and Karlsson, E. (1993) A potassium channel toxin from the secretion of the sea anemone *Bunodosoma granulifera*. Isolation, amino acid sequence and biological activity. *Biochim. Biophys. Acta.* **1157**, 86–92.

[75] Alessandri-Haber, N., Lecoq, A., Gasparini, S., Grangier-Macmath, G., Jacquet, G., Harvey, A.L., de Medeiros, C.L.C., Rowan, E.G., Gola, M., Ménez, A. and Crest, M. (1999) Mapping the functional anatomy of BgK on Kv1.1, Kv1.2, and Kv1.3. Clues to design analogs with enhanced selectivity. *J. Biol. Chem.* **274**, 35653–61.

[76] Karlsson, E., Arnberg, H. and Eaker, D. (1971) Isolation of the principal neurotoxins of two *Naja naja* subspecies. *Eur. J. Biochem.* **21**, 1–16.

[77] Low, B.W., Preston, H.S., Sato, A., Rosen, L.S., Searl, J.E., Rudko, A.D. and Richardson, J.S. (1976) Three-dimensional structure of erabutoxin b neurotoxic protein: inhibitor of acetylcholine receptor. *Proc. Natl. Acad. Sci.* **78**, 2991–4.

[78] Tsernoglou, D. and Petsko, G.A. (1976) The crystal structure of a post-synaptic neurotoxin from sea snake at 2.2 Å resolution. *FEBS Lett.* **68**, 1–4.

[79] Basus, V.J., Billeter, M., Love, R.A., Stroud, R.M. and Kuntz, I.D. (1988) Structural Studies of α-Bungarotoxin. 1. Sequence-Specific 1H NMR Resonance Assignments. *Biochemistry* **27**, 2763–71.

[80] Walkinshaw, M.D., Saenger, W. and Maelicke, A. (1980) Three-dimensional structure of the 'long' neurotoxin from cobra venom. *Proc. Natl. Acad. Sci. USA* **77**, 2400–4.

[81] Corfield, P.W., Lee, T.J. and Low, B.W. (1989) The crystal structure of erabutoxin a at 2.0-Å resolution. *J. Biol. Chem.* **264**, 9239–42.

[82] Zinn Justin, S., Roumestand, C., Gilquin, B., Bontems, F., Ménez, A. and Toma, F. (1992) Three-dimensional solution structure of a curaremimetic toxin from *Naja nigricollis* venom: a proton NMR and molecular modeling study. *Biochemistry* **31**, 11335–47.

[83] Arnoux, B., Ménez, R., Drevet, P., Boulain, J.-C., Ducruix, A. and Ménez, A. (1994) Three-dimensional crystal structure of recombinant erabutoxin a at 2.0 Å resolution. *FEBS Lett.* **342**, 12–14.

[84] Hopkins, W.K., Miller, J.L. and Miljanich, G.P. (1996) Voltage-gated potassium channels inhibitors. *Current Pharmac. Design.* **2**, 389–96.

[85] Kaczorowski, G.J. and Garcia, M.L. (1999) Pharmacology of voltage-gated and calcium-activated potassium channels. *Current Opinion Chem. Biol.* **3**, 448–58.

[86] MacKinnon, R., Heginbotham, L. and Abramson, T. (1990) Mapping the receptor site for charybdotoxin, a pore-blocking potassium channel inhibitor. *Neuron* **5**, 767–71.

[87] Gross, A., Abramson, T. and MacKinnon, R. (1994) Transfer of the scorpion toxin receptor to an insensitive potassium channel. *Neuron* **13**, 961–6.

[88] Hurst, R.S., Busch, A.E., Kavanaugh, M.P., Osborne, P.B., North, R.A. and Adelman, J.P. (1991) Identification of amino acid residues involved in dendro-toxin block of rat voltage-dependent potassium channels. *Mol. Pharmacol.* **40**, 572–6.

[89] Shon, K.J., Stocker, M., Terlau, H., Stühmer, W., Jacobsen, R., Walkern, C., Grilley, M., Watkins, M., Hillyard, D.R., Gray, W.R. and Olivera, B. (1998) kappa-conotoxin PVIIA is a peptide inhibiting the shaker K$^+$ channel. *J. Biol. Chem.* **273**, 33–8.

[90] Kim, M., Baro, D.J., Lanning, C.C., Doshi, M., Farnham, J., Moskowitz, H.S., Peck, J.H., Olivera, B.M. and Harris-Warrick, R.M. (1997) Alternative splicing in the pore-forming region of shaker potassium channels. *J. Neurosci.* **17**, 8213–24.

[91] Carbone, E., Wanke, E., Prestipino, G., Possani, L.D. and Maelicke, A. (1982) Selective blockage of voltage-dependent K$^+$ channels by a novel scorpion toxin. *Nature* **296**, 90–1.

[92] Garcia, M.L., Gao, Y.D., McManus, O.B. and Kaczorowski, G.J. (2001) Potassium channels: from scorpion venoms to high-resolution structure. *Toxicon* **39**, 739–48.

[93] Garcia, E., Scanlon, M. and Naranjo, D. (1999) A marine snail neurotoxin shares with scorpion toxins a convergent mechanism of blockade on the pore of voltage-gated K channels. *J. Gen. Physiol.* **114**, 141–57.

[94] Ranganathan, R., Lewis, J.H. and MacKinnon, R. (1996) Spatial localization of the K$^+$ channel selectivity filter by mutant cycle-based structure analysis. *Neuron* **16**, 131–9.

[95] Bednarek, M.A., Bugianesi, R.M., Leonard, R.J. and Felix, J.P. (1994) Chemical synthesis and structure–function studies of margatoxin, a potent inhibitor of voltage-dependent potassium channel in human T lymphocytes. *Biochem. Biophys. Res. Com.* **198**, 619–25.

[96] Aiyar, J., Withka, J.N., Rizzi, J.P., Singleton, D.H., Andrews, G.C., Lin, W., Boyd, J., Hanson, D.C., Simon, M., Dethlefs, B., Lee, C.L., Hall, J.E., Gutman, G.A. and Chandy, K.G. (1995) Topology of the pore-region of a K$^+$ channel revealed by the NMR-derived structures of scorpion toxins. *Neuron* **15**, 1169–81.

[97] Pennington, M.W., Mahnir, V.M., Khaytin, I., Zaydenberg, I., Byrnes, M.E. and Kem, W.R. (1996) An essential binding surface for ShK toxin interaction with rat brain potassium channels. *Biochemistry* **35**, 16407–11.

[98] Rauer, H., Pennington, M., Cahalan, M. and Chandy, K.G. (1999) Structural conservation of the pores of calcium-activated and voltage-gated potassium channels determined by a sea anemone toxin. *J. Biol. Chem.* **274**, 21885–92.

[99] Imredy, J.P. and MacKinnon, R. (2000) Energetic and structural interactions between delta-dendrotoxin and a voltage-gated potassium channel. *J. Mol. Biol.* **296**, 1283–94.

[100] Scanlon, M.J., Naranjo, D., Thomas, L., Alewood, P.F., Lewis, R.J. and Craik, D.J. (1997) Solution structure and proposed binding mechanism of a novel potassium channel toxin kappa-conotoxin PVIIA. *Structure* **5**, 1585–97.

[101] Savarin, P., Guenneugues, M., Gilquin, B., Lamthanh, H., Gasparini, S., Zinn-Justin, S. and Ménez, A. (1998). Three-dimensional structure of kappa-conotoxin PVIIA, a novel potassium channel-blocking toxin from cone snails. *Biochemistry* **37**, 5407–16.

[102] Jacobsen, R.B., Dietlind Koch, E., Lange-Malecki, B., Stocker, M.,Verhey, J., Van Wagoner, R.M., Vyazovkina, A., Olivera, B.M. and Terlau, H. (2000) Single amino acid substitutions in kappa-conotoxin PVIIA disrupt interaction with the shaker K$^+$ channel. *J. Biol. Chem.* **275**, 24639–44.

[103] Terlau, H., Shon, M., Grilley, M., Stöcker, W., Stühmer, W. and Olivera, B.M. (1996). Strategy for rapid immobilization of prey by a fish-hunting marine snail. *Nature* **381**, 148–51.

[104] Park, C.S. and Miller, C. (1992) Mapping function to structure in a channel-blocking peptide: electrostatic mutants of charybdotoxin. *Biochemistry* **31**, 7749–55.

[105] Goldstein, S.A.N. and Miller, C. (1993) Mechanism of charybdotoxin block of a voltage-gated K$^+$ channel.*Biophys. J.* **65**, 1613–19.

[106] Aiyar, J., Rizzi, J.P., Gutman, G.A. and Chandy, K.G. (1996) The signature sequence of voltage-gated potassium channels projects into the external vestibule. *J. Biol. Chem.* **271**, 31013–16.

[107] Naranjo, D. and Miller, C. (1996) A strongly interacting pair of residues on the contact surface of charybdotoxin and a Shaker K$^+$ channel. *Neuron* **16**, 123–30.

[108] Naini, A.A. and Miller, C. (1996) A symmetry-driven search for electrostatic interaction partners in charybdotoxin and a voltage-gated K$^+$ channel. *Biochemistry* **35**, 6181–7.

[109] Kalman, K., Pennington, M.W., Lanigan, M.D., Nguyen, A., Rauer, H., Mahnir, V., Paschetto, K., Kem, W.R., Grissmer, S., Gutman, G.A., Christian, E.P., Cahalan, M.D., Norton, S. and Chandy, K.G. (1998) ShK-Dap22, a potent Kv1.3–specific immunosuppressive polypeptide. *J. Biol. Chem.* **273**, 32697–707.

[110] Wrisch, A. and Grissmer, S. (2000) Structural differences of bacterial and mammalian K$^+$ channels. *J. Biol. Chem.* **275**, 39345–53.

[111] Legros, C., Pollmann, V., Knaus, H.G., Farrell, A.M., Darbon, H., Bougis, P.E., Martin-Eauclaire, M.F. and Pongs, O. (2000) Generating a high affinity scorpion toxin receptor in KcsA-Kv1.3 chimeric potassium channels. *J. Biol. Chem.* **275**, 16918–24.

[112] McIntosh, J.M., Yoshikami, D., Mahe, E., Nielsen, D.B., Rivier, J.E., Gray, W.R. and Olivera, B.M. (1994) A nicotinic acetylcholine receptor ligand of unique specificity, alpha-conotoxin ImI. *J. Biol. Chem.* **269**, 16733–9.

[113] Quiram, P.A. and Sine, S.M.. (1998) Structural elements in alpha-conotoxin ImI essential for binding to neuronal alpha(7) receptors. *J. Biol. Chem.* **273**, 11007–11.

[114] Johnson, D.S., Martinez, J., Elgoyhen, A.B., Heinemann, S.F. and McIntosh, J.M. (1995) alpha-conotoxin ImI exhibits subtype-specific nicotinic acetylcholine receptor blockade: preferential inhibition of homomeric alpha 7 and alpha 9 receptors. *Mol. Pharmacol.* **48**, 194–9.

[115] Servent, D., Lamthanh, H., Antil, S., Bertrand, D., Corringer, P.J., Changeux, J.-P. and Ménez, A. (1998) Functional determinants by which snake and cone snail toxins block the alpha 7 neuronal nicotinic acetylcholine receptors. *J. Physiol. (Paris)* **92**, 107–11.

[116] Lamthanh, H., Jegou-Matheron, C., Servent, D., Ménez, A. and Lancelin, J.M. (1999) Minimal conformation of the alpha-conotoxin ImI for the alpha7 neuronal nicotinic acetylcholine receptor recognition: correlated CD, NMR and binding studies. *FEBS Lett.* **454**, 293–8.

[117] Maslennikov, I.V., Shenkarev, Z.O., Zhmak, M.N., Ivanov, V.T., Methfessel, C., Tsetlin, V.I. and Arseniev, A.S. (1999) NMR spatial structure of alpha-conotoxin ImI reveals a common scaffold in snail and snake toxins recognizing neuronal nicotinic acetylcholine receptors. *FEBS Lett.* **444**, 275–80.

[118] Quiram, P.A., Jones, J.J. and Sine, S.M. (1999) Pairwise interactions between neuronal alpha7 acetylcholine receptors and alpha-conotoxin ImI. *J. Biol. Chem.* **274**, 19517–24.

[119] Dougherty, D.A. (1996) Cation-π interactions in chemistry and biology: A new view of benzene, Phe, Tyr, and Trp. *Science* **271**, 163–8.

[120] Ackermann, E.J. and Taylor, P. (1997) Nonidentity of the alpha-neurotoxin binding sites on the nicotinic acetylcholine receptor revealed by modification in alpha-neurotoxin and receptor structures. *Biochemistry* **36**, 12836–44.

[121] Ackermann, E.J., Ang, E.T.H., Kanter, J.R., Tsigelny, I. and Taylor, P. (1998) Identification of pairwise interactions in the alpha-neurotoxin-nicotinic acetylcholine receptor complex through double mutant cycles. *J. Biol. Chem.* **273**, 10958–64.

[122] Malany, S., Ackermann, E., Osaka, H. and Taylor, P. (1998) Complementary binding studies between α-neurotoxin and the nicotinic acetylcholine receptor. *Journal of Physiology (Paris)* **92**, 462–3.

[123] Garcia, K.C., Degano, M., Pease, L.R., Huang, M., Peterson, P.A., Teyton, L. and Wilson, I.A. (1998) Structural basis of plasticity in T cell receptor recognition of a self peptide-MHC antigen. *Science* **279**, 1166–72.

[124] Pawson, T. (1995) Protein modules and signalling networks. *Nature* **373**, 573–80.

[125] Gordon, D., Gilles, N., Bertrand, D., Molgo, J., Nicholson, G.M., Sauviat, M.P., Benoit, E., Shichor, I., Lotan, I., Gurevitz, M., Kallen, R.G. and Heinemann, S.H. Scorpion toxins differentiating among neuronal sodium channel subtypes: nature's guide for design of selective drugs. Chapter 12 in Section III of this book.

[126] Camps, M., Nichols, A., Arkinstall, S. (2000) Dual specificity phosphatases: a gene family for control of MAP kinase function. *FASEB J.* **14**, 6–16.

[127] Olivera, B.M. (1997) E.E. Just Lecture, 1996. Conus venom peptides, receptor and ion channel targets, and drug design: 50 million years of neuropharmacology. *Mol. Biol. Cell.* **8**, 2101–9.

[128] Ohno, M., Ménez, R., Ogawa, T., Danse, J.M., Shimohigashi, Y., Fromen, C., Ducancel, F., Zinn-Justin, S., Le Du, M.H., Boulain, J.-C., Tamiya, T. and Ménez, A. (1998) Molecular evolution of snake toxins: Is the functional diversity of snake toxins associated with a mechanism of accelerated evolution. *Progress in Nucleic Acid Research and Molecular Biology* **59**, 307–64.

[129] Ogawa, T., Oda, N., Nakashima, K., Sasaki, H., Hattori, M., Sakaki, Y., Kihara, H. and Ohno, M. (1992) Unusually high conservation of untranslated sequences in cDNAs for *Trimeresurus flavoviridis* phospholipase A2 isozymes. *Proc. Natl. Acad. Sci. USA* **89**, 8557–61.

[130] Nakashima, K., Nobuhisa, I., Deshimaru, M., Nakai, M., Ogawa, T., Shimohigashi, Y., Fukumaki, Y., Hattori, M., Sakaki, Y. and Hattori, S. (1995) Accelerated evolution in the protein-coding regions is universal in crotalinae snake venom gland phospholipase A2 isozyme genes. *Proc. Natl. Acad. Sci. USA* **92**, 5605–9.

[131] Olivera, B.M., Walker, C., Cartier, G.E., Hooper, D., Santos, A.D., Schoenfeld, R., Shetty, R., Watkins, M., Bandyopadhyay, P. and Hillyard, D.R. (1999) Speciation of cone snails and interspecific hyperdivergence of their venom peptides. Potential evolutionary significance of introns. *Ann. NY Acad. Sci.* **870**, 223–37.

[132] Duda, T.F. Jr. and Palumbi, S.R. (1999) Molecular genetics of ecological diversification: duplication and rapid evolution of toxin genes of the venomous gastropod *Conus*. *Proc. Natl. Acad. Sci. USA* **96**, 6820–3.

[133] Conticello, S.G., Pilpel, Y., Glusman, G. and Fainzilber, M. (2000) Position-specific codon conservation in hypervariable gene families. *Trends Genet.* **16**, 57–9.

[134] Conticello, S.G., Gilad, Y., Avidan, N., Ben-Asher, E., Levy, Z. and Fainzilber, M. (2001) Mechanisms for evolving hypervariability: the case of conopeptides. *Mol. Biol. Evol.* **18**, 120–31.

[135] Delano, W.L., Ultsch, M.H., de Vos, A.M. and Wells, J.A. (2000) Convergent solutions to binding at protein-protein interface. *Science* **287**, 1279–83.

[136] Hoffmüller, U., Knaute, T., Hahn, M., Hohne, W., Schneider-Mergener, J. and Kramer, A. (2000) Evolutionary transition pathways for changing peptide ligand specificity and structure. *EMBO J.* **19**, 4866–74.

[137] Crameri, A., Raillard, S.A., Bermudez, E. and Stemmer, W.P. (1998) DNA shuffling of a family of genes from diverse species accelerates directed evolution. *Nature* **391**, 288–91.

[138] Stemmer, W.P. (1994) Rapid evolution of a protein *in vitro* by DNA shuffling. *Nature* **370**, 389–91.

[139] Sieber, V., Martinez, C.A. and Arnold, F.H. (2001) Libraries of hybrid proteins from distantly related sequences. *Nat. Biotechnol.* **19**, 456–60.

[140] Chandy, K.G., Cahalan, M., Pennington, M., Norton, R.S., Wulff, H. and Gutman, G.A. (2001) Potassium channels in T lymphocytes: toxins to therapeutic immunosuppressants. *Toxicon* **39**(9), 1269–76.

[141] Cahalan, M.D. and Chandy, K.G. (1997) Ion channels in the immune system as targets for immunosuppression. *Curr. Opin. Biotechnol.* **8**, 749–56.

[142] Pennington, M.W., Mahnir, V.M., Krafte, D.S., Zaydenberg, I., Byrnes, M.E., Khaytin, I., Crowley, K. and Kem, W.R. (1996) Identification of three separate binding sites on SHK toxin, a potent inhibitor of voltage-dependent potassium channels in human T-lymphocytes and rat brain. *Biochem. Biophys. Res. Commun.* **219**, 696–701.

[143] Luo, S., Nguyen, T.A., Cartier, G.E., Olivera, B.M., Yoshikami, D. and McIntosh, J.M. (1999) Single-residue alteration in alpha-conotoxin PnIA switches its nAChR subtype selectivity. *Biochemistry* **38**, 14542–8.

[144] Hogg, R.C., Miranda, L.P., Craik, D.J., Lewis, R.J., Alewood, P.F. and Adams, D.J. (1999) Single amino acid substitutions in alpha-conotoxin PnIA shift selectivity for subtypes of the mammalian neuronal nicotinic acetylcholine receptor. *J. Biol. Chem.* **274**, 36559–64.

[145] Skerra, A. (2000).Engineered protein scaffolds for molecular recognition. *J. Mol. Recognit.* **13**, 167–87.

[146] Martin, L. and Vita, C. (2000) Engineering novel bioactive mini-proteins from small size natural and de novo designed scaffolds. *Current Prot. & Pept. Sci.* **1**, 403–30.

[147] Cedrone, F., Ménez, A. and Quémeneur, E. (2000) Tailoring new enzyme functions by rational redesign. *Curr. Opin. Struct. Biol.* **10**, 405–10.

[148] Cervenansky, C., Dajas, F., Harvey, A.L. and Karlsson, E. (1991) Fasciculins, anticholinesterase toxins from mamba venoms: biochemistry and pharmacology. In *Snake Toxins* (Harvey, A.L., ed.), pp. 303–21, Pergamon Press, Inc., New York.

[149] Ricciardi, A., Le Du, M.H., Khayati, M., Dajas, F., Boulain, J.-C., Ménez, A. and Ducancel, F. (2000) Do structural deviations between toxins adopting the same fold reflect functional differences? *J. Biol. Chem.* **275**, 18302–10.

[150] Le Du, M.H., Ricciardi, A., Khayati, M., Ménez, R., Boulain, J.-C., Ménez, A. and Ducancel, F. (2000) Stability of a structural scaffold upon activity transfer: X-ray structure of a three fingers chimeric protein. *J. Mol. Biol.* **296**, 1017–26.

[151] Vita, C., Roumestand, C., Toma, F. and Ménez, A. (1995) Scorpion toxins as natural scaffolds for protein engineering. *Proc. Natl. Acad. Sci. USA* **92**, 6404–8.

[152] Vita, C., Vizzavona, J., Drakopoulou, E., Zinn-Justin, S., Gilquin, B. and Ménez, A. (1998) Novel miniproteins engineered by the transfer of active sites to small natural scaffolds. *Biopolymers* **47**, 93–100.

11 Scorpion Genes and Peptides Specific for Potassium Channels: Structure, Function and Evolution

LOURIVAL D. POSSANI*, ENRIQUE MERINO, MIGUEL CORONA and BALTAZAR BECERRIL

11.1 INTRODUCTION

Scorpion venoms are a rich source of peptides, most of which were designed and selected by natural evolution to recognise ion-channels, permeable to Na^+, K^+, Cl^- and Ca^{2+} ions. Several reviews dealing with various aspects of these natural ligands have been published in recent years [1–10]. In the short time since the last review, there has been the very recent discovery of the Ergotoxins [11–13] and Tc1, the shortest scorpion toxin specific for the voltage sensitive K^+-channel [14]. There are considerable prospects for the discovery of novel peptides and functions among scorpion venom peptides, whose biodiversity is estimated at around 100 000, of which only about 0.02% are known [7]. Apart from the peptides active on ion-channel permeability, there are pore-forming substances [15], antibiotics [16], analgesic peptides [17], and others of unknown function [18]. Based on these results it can be speculated that the evolutionary forces governing the design, selection and maintenance of such peptides in scorpion venom were driven by the scorpion's need for tools (peptides and/or toxins) that interfere with normal ion-channel function, thereby enabling it to capture prey or defend itself against predators. Among scorpion venom peptides that affect ion-channels, a little over 200 distinct amino acid sequences are known, either by direct analysis of peptides or based on cloned genes from 30 different species of the 1500 surmised to exist in the world [9]. What is evident from the knowledge gathered thus far is that, although the length of these various peptides varies from toxin to toxin, the three-dimensional folding seems to follow the same general structural motif.

In this chapter we shall review only the peptides and genes, whose products are assumed to affect the function of K^+-channels. To abbreviate the nomenclature of these toxins, they will be called simply: K^+-peptides. A general

* Corresponding author

Perspectives in Molecular Toxinology
Edited by A Ménez © 2002 John Wiley & Sons, Ltd

classification for the K$^+$-peptides was recently proposed [6]; here we will add the newly discovered ones. The physiological action will be briefly revised and the results of a computational analysis, showing particular features of the three-dimensional structure, will be presented, together with a phylogenetic tree that groups the distinct peptides in sub-families. The possible relationship of these peptides to function, and some considerations related to the possible evolution of these peptides, will be discussed. Finally, some thoughts on the perspectives for future developments in this area will be mentioned.

11.2   ISOLATION AND PRIMARY SEQUENCES

Scorpion venoms are usually obtained by electrical stimulation of the telson and the crude extracts are fractionated by chromatographic procedures, using either a gel filtration column, followed by ion-exchange and high-performance liquid chromatographic techniques, or a combination of these methodologies, as discussed in more detail in reviews [5, 7].

The amino acid sequence of the various peptides are more often obtained by automatic Edman degradation. Fragmentation by means of specific enzymes yields full overlapping information on the primary structure of the peptides, as well as the determination of the disulphide bridges [5, 7, 9]; although for the latter, mass spectrometry analysis has also been used successfully to provide information on the disulphide arrangements of these peptides [12].

An initial attempt to classify the K$^+$-peptides was proposed by Miller [19] and amplified by Selisko et al. [20, see also [5]]. However, it was only recently that Tytgat et al. [6], jointly with an international panel of experts in the field, proposed a more rational nomenclature for the peptides specific for K$^+$-channels from scorpion venoms. About 49 different K$^+$-peptides were listed (review [6]). Several new sequences have since been reported, and if the information from genes that code for the expression of putative K$^+$-channel specific toxins are included, as shown by Possani et al. [9], the total number is now about 75, plus a couple of more recent ones [13, 14, 21], which clearly shows that the number of K$^+$-peptides is increasing steadily. Not included in this analysis is a recent publication [22] dealing with the three-dimensional structure of butantoxin, because it is practically identical to TsTXIV (see review [6]), except for the lack of the asparagine at the end.

Figure 11.1 lists representative examples of the 12 sub-families [6] taking only one sequence for each group. A new sub-family number 13, was recently defined [14]; it contains only 23 amino acid residues. Thus presently, there are 13 distinct sub-families of K$^+$-channel specific toxins. A new nomenclature is proposed for the K$^+$-peptides that affect function of the ERG K$^+$-channels, where ERG stands for the family of ether-a-go-go-related genes. The ERG channel is an inward rectifier K$^+$-channel, contrary to most of the other known K$^+$-channels, usually permeable to K$^+$ ions from the interior to the

Alpha-Subfamily		Sequence	alpha-KTx number
1	ChTx	ZFTNVSCTTSKECWSVCQRLHNTS-RGKCMNK-KCRCYS	1.1
2	NTx	TIINVKCTSPKQCSKPCKELYGSSAGAKCMNG-KCKCYNN	2.1
3	KTx	GVEINVKCSGSPQCLKPCKDA-GMR-FGKCMNR-KCHCTPK	3.1
4	TsII-9	VFINAKCRGSPECLPKCKEAIG-KAAGKCMNG-KCKCYP	4.1
5	ScyTx	AFC-NLRMCQLSCRSL-GL--LGKCIGD-KCECVKH	5.1
6	Pi1	L-VKCRGTSDCGRPCQQQTGCPNS-KCINR-MCKCYGC	6.1
7	Pi2	TI---SCTNPKQCYPHCKKETGYPNA-KCMNR-KCKCFGR	7.1
8	P01	VSC---EDCPEHCSTQ---KAQAKCDND-KCVCEPI	8.1
9	BmP02	VGC---EECPMHCKGK---NAKPTCD-DGVCNCNV	9.1
10	CoTx1	AVC-VYRTCDKDCKRR-GYR-SGKCINN-ACKCYPY	10.1
11	PBTx1	DEEPKESC-SDEMCVIYCKGE-EYS-TGVCDGPQKCKCSD	11.1
12	TsTXIV	WCSTCLDLACGASRECYDPCFKAFG-RAHGKCMNN-KCRCYTN	12.1
13	Tc1	ACG-S--CRKKCK---G---SGKCIN-GRCKCY	13.1

Beta-Sufamily		Sequence	beta-KTx number
1	CnERG1	DRDSCVDKSRCAKYGYYQECQDCCKN-AGHNGGTCMFFKCKCA	1.1
2	BeKm-1	RPT-DIK--CSES--YQ-CFPVCKSRFGKTNGRCVNGFCDCF	2.1

Figure 11.1   Representative examples of primary structure of families and sub-families of K$^+$-peptides

Two families are shown: α and β-KTx with 13 and 2 sub-families, respectively. The amino acid sequences were taken from the review [6] and recent publications [12, 13]

exterior of the cells [11]. For this new family, we propose to call them βKTx, to differentiate them from the α-KTx proposed in [5, 19, 20]. To date, there are only two different toxins known [12, 13], which fit into two sub-families: βKTx1 and βKTx2. The first one was isolated from the venom of *Centruroides noxius* and was initially reported to contain 6 half-cystines [11], but this was later corrected to 8 half-cystines [12], with a different C-terminal residue. The presence of these two almost identical isoforms in *Centruroides noxius* venom is described and discussed in a new publication to be submitted elsewhere (Gurrola, G., Pardo, L., Zamudio, F., Scaloni, A., Wanke, E. and Possani, L.D., in preparation). Thus, the real ergotoxin-1 (abbreviated here as CnERG1) has 42 amino acid residues closely packed by 4 disulphide bridges [12]. It also contains the two structural motifs: CXXXC, and CXC (where C stands for cysteine and X for any amino acid residue) shown to be an important characteristic of K$^+$-peptides, as will be discussed later. However, its three-dimensional structure is still unknown. The second sub-family is constituted by the toxin BeKm-1 [13], which is isolated from *Buthus eupeus* and contains 36 amino acid residues with only six half-cystines, thus forming three disulphide bridges. BeKm-1 does not have the double consecutive cysteines, as found in CnERG1. Nevertheless, BeKm-1 has the two segments of sequences flanked by half-cystines reported for the other K$^+$-peptides. Our unpublished data suggest that there is more than one peptide for each of these two new sub-families proposed here.

11.3   THREE-DIMENSIONAL CHARACTERISATION

The first three-dimensional structure of a scorpion toxin was obtained by X-ray diffraction in 1980 [23]. Since then, several other structures of scorpion toxins were determined either by crystallographic analysis or by nuclear magnetic resonance (NMR) technology (for review see [7]). Because the K^+-peptides are shorter than the Na^+-channel ones, they are ideal for NMR determination of their three-dimensional structure. All of them, show a similar three-dimensional structure as discussed in Selisko *et al.* [20], see also [5].

In Plate VII we show an example of the three-dimensional fold of a typical K^+-peptide (NTx, which stands for noxiustoxin, a K^+-peptide from *Centruroides noxius*), superimposed on three other peptides. The three peptides that do not belong to the K^+-channel-specific scorpion toxins group but have a conserved three-dimensional fold are Drosomycin, an antifungal protein from *Drosophila melanogaster* [24], a peptide called 1AYJ, which is also an antifungal protein, isolated from the plant *Raphanus sativus* [25], and BmkM4, a scorpion neurotoxin from *Buthus martensii* which is specific for Na^+-channels [26]. The first important observation in Plate VII is the striking three-dimensional positions of the elements that form the secondary structure, made by the α-helix segment and the antiparallel β-sheets. Equally important is the tertiary folding stabilised by two disulphide bridges, which is highly conserved around the central core of the molecules. Clear differences can be seen in the loops extending out from the α-helix and β-sheets, as well the extension of the C-terminal region of the Na^+-channel specific toxin, about 30 % longer than that of the others. It is intuitively logical to assume that these variable regions contain functional information. In other words, while the target molecules towards which these similar peptides evolved are different, such as K^+ or Na^+-channels, fungi or even microorganisms, they are expected to contain the information that determines the recognition of and affinity for the receptor molecules, hence the function of these peptides. However, this is only partially true. The studies of the group of C. Miller, R. Mackinnon, M. Garcia and others (discussed in review [5]) have demonstrated that specific amino acids, sometimes only one residue, such as Lys27 in charybdotoxin [27], or Lys28 for the case of noxiustoxin [28], is essential for the blocking of certain types of voltage-dependent K^+-channels.

Additional critical residues are situated in other sections of the molecule, such as Lys6, Thr8 and the C-terminal tripeptide Tyr–Asn–Asn in the case of NTx [28]. Our group demonstrated recently that in the case of Pi2 and Pi3, two K^+-peptides that differ only in one amino acid in position 7 (a proline for a glutamic acid) have a 17-fold difference in their affinities for the shaker K^+-channel [29] and for K^+-channels present in human lymphocytes [30]. This difference has been attributed to the formation of a salt bridge between Glu7 and Lys24 [30]. According to the group of A. Ménez in France [31], convergent evolution of animal toxins selected a diad of amino acids (Lys and Tyr) as a minimum requirement for blockage of K$^+$-channels. The extended segment of the *Buthus martensii* Na$^+$-channel-specific toxin is certainly implicated in channel recognition and in the control of its gating mechanism, judging by its similarity with other Na$^+$-peptides as shown elsewhere [32–34]. Additionally, it has been proposed that the overall charge distribution of these peptides plays a role in the binding and affinity of the K$^+$-peptides towards their receptor channels (review in [1]). In some cases a more than one million-fold difference may exist between two different toxins [5, 7, 9], which nonetheless have a strikingly similar three-dimensional structure as shown in Plate VII. The receptor molecule to which the other two short-chain peptides (drosomycin and peptide 1AYJ) bind is still not well characterised. In Plate VIID and E additional evidence is presented, taking into consideration the amino acid sequence of these four peptides. The highly conserved and highly variable regions are indicated. The elements governing the three-dimensional folding provide evidence that amino acids with similar or equivalent charge or hydrophobicity are preserved in the structurally neighbouring residues of the aligned structures.

In conclusion, the definition of similar function based on the 'α-helix plus β-sheets' structural motif is not enough to ensure that the peptide will recognise an ion-channel of the K$^+$-type. Other elements and variables are implicated, and it is sufficient to change only one amino acid to modify the function. This will foster future research and development in this area, as will be discussed in the section on perspectives.

11.4   PHYSIOLOGICAL EFFECTS

As discussed in a previous review, most of the peptides initially isolated from scorpion venoms were purified using bioassays, in which live animals were used [6]. Injection into mammals, usually mice, insects such as flies or crickets, and several kinds of crustaceans such as sweet water crayfishes, were examples of methods used to determine the physiological action of these peptides, generically called toxins. However, most components active on K$^+$-channels were never assayed in living organisms, due to limitations of available material, or if injected, showed none of the lethal effects earlier described for the Na$^+$-channel toxins [6]. For this reason, we prefer to call them just K$^+$-peptides, rather than K$^+$-toxins. The great majority of K$^+$-peptides were purified using either binding and displacement experiments with a radiolabelled known K$^+$-specific peptide, or by the use of fine electrophysiological systems [6]. The *in vitro* experiments used a variety of different systems, including squid giant axons [35], various cells in culture [11, 21, 36], insect cell lines such as Sf9 [14], human lymphocytes [30], or mRNA expressed in *Xenopus laevis* oocytes [20].

In Figure 11.2 we have redrawn from the literature examples of our own work [11, 14, 21, 36]. This figure shows several types of studies using whole

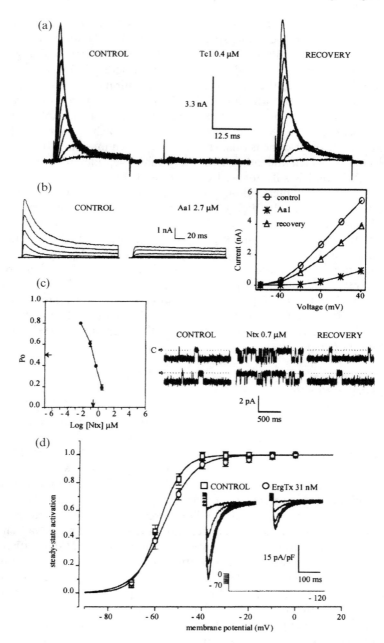

Figure 11.2 Electrophysiological effects of K$^+$-peptides
(a) Effects of Tc1 on Shaker B K$^+$-channels, expressed in Sf9 cells, showing control, under the action of 0.4 µM of toxin, and recovery (right figure). The channels were activated by

cell-clamp techniques with isolated cells or excised patches of membranes obtained from specific tissues. Figure 11.2A illustrates recent work performed with the *shaker* K$^+$-channel expressed in *Spodoptera frugiperda* (Sf9) cells. A complete blockade of K$^+$-currents is obtained with Tc1, the shortest K$^+$-channel-specific scorpion toxin known [14]. Plate VIIB is also from a recent study in which the A-type K$^+$-current, from cerebellum granular cells in culture, is completely inhibited by the addition of Aa1 scorpion toxin [21]. Figure 11.2C is an example of noxiustoxin blockade of single Ca^{2+}-activated K$^+$-channels of small conductance recorded in patches of membranes from cultured bovine aortic endothelial cells [36]. The probability of opening is reduced, with an IC$_{50}$ on the order of 300 nM. Finally, Figure 11.2D is an example of the effect of the new family of K$^+$-peptides, using ErgTX1 on F-11 neuroblastoma cells [11], in which the blocking effect, i.e. decrease in the inward current of potassium through the ERG-channel, is evident.

To date, relatively few specific bioassays involving known ion channels expressed in heterologous systems have been studied. This greatly hampers understanding of the functional relationships of these ligands to their receptor sites. It also complicates understanding of the concept of species specificity. Why are there toxins specific for insects, crustaceans and mammals? Different tissues and other possible receptor sites should be investigated. For many of these peptides we still do not know the specific target, if any, for which they evolved. This will be briefly discussed in the section on future perspectives.

11.5 STRUCTURE OF GENES ENCODING K$^+$-PEPTIDES

The cDNA and genomic regions encoding several K$^+$-peptides have been published [10, 37]. Examples of a cDNA and a genomic segment encoding

---

30 ms pulses from -30 to $+50$ mV in 10 mV increments, delivered every 20 s from the holding potential of -90 mV. Redrawn and modified from [14].

(b) Effect of toxin Aa1 on I_A type K$^+$ current of cerebellum granular cells. On the left is a family of control currents in response to a voltage step from -60 to $+40$ mV in 20 mV increments. In the middle are the records obtained in the presence of toxin. On the right is the current–voltage relationship for peak I_A under control conditions, in the presence of toxin and after washing. Redrawn and modified from [21].

(c) Blocking effect of Noxiustoxin on single Ca^{2+}-activated K$^+$-channels from patches of bovine aortic endothelial cells, incorporated in artificial bilayers. Left shows a probability of opening (Po) value plotted against toxin concentration, showing the half-inhibitory concentration (IC$_{50}$) = 310 nM. On the right, from left to right: a control trace, effect of addition of toxin and recovery after washing the system with buffer alone. Traces are unitary channel activity at -40 mV from an outside-out patch. The dotted line indicates the zero current level (baseline), arrows labelled C, show the closed level of a single channel. Modified from [36].

(d) Effect of CnERG1 on F-11 clone cells. Steady-state activation curves under control and during perfusion with 31 nM of CnERG1 (n = 3); conditioning duration of 15 s. Inset: superposition of tail currents from one of the three cells, in the control and under the effect of the toxin. Modified from [11]

(a)

```
ATG GAG GGT ATT GCC AAA ATA ACA CTA ATC CTA TTG TTT TTG TTC GTA ACA ATG CAT    57
 M   E   G   I   A   K   I   T   L   I   L   L   F   L   F   V   T   M   H     -10
ACA TTC GCT AAT TGG AAT ACC GAA GCG GCC GTG TGT GTT TAT CGC ACT TGT GAT AAG   114
 T   F   A   N   W   N   T   E   A   A   V   C   V   Y   R   T   C   D   K      10
GAT TGC AAA CGT AGG GTG TAT AGG TCG GGA AAA TGC ATT AAC AAC GCC TGC AAA TGT   171
 D   C   K   R   R   G   Y   R   S   G   K   C   I   N   N   A   C   K   C      29
TAT CCC TAT GGA AAA TAA TGTTCAGTTAACAAAAAAAAA                                  210
 Y   P   Y   G   K  END                                                        34
```

(b)

```
ATTTAATAATTGACTTTTATGGATATAATATATCTTTATTCATTCGAAA    ATG AAG GTG TTT TCC GCA    64
                                                      M   K   V   F   S   A   -17
GTT TTG ATA ATT CTC TTC GTC TGT TCA ATG Agtaattacgaatttttattaatttatatattttta   129
 V   L   I   I   L   F   V   C   S   M                                          -7
tatgtaaaaacttaataattcattaattaagcatatgttgtttaatattttagTT ATT GGA ATT AAT GCA   199
                                                          I   I   G   I   N   A  -1
GTG AGA ATT CCA GTG TCA TGT AAA CAT TCT GGT CAA TGT TTA AAA CCA TGC AAG GAT   256
 V   R   I   P   V   S   C   K   H   S   G   Q   C   L   K   P   C   K   D      19
GCT GGA ATG AGA TTT GGA AAA TGC ATG AAT GGC AAA TGC GAT TGT ACA CCA AAG TGA   313
 A   G   M   R   F   G   K   C   M   N   G   K   C   D   C   T   P   K  END     37
TTTTTTCTTCCATAAAAATATTTTCAATGTGTAATAGTT                                        352
```

Figure 11.3 cDNA or genomic nucleotide sequences encoding two K$^+$-channel scorpion toxins

(a)   Nucleotide sequence of the cDNA encoding Cobatoxin 1 from *Centruroides noxius* Hoffmann [20]. Mature peptide consists of 32 amino acid residues ending at an amidated tyrosine. The last two amino acids (Gly33 and Lys34) are processed during toxin maturation. Preceding mature toxin, a putative signal peptide (28 residues) is underlined.

(b)   Nucleotide sequence of the KTX$_2$ gene from *Androctonus australis* (modified from [38]). In this sequence a 22-amino-acid signal peptide (underlined), and a 37-residue mature toxin are shown. An intron of 87 bp interrupting the region encoding the signal peptide is present. Numerals on the right indicate nucleotide and amino acid numbering. Negative digits correspond to amino acid positions at the signal peptide

two different toxins are shown in Figure 11.3. The organisation of these coding regions is very similar to the corresponding fragments encoding Na$^+$-channel toxins [7]. These peptides are synthesised as precursors in which a signal peptide is removed by a signal peptidase. Processing at the carboxy-terminus seems to follow rules similar to those described for Na$^+$-channel toxins [7]. The cDNA sequences have been reported for a new family of long-chain (60–64 amino acid residues), K$^+$-channel toxins stabilised by three disulphide bridges [10]. The genomic organisation of the DNA segments encoding K$^+$-channel toxins is similar to that reported for Na$^+$-channel toxins: two exons and one intron which interrupts the region that codes for the signal peptide (Figure 11.2B) [10]. A similar structure has also been found for Chlorotoxin genes [39, 40]. Smaller introns and longer signal peptides are associated with K$^+$-or Cl$^-$-channel toxin genes as compared to Na$^+$-channel toxin genes.

11.6   PHYLOGENETIC TREE AND EVOLUTION

For this study we analysed 75 scorpion toxins specific for K$^+$-channels and three other related toxins from other sources, obtained from the Swiss-Prot,

GenBank and PIR protein databases. Multiple alignment of these sequences was done using the CLUSTALX program [41] and further refined manually. Based on this initial alignment, 1000 bootstrapped data were resampled using the SEQBOOT program of J. Felsenstein's PHYLIP phylogeny inference package program [42]. Genetic distances of these alignments were calculated using the Dayhoff PAM matrix with the PROTDIST program [42]. Subsequently, the trees were constructed by successive clustering of lineages using the Neighbor-Joining algorithm [43] as implemented in the NEIGHBOR program [42]. The strict consensus tree was obtained using the CONSENSUS program [42] and the unrooted tree diagram was generated with the DRAWTREE program [42]. The three-dimensional structures, alignment data and drawing program Cn3D were obtained from the National Center of Biotechnology Information (http://www.ncbi.nlm.nih.gov/Structure). Finally, the art work was done using the drawing program CANVAS.

The unrooted phylogenetic tree of K$^+$-peptides is shown in Figure 11.4. There is a substantial amount of information in this tree, which merits long analysis and discussion. Here, we will limit ourselves to the most significant findings. The first observation is that although all these different peptides (almost 80) have quite similar three-dimensional folding (Plate VII), they fall into a broad number of clusters when analysed by the computational methods described here. On the order of 15 or more different groups are evident. The α-K$^+$-peptides are all related to each other, justifying the nomenclature proposed by Tytgat [6]. The sub-families of Charybdotoxin (ChTx), Noxiustoxin (NTx, 1SX, in box), Kaliotoxin, two sub-families of *Pandinus imperator* toxins (Pi1, Pi2 plus Pi3), P01 and leiuropeptides, scyllatoxin, cobatoxins, *Parabuthus* toxins, *Buthus martensii* peptides and two sub-families of *Tityus* toxins are all clustered separately with their homologues, as expected from the primary sequence analysis [6]. However, it is worth noticing that all three different non-K$^+$-channel toxins used in Plate VII (1SN4, 1MYN, 1AYJ) are separated. Equally well separated are the ERG-channel toxins (CnERG1 from *Centruroides noxius*) and BeKm-1, which although quite different from CnERG1 is closely related to BmTx1, another toxin of *Buthus martensii*.

An additional important observation from this phylogenetic tree is the suggestion that all these peptides came from a common ancestor. The distinct peptides with different affinities toward distinct receptor targets but basically similar structural motifs evolved from a common scaffold, which served as a prototype for the selection and maturation of these peptides. This process is certainly still occurring today.

11.7 PERSPECTIVES AND CONCLUDING REMARKS

Future work is likely to focus on at least two important issues: the biodiversity of these natural ligands, many of which have not yet been described (only 0.2 %

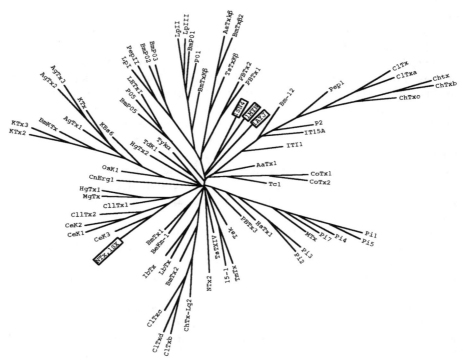

Figure 11.4   Unrooted phylogenetic tree of scorpion toxins specific for K⁺-channels

A multiple sequence alignment of 75 scorpion toxins specific for K⁺-channels were used to calculate a matrix with the genetic distances for each pair of the sequences. Based on this matrix, successive clustering of lineages were used to construct the unrooted tree with the Neighbour-Joining algorithm [43]. This analysis also includes the sequence of three toxins (1SN4, 1MYN, 1AYJ, inside a box) that do not belong to K⁺-channel-specific scorpion toxins but have a common three-dimensional fold as NTx, shown in Plate VII. Also included are peptides ClTxa, ClTxb, ClTxc, ClTxd, putative chloride-channel-specific peptides [39], whose sequence similarities with K⁺-peptides are considerable. Abbreviated names of the K⁺-peptides were taken from [6, 9 and 39]. CnERG1 [12] and BeKm-1 [13] are defined in the text of this chapter. AaTX1 is from [21], whereas KTx3 is from [44]. CeK1, CeK2, CeK3, Pi5 are novel K⁺-peptides from our group (Olamendi-Portugal, T. *et al.*, unpublished)

of the possible universe is known), and the identification of their target molecules.

In terms of biodiversity, it has been shown that a change in only one amino acid residue in the primary structure of these peptides can alter by one to three orders of magnitude the fitness of the ligand to the receptor site [19, 27, 28, 30], meaning that it is worth pursuing the isolation and determination of the amino acid sequences of the 99.8% remaining putative peptides in scorpion venoms. The issue of target molecules is even more important, because we really do not know how many more different receptor sites exist for the biological effects of these peptides. As discussed before by our group [5, 7, 9], some peptides still have an unknown function. They have been tested in living organisms, tissues or specific channels expressed in cell culture systems, with no observable effects. The question is, are there other target receptor sites? Furthermore, many of these peptides were assayed in different systems and the relative potency and fitness can not be ascertained, because there are no systematic studies done with a given toxin against all the possible target molecules [5, 7, 9, 10]. Some of the experiments done in order to assay specificity of known peptides were performed in heterologous systems, and we know now that many of these channel molecules have other internal molecules that work as modulators of the channel activity, as is the case for the β-subunits of K$^+$-channels [45]. The perspectives are therefore wide-ranging and many more interesting and novel peptides and mechanisms are likely to be discovered in the future. It is also possible that other, as yet unknown, types of structure foldings could be present in these K$^+$-peptides, as with a K$^+$-channel-specific peptide from a sea anemone [31]. No less appealing is the fact that using knowledge of these various peptide structures, rational structural modelling should help in the future the design of new drugs [5].

ACKNOWLEDGEMENTS

Partially supported by grants: 55000574 of the Howard Hughes Medical Institute; IN216900 from the *Direccion General de Asuntos del Personal Academico – UNAM*; 31691-N and Z-005 from the National Council of Science and Technology, Mexican Government.

REFERENCES

[1] Garcia, M.L., Hanner, M., Knaus, H.G., Koch, R., Schmalhofer, W., Slaughter, R.S. and Kaczorowski, G.J. (1997) Pharmacology of potassium channels. *Adv. Pharmacol.* **39**, 425–71.
[2] Legros, C. and Martin-Eauclaire, M-F. (1997) Les toxines de scorpion. *Comptes Rendus Soc. Biol. (France)* **191**, 345–80.
[3] Gordon, D., Savarin, P., Gurevitz, M. and Zinn-Justin, S. (1998) Functional anatomy of scorpion toxins affecting sodium channels. *J. Toxicol. Toxin. Rev.* **17**, 131–59.
[4] Valdivia, H. and Possani, L.D. (1998) Peptide toxins as probes of Ryanodine receptor. *Trends Cardiovascular Med.* **8**, 111–18.
[5] Possani, L.D., Selisko, B. and Gurrola, G. (1999) Structure and function of scorpion toxins affecting K$^+$-channels. *Perspective in Drug Discovery and Design* **15/16**, 15–40.
[6] Tytgat, J., Chandy, K.G., Garcia, L.M., Gutman, G.A., Martin-Eauclaire, M-F., van del Walt, J.J. and Possani, L.D. (1999) A unified nomenclature for short chain peptides isolated from scorpion venoms: α-KTx molecular subfamilies. *Trends in Pharmacol. Sciences* **20**, 445–47.
[7] Possani, L.D., Becerril, B., Delepierre, M. and Tytgat, J. (1999) Scorpion toxins specific for Na$^+$-channels. *Eur. J. Biochem.* **264**, 287–300.

[8] Legros, C., Bougis, P.E. and Martin-Eauclaire, M-F. (1999) Molecular biology of scorpion toxins active on potassium channels. *Perspectives in Drug Discovery and Design* **15/16**, 1–14.

[9] Possani, L.D., Merino, E., Corona, M., Bolivar, F. and Becerril, B. (2000) Peptides and genes coding for scorpion toxins that affect ion-channels. *Biochimie (France)* **82**, 861–8.

[10] Possani, L.D., Becerril, B., Tytgat, J. and Delepierre, M. (2001) High affinity scorpion toxins for studying potassium and sodium channels. In: *Ion Channel Localization Methods and Protocols* (eds: Nichols C. and Lopatin A.). Humana Press, Totowa NJ, USA.

[11] Gurrola, G.B., Rosati, B., Rocchetti, M., Pimienta, G., Zaza, A., Arcangeli, A., Olivotto, M., Possani, L.D. and Wanke, E. (1999) A toxin to nervous, cardiac, and endocrine ERG K$^+$ channels isolated from *Centruroides noxius* scorpion venom. *FASEB J.* **13**, 953–62.

[12] Scaloni, A., Bottiglieri, C., Ferrera, L., Corona, M., Gurrola, G.B., Batista, C., Wanke, E. and Possani, L.D. (2000) Disulfide bridges of Ergotoxin, a member of a new sub-family of peptide blockers of the *ether-a-go-go*-related K$^+$ channel. *FEBS Letters* **479**, 156–7.

[13] Korolkova, Y.V., Kozlov, S.A., Lipkin, A.V., Pluzhnikov, K.A., Hadley, J.K., Filippov, A.K., Brown, D.A., Angelo, K., Strobak, D., Jespersen, T., Olesen, S.P., Jensen, B.S. and Grishin, E.V. (2001) An ERG channel inhibitor from the scorpion Buthus eupeus. *J. Biol. Chem.* **276**, 9868–76.

[14] Batista, C.V.F., Gómez-Lagunas, F., Lucas, S. and Possani, L.D. (2000) Tc1, from *Tityus cambridgei*, is the first member of a new sub-family of scorpion toxin that blocks K$^+$-channels. *FEBS Letters* **486**, 117–20.

[15] Verdonck, F., Bosteels, S., Desmet, J., Moerman, L., Noppe, W., Willems, J., Tytgat, J. and van der Walt, J. (2000) A novel class of pore-forming peptides in the venom of *Parabuthus schlechteri* Purcell (Scorpions: Buthidae), *Cimbebasia* **16**, 247–60.

[16] Torres-Larios, A., Gurrola, G.B., Zamudio, F.Z. and Possani, L.D. (2000) Hadrurin, a new antimicrobial peptide from the venom of the scorpion *Hadrurus aztecus*. *Eur. J. Biochem.* **267**, 5023–31.

[17] Wang, C-Y., Tan, Z-Y., Chen, B., Zhao, Z.Q. and Ji, Y-H. (2000) Antihyperalgesia effect of BmK IT2, a depressant insect-selective scorpion toxin in rat by peripheral administration. *Brain Res. Bull.* **53**, 335–8.

[18] Delepierre, M., Prochnicka-Chalufour, A., Boisbouvier, J. and Possani, L.D. (1999) Pi7, an orphan peptide isolated from the scorpion *Pandinus imperator*: [1]N-NMR analysis using a nano-nmr probe *Biochem.* **38**, 16756–65.

[19] Miller, C. (1995) The charybdotoxin family of K$^+$-channel-blocking peptides. *Neuron* **15**, 5–10.

[20] Selisko, B., García, C., Becerril, B., Gomez-Lagunas, F., Garay, R. and Possani, L.D. (1998) Cobatoxins 1 and 2 from *Centruroides noxius* Hoffmann constitute a new subfamily of potassium-channel-blocking scorpion toxins. *Eur. J. Biochem.* **254**, 468–79.

[21] Pisciotta, M., Coronas, F.I., Bloch, C., Prestipino, G. and Possani, L.D. (2000) Fast K$^+$ currents from cerebellum granular cells are completely blocked by a peptide from *Androctonus australis* Garzoni scorpion venom. *Biochim. Biophys. Acta* **1468**, 203–12.

[22] Holaday, S.K., Martin, B.M., Fletcher, P.L. and Krishna, N.R. (2000) NMR solution structure of butantoxin. *Arch. Biochem. & Biophys.* **379**, 18–27.

[23] Fontecilla-Camps, J-C., Almassy, R.J., Suddath, F.L., Watt, D.D. and Bugg, C. (1980) Three-dimensional structure of a protein from scorpion venom: a new structural class of neurotoxins. *Proc. Natl. Acad. Sci. (USA)* **77**, 6496–500.

[24] Landon, C., Sodano, P., Hetru, C., Hoffmann, J. and Ptak, M. (1997) Solution structure of drosomycin, the first inducible antifungal protein from insects. *Protein Sci.* **6**, 1878–84.

[25] Fant, F., Vranken, W., Broekaert, W. and Borremans, F. (1998) Determination of the three-dimensional solution structure of *Raphanus sativus* antifungal protein 1 by ^{1}HNMR. *J. Mol. Biol.* **279**, 257–70.

[26] He, X.L., Li, H.M., Zeng, Z.H., Liu, X.Q., Wang, M. and Wang, D.C. (1999) Crystal structures of two α-like scorpion toxins: non-proline cis peptide bonds and implications for new binding site selectivity on the sodium channel. *J. Mol. Biol.* **292**, 125–35.

[27] Park, C.S. and Miller, C. (1992) Interaction of charybdotoxin with permeant ions inside the pore of a K$^+$-channel. *Neuron* **9**, 307–13.

[28] Martinez, F., Muñoz-Garay, C., Gurrola, G., Darszon, A., Possani, L.D. and Becerril, B. (1998) Site directed mutants of Noxiustoxin reveal specific interactions with potassium channels *FEBS Letters* **429**, 381–4.

[29] Gómez-Lagunas, F., Olamendi-Portugal, T., Zamudio, F.Z. and Possani, L.D. (1996) Two novel toxins from the venom of the scorpion *Pandinus imperator* show that the N-terminal amino acid sequence is important for their affinities toward the Shaker B K$^+$-channels. *J. Membr. Biol.* **152**, 49–56.

[30] Péter, M. Jr, Varga, Z., Gáspár, R. Jr, Damjanovich, S., Horjales, E., Possani, L.D. and Panyi, G. (2001) Effects of toxins Pi2 and Pi3 on human T lymphocyte Kv1.3 channels: the role of Glu7 and Lys24. *J. Membr. Biol.* **179**, 13–25.

[31] Dauplais, M., Lecoq, A., Song, J., Cotton, J., Jamin, N., Gilquin, B., Roumestand, C., Vita, C., de Medeiros, C.L., Rowan, E.G., Harvey, A.L. and Ménez, A. (1997) On the convergent evolution of animal toxins. Conservation of a diad of functional residues in potassium channel-blocking toxins with unrelated structures. *J. Biol. Chem.* **272**, 4302–9.

[32] Darbon, H., Weber, C. and Braun, W. (1991) Two-dimensional ^{1}H nuclear magnetic resonance study of AaH IT, an anti-insect toxin from the scorpion Androctonus austalis Hector. Sequential resonance assignments and folding of the polypeptide chain. *Biochem.* **30**, 1836–45.

[33] Oren, D.A., Froy, O., Amit, E., Kleinberger-Doron, N., Gurevitz, M. and Shaanan, B. (1998) An excitatory scorpion toxin with a distinctive feature: an additional α helix at the C terminus and its implication for interaction with insect sodium channels. *Structure* **6**, 1095–103.

[34] Tugarinov, V., Kustanovich, I., Zilberberg, N., Gurevitz, M. and Anglister, J. (1997) Solution structures of a highly insecticidal recombinant scorpion α-toxin and a mutant with increased activity. *Biochem.* **36**, 2414–24.

[35] Carbone, E., Prestipino, G., Spadavecchia, L., Franciolini, F. and Possani, L.D. (1987) Blocking of the squid axon K$^+$ channel by Noxiustoxin: a toxin from the venom of the scorpion *Centruroides noxius*. *Pfluegers Arch. (Eur. J. Physiol.)* **408**, 423–31.

[36] Vaca, L., Gurrola, G.B., Possani, L.D. and Kunze, D.L. (1993) Blockade of an endothelial K$_{Ca}$ channel with synthetic peptides corresponding to the amino acid sequence of Noxiustoxin: a K$^+$ channel blocker. *J. Membr. Biol.* **134**, 123–29.

[37] Dai, L., Wu, J.J., Gu, Y.H., Lan, Z.D., Ling, M.H. and Chi, C.W. (2000) Genomic organization of three novel toxins from the scorpion *Buthus martensi* Karsch that are active on potassium channels. *Biochem. J.* **346**, 805–9.

[38] Legros, C., Bougis, P.E. and Martin-Eauclaire, M-F. (1997) Genomic organization of the KTX$_2$ gene, encoding a 'short' scorpion active on K$^+$ channels. *FEBS Lett.* **402**, 45–9.

[39] Froy, O., Sagiv, T., Poreh, M., Urbach, D., Zilberberg, N. and Gurevitz, M. (1999) Dynamic diversification from a putative common ancestor of scorpion toxins affecting sodium, potassium and chloride channels. *J. Mol. Evol.* **48**, 187–96.

[40] Wu, J.J., Dai, L., Lan, Z.D. and Chi, C.W. (2000) The gene cloning and sequencing of Bm-12, a chlorotoxin-like peptide from the scorpion *Buthus martensi* Karsch. *Toxicon* **38**, 661–8.

[41] Thompson, J.D., Gibson, T.J., Plewinak, F., Jeanmougin, F., Higins, D.G. (1997) The CLUSTAL_X window interface; flexible strategies for multiple sequence alignment aided by quality analysis tools. *Nucleic Acids Res.* **25**, 4876–82.

[42] Felsenstein, J. (1995) PHYLIP (Phylogeny Interface Package) version 3.57c., Department of Genetics, University of Washington, Seattle USA.

[43] Zang, J. and Nei, M. (1997) Accuracies of ancestral amino acid sequences inferred by parsimony, likelihood, and distance methods. *J. Mol. Evol.* **4**, S129–46.

[44] Meki, A., Mansuelle, P., Laraba-Djebari, F., Oughideni, R., Rochat, H., Martin-Eauclaire, M.F. (2000) KTX3, the kaliotoxin from *Buthus occitanus tunetanus* scorpion venom: one of an extensive family of peptidyl ligands of potassium channels. *Toxicon* **38**, 105–11.

[45] Meera, P., Wallner, M. and Toro, L. (2000) A neuronal β subunit (KCNMB4) makes the large conductance, voltage-and Ca^{2+}-activated K^+ channel resistant to charybdotoxin and iberiotoxin. *Proc. Natl. Acad. Sci. USA* **97**, 5562–7.

12 Scorpion Toxins Differentiating among Neuronal Sodium Channel Subtypes: Nature's Guide for Design of Selective Drugs

DALIA GORDON*, NICOLAS GILLES, DANIEL BERTRAND, JORDI MOLGO, GRAHAM M. NICHOLSON, MARTIN P. SAUVIAT, EVELYNE BENOIT, IRIS SHICHOR, ILANA LOTAN, MICHAEL GUREVITZ, ROLAND G. KALLEN and STEFAN H. HEINEMANN

12.1 INTRODUCTION

The need for selective drugs

Many drugs used in medicine and insecticides used in pest control act as modifiers of voltage-gated sodium channels, which are essential for cardiac, muscular and nervous system activity. Unfortunately, most modifiers are rather nonselective as they are unable to distinguish between channel subtypes throughout the body and among species, probably due to the general conservation of sodium channel structure in the animal kingdom. The lack of specificity of drugs, such as those controlling epilepsy, pain sensation, and cardiac arrhythmias, often renders adverse side effects. Likewise, the present generation of insecticides (e.g. pyrethroids) shows little preference for insect over noninsect sodium channels, posing risks to humans and other animals. However, sodium channels do show differential sensitivities to various natural toxins (e.g. scorpion, cone snail, spider and sea anemone neurotoxins) indicating that important structural differences exist, which might be exploited in the design of highly specific drugs and insecticides. During recent years, we have characterised a number of scorpion toxins able to differentiate between distinctive sodium channel subtypes in various animals (mammals and insects) and between different mammalian neuronal channels [1–7]. This has established an experimental basis for elucidating structural elements in both toxin ligands and in sodium channel receptor sites that are involved in this specificity (see Chapter 13 by Gurevitz *et al.*). Such

---

* Corresponding author

Perspectives in Molecular Toxinology
Edited by A Ménez © 2002 John Wiley & Sons, Ltd

information not only may broaden our knowledge about structural motifs that affect sodium channel function, but also provide new insights and strategies for development of clinically relevant drugs and a new generation of insecticides.

12.1.1 MAMMALIAN EXCITABLE TISSUES DISPLAY DIFFERENT SODIUM CHANNEL SUBTYPES

Vertebrate and invertebrate voltage-gated sodium channels are made of one large, pore-forming α-subunit (~ 260 kDa) which provides a functional channel upon heterologous expression with conductance properties much like those of the respective native channel [1, 8, 9]. The α-subunit consists of four repeat domains (D1–D4), arranged around a central pore. Each domain contains six transmembrane segments (S1–S6) and a membrane-associated re-entrant loop (SS1–SS2; see Figures 12.1 and 12.2). Movement of the positively charged voltage-sensing S4 segments activates the channel in response to membrane depolarisation [1, 8, 10]. Multiple voltage-gated sodium channel subtypes with different kinetic properties and pharmacological features have been described in various excitable tissues [8, 9]. This variability is generated by a multigene family encoding the α-subunit of these channels, some of which are broadly expressed in the central nervous system (CNS: brain subtypes, rBI-III (rNa$_v$1.1–1.3), Na6 (rNa$_v$1.6, PN4) [11–13]. Conversely, the expression of other subtypes is restricted to specific tissues such as the peripheral nervous system (PN1, rNa$_v$1.7) [14], muscle (skeletal muscle SkM1 (μ1, rNa$_v$1.4) and cardiac, H1/SkM2 (rNa$_v$1.5) subtypes [15–18], and channels specific to sensory neurons such as dorsal root ganglia (DRGs), which are involved in transmission of pain signals (PN3/SNS, rNa$_v$1.8) [17–20], and SNS2/NaN (rNa$_v$1.9) [21, 22]. The PN3 and SNS2 sodium channels give rise to TTX-resistant sodium currents, which typify some pain-sensing neurons [19, 20], whereas the TTX-sensitive sodium currents in peripheral neurons are mediated by several sodium channel subtypes, mostly PN1, Na6 and brain I expressed in most DRG neurons [23]. Evidently, the different channel subtypes have specialised in mediating various physiological functions [1, 8, 13–15, 17, 23–28]. However, in the absence of ligands capable of discriminating between sodium channel subtypes, little is known about the distribution and function of distinct sodium channels in different tissues and sub-cellular regions (neural cell bodies, dendrites, axons and synapses).

12.1.2 SCORPION TOXINS AFFECTING SODIUM CHANNELS

Neurotoxins from plants, animals and microorganisms that target sodium channels bind to at least seven topologically distinct receptor sites on the α-subunit and affect either gating properties or ion permeation through the pore (Figure 12.1; reviewed by [1, 2]). Polypeptide toxins from various animals, such

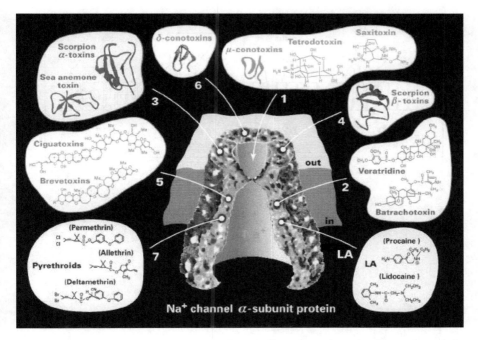

Figure 12.1 A schematic representation of the sodium channel α-subunit protein and the major groups of ligands that modify its function. Receptor sites of each ligands group are indicated by a number and the putative localisation in the channel is shown. Receptors of the lipid-soluble sodium channel modifiers, such as alkaloid toxins (veratridine, batrachotoxin – receptor site-2), marine cyclic polyether toxins that bind to receptor site-5 (e.g. brevetoxins and ciguatoxins), synthetic pyrethroids (e.g. permethrin, deltamethrin – receptor site-7) and local anesthetics (LA), such as lidocaine, are suggested to be within the hydrophobic protein core. Receptor sites of water-soluble polypeptide toxins, such as scorpion α- and β-toxins (that bind to receptor sites 3 and 4, respectively), and δ-conotoxins (receptor site-6), are located at the extracellular side of the channel protein. The external vestibule of the ion conducting pore, at the centre of the protein, contains receptor site-1, which binds the sodium channel blockers μ-conotoxins, tetrodotoxin (TTX) and saxitoxin. For details see Catterall, 1992 and Gordon, 1997 [1, 4, 10]

as scorpions, cone snails, spiders and sea anemones, modify voltage-dependent activation and/or inactivation of sodium channels by binding to at least three distinct receptor sites on their extra-cellular surface (sites 3, 4 and 6, see Figure 12.1; [1, 2]). Scorpion toxins are divided into α and β classes according to their mode of action and binding properties. Alpha toxins inhibit sodium current inactivation upon binding to receptor site-3, which is mapped to the extra-cellular loop S3–S4 in Domain-4 (D4) of the sodium channel (Figure 12.2; [10, 29]). Beta toxins (2) shift the voltage dependence of activation to a more negative membrane potential [30–35] upon binding to receptor site-4 [36, 37], which is assigned to the S3–S4 external loop in D2 of sodium channels [38–40]; see Figure 12.2).

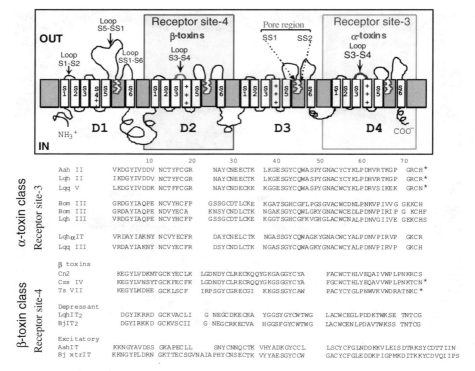

Figure 12.2 **Upper panel** Topology of the pore-forming α-subunit of Na⁺ channels and toxin receptor sites 3 and 4. Segments SS1 and SS2 form the pore region, around which the four homologous domains (D1–D4) assemble. Loop S3–S4 of D2, is involved in β-toxin binding to site-4 [38]. The loop S3–S4 in D4 shows α-toxin receptor site-3 [29]

Lower panel Sequence alignment of several scorpion toxins affecting Na⁺ channels (based on [2, 42, 69]). B, β-strand; T, turn; H, α-helix. Aah, *Androctonus australis hector*; Bj, *Buthotus judaicus*; Bom, *Buthus occitanus mardochei*: Lqh, *Leiurus quinquestriatus hebraeus*; Lqq, *L. q. quinquestriatus*; Cn, *Centruroides noxius*; Css, *C. suffusus suffusus*; Ts, *Tityus serrulatus*. For references, see [2, 42, 69]

12.2 SCORPION α-TOXINS AND RECEPTOR SITE-3 ON SODIUM CHANNEL SUBTYPES

A variety of peptide toxins from sea anemones, spiders and scorpions, that share little sequence homology, compete for binding to voltage-gated sodium channels at similar or overlapping sites named receptor site-3 [1, 10, 41]. All these toxins prolong the sodium current under voltage-clamp conditions by slowing its fast inactivation phase in different excitable membranes of insects and mammals [10, 42–46]. Despite the commonality in mode of action and site of recognition (receptor site-3), scorpion α-toxins show vast differences in preference for insects and mammals. Accordingly, they have been divided into three major groups:

classical α-toxins that are highly active in the mammalian brain (e.g. *Androctonus australis hector* (Aah) and *Leiurus quinquestriatus hebraeus* (Lqh) toxins, Aah-II and Lqh-II); α-toxins that are very active in insects (e.g. LqhαIT and the *L. q. quinquestriatus* toxin, Lqq-III); and α-like toxins that are active in both mammalian and insect CNS (e.g. Lqh-III, and the *Buthus occitanus mardochei* (Bom) toxins, Bom-III and Bom-IV; see Figure 12.2 and Table 12.1). Receptor site-3 is homologous, yet non-identical in insect and rat brain sodium channels. This is inferred from the ability of all site-3 toxins to compete for the α-toxins binding-site in insect sodium channels. However, their binding affinity for insect and mammalian sodium channels varies greatly [6, 29, 41–51].

12.2.1   LOCALISATION OF RECEPTOR SITE-3 ON THE SODIUM CHANNEL

Current information about receptor site-3 on the sodium channel is still rudimentary. Two distinct extra-cellular regions of the sodium channel α-subunit have been implicated in scorpion α-toxin binding. The S5–SS1 loop in Domain-1 (D1), identified by photo-affinity labelling, and the S5–S6 loops of D1 and D4, highlighted by inhibition of toxin binding with site-directed antibodies ([52, 53]; see Figure 12.2). Evidence for the involvement of the S3–S4 extracellular region of D4 in toxin binding was provided by site-specific mutagenesis ([28, 54]; Figure 12.3). Inversion of the negative charge of glutamate 1613 (E1613R) of this region in the rat brain channel, rBIIA, reduced 100-fold the binding affinity of a scorpion α-toxin [28]. Binding of sea anemone *Anthopleurin* toxins, ApA and ApB, to the cardiac channel, hH1 [17], and chimeric hH1 bearing D4 of rSkM1, pointed out an aspartate residue involved in binding of site-3 toxins. This Asp is located at a site homologous to Glu1613 in S3–S4 in D4 of rBIIA [54]. It is important to note that only the affinity of toxin binding has changed whereas other channel properties have not been affected by this mutagenesis. Overall, site-3 on the extracellular surface of these channels is the region interacting with various toxins and, therefore, any attempt to rationalise selective drug design at this site requires details about its spatial organisation in channel subtypes.

12.3   SELECTIVE INTERACTION OF α-TOXINS WITH DISTINCT MAMMALIAN SODIUM CHANNEL SUBTYPES

12.3.1   DIFFERENCES IN TOXICITY OF α-TOXINS IN CNS AND PERIPHERY

Intracerebroventricular (*icv*) injection of α-toxins in mice revealed vast differences in toxicity between α-toxins (Table 12.1). The toxins that are very active on insects were ∼2000 times less toxic to mice compared to the classical α-toxins

Table 12.1 Activity of some scorpion α-toxins on mammals and insects.

α-Toxin Group	Toxin	Activity on mammals			Activity on insect	
		LD$_{50}$ Mice (i.c.v.)[e] pmol/g	LD$_{50}$ Mice (s.c)[d] pmol/g	Ki, nM Rat brain[g] (^{125}I-α-Toxin)	LD$_{50}$ Cockroach[f] pmol/g	Ki, nM Cockroach[k] (^{125}I-LqhαIT)
Classical α-Toxins	Aah-II	0.004[a]	1.7[a]	0.3[a]	897[c]	59[c]
	Lqh-II	0.014[b]	8.8[g]	0.3±0.1[b]	280[b]	45[b]
	Lqq-V	0.018[a]	3.4[a]	1.0[a]	2317[c]	120[c]
	Aah-I	0.071[a]	2.4[a]	4.5[a]	276[c]	160[c]
α-Like	Lqh-III	0.36[b]	23[b]	2004±500[i]	28[b]	0.4[b]
	Bom-III	0.16[a]	19.7[a]	≫10000[a,h]	52.6[c]	29[c]
	Bom-IV	0.16[a]	5.5[a]	≫10000[a,h]	19.7[c]	5[c]
α-Insect	LqhαIT	7.9[b]	8.3[c]	2760±470[i]	2.5[c]	0.03[c]
	Lqq-III	7.9[a]	6.9[a]	700[a]	8.3[c]	0.03[c]

a see [42] for details; *b* [55]; *c* [43]; *d* subcutaneous injection to mice; *e* Intracerebroventricular injection to 20 g mouse; *f* Injection to abdominal segments of cockroach *Blatella germanica*, 50 ± 2 mg body weight; *g* competition for ^{125}I-Aah-II or ^{125}I-Lqh-II binding in rat brain synaptosomes; *h* no significant inhibition was detected at 10 μM toxin; *i* [6]; *k* competition for ^{125}I-LqhαIT binding to cockroach (*Periplaneta americana*) neuronal membranes

S3-S4-Domain-4

```
---S3---|-------- Loop region --------|---S4---
             *         *

rSkM1    S  D  L  I  Q  K  Y  -  -  -  F  V  S  P  T  L  F  R

rBIIA    A  E  L  I  E  K  Y  -  -  -  F  V  S  P  T  L  F  R

rBI      A  E  L  I  E  K  Y  -  -  -  F  V  S  P  T  L  F  R

rBIII    A  E  L  I  E  K  Y  -  -  -  F  V  S  P  T  L  F  R

rPN1     A  E  N  I  E  K  Y  -  -  -  F  V  S  P  T  L  F  R

hH1      S  D  I  I  Q  K  Y  -  -  -  F  F  S  P  T  L  F  R

rPN3     F  S  A  I  L  K  S  L  E  N  Y  F  S  P  T  L  F  R

Para     S  D  I  I  E  K  Y  -  -  -  F  V  S  P  T  L  L  R
```

Figure 12.3 Comparison of amino acid sequences within the S3–S4 extracellular loop of Domain-4 in several mammalian and insect sodium channel subtypes. r designates rat; h stands for human; Para is from *Drosophila*. Asterisks indicate Asp1428 and Lys1432 in rSkM1, which are important for LqhαIT binding. Mutated residues are underlined (see Table 12.2). For references of sequences see [9]

Aah-II and Lqh-II (Table 12.1, Figure 12.2). The α-like toxins Lqh-III, Bom-III and Bom-IV, were toxic in mouse brain (Table 12.1) but neither bound, nor competed with classical α-toxins for the rat brain synaptosomal site [6, 43, 44, 55]. This peculiarity could have suggested a binding site distinct from the receptor site-3 that is recognised by classical α-toxins, or different targets in the CNS. An additional peculiar effect of α-toxins in mammals is their similar toxicity when administered subcutaneously (*s.c.*) to mice, in contrast to the differences noted upon *icv* injection (Table 12.1). To gain insight into this complex issue, we compared the binding and effects of the α-like toxin Lqh-III and the classical α-toxin, Lqh-II, in central and peripheral tissues and on various sodium channel subtypes.

12.3.2 α-TOXINS DIFFERENTIATE BETWEEN NEURONAL SODIUM CHANNEL SUBTYPES IN CNS

Despite its prominent toxicity to both insects and mice, the α-like toxin, Lqh-III, binds with high affinity to cockroach neuronal membranes [6, 45, 51] but competes only at high concentration for the high-affinity binding sites of Lqh-II in rat brain synaptosomes (Figure 12.4A; [6, 56]). This result suggests a very low affinity of Lqh-III for the Lqh-II binding site in rat brain [6]. Furthermore, Lqh-III has no effect on either sodium channels of cultured embryonic chick central neurons or rat brain sodium channel subtype II (rBII) expressed in

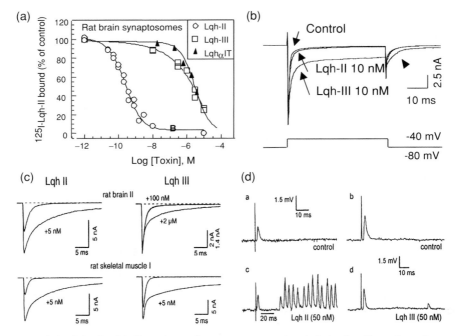

Figure 12.4 Selective binding of α-toxins to rat neuronal and skeletal sodium channels.

(a) Competition for the classical α-toxin binding, ^{125}I-Lqh-II. Rat brain synaptosomes were incubated with 0.08 nM ^{125}I-Lqh-II with increasing concentrations of Lqh-II, Lqh-III or LqhαIT. Inhibition of specific binding is presented as a percentage of the control in the absence of competitors. The Ki values are: Lqh-II, 0.3 ± 0.1 nM; Lqh-III, 2.0 ± 0.5 μM and LqhαIT, 2.8 ± 0.5 μM (for details see [6, 56]).

(b) Effects of Lqh-II and Lqh-III in rat CA1 pyramidal cells recorded in an acute hippocampal slice. Lqh-III, but not Lqh-II, strongly inhibit sodium current inactivation under voltage clamp conditions. See [6] for details.

(c) Effects of Lqh-II and Lqh-III on rat brain II (rBII) and skeletal muscle, rSkM1, sodium channel subtypes expressed in mammalian HEK293 cells. Superimposed current traces through rBII (top) and rSkM1 (bottom) under voltage clamp conditions. In rBII channels, Lqh-II inhibits sodium current inactivation at 5 nM, whereas Lqh-III had a only a partial effect at 2 μM. In contrast, both toxins affected the currents in skeletal muscle channels. See details in [56].

(d) Effect of Lqh-II and Lqh-III on neuromuscular transmission. Endplate potentials (EPPs) recorded intra-cellularly on two different nerve-hemidiaphragm muscles before (a,b) and 30 min after addition of 50 nM Lqh-II (c) or Lqh-III (d). Lqh-II induced repetitive EPPs due to increased neurotransmitter release from the motor nerve terminal, whereas Lqh-III had no effect on the nerve but affected the muscle membrane directly [56]

Xenopus oocytes [6] or in mammalian cells [56], whereas site-3 toxins, such as the classical α-toxin, Lqh-II, and the sea anemone toxin, ATXII, are very active on these targets ([6, 56]; Figure 12.4C). Characterisation of Lqh-III effects on rat brain slices provided an unexpected explanation for the peculiar behaviour of α-like toxins. Lqh-III strongly inhibited sodium current inactivation of rat CA1 pyramidal neurons in acute hippocampal slices, whereas Lqh-II had almost no effect (Figure 12.4B; [6]). These results have demonstrated that Lqh-II and Lqh-III (representing scorpion classical α-toxins and α-like toxins) discriminate between sodium channel subtypes in rat brain.

12.3.3 DIFFERENCE IN SUB-CELLULAR LOCALISATION OF LQH-II AND LQH-III SODIUM CHANNEL TARGETS

The riddle of α-like toxins that kill mice upon *icv* injection but are incapable of specific binding to rat brain synaptosomes may now be explained [6, 56]. Rat brain synaptosomes contain nerve terminal membranes but are usually devoid of contamination by cell bodies [57]. The main sodium channel subtype in rat brain synaptosomes is II/IIA (rBII/IIA; about 80%) and another 20% is occupied by rBI [12, 58]. It has been shown that the α-like toxin, Bom-IV, is incapable of specific binding to rat caudal brain regions and spinal cord membranes [44], which contain a high level of rBI [25, 58, 59]. Thus, neither rBII nor, most likely, rBI are targets for α-like toxins in rat CNS. Experiments with rat CA1 pyramidal cells have indicated, however, that at least one sodium channel subtype that is sensitive to Lqh-III but not to Lqh-II must exist in the brain [6]. Therefore, α-like toxins affect sodium channel subtypes that are most likely present exclusively on neuronal somata and are absent in synaptosomes. The sodium channel target for Lqh-II, i.e. rBII/IIA, has been localised in axons, but not the somata of brain neurons [60]. Inhibition of sodium current inactivation by α-like toxins using patch clamp analysis of sodium currents in rat CA1 pyramidal (Figure 12.4B, [6]) and cultured cerebellar granule neurons [43] provided further indication for the localisation of sodium channels that were targets for these toxins. Due to space limitation of the clamp, this method records only the electrophysiological properties of the soma and proximal processes thus suggesting that the target sodium channels for α-like toxins exist on the somata.

Neurons of the mammalian brain express multiple subtypes of sodium channels encoded by at least six distinct genes [8, 12, 13, 61]. These channels differ in sub-cellular localisation, developmental pattern of expression and abundance in different brain regions [13, 23–26, 58–61]. Although these variations may suggest functional differences, no evidence for distinctive physiological roles has been adduced for any of the sodium channel subtypes. However, due to their variable sensitivity to α-classical and α-like toxins, we were able to discriminate pharmacologically between sodium channel subtypes in sub-cellular regions of mammalian brain neurons. Thus, new tools are now available for studying the functional role and distribution of these channels.

12.3.4   EFFECTS OF SCORPION α-TOXINS ON PERIPHERAL SODIUM CHANNELS

The similar toxicity of Lqh-II and Lqh-III upon peripheral injection was resolved by examination of the toxins' effects on several peripheral preparations. The neuromuscular junction is useful to examine separately the effects of a toxin on motor nerves and on skeletal muscle. Fifty nanomolar Lqh-II applied onto the nerve terminal elicited repetitive end plate potentials (EPPs) in response to a single nerve stimulus (Figure 12.4D; [56]). Repetitive EPPs may result from inhibition of sodium channel inactivation at both the nodes of Ranvier and nerve terminals, which is expected to cause membrane depolarisation and repetitive firing of nerve terminals, as was previously described with another site-3 toxin, ATXII [62]. Conversely, Lqh-III had no apparent effect on the motor nerve even at much higher concentrations (up to 1 μM, Figure 12.4D; [56]), but did affect the muscle, which showed tetanic contraction in the presence of 50 nM toxin. This result demonstrated how Lqh-III differentiates between sodium channels of the muscle versus those in peripheral motor nerves. On the one hand, Lqh-II and Lqh-III have a similar effect on native mammalian tissues of skeletal and cardiac muscles as well as on recombinant rSkM1 and hH1 channels expressed in mammalian cells (Figure 12.4C; [56, 63, 64; unpublished results]). On the other hand, although both toxins inhibit tetrodotoxin-sensitive sodium current inactivation in dorsal root ganglion (DRG) neurons, they affect the currents in different ways and Lqh-II is active at lower concentrations than Lqh-III. The different effects induced by Lqh-III, such as the larger shift in activation and more profound slowing of inactivation compared to Lqh-II, suggests that the two toxins do not affect identical sodium channel subtypes in DRGs [56]. This difference may be explained by heterogeneity of sodium channels in DRG neurons. Indeed, such heterogeneity has been reported at the level of multiple transcripts encoding sodium channel proteins that produce various Na currents, as well as by demonstrating TTX-sensitive versus TTX-resistant channels [14, 23, 61, 65, 66]. We have also shown that scorpion α-toxins affect the TTX-sensitive but not the TTX-resistant sodium channels in DRG neurons [56]. On the basis of these results, it is not yet possible to define the sodium channel subtype that is the main target for modification by the α-like toxin Lqh-III in mammalian brain and PNS, but rBII and perhaps also rBI can most likely be excluded [6, 56]. Thus, our study elucidates for the first time the differences in neuronal targets of scorpion α-toxins in mammalians. The similarity in peripheral toxicity is attributed to the similar effects of all α-toxins on sodium channels in skeletal (and probably also cardiac) muscles.

12.3.5   SODIUM CHANNELS OF SIMILAR TISSUES DIFFER IN VARIOUS VERTEBRATES

The sensitivity of sodium channels for scorpion α-toxins differs in motor nerves, central neurons or skeletal and cardiac muscles of rat, chick and frog. For example, Lqh-III is inactive on rat motor nerves but is very active on the frog myelinated axons at equivalent concentrations [56, 67; unpublished results]. Likewise, the neuronal sodium channel subtype of rat brain, which is sensitive to Lqh-III but not to Lqh-II, is absent from chick brain [6]. Furthermore, Lqh-II affects differently sodium currents produced by cardiac sodium channels in mammals and in frogs (unpublished results). These findings correlate with the absence of channel subtypes I and II in frog and chick brains, as was suggested by immuno-precipitation using subtype-specific antibodies [58]. Thus, not only do scorpion toxins discern between sodium channel subtypes in the same animal, but they also differentiate between sodium channels of the same tissue but in another organism. It would be of interest to examine whether these pharmacological differences are attributed to variations in the corresponding genes or to post-transcriptional modifications of their products.

12.4   DO THE VARIOUS α-TOXINS INTERACT IDENTICALLY WITH RECEPTOR SITE-3?

In contrast to neuronal sodium channel subtypes, the rSkM1 sodium channel is equally sensitive to the three different scorpion α-toxins, Lqh-II, Lqh-III and LqhαIT [63, 68]. However, these toxins have no effect on the TTX-resistant sodium channels in cultured DRG neurons [56], implying that the SNS/PN3 channels [19, 20] are insensitive to scorpion α-toxins. This result is in concert with the salient difference in the sequence of the S3–S4 loop in D4 of PN3 compared to all other sodium channel subtypes (see Figure 12.3), and provides further indirect evidence for the importance of the S3–S4 loop in D4 for scorpion α-toxin binding [28]. Direct evidence was then provided by mutagenesis of the S3–S4 external loop in D4 of the rSkM1 channel that was expressed in mammalian cells (Figure 12.4C; [68]). This approach highlighted two residues, Asp1428 and Lys1432, of prime importance for toxin binding. Surprisingly, despite the similar affinity of the three toxins, Lqh-II, Lqh-III, and LqhαIT, for rSkM1, each substitution differentially affected the binding affinity of each of the toxins (see Table 12.2; [68]). The properties of toxin binding to the modified channels were studied by quantifying the magnitude of the fast and slow exponential current decay kinetics, attributed to free and toxin-bound channels, respectively. The dissociation constants (K_d) obtained for each toxin with the rSkM1 channel mutants are presented in Table 12.2 [68].

   The results imply that the three toxins bind overlapping sites that include the S3–S4 loop of D4 and that Asp1428 within this site is most crucial for their

Table 12.2 Apparent K_d (nM) of scorpion α-toxins with rSkM1 expressed in HEK293 cells at -120 mV holding potential (68). n = number of cells.

rSkM1 Mutants	Lqh-II K_d (nM)	Change in K_d (mut/wt)	Lqh-III K_d (nM)	Change in K_d (mut/wt)	LqhαIT K_d (nM)	Change in K_d (mut/wt)
WT	5.6 ± 0.8 (n = 10)	1	4.6 ± 0.7 (n = 7)	1	4.8 ± 0.9 (n = 7)	1
D1428R	591 ± 29 (n = 4)	**106**	2370 ± 173 (n = 4)	**515**	4176 ± 931 (n = 3)	**865**
D1428N	37.4 ± 5 (n = 7)	**6.7**	481 ± 15 (n = 5)	**104.6**	2955 ± 745 (n = 3)	**612**
Q1431E	13.3 ± 1.1 (n = 6)	2.4	7.0 ± 0.8 (n = 7)	1.5	13.7 ± 2.6 (n = 6)	2.8
K1432N	19 ± 1.8 (n = 6)	3.4	4.6 ± 0.4 (n = 7)	1	208 ± 38 (n = 4)	**43**
S1436A	6.5 ± 0.6 (n = 6)	1.2	4.8 ± 0.6 (n = 7)	1	7.2 ± 1.4 (n = 6)	1.5
L1439A	8.7 ± 1.2 (n = 6)	1.6	5.9 ± 1.1 (n = 6)	1.3	6.3 ± 1.4 (n = 6)	1.3
F1440L	9.3 ± 0.8 (n = 7)	1.7	5.2 ± 0.9 (n = 6)	1.1	4.6 ± 1 (n = 6)	1

Plate I Computer-generated comparison of the 2-D gels of *C. atrox* and *B. jararaca* Venoms. Digitised images of the 2-D gels were made and analysed with Compugen Z3 analysis software. The software analysed the images, identified spots in each image and determined which spots shared identical migration patterns. The spots from the *C. atrox* venom gel are represented in green and those from the *B. jararaca* venom gel are represented in magenta. Spots in both gels that were determined to have identical migration patterns are denoted in black

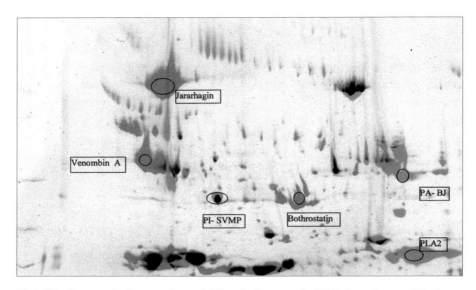

Plate II Computerised comparison of 2-D gels from two individual specimens of *Bothrops jararaca*. Digitised images of the 2-D gels were made and analysed with Compugen Z3 analysis software. The software analysed the images, identified spots in each image and determined which spots shared identical migration patterns. Digitised images in magenta are from 1st gel in upper left corner of Figure 5.5. Digitised images in green are from 2nd gel in upper left of Figure 5.5. Spots determined to be identical in both images are represented in black. Spots that were identified by LC/MS/MS sequence analysis are indicated on the image. Proteins identified were: PA-BJ, platelet-aggregating serine peptidase [23]; Jararhagin, hemorrhagic P-III SVMP [24]; PLA2, phospholipase A2, (BJ-PLA2) [25]; Venombin, (bothrombin) fibrinogen-clotting serine peptidase [26] [Reference numbers relate to Chapter 5.]

Plate III Different foraging strategies of two fish-hunting cone snails. The figure shows the behaviour elicited in two piscivorous *Conus* species by a fish introduced into an aquarium. Top: *Conus striatus*, the striated cone, has both its siphon and its proboscis extended. The siphon which detects the presence of fish by a chemosensory mechanism is the larger tube; the extended proboscis (lower, thinner tube) is where the harpoon is ejected, through which venom is injected into prey. The fish is tethered by the barbed, harpoon-like tooth and drawn into the mouth. Bottom: The geography cone, *Conus geographus* has its siphon extended and its rostrum (mouth) wide open with the proboscis contracted within the snail's mouth. The distended mouth is capable of engulfing many fish at once.
It has been postulated that harpoon-and-line snails like *Conus striatus* (top) approach fish hiding in crevices at night, harpoon them from a distance, and reel them in after the fish has been envenomated and immobilised. In contrast, net strategists like *Conus geographus* (bottom) probably crawl out at night, stalk schools of fish hiding in holes in the reef and try to bag multiple fish from the school in their rostrum nets. Once engulfed in the snail's mouth, the fish are harpooned through the snail's proboscis one-by-one and drawn further down into the digestive tract

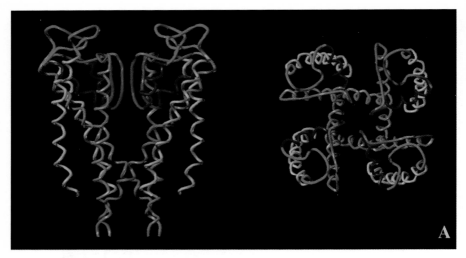

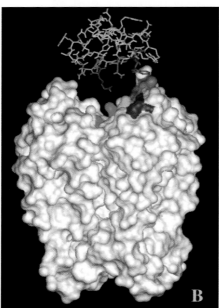

Plate IV Calculated model of the structure of a voltage-gated potassium channel.
A Backbone of a 3-D model of the Kv1.1 channel, derived from the crystallographic structure of the bacterial channel KcsA [14]. Transverse and top views of the structure are shown left and right. The filter of selectivity is coloured red, the pore helix is coloured blue and the turrets are coloured green.

B A transverse view of the channel oriented as in A (left) is shown with a sea anemone toxin (BgK) ready to plug in its binding site. The critical lysine of BgK protrudes toward the pore of the channel and is coloured red. The other coloured residues of the toxin interact with those of the channel that are similarly coloured [Reference numbers relate to Chapter 10.]

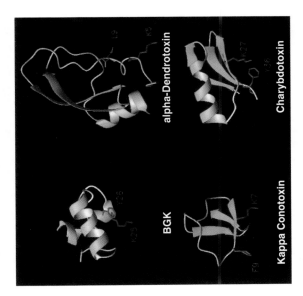

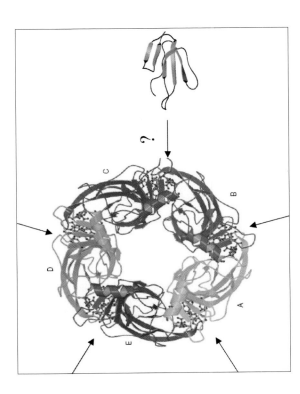

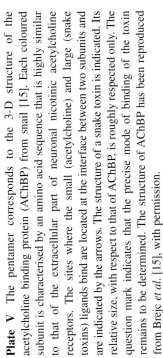

Plate V The pentamer corresponds to the 3-D structure of the acetylcholine binding protein (AChBP) from snail [15]. Each coloured subunit is characterised by an amino acid sequence that is highly similar to that of the extracellular part of neuronal nicotinic acetylcholine receptors. The sites where the small (acetylcholine) and large (snake toxins) ligands bind are located at the interface between two subunits and are indicated by the arrows. The structure of a snake toxin is indicated. Its relative size, with respect to that of AChBP, is roughly respected only. The question mark indicates that the precise mode of binding of the toxin remains to be determined. The structure of AChBP has been reproduced from Brejc *et al.* [15], with permission.

Plate VI The 3-D structure of four animal toxins that bind competitively to voltage-gated channels in rat brain [53]. BgK is a toxin from the sea anemone *Bunodosoma granulifera*. Its 3-D structure was solved by NMR [46]. α-dendrotoxin is from the snake *Dendroaspis angusticeps*. Its 3-D structure was solved by X-ray crystallography [15]. κ-conotoxin is from the cone snail *Conus purpurascens*. Its 3-D structure was solved by NMR [100, 101]. Charybdotoxin is from the scorpion *Leiurus quinquestriatus hebraeus*. Its 3-D structure was solved by NMR [64]. The critical residues that form a common dyad for the four toxins to bind competitively to Kv1 channels are coloured red (lysine) and blue (hydrophobic residue) [Reference numbers relate to Chapter 10.]

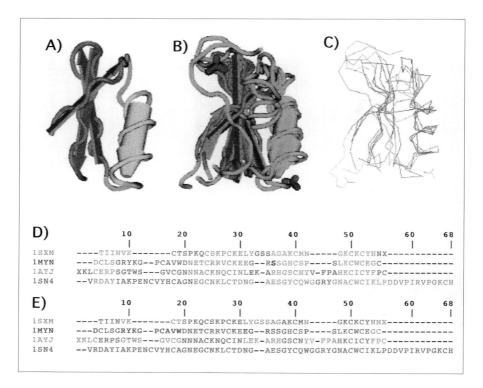

Plate VII   Common three-dimensional fold of scorpion toxins specific for K⁺-channels and related toxins

Superposition of overall folds of: Noxiustoxin, a K⁺-channel specific toxin from *Centruroides noxius*; Drosomycin, an antifungal protein from *Drosophila melanogaster*, protein 1, an antifungal protein from *Raphanus sativus* and BmkM4, a scorpion neurotoxin specific for Na⁺-channels from *Buthus martensii* (PDB accession numbers are: 1SXN, 1MYN, 1AYJ and 1SN4A, respectively).

A The α-helix segment is coloured blue, β-strand structures are shown in yellow and coils in green.

B Secondary structures in a three-dimensional alignment of NTx and related peptides. α-helix, β-strand and coil segments in the alignment structures are shown using the same colour code as in A.

C Structurally neighbouring residues of the aligned structures are coloured red. Residues with no neighbours are shown as follows; 1SXM: pink; 1MYN: dark blue; 1AYJ: brown; 1SN4: green.

D Amino acid alignment with residues shown according to the secondary structure colour code used in A.

E Amino acid alignment with residues shown according to the three-dimensional neighbours colour code used in C. The three-dimensional structure, alignment data and Cn3-D drawing program were obtained from the National Center of Biotechnology Information (http://www.ncbi.nlm.nih.gov/Structure)

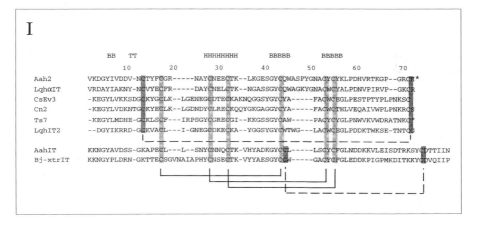

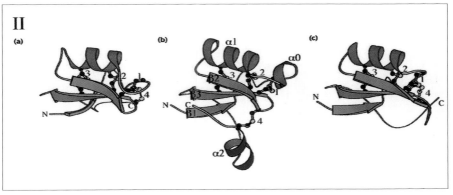

Plate VIII   Alignment of scorpion 'long' toxin representatives that affect sodium channels. I, Sequence alignment based on published data (15,20,30). Dashes indicate gaps. Asterisks designate α-amidation of carboxy-terminal residues. The conserved (solid lines) and the non-conserved (dashed line) disulphide bonds are depicted at the bottom. B, β-strand; T, turn; H, α-helix. Abbreviations: Aah, *Androctonus australis hector*; Bj, *Buthotus judaicus*; Lqh, *Leiurus quinquestriatus hebraeus*; Cn, *Centruroides noxius*; CsE, *Centruroides sculpturatus Ewing*; Ts, *Tityus serrulatus*. II, Overall three-dimensional carbon backbone of (a) CsEv3 [42], (b) Bj-xtrIT [53], and (c) Aah2 [43], viewed in the same orientation [Reference numbers relate to Chapter 13.]

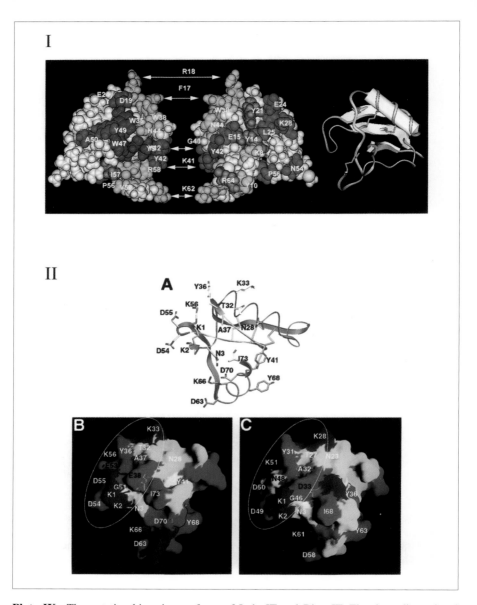

Plate IX The putative bioactive surfaces of LqhαIT and Bj-xtrIT. The three-dimensional models are based on the solution [46] and crystal [53] structures, respectively. I, LqhαIT with identical orientation of the carbon backbone (right) and the space-filling model (middle). Modified residues affecting bioactivity are in yellow and orange. Residues, whose substitution has not affected bioactivity are in green. Residues, which belong to the 'conserved hydrophobic surface' are in blue. II, Putative bioactive surface of scorpion excitatory toxins. **A** a ribbon diagram of Bj-xtrIT structure [53] with the side-chain residues indicated in panel B. **B** and **C** the putative bioactive surfaces of Bj-xtrIT and AahIT, respectively, highlighted by the ellipsoid (for details see [10]) [Reference numbers relate to Chapter 13.]

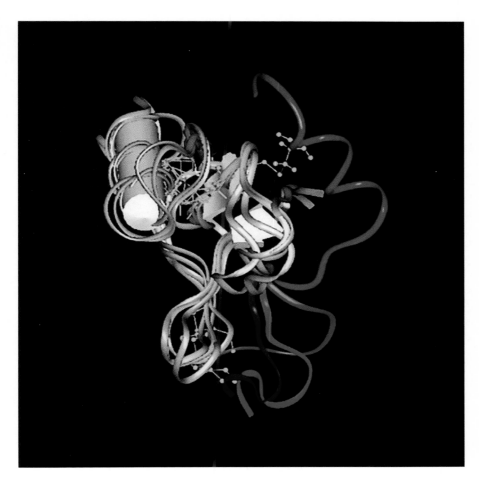

Plate X Superposition of the Cα backbone of scorpion toxins affecting sodium channels. Four toxins representing distinct pharmacological groups are delineated in ribbon: the alpha toxin, LqhαIT [46], the beta toxin Cn2 [63], and the excitatory toxin, Bj-xtrIT [53] have a blue, magenta, and red C-tails, respectively. Note: the last three residues could not be seen in the X-ray structure of Bj-xtrIT, but they cover, most likely, residue 38 in the centre of the bioactive surface [53]. In addition, a model of a depressant toxin, LqhIT2, is presented with a green C-tail (the model was constructed by alignment with Cn2 using the MODELLER and HOMOLOGY modules of Insight II, MSI, Cambridge, UK). The α-helix and β-strands are highlighted (barrel and arrows, respectively). The conserved three disulphide bonds are depicted in yellow lines, whereas the fourth disulphide bond is pinpointed in yellow balls and sticks [Reference numbers relate to Chapter 13.]

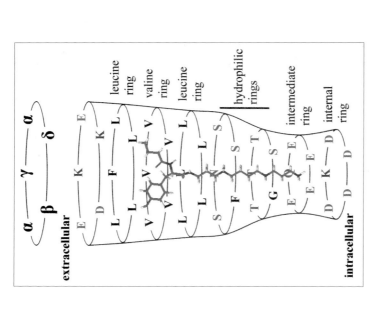

Plate XI Steric antagonism of nAChR. Cartoon of the immature mammalian skeletal muscle nAChR pore showing the rings of residues which form its lining and its internal and external mouths (Red = acidic, Blue = basic, Green = hydrophilic, Black = hydrophobic). PhTX-343 is superimposed showing a potential binding site in the pore of the ion channel where it sterically inhibits ion flow (Cyan = C, Blue = N, Red = O, Grey = H)

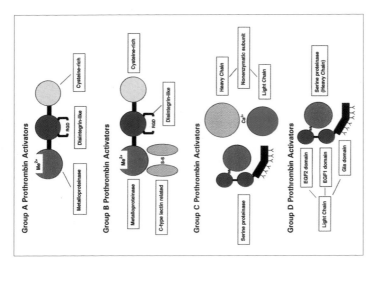

Plate XII Schematic representation of subunit structures of prothrombin activators from snake venoms

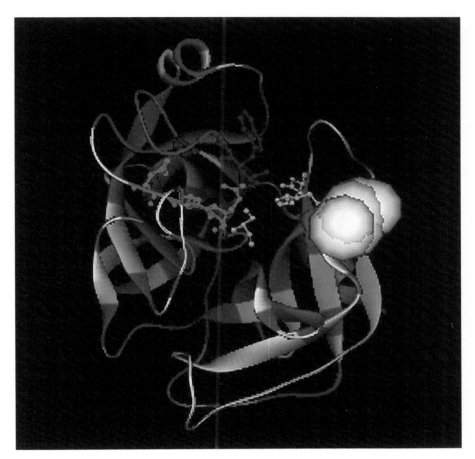

Plate XIII Molecular model of trocarin. Trocarin was modelled based on factor Xa structure (PDB file: 1XKA). Segments that are identical or conserved are shown in red or yellow, respectively. White segments indicate no structural similarity between trocarin and human factor Xa. Active site residues, His 42, Ser 185 and Asp 88 are shown in light blue. The substrate-binding pocket (Asp 179, Ala180, Cys 181, Gln 182) colored green and residues involved in forming the hydrogen-bonded structure antiparallel to the substrate (Ser 204, Trp 205, Gly 206, Gly 216) colored blue. A schematic representation of the carbohydrate moiety located at the lip to the entrance of the active site is shown as CPK model. The actual size of this moiety is expected to be at least five times the size shown

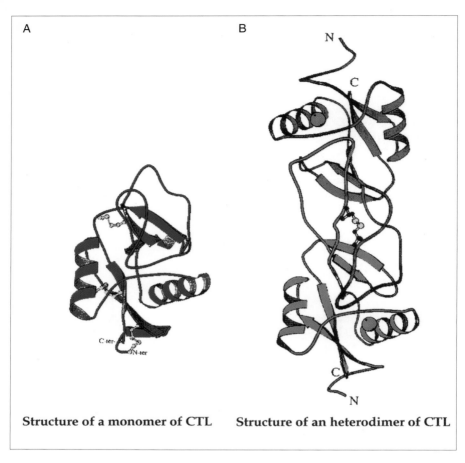

A

B

N

C

C-ter

N-ter

C

N

Structure of a monomer of CTL   **Structure of an heterodimer of CTL**

Plate XIV   Crystal structures of monomeric (**A**) and heterodimeric (**B**) of CTL

binding. However, relative differences in affinity caused by charge-neutralising mutations at Asp1428 and the neighbouring Lys1432 suggest that the interacting surface of these toxins is distinguishable. Since the high-affinity binding of Lqh-II is likely to involve more than one residue (Asp1428), additional residues on the channel that may belong to other external loops may contribute to toxin binding. Moreover, these results also indicate that the differential binding of the three toxins to the brain and skeletal muscle channel subtypes must involve extracellular loops other than the relatively conserved S3–S4 loop in D4. It seems though that further channel mutations and chimeras are required to resolve the selectivity issue. From the unaffected affinity of binding of Lqh-II and LqhαIT to a rat brain rBIIA chimera, in which D2 has been replaced with that of *Drosophila* Para sodium channel (see Figure 12.5; unpublished results), it has become clear that D2 is not the channel region responsible for the differential binding of α-toxins to receptor site-3 on rat and insect channels. Several more extra-cellular regions of sodium channels still remain to be explored to resolve this enigma.

12.5 SCORPION β-TOXINS THAT BIND TO RECEPTOR SITE-4

β-toxins bind to receptor site-4 of the sodium channel [36, 37] and generate a shift in the voltage dependence of activation to a more negative membrane potential [30–35]. A few β-toxins have been shown to be much less effective on rat cardiac channels compared to brain and skeletal muscle channel subtypes [39, 40, 69]. Like most α-toxins, β-toxins affect both insects and mammals in a variety of preferences. The absolute preference of two β-toxin groups, the excitatory and the depressant, for insect sodium channels is most intriguing [2, 70, 71]. Although their specificity for insects was first described in 1971 [71, 72], the molecular basis for this unusual phenomenon is still unknown. Yet this specificity has led to the design of baculoviruses expressing anti-insect specific toxins as a means of insect pest control [3, 73, 74, 75, 76].

12.5.1 EXCITATORY AND DEPRESSANT TOXINS EXCLUSIVELY BIND TO INSECT SODIUM CHANNELS

Insect sodium channels are closely related to their mammalian counterparts, as implied by biochemical, pharmacological and molecular studies [2, 4, 43, 77, 78]. Yet scorpion excitatory and depressant toxins show absolute preference for insect sodium channels, indicating that structural elements unique to insect sodium channels predominate at the toxin interacting surface. Excitatory toxins, such as AahIT, LqqIT1, and Bj-xtrIT (from *Buthotus judaicus*) (Figure 12.2), constitute a unique group among long-chain scorpion toxins. They are 70- to 76-amino-acid polypeptides displaying a shift in the otherwise conserved

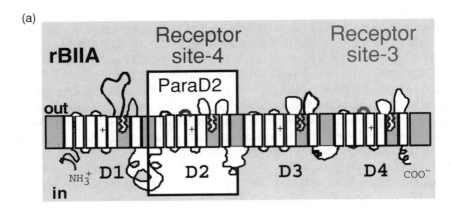

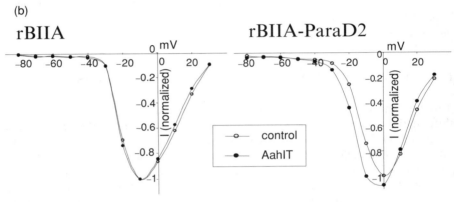

Figure 12.5 Na$^+$ currents mediated by rBIIA and rBIIA-ParaD2 channels (model in (a)) in *Xenopus* oocytes. cRNAs of α and β1-subunits were co-injected into oocytes. (b) Similar currents were produced by rBIIA and rBIIA-ParaD2. Current–voltage (I–V) curves of rBIIA ((b), left) before and after application of 2.3 μM AahIT, and of rBIIA-ParaD2 chimera ((b), right) before (empty symbols) and after (black symbols) addition of 1 μM AahIT. Note the shift to a more negative potential produced by AahIT in the chimera. Both channels are similarly sensitive to Lqh-II and insensitive to LqhαIT indicating unchanged selectivity for α-toxins (unpublished)

position of the fourth disulphide bridge [2, 79]. The only excitatory toxin with known 3-D structure is Bj-xtrIT, whose recombinant form was analysed by X-ray crystallography ([79]; see Chapter 13 by Gurevitz *et al.*). Excitatory toxins shift the threshold of insect sodium channel activation to a more negative membrane potential and increase peak sodium currents [4, 5, 71, 72, 80]. Their specificity to insects is also shown in radio-labelled toxin-receptor binding assays, which reveal a single, high-affinity binding site on insect sodium channels [51, 81, 82]. The β-toxin Ts-VII, which binds with high affinity to both mammalian and insect sodium channels, competes with the excitatory toxin

AahIT as well as with the depressant toxin LqhIT2 for binding to insect neuronal membranes [78, 83, 84]. This suggests that excitatory and depressant toxins bind to the equivalent of receptor site-4 on insect sodium channels [2]. The natural capacity of excitatory and depressant anti-insect toxins to discern insect from noninsect sodium channels may be exploited for the design of new selective insecticides. However, a prerequisite for such a design is information on the bioactive surface and receptor site of excitatory toxins on the insect channel.

12.5.2 LOCALISATION OF RECEPTOR SITE-4 ON THE SODIUM CHANNEL

To identify the receptor site for β-toxins, a chimeric rat skeletal muscle channel (rSkM1) has been constructed, in which domain-2 (D2) was replaced by its counterpart domain from the heart sodium channel (rH1) [39, 40]. The loss of sensitivity of the chimera for Ts-VII, much like the insensitivity of the intact rH1 channel to this toxin, implied that D2 has a critical role in recognition of β-toxins. Indeed, a single point mutation (G845N) in S3–S4 of D2 of the rat brain sodium channel, rBIIA, reduced the binding affinity of the β-toxin, Css-IV, almost 20-fold [38]. An Asn residue, which appears in this position in the less sensitive cardiac sodium channel subtype, rH1 (region S3–S4 in Figure 12.6; [38]), implies that this residue plays a role in binding of β-toxins. In addition, other external loops in D2 (S1–S2), D1 (S5–SS1), and D3 (SS2–S6) were suggested to contribute to β-toxin binding (Figure 12.2; [38]).

Domain-2 loop S3-S4

```
              ----S3-----        *      -----S4----
rSkM1         L V E L G L A N V Q G L S V L R S F R
rBIIA         - M - - - - - - E - - - - - - - -
rBI           - - - - - - - - E - - - - - - - -
rH1           - M - - - - S R M G N - - - - - - -
hH1           - M - - - - S R M S N - - - - - - -
Drosophila    - L - - - - E G - - - - - - - - -
Musca         - L - - - - E G - - - - - - - - -
Cockroach     - L - - - - E G - - - - - - - - -
```

Figure 12.6 Comparison of S3–S4–D2 sequences of various mammalian and insect channels. Gly845 of rBIIA is highlighted by an asterisk [38]. Residues in bold are equivalents of Gly845 in other channel subtypes: hSkM1 and rSkM1, skeletal muscle from human and rat; rBII and rBI, rat brain; hH1 and rH1, cardiac channels of human and rat [9]. *Drosophila, Musca* and cockroach, Para channel from *D. melanogaster* [85], *M. domestica* [86], and *Blatella germanica* [93]

12.6   SODIUM CHANNEL DOMAIN-2 CONFERS SELECTIVE
BINDING OF EXCITATORY TOXINS

Only limited information on molecular aspects of insect sodium channels is available and only a few have been cloned. Thus far, two insect channels (*Drosophila melanogaster* [85] and *Musca domestica*, [86]) have been functionally expressed in *Xenopus* oocytes [47, 87]. Nonetheless, insect neuronal sodium channels have been analysed biochemically and immuno-chemically [2, 77] and by binding studies using anti-insect specific scorpion toxins. These analyses have indicated not only that insect neuronal sodium channels differ structurally and pharmacologically from their mammalian counterparts, but that they also differ among various insect species (e.g. locust, cockroach, blowfly and moth; [78, 82, 88, 89]). Clarifying the structural basis of these differences may be exploited in future design of highly selective pesticides.

On the basis of previous reports assigning S3–S4 of D2 in rBIIA to binding of β-toxins [38], D2 of *Drosophila*-Para sodium channel was exchanged with its counterpart domain of rBIIA and the chimeric channel was analysed following expression in *Xenopus* oocytes (Figure 12.5; unpublished results). The recombinant chimera, rBIIA-ParaD2, produced TTX-sensitive voltage-activated Na currents and its sensitivity to the classical α-toxin, Lqh-II, was unaffected. However, in contrast to the insensitivity of intact rBIIA to the excitatory toxin AahIT (Figure 12.5), the chimeric channel acquired sensitivity to this toxin (unpublished results). The toxin effect was similar to that on insect neurons, namely the Na current was activated at more negative potentials ([3, 71, 80, unpublished results]). Since the S3–S4–D2 loop is highly conserved in insect Para, rat brain, and skeletal muscle sodium channels (Figure 12.6), an additional region in ParaD2 is most likely involved in the selective recognition of insect sodium channels by excitatory toxins, yet a single domain, D2, dictates the specificity of AahIT for insects.

12.7   CONCLUDING REMARKS AND PERSPECTIVES

The recent study of scorpion toxins and sodium channels of various origins (animals and tissues) highlighted two significant aspects of this field of research, namely the existence of sodium channel subtypes and the selectivity of various toxins that have developed a capability to distinguish between various channels.

12.7.1   STRUCTURAL DIFFERENCES AT THE TOXIN–RECEPTOR INTERACTING SURFACES OF VARIOUS SODIUM CHANNELS

Scorpion toxins of the α-class that are able to discern sodium channel subtypes in the mammalian nervous system enable us to unravel the distribution and functional role of distinct sodium channels in different tissues and sub-cellular regions (neural cell bodies, dendrites, axons and synapses; [6, 56]). Fluorescent labelling of such toxins, e.g., Lqh-II and Lqh-III, may allow a simultaneous follow-up of toxin binding and effect as well as localisation of the recognition site in the cell and tissue, using various optical and electro-physiological techniques.

A major goal in future studies is to elucidate the bioactive surface of different toxins (see Chapter 13 by Gurevitz *et al.*) as well as the mirror face of their receptor site in the channel. Although the bioactive surface of the alpha toxin, LqhαIT, has been clarified [90, 91] and a few channel mutants have highlighted putative residues involved in toxin binding [29, 68], further analysis of toxin-receptor amino acid interacting pairs is required for precise delineation of the interacting face. This goal is approachable by mutant thermodynamic cycle analysis applied on recombinant toxins and expressed channels, for which initial experiments have been described recently [68]. Despite this experimental approach, it should be borne in mind that selectivity of toxins for sodium channel subtypes might be imposed by structural elements that are not necessarily part of the functional interacting surface. Elucidation of such elements is far more problematic than identification of residues involved in toxin binding and requires detailed description of channel external architecture, in other words, the three-dimensional determination of sodium channel structures. Although initial structural data on sodium channels have been recently reported [92], an alternative approach to resolving the selectivity issue is to create a large repertoire of functional channel chimeras whose function and binding capacity can be tested against the various toxins and mutants. Together with the known structure of the toxin, such an approach may assist in describing the counterpart surface of the channel that dictates selectivity.

12.7.2 SELECTIVE RECOGNITION

On the basis of our recent results, we suggest that toxin selectivity for specific receptor sites may be conferred not only by structural variables at the interacting surface, but also by external channel elements that affect toxin access to the binding site. We have shown in binding competition studies that toxins with slightly different shape or electrostatic surface potential are capable of reaching the same or overlapping receptor sites [2, 5, 6, 41, 43–45, 49–51, 78, 82, 84]. Furthermore, the sequences of channel regions involved in toxin binding do not always vary sufficiently to explain selective recognition by toxins. Thus, structural differences in channel regions that flank the binding site could be involved in toxin binding. This assumption predicts that toxin binding to various channel subtypes as well as mutations at channel external regions (outside the interacting site) are more likely to affect the association rather than dissociation rate of toxin binding, and is susceptible to experimental confirmation. This notion was supported by the differences in association rates of Lqh-II, Lqh-III and LqhαIT observed for binding to rSkM1 channels expressed in mammalian cells [63], and the change in association rate of Lqh-II detected in synaptosomes under depolarised conditions [94, 95]. The rational design of subtype-specific compounds will take into account such architectural considerations, which may facilitate the design of selective ligands.

ACKNOWLEDGEMENTS

Parts of this research were supported by grants from the Israeli Science Foundation (no. 508/00 to D. Gordon) and BARD, the United States – Israel Binational Agricultural Research and Development (IS-3259-01 to D. Gordon and M. Gurevitz).

REFERENCES

[1] Gordon, D. (1997) Sodium channels as targets for neurotoxins: mode of action and interaction of neurotoxins with receptor sites on sodium channels. In *Toxins and Signal Transduction* (P. Lazarowici and Y. Gutman, eds.), pp. 119–49, Harwood Press, Amsterdam.

[2] Gordon, D., Savarin, P., Gurevitz, M. and Zinn-Justin, S. (1998) Functional anatomy of scorpion toxins affecting sodium channels. *J. Toxicol. Toxin. Rev.* **17**, 131–58.

[3] Gurevitz, M., Froy, O., Zilberberg, N., Turkov, M., Strugatzky, D., Gershburg, E., Lee, D., Adams, M.E., Tugarinov, V., Anglister, J., Shaanan, B., Loret, E., Stankiewicz, M., Pelhate, M., Gordon, D. and Chejanovsky, N. (1998) Sodium channel modifiers from scorpion venom: Structure–activity relationship, mode of action, and application. *Toxicon* **36**, 1671–82.

[4] Gordon, D. (1997) A new approach to insect-pest control – combination of neurotoxins interacting with voltage sensitive sodium channels to increase selectivity and specificity. *Invertebr. Neurosci.* **3**, 103–16.

[5] Froy, O., Zilberberg, N., Gordon, D., Turkov, D., Gilles, N., Stankiewicz, M., Pelhate, M., Loret, E., Oren, D., Shaanan, B. and Gurevitz, M. (1999) The putative bioactive surface of insect-selective scorpion excitatory neurotoxins. *J. Biol. Chem.* **274**, 5769–76.

[6] Gilles, N., Blanchet, B., Shichor, I., Zaninetti, M., Lotan, I., Bertrand, D. and Gordon, D. (1999) A scorpion α-like toxin active on insects and mammals reveals an unexpected specificity and distribution of sodium channel subtypes in rat brain neurons. *J. Neurosci.* **19**, 8730–9.

[7] Turkov, M., Rashi, S., Zilberberg, N., Gordon, D., Ben Khalifa, R., Stankiewicz, M., Pelhate, M. and Gurevitz, M. (1997) In vitro folding and functional analysis of an anti-insect selective scorpion depressant neurotoxin produced in *Escherichia coli*. *Prot. Expr. Purific.* **9**, 123–31.

[8] Goldin, A.L. (1999) Diversity of mammalian voltage-gated sodium channels. *Ann. NY Acad. Sci.* **868**, 38–50.

[9] Goldin, A.L., Barchi, R.L., Caldwell, J.H., Hofmann, F., Howe, J.R., Hunter, J.C., Kallen, R.G., Mandel, G., Meisler, M.H., Netter, Y.B., Noda, M., Tamkun, M.M.,

Waxman, S.G., Wood, J.N. and Catterall, W.A. (2001) Nomenclature of voltage-gated sodium channels. *Neuron* **28**, 365–8.

[10] Catterall, W.A. (1992) Cellular and molecular biology of voltage-gated sodium channels. *Physiol. Rev.* **72**, S15–S48.

[11] Auld, V.J., Goldin, A.L., Krafte, D.S., Marshall, J., Dunn, J.M., Catterall, W.AA, Lester, H.A., Davidson, N. and Dunn, R.J. (1988) A rat brain Na$^+$ channel alpha subunit with novel gating properties. *Neuron* **1**, 449–61.

[12] Noda, M., Ikeda, T., Kayano, T., Suzuki, H., Takeshima, H., Kurasaki, M., Takahashi, H. and Numa, S. (1986) Existence of distinct sodium channel messenger RNAs in rat brain. *Nature* **320**, 188–92.

[13] Schaller, K.L., Krzemien, D.M., Yarowsky, P.J., Krueger, B.K. and Caldwell, J.H. (1995) A novel, abundant sodium channel expressed in neurons and glia. *J. Neurosci.* **15**, 3231–42.

[14] Toledo-Aral, J.J., Moss, B.L., He, Z-J., Koszowski, A.G., Whisenand, T., Levinson, S.R., Wolf, J.J., Silos-Santiago, I., Halegoua, S. and Mandel, G. (1997) Identification of PN1, a predominant voltage dependent sodium channel expressed principally in peripheral neurons. *Proc. Natl. Acad. Sci. USA* **94**, 1527–32.

[15] Trimmer, J.S., Cooperman, S.S., Tomiko, S.A., Zhou, J., Crean, S.M., Boyle, M. B., Kallen, R.G., Sheng, Z., Barchi, R.L., Sigworth, F.J., Goodman, R.H., Agnew, W.S. and Mandel, G. (1989) Primary structure and functional expression of a mammalian skeletal muscle sodium channel. *Neuron* **3**, 33–49.

[16] Rogart, R.B., Cribbs, L.L., Muglia, L.K., Kephart, D.D. and Kaiser, M.W. (1989) Molecular cloning of a putative tetrodotoxin-resistant rat heart sodium channel isoform. *Proc. Natl. Acad. Sci. USA*, **86**, 8170–4.

[17] Kallen, R.G., Sheng Z-H., Yang, J., Chen, L., Rogart, R.B. and Barchi, R.L. (1990) Primary structure and expression of a sodium channel characteristic of denervated and immature rat skeletal muscle. *Neuron* **4**, 233–42.

[18] Akopian, A.N., Souslova, V., England, S., Okuse, K., Ogata, N., Ure, J., Smith, A., Kerr, B.J., McMahon, S.B., Boyce, S., Hill, R., Stanfa, L.C., Dickenson, A.H. and Wood, J.N. (1999) The tetrodotoxin-resistant sodium channel SNS has a specialized function in pain pathways. *Nature Neurosci.* **2**, 541–8.

[19] Akopian, A.N., Sivilotti, L. and Wood, J.N. (1996) A tetrodotoxin-resistant voltage-gated sodium channel expressed by sensory neurons. *Nature* **379**, 257–62.

[20] Sangameswaran, L., Delgado, S.G., Fish, L.M., Koch, B.D., Jakeman, L.B., Stewart, G.R., Sze, P., Hunter, J.C., Eglen, R.M. and Herman, R.C. (1996) Structure and function of a novel voltage-gated tetrodotoxin-resistant sodium channel specific to sensory neurons. *J. Biol. Chem.* **271**, 5953–6.

[21] Dib-Hajj, S.D., Tyrrell, L., Black, J.A. and Waxman, S.G. (1998) NaN, a novel voltage-gated Na channel, is expressed preferentially in peripheral sensory neurons and down-regulated after axotomy. *Proc. Natl. Acad. Sci. USA* **95**, 8963–8.

[22] Tate, S., Benn, S., Hick, C., Trezise, D., John, V., Mannion, R.J., Costigan, M., Plumpton, C., Grose, D., Cladwell, Z., Kendall, G., Dale, K., Bountra, C. and Woolf, C.J. (1998) Two sodium channels contribute to the TTX-R sodium current in primary sensory neurons. *Nature Neurosci.* **1**, 653–5.

[23] Felts, P.A., Yokoyama, S., Dib-Hajj, S., Black, J.A. and Waxman, S.G. (1997) Sodium channel α subunit mRNAs I, II, III, NaG, Na6 and hNE (PN1): different expression patterns in developing rat nervous system. *Molec. Brain. Res.* **45**, 71–82.

[24] Beckh, S. (1990) Differential expression of sodium channel mRNA in rat peripheral nervous system and innervated tissues. *FEBS Lett.* **262**, 317–22.

[25] Beckh, S., Noda, M., Lubbert, H. and Numa, S. (1989) Differential regulation of three sodium channel RNAs in the rat central nervous system during development. *EMBO J.* **8**, 3611–16.

[26] Black, J.A., Yokoyama, S., Higashida, H., Ransom, B.R. and Waxman, S.G. (1994) Sodium channel mRNAs I, II and III in the CNS: cell-specific expression. *Molec. Brain. Res.* **22**, 275–89.

[27] Furuyama, T., Morita, Y., Inagaki, S. and Tagaki, H. (1993) Distribution of I, II and III subtypes of voltage-sensitive Na$^+$ channel mRNA in the rat brain. *Molec. Brain. Res.* **17**, 169–73.

[28] Gautron, S., Dos Santos, G., Pinto-Henrique, D., Koulakoff, A., Gros, F. and Berwald Netter, Y. (1992) The glial voltage-gated sodium channel: Cell- and tissue-specific mRNA expression. *Proc. Natl. Acad. Sci. USA* **89**, 7272–6.

[29] Rogers, J.C., Qu, Y., Tanada, T.N., Scheuer, T. and Catterall, W.A. (1996) Molecular determinants of high affinity binding of α-scorpion toxin and sea anemone toxin in the S3–S4 extracellular loop in domain IV of the sodium channel α subunit. *J. Biol. Chem.* **271**, 15950–62.

[30] Cahalan, M.D. (1975) Modification of sodium channel gating in frog myelinated nerve fibers by *Centruroides sculpturatus* scorpion venom. *J. Physiol.* **244**, 511–34.

[31] Jaimovich, E., Ildefonse, M., Barhanin, J., Rougier, O. and Lazdunski, M. (1982) *Centruroides* toxin, a selective blocker of surface Na channels in skeletal muscle: Voltage clamp analysis and biochemical characterization of the receptor. *Proc. Natl. Acad. Sci. USA* **79**, 3896–900.

[32] Jonas, P., Vogel, W., Arantes, E.C. and Giglio, J.R. (1986) Toxin γ of the scorpion *Tityus serrulatus* modifies both activation and inactivation of sodium permeability of nerve membrane. *Pflügers Archiv.* **407**, 92–9.

[33] Meves, H., Rubly, N. and Watt, D.D. (1982) Effect of toxins isolated from the venom of the scorpion *Centruroides sculpturatus* on the Na currents of the Node of Ranvier. *Eur. J. Physiol. Pflügers Archiv.* **393**, 56–62.

[34] Vijverberg, H.P., Pauron, D. and Lazdunski, M. (1984) The effect of *Tityus serrulatus* scorpion toxin γ on sodium channel in neuroblastoma cells. *Pflügers Archiv.* **40**, 297–303.

[35] Wang, G.K. and Strichartz, G. (1983) Purification and physiological characterization of neurotoxins from venom of the scorpion *Centruroides sculpturatus* and *Leiurus quinquestriatus*. *Molec. Pharmacol.* **23**, 519–33.

[36] Jover, E., Couraud, F. and Rochat, H. (1980) Two types of scorpion neurotoxins characterized by their binding to two separate receptor-sites on rat brain synaptosomes. *Biochem. Biophys. Res. Commun.* **95**, 1607–14.

[37] Barhanin, J., Giglio, J.R., Leopold, P., Schmid, A., Sampaio, S.V. and Lazdunski, M. (1982) *Tityus serrulatus* venom contains two classes of toxins. *J. Biol. Chem.* **257**, 12553–8.

[38] Cestèle, S., Qu, Y., Rogers, J.C., Rochat, H. and Catterall, W.A. (1998) Voltage sensor trapping: enhanced activation of sodium channels by β-scorpion toxin bound to the S3–S4 loop in domain II. *Neuron* **21**, 919–31.

[39] Marcotte, P., Chen, L-Q., Kallen, R.G. and Chahnin, M. (1997) Effects of *Tityus serrulatus* scorpion toxin γ on voltage-gated Na$^+$ channels. *Circ. Res.* **80**, 363–9.

[40] Tsushima, R.G., Borges, A. and Backx, P.H. (1999) Inactivated state dependence of sodium channel modulation by β-scorpion toxin. *Pflügers Archiv.* **437**, 661–8.

[41] Little, M.J., Zappia, C., Gilles, N., Connor, M., Tyler, M.F., Martin-Eauclaire, M.F., Gordon, D. and Nicholson, G.M. (1998) δ-Atracotoxins from Australian funnel-web spiders compete with scorpion α-toxin binding but differentially modulate alkaloid toxin activation of voltage-gated sodium channels. *J. Biol. Chem.* **273**, 27076–83.

[42] Martin-Eauclaire, M.F. and Couraud, F. (1995) Scorpion neurotoxins: Effects and mechanisms. In *Handb. Neurotoxicology* (LW Chang and RS Dyer, eds.), pp. 683–716, Marcel Dekker, NY.

[43] Gordon, D., Martin-Eauclaire, M.F., Cestèle, S., Kopeyan, C., Carlier, E., Ben Khalifa, R., Pelhate, M. and Rochat, H. (1996) Scorpion toxins affecting sodium current inactivation bind to distinct homologous receptor sites on rat brain and insect sodium channels. *J. Biol. Chem.* **271**, 8034–45.

[44] Cestèle, S., Stankiewicz, M., Mansuelle, P., Dargent, B., De Waard, M., Gilles, N., Pelhate, M., Rochat, H., Martin-Eauclaire, M.F. and Gordon, D. (1999) Scorpion α-like toxins, toxic to both mammals and insects, differentially interact with receptor site 3 on voltage-gated sodium channels in mammals and insects. *Eur. J. Neurosci.* **11**, 975–85.

[45] Krimm, I., Gilles, N., Sautiere, P., Stankiewicz, M., Pelhate, M., Gordon, D. and Lancelin, J-M. (1999) NMR structures and activity of a novel α-like toxin from the scorpion *Leiurus quinquestriatus hebraeus*. *J. Mol. Biol.* **285**, 1749–63.

[46] Strichartz, G., Rando, T. and Wang, G.K. (1987) An integrated view of the molecular toxicology of sodium channel gating in excitable cells. *Annu. Rev. Neurosci.* **10**, 237–67.

[47] Warmke, J.W., Reenan, R.A.G., Wang, P., Qian, S., Arena, J.P., Wang, J., Wunderler, D., Liu, K., Kaczorowski, G.J., Van der Ploeg, L.H.T., Ganetzky, B. and Cohen, C.J. (1997) Functional expression of *Drosophila* para sodium channels. Modulation by the membrane protein TipE and toxinpharmacology. *J. Gen. Physiol.* **110**, 119–33.

[48] Vargas, O., Martin, M.F. and Rochat, H. (1987) Characterization of six toxins from the venom of the Moroccan scorpion *Buthus occitanus mardochei*. *Eur. J. Biochem.* **162**, 589–99.

[49] Gordon, D. and Zlotkin, E. (1993) Binding of an alpha scorpion toxin to insect sodium channels is not dependent on membrane potential. *FEBS Lett.* **315**, 125–9.

[50] Little, M.J., Wilson, H., Zappia, C., Cestèle, S., Tyler, M.I., Martin-Eauclaire, M.F., Gordon, D. and Nicholson, G.M. (1998) δ-Atracotoxins from Australian funnel-web spiders compete with scorpion α-toxin binding on both rat brain and insect sodium channels. *FEBS Lett.* **439**, 246–52.

[51] Gilles, N., Krimm, I., Bouet, F., Froy, O., Gurevitz, M., Lancelin, J.M. and Gordon, D. (2000) Structural implications on the interaction of scorpion α-like toxins with the sodium channel receptor site inferred from toxin iodination and pH-dependent binding. *J. Neurochem.* **75**, 1735–45.

[52] Tejedor, F. and Catterall, W.A. (1990) Photoaffinity labeling of the receptor site for α scorpion toxins on purified and reconstituted sodium channels by a new toxin derivative. *Cellular. Molec. Neurobiol.* **10**, 257–65.

[53] Thomsen, W.J. and Catterall, W.A. (1989) Localization of the receptor site for α-scorpion toxins by antibody mapping implications for sodium channel topology. *Proc. Natl. Acad. Sci. USA* **86**, 10161–5.

[54] Benzinger, G.R., Kyle, J.W., Blumenthal, K.M. and Hanck, D.A. (1998) A specific interaction between the cardiac sodium channel and site-3 toxin Anthopleurin B. *J. Biol. Chem.* **273**, 80–4.

[55] Sautiere, P., Cestèle, S., Kopeyan, C., Martinage, A., Drobecq, H., Doljansky, Y. and Gordon, D. (1998) New toxins acting on sodium channels from the scorpion *Leiurus Quinquestriatus hebraeus* suggest a clue to mammalian versus insect selectivity. *Toxicon* **36**, 1141–54.

[56] Gilles, N., Chen, H., Wilson, H., Legall, F., Montoya, G., Molgo, J., Schönherr, R., Nicholson, G., Heinemann, S.H. and Gordon, D. (2000) Scorpion α- and α-like toxins differentially interact with sodium channels in mammalian CNS and periphery. *Eur. J. Neurosci.* **12**, 2823–32.

[57] Gray, E.G. and Whittaker, V.P. (1962) The isolation of nerve ending from brain: an electron microscopic study of cell fragments derived by homogenization and centrifugation. *J. Anatomy* **96**, 79–96.

[58] Gordon, D., Merrick, D., Auld, V., Dunn, R., Goldin, A.L., Davidson, N. and Catterall, W.A. (1987) Tissue-specific expression of RI and RII sodium channel subtypes. *Proc. Natl. Acad. Sci. USA* **84**, 8682–6.

[59] Furuyama, T., Morita, Y., Inagaki, S. and Tagaki, H. (1993) Distribution of I, II and III subtypes of voltage-sensitive Na^+ channel mRNA in the rat brain. *Molec. Brain Res.* **17**, 169–73.

[60] Westenbroek, R.E., Merrick, D.K., Catterall, W.A. (1989) Differential subcellular localization of the R_I and $R_{II}Na^+$ channel subtypes in central neurons. *Neuron* **3**, 695–704.

[61] Sangameswaran, L., Fish, L.M., Koch, B.D., Rabert, D.K., Delgado, S.G., Ilnicka, M., Jakeman, L.B., Novakovic, S., Wong, K., Sze, P., Tzoumaka, E., Stewart, G.R., Herman, R.C., Chan, H., Eglen, R.M. and Hunter, J.C. (1997) A novel tetrodotoxin-sensitive, voltage-gated sodium channel expressed in rat and human dorsal root ganglia. *J. Biol. Chem.* **272**, 14805–9.

[62] Molgo, J. and Mallart, A. (1985) Effects of Anemonia sulcata toxin II on presynaptic currents and evoked transmitter release at neuromuscular junctions of the mouse. *Pflügers Archiv.* **405**, 349–53.

[63] Chen, H., Gordon, D. and Heinemann, S. (2000) Modulation of cloned skeletal muscle sodium channel by the scorpion toxins Lqh II, Lqh III and LqhαIT. *Pflügers Archiv.* **439**, 423–32.

[64] Chen, H. and Heinemann, S.H. (2001) Interaction of scorpion α-toxins with cardiac sodium channels: binding properties and enhancement of slow inactivation. *J. Gen. Physiol.*, **117**, 505–18.

[65] Roy, M.L. and Narahashi, T. (1992) Differential properties of tetrodotoxin-sensitive and tetrodotoxin-resistant sodium channels in rat dorsal root ganglion neurons. *J. Neurosci.* **12**, 2104–11.

[66] Black, J.A., Dib-Hajj, S., McNabola, K., Jeste, S., Rizzo, M.A., Kocsis, J.D. and Waxman, S.G. (1996) Spinal sensory neurons express multiple sodium channel α-subunit mRNAs. *Mol. Brain Res.* **43**, 117–31.

[67] Benoit, E. and Gordon, D. (2001) The scorpion alpha-like toxin Lqh III specifically alters sodium channel inactivation in myelinated axons. *Neuroscience* **104**, 551–9.

[68] Ma, Z., Tang, L., Lu, S., Kong, J., Gordon, D. and Kallen, R.G. (2000) The domain 4 S3–S4 extracellular loop provides molecular determinants for binding of α-scorpion toxins (Lqh II, Lqh III and LqhαIT) to the voltage-gated rat skeletal muscle Na^+ channel (rSkM1). *Biophys. J.* **78**, A1011.

[69] Possani, L.D., Becerril, B., Delepierre, M. and Tytgat, J. (1999) Scorpion toxins specific for Na^+-Channels. *Eur. J. Biochem.* **264**, 287–300.

[70] Dianous, S., Hoarau, F. and Rochat, H. (1987) Re-examination of the specificity of the scorpion *Androctonus australis* Hector insect toxin towards arthropods. *Toxicon* **25**, 411–17.

[71] Zlotkin, E., Miranda, F. and Rochat, H. (1978) Chemistry and pharmacology of Buthinae scorpion venoms. In: *Arthropods venoms*, (Bettini S, ed.), Springer Verlag, New York, NY, pp. 317–69.

[72] Zlotkin, E., Rochat, H., Kopeyan, C., Miranda, F. and Lissitzky, S. (1971) Purification and properties of the insect toxin from the venom of the scorpion *Androctonus australis* Hector. *Biochimie* **53**, 1073–8.

[73] Chejanovsky, N., Zilberberg, N., Rivkin, H., Zlotkin, E. and Gurevitz, M. (1994) Expression of a functional alpha scorpion neurotoxin in insect cells and in insect larvae. *FEBS Lett.* **376**, 181–4.

[74] Gershburg, E., Stockholm, D., Froy, O., Rashi, S., Gurevitz, M. and Chejanovsky, N. (1998) Baculovirus-mediated expression of a scorpion depressant toxin improves the insecticidal efficacy achieved with excitatory toxins. *FEBS Lett.* **422**, 132–36.

[75] Maeda, S., Volrath, S.L., Hanzlik, T.N., Harper, S.A., Majuma, K., Maddox, D.W., Hammock, B.D. and Fowler, E. (1991) Insecticidal effects of an insect-specific neurotoxin expressed by a recombinant baculovirus. *Virology* **164**, 777–80.

[76] Stewart, L.M.D., Hirst, M., Ferber, M.L., Merryweather, A.T., Cayley, P.J. and Possee, R.D. (1991) Construction of an improved baculovirus insecticide containing an insect-specific toxin gene. *Nature* **352**, 85–8.

[77] Gordon, D., Moskowitz, H. and Zlotkin, E. (1990) Sodium channel polypeptides in central nervous systems of various insects identified with site-directed antibodies. *Biochim. Biophys. Acta.* **1026**, 80–6.

[78] Gordon, D., Moskowitz, H., Eitan, M., Warner, C., Catterall, W.A. and Zlotkin, E. (1992) Localization of receptor sites for insect-selective toxins on sodium channels by site-directed antibodies. *Biochemistry* **31**, 7622–8.

[79] Oren, D.A., Froy, O., Amit, E., Kleinberger-Doron, N., Gurevitz, M. and Shaanan, B. (1998) An excitatory scorpion toxin with a distinctive feature: an additional α-helix at the C-terminus and its implications for interaction with insect sodium channels. *Structure* **6**, 1095–103.

[80] Lee, D. and Adams, M.E. (2000) Sodium channels in central neurons of the tobacco budworm, *Heliothis virescens*: basic properties and modification by scorpion toxins. *Insect. Physiol.* **46**, 499–508.

[81] Gordon, D., Jover, E., Couraud, F., and Zlotkin, E. (1984) The binding of the insect selective neurotoxin (AaIT) from scorpion venom to locust synaptosomal membranes. *Biochim. Biophys. Acta.* **778**, 349–58.

[82] Moskowitz, H., Herrmann, R., Zlotkin, E. and Gordon, D. (1994) Variability among insect sodium channels revealed by binding of selective neurotoxins. *Insect. Biochem. Molec. Biol.* **24**, 13–19.

[83] Lima, M.E., Martin, M-F., Diniz, C.R. and Rochat, H. (1986). *Tityus serrulatus* toxin VI bears pharmacological properties of both β-toxin and insect toxin from scorpion venoms. *Biochem. Biophys. Res. Commun.* **139**, 296–302.

[84] Lima, M.E., Martin-Eauclaire, M-F., Hue, B., Loret, E., Diniz, C.R. and Rochat, H. (1989) On the binding of two scorpion toxins to the central nervous system of the cockroach *Periplaneta americana*. *Insect. Biochem.* **19**, 413–22.

[85] Loughney, K., Kreber, R. and Ganetzky, B. (1989) Molecular analysis of the *para* locus, a sodium channel gene in *Drosophila*. *Cell* **58**, 1143–54.

[86] Williamson, M.S., Martinez-Torres, D., Hick, C.A. and Devonshire, A.L. (1996) Identification of mutations in the housefly para-type sodium channel gene associated with knockdown resistance (*kdr*) to pyrethroid insecticides. *Mol. Gen. Genet.* **252**, 51–60.

[87] Smith, T.J., Lee, S.H., Ingles, P.J., Knipple, D.C. and Soderlund, D.M. (1997) The L1014F point mutation in the house fly Vssc1 sodium channel confers knockdown resistance to pyrethroids. *Insect Biochem. Molec. Biol.* **27**, 807–12.

[88] Cestèle, S., Gordon, D., Kopeyan, C. and Rochat, H. (1997) Toxin III from *Leiurus quinquestriatus quinquestriatus:* A specific probe for receptor site 3 on insect sodium channels. *Insect. Biochem. Molec. Biol.* **27**, 523–8.

[89] Fishman, L., Herrmann, R., Gordon, D. and Zlotkin, E. (1997) Insect tolerance to neurotoxic polypeptide: pharmacokinetic and pharmacodynamic aspects. *J. Exp. Biol.* **200**, 1115–23.

[90] Zilberberg, N., Froy, O., Loret, E., Cestèle, S., Arad, D., Gordon, D. and Gurevitz, M. (1997) Identification of structural elements of a scorpion α-neurotoxin important for receptor site recognition. *J. Biol. Chem.* **272**, 14810–16.

[91] Gurevitz, M., Gordon, D., Ben-Natan, S., Turkov, M. and Froy, O. (2001) Diversification of neurotoxins by C-tail 'wiggling' – a scorpion recipe for survival. *FASEB J.* **15**, 1201–5.

[92] Sato, C., Ueno, Y., Asal, K., Takahashi, K., Sato, M., Engel, A. and Fujiyoshi, Y. (2001) The voltage sensitive sodium channel is a bell-shaped molecule with several cavities. *Nature* **409**, 1047–51.

[93] Dong, K. (1997) A single amino acid change in the Para sodium channel protein is associated with knockdown-resistance (kdr) to pyrethroid insecticides in German cockroach. *Insect. Biochem. Molec. Biol.* **27**, 93–100.

[94] Gilles, N., Leipold, E., Chen, H., Heinemann, S.H. and Gordon, D. (2001) Effects of depolarization on binding kinetics of scorpion α-toxins highlights conformational changes of rat brain sodium channels. *Biochemistry* **40**, 14576–84.

[95] Gilles, N., Harrison, G., Karbat, I., Gurevitz, M., Nicholson, G. and Gordon, D. (2002) Variations in receptor site-3 on rat brain and insect sodium channels highlighted by binding of a funnel-web spider δ-atracotoxin. *Eur. J. Biochem., in press.*

13 Diversification of Toxic Sites on a Conserved Protein Scaffold – a Scorpion Recipe for Survival

MICHAEL GUREVITZ*, NOAM ZILBERBERG, OREN FROY,
MICHAEL TURKOV, RUTHIE WILUNSKY, IZHAR KARBAT,
JACOB ANGLISTER, BOAZ SHAANAN, MARCEL PELHATE,
MICHAEL E. ADAMS, NICOLAS GILLES and DALIA GORDON

13.1 DIVERSE PHARMACOLOGY OF SCORPION TOXINS AFFECTING SODIUM CHANNELS

Scorpion toxins affecting sodium channels are polypeptides of 61–76 amino acid residues ('long chain' toxins) that have been divided into two major classes, α and β, according to their mode of action and binding properties [1, 2]. Both classes modulate the sodium current upon binding at two distinct receptor sites on the sodium channel. α-Toxins inhibit the inactivation of the sodium current by binding to receptor site-3 [3] and are subdivided into distinct pharmacological groups according to their preference for phylogenetically distinct target sites ([4, 5]; see Chapter 12 by Gordon *et al.*). Conversely, β-toxins bind to receptor site-4 and shift the activation of sodium current to more negative membrane potentials (reviewed in [4, 6]). Most α- and β-toxins affect mammalian and insect sodium channels with various affinities. However, two additional pharmacologically distinct toxin groups, excitatory and depressant, which affect exclusively sodium channels of insects, have been characterised [7–11]. Both groups induce opposite effects on the sodium current by affecting the activation process of the sodium channel and bind to distinct, yet closely related receptor sites [12]. For these reasons and due to their capability to compete with β-toxins for the receptor binding-site, excitatory and depressant toxins have been suggested to belong to the β-class [2, 13]. Sequence comparison among the various groups reveals approximately 30% similarity, whereas the toxins within each group may differ by up to 50% [2, 14] (Plate VIII). Much information has been obtained from chemical modifications introduced at specific residues of alpha toxins, e.g. Aah2 and Aah3 from the scorpion

---

* Corresponding author

Perspectives in Molecular Toxinology
Edited by A Ménez © 2002 John Wiley & Sons, Ltd

Androctonus australis hector, Lqq5 from *Leiurus quinquestriatus quinquestriatus*, and Bom3 from *Buthus occitanus mardochei* [15–17], and from the interaction of Aah2 with its receptor site using antibodies produced against toxin fragments [18, 19]. However, clarification of the molecular basis of toxin preference for various sodium channels requires a reliable quantitative expression system that would enable extensive mutagenesis for elucidation of bioactive residues.

13.2 STRUCTURE–ACTIVITY RELATIONSHIP

13.2.1 CLONING AND FUNCTIONAL EXPRESSION

Expression in eukaryotic systems of several 'long chain' toxins has been documented [20–23], but the minute quantities obtained limited further studies. The expression of scorpion toxins in *Escherichia coli*, a preferred system for genetic manipulations, was not achieved, possibly due to the reducing conditions in the cytosol, lack of recognition of regulatory sequences, and rapid degradation [24–26]. These difficulties were overcome [27–29] by enforcing the accumulation of toxins in *E. coli* within inclusion bodies and then exploiting the competence of small denatured polypeptides to fold into their functional conformation *in vitro* [30, 31]. Another approach was to express a scorpion toxin by fusion to an *E. coli* resident protein, which accumulated in the periplasmic space. Purification and proteolytic cleavage of the fusion protein resulted in an active scorpion toxin [32].

Genes encoding toxins from the alpha, excitatory, and depressant groups were isolated from cDNA libraries prepared from mRNA extracted from telsons of *Leiurus quinquestriatus hebraeus* and *Buthotus judaicus* [33–37]. Selected cDNA clones encoding the strongest anti-insect α-toxin, LqhαIT [38], the anti-insect selective depressant toxin, LqhIT2 [39], and the excitatory toxin, Bj-xtrIT [10], were used for production of recombinant insoluble toxins in *E. coli* [40, 28]. The recombinant toxins, isolated from inclusion bodies and renatured *in vitro*, were functionally indistinguishable from the authentic native toxins purified from scorpion venom, as revealed in toxicity assays, binding and electrophysiological studies [28, 41, 10]. The efficient production of recombinant active toxins paved the way for site-directed mutagenesis and structural analyses, which enabled elucidation of bioactive surfaces.

13.2.2 DETERMINATION OF THREE-DIMENSIONAL STRUCTURES

The 3-D structure of several α-toxins has been known for more than a decade [42–44], and yet, investigation of toxin bioactive surfaces was unapproachable due to the lack of a reliable expression system. The first toxin that was amenable to mutagenic dissection was the alpha anti-insect toxin, LqhαIT, which was expressed in milligram amounts using bacterial expression and *in vitro* folding protocols [28, 45]. To distinguish between amino acid residues that were exposed to the solvent, and hence participate in bioactivity, from residues buried within the molecule, the solution structure of recombinant LqhαIT was determined by ^{1}H-2D-NMR spectroscopy [46] (Plate IX-I). The resolved structure resembled those of other scorpion α-toxins consisting of an α-helix, three-strand antiparallel β-sheet, three type I tight turns, five-residue turn, and a surface-exposed hydrophobic patch [43, 47–50]. The conformation of the polypeptide carbon backbone of LqhαIT was compared to that of the most active anti-mammalian scorpion α-toxin, Aah2 [51, 1]. Both structures were found to be very similar (the rmsd for the backbone atoms between both toxins was only 1.1 Å) despite only 60 % sequence homology. The most salient differences between both structures were observed in the five-residue turn and C-tail regions and in their mutual disposition [46]. Thus, the similar scaffold of α-toxins, on the one hand, and diverse binding properties and preference for various sodium channels, on the other hand, provided an ideal model for clarifying structural features that correlate with the different pharmacologies (see below).

Another group of scorpion toxins with unique structural features and disulphide bond arrangement are the excitatory toxins [7, 52, 53] (Plate VIII-IIb). Due to their selectivity for insect sodium channels [54], these toxins may serve as leads for generating selective insecticides. Indeed, the heralded representative, AahIT [55], was engineered into the genome of the entomopathogenic baculovirus *Autographa californica* nucleopolyhedrovirus (AcMNPV) and shown to increase their insecticidal efficacy [56, 22, 57]. However, expression of AahIT [20] or its counterpart from *L. q. hebraeus*, LqhIT1 ([58]; unpublished results), in heterologous systems was extremely limited. Conversely, the excitatory toxin, Bj-xtrIT, of *Buthotus judaicus*, sharing only 49 % sequence homology with AahIT (Plate VIII-I), folded easily to provide milligram amounts of recombinant active protein [10]. Bj-xtrIT was very similar to AahIT in toxicity assays, electrophysiological effects on a cockroach axon, and competition for ^{125}I-AahIT and ^{125}I-LqhIT2 binding to cockroach neuronal membranes [10]. Moreover, ^{125}I-Bj-xtrIT bound with high affinity ($K_d = 0.15 \pm 0.03$ nM; n = 4; unpublished results) to its receptor site and was displaced by LqhIT2 and by AahIT [59, 12, 60]. Therefore, Bj-xtrIT has become the model of choice to study the structure–activity relationships and three-dimensional structures of excitatory toxins. A Bj-xtrIT crystal was easily obtained, which enabled determination of its 3-D structure by X-ray diffraction [53]; Plates VIII-IIb and IX-II). Like other scorpion 'long chain' toxins, the scaffold of Bj-xtrIT is composed of a closely packed module comprising an α-helix linked to a three-strand antiparallel β-sheet via two disulphide bonds. However, the last 17 residues form a smaller structural entity containing a short α-helix, anchored to the large module via a disulphide bridge, and a terminal mobile stretch of seven residues, of which only four are seen in the electron-density map. The C-tail covers part of the conserved hydrophobic patch and was later shown to be very important for bioactivity (see below).

   Scorpion depressant toxins exhibit unique pharmacology and are also very important due to their preference for insect sodium channels [61, 9] and prominent insecticidal potential when expressed by AcMNPV [62]. Despite the efficient expression of a functional depressant toxin, LqhIT2 [41], all attempts to determine its 3-D structure have failed thus far, probably due to hydrophobicity and tendency to aggregate. Therefore, in order to enable a mutagenic dissection, solvent-exposed residues that may be involved in bioactivity were highlighted in a computer model constructed using the coordinates of the toxin, CsE v3, from the scorpion *Centruroides sculpturatus* Ewing [42] and later those of the β-toxin, Cn2 (from *Centruroides noxius*; [63]; Plates VIII and X). As depressant and β-toxins resemble each other in sequence, neurophysiological action, and competition for receptor-binding site [2], selection of sites for mutagenesis could be rationalised by relying on this model.

   These structural studies have indicated that despite the diverse bioactivity and sequence, scorpion 'long chain' toxins share a common α/β scaffold that occupies most of their globular structure [2, 14, 64]. The scaffold is stabilised by three, spatially conserved, disulphide bonds and a fourth bond is conserved in all but the excitatory toxins [65, 53] (Plate VIII). Moreover, all these toxins contain a solvent-exposed patch of aromatic and hydrophobic residues, known as the 'conserved hydrophobic surface' [43], whose function is still unknown. The mutual disposition of the α-helix and the β-sheet in all toxins is similar. However, the carboxy-terminal stretch and the regions connecting the secondary structure elements differ in the various toxins [2, 44, 66]. Superposition of the four toxins indicates quite explicitly a common starting point after the third β-strand, from which all C-tails diverged (see below). Hence, all toxins constitute on the one hand a rigid α/β scaffold, and on the other hand a highly variable carboxy-terminal tail confined mainly by the fourth disulphide bridge [66].

13.2.3   ELUCIDATION OF BIOACTIVE SURFACES

The expression system developed for production of recombinant LqhαIT, Bj-xtrIT, and LqhIT2 was used for extensive mutagenesis of amino acid residues that appear on the molecule's surface. Modifications were introduced at sites that either differ or were conserved among toxins of the same pharmacological group. All toxin variants were purified by C_{18}-RP-HPLC and analysed by (i) toxicity assays on blowfly larvae and mice; (ii) binding-competition studies using cockroach neuronal membrane preparations; and (iii) CD-spectroscopy in order to discern mutations that may have affected the structure or the bioactive surface. More than 40 point mutations introduced to LqhαIT highlighted Lys8, Tyr10, Phe17, Arg18, Arg58, Val59 and Lys62. These residues are organised in a belt shape surrounding the toxin molecule and are important for bioactivity of LqhαIT as their substitution reduced toxicity at least five-fold compared to the unmodified toxin [45, 29; unpublished results). Three of them, Arg58, Val59, and Lys62, belong to the C-tail, and the side chains of the rest, Lys8, Tyr10, Phe17, and Arg18 are located on the outer surface of the α/β scaffold (Plate IX-I). Out of 26 mutations introduced to the excitatory toxin, Bj-xtrIT, substitution of Asp55, Asp70, Ser76, Asn28, and Tyr36 reduced toxicity more than five-fold while substitution of Ile73 and Ile74 had a much larger effect of over 100-fold ([10]; Plate IX-II; unpublished results). More than 50 amino acid substitutions were performed at 37 sites of LqhIT2. The residues with more than five-fold effect on toxicity were Asn58 and Gly61 of the C-tail, and Asp8, Lys11 and Lys26 of its vicinal surface. These residues form a bioactive surface encompassing the fourth disulphide bridge that anchors the C-tail onto the molecule (unpublished results).

Superposition of the Cα-backbones of the α-toxin LqhαIT [46], β-toxin Cn2 [63], excitatory toxin Bj-xtrIT [53], and LqhIT2 model [66] highlighted the 180° positional shift of the C-tail of Bj-xtrIT and its fourth disulphide bridge relative to the other toxins (Plate X). The orientation of the bioactive surface, with respect to the α/β scaffold, differs between Bj-xtrIT, LqhαIT, and LqhIT2 and seems to correspond to the molecular exterior onto which the C-tail is anchored. From these analyses it has become evident that the bioactive surface of each toxin examined thus far includes the distal region of the C-tail and spatially vicinal residues. Furthermore, substitution of the residues contributed by the distal part of the C-tail had the most effect on bioactivity.

13.3   DIVERSIFICATION OF TOXINS

A variety of animals of distant phylogenetic origin, such as sea anemones (Coelenterata), cone snails (Mollusca), scorpions and spiders (Arthropoda), and snakes (Chordata), have developed venomous competence for predation and defense that compensates for their physical limitations. They produce within a special gland a mixture of deadly neurotoxins that are injected by a stinging device into a fast-moving prey or enemy in order to achieve fast paralysis. The continuous evolution of sodium channels may have caused changes in toxin target sites, which could, however, be compensated by developing new toxins or by accelerated reshaping of existing toxins.

13.3.1   CONVERGENT AND DIVERGENT TOXIN EVOLUTION

Toxins of a given animal group may diverge from a common progenitor and provide a battery of functionally distinct polypeptides (divergent evolution) [67–72, 58]. However, genetically unrelated toxins, whose bioactive site has been formed independently, may recognise similar target sites (convergent evolution; e.g. potassium channel blockers, such as α-dendrotoxins, from the green mamba snake *Dendroaspis angusticeps*; charybdotoxin, from the scorpion *Leiurus quin-questriatus hebraeus*; BgK sea anemone toxin, from *Bunodosoma granulifera*; PVIIA κ-conotoxin, from *Conus purpurascens*) (reviewed in [71]). Snake toxins, such as curaremimetics and fasciculins (reviewed in [70]), and dendrotoxins and bovine pancreatic trypsin inhibitors (BPTI), display different bioactive sites that recognise two distinct targets. These bioactive sites occupy topographically unrelated regions on conserved protein scaffolds [73–75]. Cone snails have adopted an alternative route of toxin divergence by producing a large variety of small peptides (conotoxins), via gene duplication, accelerated mutagenesis, insertion of unusual amino acid residues, and a variety of post-translational modifications [72, 76, 77]. A unique route of diversification of 'long chain' scorpion toxins has recently been suggested on the basis of genetic studies combined with determination of the bioactive surface of toxin representatives of various pharmacological groups [58, 66].

13.3.2   SCORPION TOXIN GENE FAMILIES AND GENOMIC ORGANISATION

Analysis of cDNA and genomic clones of representative scorpion 'short chain' (mainly potassium channel blockers) and 'long chain' (sodium channel modi-fiers) neurotoxins from *Leiurus quinquestriatus hebraeus*, *Buthotus judaicus*, and *Buthus occitanus mardochei* revealed in all instances gene families. This poly-morphism is not due to population variations since it was found in individual scorpion segments and could be explained by gene duplication and a high rate of mutations [58, 78, 66]. Comparison of the various genomic clones uncovered similar gene organisation of two exons and one intron. The introns vary in length and are located 43–55 nucleotides downstream of the ATG initiation codon of the leader sequence [79–83, 58], thereby suggesting the formation of a pro-protein during toxin maturation and secretion [58].

The prominent polymorphism together with the conserved gene organisa-tion, intron features, common cysteine-stabilised α-helical (CSH) core connect-ing the α-helix to the three-stranded β-sheet (Plate VIII and Figure 13.1), and the modulation of membrane potential upon binding to ion channels suggest that all scorpion toxins evolved from an ancestral common progenitor. The vast diversity found among genomic copies, cDNAs, and their protein products for each toxin suggests an extensive evolutionary process of the scorpion 'pharma-ceutical factory' whose success is due, most likely, to the inherent permissive-ness of the toxin exterior to structural alterations [58].

Specific features of the scorpion toxins that suggest common ancestry were, however, detected in other polypeptides of enormous phylogenetic distance. A group of membrane potential modulators including plant γ-thionines, scorpion toxins, insect and scorpion defensins, bee venom apamin and MCD peptide, snake sarafotoxins, and human endothelins, have been shown to share a similar

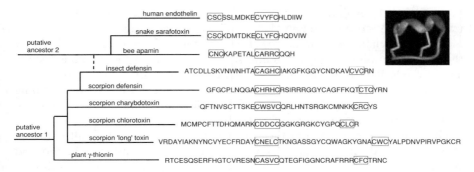

Figure 13.1 Sequence comparison and a putative evolutionary scheme for various membrane potential modulators encompassing plant γ-thionines, arthropod toxins and defensins, snake toxins, and human endothelin (for details see [69]). The CSH motif is boxed in the sequences and is also shown three-dimensionally by the endothelin 1 structure (PDB accession number 1EDP). The α-helix and β-strand (blue) are connected by two disulphide bonds (yellow)

CSH motif [84] and gene organisation [69]. This commonality suggested that these polypeptides might have diverged from a common ancestor (Figure 13.1).

13.3.3   A PUTATIVE MECHANISM FOR DIVERSIFICATION OF BIOACTIVE SURFACES IN SCORPION 'LONG' TOXINS

On the basis of common ancestry, conservation of core structure, and divergence of the C-tail from a common starting point, which coincides with the bioactive surface, a putative accelerated evolutionary route for diversification of bioactive surfaces has been suggested [66]. According to this suggestion, the C-tail, starting after the third β-strand and protruding out of the α/β scaffold, has undergone structural rearrangements so that new molecular exteriors were formed. The evolutionary 'wiggling' of the C-tail has been confined by the fourth disulphide bridge, whose position in excitatory toxins is remote from the conserved position in all other toxins (Plates VIII and X). Thus, it was hypothesised that alterations in a short hypervariable region, having a limited degree of structural freedom due to the fourth disulphide bond, induced the formation of a variety of bioactive surfaces (Plate X). This process may have occurred in parallel to the evolutionary changes of target sites in sodium channels. The molecular surfaces exerting novel bioactivities with an advantage for predation or defense have, most likely, become the progenitors of the existing pharmacological groups, e.g., alpha and its subgroups, beta, excitatory, and depressant toxins. Apparently, additional alterations of residues in these surfaces enabled increases in binding affinity to the various receptor sites. In this respect, the vast genetic polymorphism of all scorpion neurotoxins [58] suggests that the evolutionary process has not ceased and that the 'pharmaceutical factory' in scorpion venom glands still seeks new bioactive sites.

From a mechanistic point of view, this scorpion toxin model is reminiscent of the diversification of antigen binding sites that occurs in the immune system. The variable light (VL) and heavy (VH) chains of antibodies constitute two domains, a hypervariable unit comprising six complementarity-determining regions (CDRs) and a framework region (FR) exhibiting far less variation. The framework region serves as a scaffold for structural disposition of the CDR loops of the heavy and the light chains. Superposition of all antibody structures reveals differences in CDR loop orientation, length, and amino acid composition, thereby generating a wide array of specificities [85]. Although scorpion toxins and antibodies are not by any means comparable, it is interesting to observe a strategic commonality in their diversification where only a section of the protein displays structural variability, thereby inducing changes in an interactive molecular face to cope with ever-changing targets at minimal energy expense.

13.4   PERSPECTIVES IN TOXIN DESIGN

13.4.1   MOBILISATION OF BIOACTIVE SITES ON A CONSERVED SCAFFOLD

The diversification of 'long chain' scorpion toxins from a common progenitor ([69]; Figure 13.1) and the putative route of active site diversification on a conserved α/β scaffold ([66]; Plate X) raised various possibilities of toxin engineering and active site manipulation. One such approach examined the feasibility of active site exchange between two scorpion α-toxins from *L. q. hebraeus* that differ greatly in preference for insect or mammalian sodium channels. The five-residue turn (residues 8–12), C-tail (residues 56–64), and Phe17, that have been shown to be important for bioactivity of LqhαIT in insects ([45, 66]; unpublished results), were mobilised onto the Lqh2 scaffold. The recombinant chimera was unable to bind rat brain synaptosomes (the original Lqh2 activity) but gained anti-insect toxicity and the ability to bind to cockroach neuronal membranes much like LqhαIT. A reciprocal construction provided an LqhαIT scaffold with bioactive surface much like that of Lqh2 as determined in toxicity assays in mice and binding assays using rat brain synaptosomes (unpublished results). This experiment has confirmed that both bioactive surfaces occupy similar molecular exteriors, which suggests that subtle structural differences of the bioactive surface dictate selectivity. This experiment does not imply that the bioactive surface of toxins from other pharmacological groups, e.g. β, excitatory, and depressant, would be oriented identically. However, molecular dissection of LqhIT2 suggests that the bioactive face is oriented around a similar molecular exterior, whereas that of the excitatory toxin, Bj-xtrIT, occupies another region, which is located around the C-tail anchoring site [66].

13.4.2 FUTURE PROSPECTS IN DESIGN OF TARGET-SELECTIVE TOXINS

Beyond the importance of scorpion toxins in studying sodium channel structure and function, the selectivity of certain toxins for insect or mammalian sodium channel subtypes offers a potential model for the design of a new generation of insecticides [56, 57, 22, 86, 62] or selective drugs [87, 88; Gordon *et al.*, Chapter 12]. The most significant information required for such applications is the details of how the surface of each toxin interacts with its receptor binding-site as well as the structural elements in both toxins and receptors that dictate their selective encounter. However, to enable such a study, toxin structures and active sites as well as channel outer regions that interact with the toxins should be determined. In this respect, the details on the bioactive surface and 3-D structure of the 'long chain' scorpion toxin representatives achieved thus far pave the way in what seems a tedious, yet rewarding 'plan of operation'. The expression of functional sodium channels in heterologous systems and preliminary data on channel receptor sites that bind scorpion toxins [89–91; unpublished results; Gordon *et al.*, Chapter 12) as well as emerging details on channel structure [92] suggest that elucidation of toxin-channel interacting surfaces is imminent. Moreover, the elucidation of toxins that differentiate between sodium channel subtypes (Gordon *et al.*, Chapter 12) and the successful mobilisation of α-toxin bioactive surfaces on a conserved protein scaffold (unpublished results) raise future possibilities for toxin design. An example of such a useful design would be the toxin, LqhαIT, which is active in both insects and mammals. Restriction of LqhαIT toxicity to insects by molecular design could be of high value in insect control management due to its prominent synergism with pyrethroids [93; unpublished results]. Another α-toxin, Lqh3, which has been shown to recognise specifically a distinct sodium channel subtype in neuronal cell bodies of the mammalian nervous system, also affects other sodium channel subtypes, e.g. in skeletal and cardiac muscles [87, 88] and in insects [94]. Restriction of the activity of Lqh3, whose structure resembles that of other α-toxins [95], to a sodium channel subtype specific to cell bodies in brain neurons may provide a model for design of selective drugs [87].

   Another approach to the design of selective sodium channel modifiers might capitalise on elucidated toxic surfaces of certain toxins with the aim of creating miniproteins that mimic the function of their large templates [71], or even synthesise chemicals capable of producing similar effects in a selective mode. Such an approach has already been initiated with 'short chain' scorpion toxins [71] and due to the recent advances in the study of structure–activity relationships has now become imminent with 'long chain' toxins.

ACKNOWLEDGEMENTS

Our research was supported by Grants: IS-2901–97C from BARD, The United States–Israel Binational Agricultural Research & Development (MG); 466/97 (MG) and 508/01 (DG) from The Israel Academy of Sciences and Humanities.

REFERENCES

[1] Martin-Eauclaire, M-F. and Couraud, F. (1995) in *Handbook of Neurotoxicology* (Chang, L.W. and Dyer, R.S., eds.) pp. 683–716, Marcel Dekker, New York.

[2] Gordon, D., Savarin, P., Gurevitz, M. and Zinn-Justin, S. (1998) Functional anatomy of scorpion toxins affecting sodium channels. *J. Toxicol. Toxin. Rev.* **17**, 131–59.

[3] Catterall, W.A. (1986) Molecular properties of voltage-sensitive sodium channels. *Annu. Rev. Biochem.* **55**, 953–85.

[4] Catterall, W.A. (1992) Cellular and molecular biology of voltage-gated sodium channels. *Physiol. Rev.* **72**, S15–S48.

[5] Gordon, D., Martin-Eauclaire, M-F., Cestele, S., Kopeyan, C., Carlier, E., Ben Khalifa, R., Pelhate, M. and Rochat, H. (1996) Scorpion toxins affecting sodium current inactivation bind to distinct homologous receptor sites on rat brain and insect sodium channels. *J. Biol. Chem.* **271**, 8034–45.

[6] Gordon, D. (1997) Sodium channels as targets for neurotoxins: mode of action and interaction of neurotoxins with receptor sites on sodium channels. in *Toxins and Signal Transduction*, (Lazarovici, P. and Gutman, Y., eds.) pp. 119–49, Harwood Press, Amsterdam.

[7] Zlotkin, E., Miranda, F. and Rochat, H. (1978) Chemistry and pharmacology of Buthinae scorpion venoms. In *Arthropod venoms* (Bettini, S., ed.), Springer Verlag, New York, NY, pp. 317–69.

[8] Zlotkin, E. (1997) Insect selective neurotoxins from scorpion venoms affecting sodium conductance. in *Toxins and Signal Transduction* (Lazarovici, P. and Gutman, Y., eds.), pp. 95–117, Harwood Press, Amsterdam.

[9] Ben-Khalifa, R., Stankiewicz, M., Lapied, B., Turkov, M., Zilberberg, N., Gurevitz, M. and Pelhate, M. (1997) Refined electrophysiological analysis suggests that a depressant toxin is a sodium channel opener rather than a blocker. *Life Sci.* **61**, 819–30.

[10] Froy, O., Zilberberg, N., Gordon, D., Turkov, M., Gilles, N., Stankiewicz, M., Pelhate, M., Loret, E., Oren, D.A., Shaanan, B. and Gurevitz, M. (1999) The putative bioactive surface of insect-selective scorpion excitatory neurotoxins. *J. Biol. Chem.* **274**, 5769–76.

[11] Gordon, D. (1997) A new approach to insect-pest control – combination of neurotoxins interacting with voltage sensitive sodium channels to increase selectivity and specificity. *Invertebr. Neurosci.* **3**, 103–16.

[12] Gordon, D., Moskowitz, H., Eitan, M., Warner, C., Catterall, W.A. and Zlotkin, E. (1992) Localization of receptor sites for insect-selective toxins on sodium channels by site-directed antibodies. *Biochemistry* **31**, 7622–8.

[13] De Lima, M.E., Martin-Eauclaire, M-F., Hue, B., Loret, E., Diniz, C.R. and Rochat, H. (1989) On the binding of two scorpion toxins to the central nervous system of the cockroach *Periplaneta americana. Insect Biochem.* **19**, 413–22.

[14] Possani, L.D., Becerril, B., Delepierre, M. and Tytgat, J. (1999) Scorpion toxins specific for Na$^+$-channels. *Eur. J. Biochem.* **264**, 287–300.

[15] Darbon, H., Jover, E., Couraud, F. and Rochat, H. (1983) α-Scorpion neurotoxin derivatives suitable as potential markers of sodium channels. *Int. J. Peptide Protein Res.* **22**, 179–86.
[16] Kharrat, R., Darbon, H., Rochat, H. and Granier, C. (1989) Structure–activity relationships of scorpion α-toxins. Multiple residues contribute to the interaction with receptors. *Eur. J. Biochem.* **181**, 381–90.
[17] El-Ayeb, M., Darbon, H., Bahraoui, E-M., Vargas, O. and Rochat, H. (1986) Differential effect of defined chemical modifications of antigenic and pharmacological activities of scorpion α and β toxins. *Eur. J. Biochem.* **155**, 289–94.
[18] Bahraoui, E., Granier, C., van Rietschoten, J., Rochat, H. and El Ayeb, M. (1986) Specifity and neutralizing capacity of antibodies elicited by a synthetic peptide of scorpion toxin. *J. Immunol.* **136**, 3371–7.
[19] El-Ayeb, M., Bahraoui, E-M., Granier, C. and Rochat, H. (1986) Use of antibodies specific to defined regions of scorpion α-toxin to study its interaction with its receptor site on the sodium channel. *Biochemistry* **25**, 6671–8.
[20] Bougis, P.E., Rochat, H. and Smith, L.A. (1989) Precursors of *Androctonus australis* scorpion neurotoxins: Structures of precursors, processing outcomes, and expression and functional recombinant toxin II. *J. Biol. Chem.* **264**, 19259–65.
[21] Dee, A., Belagaje, R.M., Ward, K., Chio, E. and Lai, M-H.T. (1990) Expression and secretion of a functional scorpion insecticidal toxin in cultured mouse cells. *Bio/ Technology* **8**, 339–42.
[22] Stewart, L.M.D., Hirst, M., Ferber, M.L., Merryweather, A.T., Cayley, P.J. and Possee, R.D. (1991) Construction of an improved baculovirus insecticide containing an insect-specific toxin gene. *Nature* **352**, 85–8.
[23] Tomalski, M.D. and Miller, L.K. (1991) Insect paralysis by baculovirus-mediated expression of a mite neurotoxin gene. *Nature* **352**, 82–5.
[24] Pang, S-Z., Oberhouse, S.M., Rasmussen, J.L., Knipple, D.C., Bloomquist, J.R., Dean, D.H., Bowman, K.D. and Sanford, J.C. (1992) Expression of a gene encoding a scorpion insectotoxin peptide in yeast, bacteria and plants. *Gene* **116**, 165–72.
[25] Howell, M.L. and Blumenthal, K.M. (1989) Cloning and expression of synthetic gene for *Cerebratulus lacteus* neurotoxin B-IV. *J. Biol. Chem.* **264**, 15268–73.
[26] Fiordalisi, J.J., Fetter, C.H., TenHarmsel, A., Gigowski, R., Chiappinelli, V.A. and Grant, G.A. (1991) Synthesis and expression in *Escherichia coli* for k-bungarotoxin. *Biochemistry* **30**, 10337–43.
[27] Gurevitz, M. and Zilberberg, N. (1994) Advances in molecular genetics of scorpion neurotoxins. *J. Toxicol. Toxin Rev.* **13**, 65–100.
[28] Zilberberg, N., Gordon, D., Zlotkin, E., Pelhate, M., Adams, M.E., Norris, T. and Gurevitz, M. (1996) Functional expression and genetic alteration of an alpha scorpion neurotoxin. *Biochemistry* **35**, 10215–22.
[29] Gurevitz, M., Froy, O., Zilberberg, N., Turkov, M., Strugatzky, D., Gershburg, E., Lee, D., Adams, M.E., Tugarinov, V., Anglister, J., Shaanan, B., Loret, E., Stankiewicz, M., Pelhate, M., Gordon, D. and Chejanovsky, N. (1998) Sodium channel modifiers from scorpion venom: Structure–activity relationship, mode of action, and application. *Toxicon* **36**, 1671–82.
[30] Marston, F.A.O. (1986) The purification of eukaryotic polypeptides synthesized in *Escherichia coli*. *Biochem. J.* **240**, 1–12.
[31] Sabatier, J.M., Darbon, H., Fourquet, P., Rochat, H. and van Rietschoten, J. (1987) Reduction and reoxidation of the neurotoxin II from the scorpion *Androctonus australis* Hector. *Int. J. Pept. Protein Res.* **30**, 125–34.
[32] Bouhaoula-Zahar, B., Ducancel, F, Zenouaki, I., Ben Khalifa, R., Borchani, L., Pelhate, M., Boulain, J.C., El-Ayeb, M., Menez, A. and Karoui, H. (1996) A recombinant insect-specific α-toxin of *Buthus occitanus tunetanus* scorpion confers protection against homologous mammal toxins. *Eur. J. Biochem.* **238**, 653–60.

[33] Zilberberg, N., Zlotkin, E. and Gurevitz, M. (1990) Characterization of the transcript for a depressant insect selective neurotoxin gene with an isolated cDNA clone from the scorpion *Buthotus judaicus. FEBS Lett.* **269**, 229–32.

[34] Zilberberg, N., Zlotkin, E. and Gurevitz, M. (1991) The cDNA sequence of a depressant insect selective neurotoxin from the scorpion *Buthotus judaicus. Toxicon* **29**, 1155–8.

[35] Gurevitz, M., Urbach, D., Zlotkin, E. and Zilberberg, N. (1991) Nucleotide sequence and structure analysis of a cDNA encoding an alpha insect toxin from the scorpion *Leiurus quinquestriatus hebraeus. Toxicon* **29**, 1270–2.

[36] Zilberberg, N., Zlotkin, E. and Gurevitz, M. (1992) Molecular analysis of cDNA and transcript encoding the depressant insect neurotoxin of the scorpion *Leiurus quinquestriatus hebraeus. Insect Biochem. Molec. Biol.* **22**, 199–203.

[37] Zilberberg, N. and Gurevitz, M. (1993) Rapid isolation of full length cDNA clones by 'inverse PCR': Purification of a scorpion cDNA family encoding α-neurotoxins. *Anal. Biochem.* **209**, 203–5.

[38] Eitan, M., Fowler, E., Herrmann, R., Duval, A., Pelhate, M. and Zlotkin, E. (1990) A scorpion venom neurotoxin paralytic to insects that affects sodium current inactivation: Purification, primary structure, and mode of action. Biochemistry **29**, 5941–7.

[39] Zlotkin, E., Eitan, M., Bindokas, V., Adams, M.E., Moyer, M., Brukhart, W. and Fowler, E. (1991) Functional duality and structural uniquness of depressant insect-selective neurotoxins. *Biochemistry* **30**, 4814–20.

[40] Rosenberg, A.H., Lade, B.N., Chui D., Lin, S., Dunn, J.J. and Studier, F.W. (1987) Vectors for selective expression of cloned DNAs by T7 RNA polymerase. *Gene* **56**, 125–35.

[41] Turkov, M., Rashi, S., Zilberberg, N., Gordon, D., BenKhalifa, R., Stankiewicz, M., Pelhate, M. and Gurevitz, M. (1997) Expression in *Escherichia coli* and reconstitution of a functional recombinant scorpion depressant neurotoxin specific against insects. *Prot. Express. Purific.* **10**, 123–31.

[42] Fontecilla-Camps, J.C., Almassy, R.J., Suddath, F.L., Watt, D.D. and Bugg, C.E. (1980) Three-dimensional structure of a protein from scorpion venom: a new structural class of neurotoxins. *Proc. Natl. Acad. Sci. USA* **77**, 6496–500.

[43] Fontecilla-Camps, J.L., Habersetzer-Rochat, C. and Rochat, H. (1988) Orthorhombic crystals and three dimensional structure of the potent toxin II from the scorpion *Androctonus australis Hector. Proc. Natl. Acad. Sci. USA* **85**, 7443–7.

[44] Fontecilla-Camps, J.C. (1989) Three-dimentional model of the insect-directed scorpion toxin from *Androctonus australis* Hector and its implication for the evolution of scorpion toxins in general. *J. Mol. Evol.* **29**, 63–7.

[45] Zilberberg, N., Froy, O., Cestele, S., Loret, E., Arad, D., Gordon, D. and Gurevitz, M. (1997) Elucidation of the putative toxic-surface of a highly insecticidal scorpion α-neurotoxin affecting voltage-sensitive sodium channels. *J. Biol. Chem.* **272**, 14810–16.

[46] Tugarinov, V., Kustanovitz, I., Zilberberg, N., Gurevitz, M. and Anglister, Y. (1997) Solution structure of a highly insecticidal recombinant scorpion α-toxin and a mutant with increased activity. *Biochemistry* **36**, 2414–24.

[47] Pashkov, V.S., Maiorov, V.N., Bystrov, V.F., Hoang, A.N., Volkova, T.M. and Grishin, E.V. (1988) Solution spatial structure of 'long' neurotoxin M9 from the scorpion *Buthus eupeus* by ^{1}H-NMR spectroscopy. *Biophys. Chem.* **31**, 121–31.

[48] Zhao, B., Carson, M., Ealick, S.E. and Bugg, C.E. (1992) Structure of scorpion toxin variant-3 at 1.2 Å resolution. *J. Mol. Biol.* **227**, 239–52.

[49] Jablonsky, M.J., Watt, D.D. and Krishna, N.R. (1995) Solution structure of an Old World-like neurotoxin from the venom of the New World scorpion *Centruroides sculpturatus* Ewing. *J. Mol. Biol.* **248**, 449–58.

[50] Landon, C., Sodano, P., Cornet, B., Bonmatin, J-M., Kopeyan, C., Rochat, H., Vovelle, F. and Ptak, M. (1997) Refined solution structure of the anti-mammal and anti-insect Lqq III scorpion toxin: comparison with other scorpion toxins. *Proteins* **28**, 360–74.

[51] Miranda, F., Kupeyan, C., Rochat, H., Rochat, C. and Lissitzky, S. (1970) Purification of animal neurotoxins. Isolation and characterization of eleven neurotoxins from the venoms of the scorpion *Androctonus australis* Hector, *Buthus occitanus tunetanus*, and *Leiurus quinquestriatus quinquestriatus*. *Eur. J. Biochem.* **16**, 514–23.

[52] Darbon, H., Zlotkin, E., Kopeyan, C., van Rietschoten, J. and Rochat, H. (1982) Covalent structure of the insect toxin of the North African scorpion *Androctonus australis* Hector. *Int. J. Pept. Protein Res.* **20**, 320–30.

[53] Oren, D.A., Froy, O., Amit, E., Kleinberger-Doron, N., Gurevitz, M. and Shaanan, B. (1998) An excitatory scorpion toxin with a distinctive feature: an additional α-helix at the C-terminus and its implications for interaction with insect sodium channels. *Structure* **6**, 1095–103.

[54] Dianous, S., Hoarau, F. and Rochat, H. (1987) Re-examination of the specificity of the scorpion *Androctonus australis* Hector insect toxin towards arthropods. *Toxicon* **25**, 411–17.

[55] Zlotkin, E., Rochat, H., Kupeyan, C., Miranda, F. and Lissitzky, S. (1971) Purification and properties of the insect toxin from the venom of the scorpion *Androctonus australis* Hector. *Biochimie* **53**, 1073–8.

[56] Maeda, S., Volrath, S.L., Hanzlik, T.N., Harper, S.A., Majuma, K., Maddox, D.W., Hammock, B.D. and Fowler, E. (1991) Insecticidal effects of an insect-specific neurotoxin expressed by a recombinant baculovirus. *Virology* **164**, 777–80.

[57] McCutchen, B.F., Hoover, K., Preisler, H.K., Betana, M.D., Herrmann, R., Robertson, J.L. and Hammock, B.D. (1997) Interactions of recombinant of wild-type baculoviruses with classical insecticides and pyrethroid-resistant tobacco budworm (Lepidoptera: Noctuidae). *J. Econ. Entomol.* **90**, 1170–80.

[58] Froy, O., Sagiv, T., Poreh, M., Urbach, D., Zilberberg, N. and Gurevitz, M. (1999b) Dynamic diversification from a putative common ancestor of scorpion toxins affecting sodium, potassium and chloride channels. *J. Mol. Evol.* **48**, 187–96.

[59] Gordon, D., Jover, E., Couraud, F. and Zlotkin, E. (1984) The binding of the insect selective neurotoxin (AaIT) from scorpion venom to locust synaptosomal membranes. *Biochim. Biophys. Acta* **778**, 349–58.

[60] Moskowitz, H., Herrmann, R., Zlotkin, E. and Gordon, D. (1994) Variability among insect sodium channels revealed by binding of selective neurotoxins. *Insect Biochem. Molec. Biol.* **24**, 13–19.

[61] Lester, D., Lazarovici, P., Pelhate, M. and Zlotkin, E. (1982) Two insect toxins from the venom of the scorpion *Buthotus judaicus*. Purification, characterization and action. *Biochim. Biophys. Acta* **701**, 370–81.

[62] Gershburg, E., Stockholm, D., Froy, O., Rashi, S., Gurevitz, M. and Chejanovsky, N. (1998) Baculovirus-mediated expression of a scorpion depressant toxin improves the insecticidal efficacy achieved with excitatory toxins. *FEBS Lett.* **422**, 132–6.

[63] Pintar, A., Possani, L.D. and Delepierre, M. (1999) Solution structure of toxin 2 from *Centruroides noxius* Hoffmann, a beta-scorpion neurotoxin acting on sodium channels. *J. Mol. Biol.* **287**, 359–67.

[64] Polikarpov, I., Sanches Matilde Jr., M., Marangoni, S., Toyama, M.H. and Teplyakov, A. (1999) Crystal structure of neurotoxin Ts1 from *Tityus serrulatus* provides insights into the specificity and toxicity of scorpion toxins. *J. Mol. Biol.* **290**, 175–84.

[65] Darbon, H., Weber, C. and Braun, W. (1991) Two-dimensional ¹H nuclear magnetic resonance study of AaHIT, an anti-insect toxin from the scorpion *Androctonus australis* Hector. Sequential resonance assignments and folding of the polypeptide chain. *Biochemistry* **30**, 1836–45.

[66] Gurevitz, M., Gordon, D., Ben-Natan, S., Turkov, M. and Froy, O. (2001) Diversification of neurotoxins by C-tail 'wiggling' – a scorpion recipe for survival. *FASEB J.* **15**, 1201–5.

[67] Norton, R.S. (1991) Structure and structure–function relationships of sea anemone proteins that interact with the sodium channel. *Toxicon* **29**, 1051–94.

[68] Norton, R.S. and Pallaghy, P.K. (1998) The cystine knot structure of ion channel toxins and related polypeptides. *Toxicon* **36**, 1573–83.

[69] Froy, O. and Gurevitz, M. (1998) Membrane potential modulators – a thread of scarlet from plants to humans. *FASEB J.* **12**, 1793–6.

[70] Ohno, M., Ménez, R., Ogawa, T., Danse, J.M., Shimohigashi, Y., Fromen, C., Ducancel, F., Zinn-Justin, S., Le Du, M.H., Boulain, J-C., Tamiya, T. and Menez, A. (1998) Molecular evolution of snake toxins: Is the functional diversity of snake toxins associated with a mechanism of accelerated evolution? in *Progress in Nucleic Acid Research and Molecular Biology* (Moldave, K., ed.) Vol. 59, pp. 307–64, Academic Press, NY.

[71] Ménez, A. (1998) Functional architectures of animal toxins: a clue to drug design. *Toxicon* **36**, 1557–72.

[72] Olivera, B.M., Rivier, J., Scott, J.K., Hillyard, D.R. and Cruz, L.J. (1999) Conotoxins. *J. Biol. Chem.* **266**, 22067–70.

[73] Berndt, K.D., Guntert, P., Orbons, L.P. and Wuthrich, K. (1992) Determination of a high-quality nuclear magnetic resonance solution structure of the bovine pancreatic trypsin inhibitor and comparison with three crystal structures. *J. Mol. Biol.* **227**, 757–75.

[74] Harvey, A.L. (1997) Recent studies on dendrotoxins and potassium ion channels. *Gen. Pharmacol.* **28**, 7–12.

[75] Gasparini, S., Danse, J.M., Lecoq, A., Pinkasfeld, S., Zinn-Justin, S., Young, L.C., de Medeiros, C.C., Rowan, E.G., Harvey, A.L. and Ménez, A. (1998) Delineation of the functional site of α-dendrotoxin: the functional topographies of dendrotoxins are different but share a conserved core with those of other Kv1 potassium channel blocking toxins. *J. Biol. Chem.* **273**, 25393–403.

[76] Duda, T.F. and Palumbi, S.R. (1999) Molecular genetics of ecological diversification: duplication and rapid evolution of toxin genes of the venomous gastropod conus. *Proc. Natl. Acad. Sci. USA* **96**, 6820–3.

[77] Conticelo, S.G., Gilad, Y., Avidan, N., Ben-Asher, E., Levy, Z. and Fainzilber, M. (2001) Mechanisms for evolving hypervariability: the case of conopeptides. *Mol. Biol. Evol.* **18**, 120–31.

[78] Vita, C., Roumestand, C., Toma, F. and Ménez, A. (1995) Scorpion toxins as natural scaffolds for protein engineering. *Proc. Natl. Acad. Sci. USA* **92**, 6404–8.

[79] Becerill, B., Corona, M., Mejia, M.C., Martin, B.M., Lucas, S., Bolivar, F. and Possani, L.D. (1993) The genomic region encoding toxin gamma from the scorpion *Tityus serrulatus* contains an intron. *FEBS Lett.* **335**, 6–8.

[80] Delabre, M.L., Pasero, P., Marilley, M. and Bougis, P.E. (1995) Promoter structure and intron-exon organization of a scorpion α-toxin gene. *Biochemistry* **34**, 6729–36.

[81] Becerill, B., Corona, M., Coronas, F.I.V., Zamudio, F., Calderon-Aranda, E.S., Fletcher, P.L., Martin, B.M. and Possani, L.D. (1996) Toxic peptides and genes encoding toxin γ of the Brazilian scorpions *Tityus bahiensis* and *Tityus stigmurus*. *Biochem. J.* **313**, 753–60.

[82] Corona, M., Zurita, M., Possani, L.D. and Becerill, B. (1996) Cloning and characterization of the genomic region encoding toxin IV-5 from the scorpion *Tityus serrulatus* Lutz and Mello. *Toxicon* **34**, 251–6.

[83] Legros, C., Bougis, P.E. and Martin-Eauclaire M-F. (1997) Genome organization of the KTX₂ gene, encoding a 'short' scorpion toxin active on K⁺ channels. *FEBS Lett.* **402**, 45–9.

[84] Kobayashi, Y., Takashima, H., Tamaoki, H., Kyogoku, Y., Lambert, P., Kuroda, H., Chino, N., Watanabe, T.X., Kimura, T., Sakakibara, S. and Moroder, L. (1991) The cysteine-stabilized α-helix: A common structural motif of ion-channel blocking neurotoxic peptides. *Biopolymers* **31**, 1213–20.

[85] Kuby, J. (1994) *Immunology*, Chapter 5, 2nd Edition, pp. 109–34, Freeman & Company, New York.

[86] Chejanovsky, N., Zilberberg, N., Rivkin, H., Zlotkin, E. and Gurevitz, M. (1995) Functional expression in insect cells and in lepidopterous larvae of a scorpion anti-insect alpha neurotoxin. *FEBS Lett.* **376**, 181–4.

[87] Gilles, N., Blanchet, C., Shichor, I., Zaninetti, M., Lotan, I., Bertrand, D. and Gordon, D. (1999) A scorpion α-like toxin that is active on insects and mammals reveals an unexpected specificity and distribution of sodium channel subtypes in rat brain neurons. *J. Neurosci.* **19**, 8730–9.

[88] Gilles, N., Chen, H., Wilson, H., Legall, F., Montoya, G., Molgo, J., Schönherr, R., Nicholson, G., Heinemann, S.H. and Gordon, D. (2000) Scorpion α- and α-like toxins differentially interact with sodium channels in mammalian CNS and periphery. *Eur. J. Neurosci.* **12**, 2823–32.

[89] Rogers, J.C., Qu, Y., Tanada, T.N., Scheuer, T. and Catterall, W.A. (1996) Molecular determinants of high affinity binding of α-scorpion toxin and sea anemone toxin in the S3–S4 extracellular loop in domain IV of the sodium channel α subunit. *J. Biol. Chem.* **271**, 15950–62.

[90] Cestele, S., Qu, Y., Rogers, J.C., Rochat, H. and Catterall, W.A. (1998) Voltage sensor-trapping: enhanced activation of sodium channels by α-scorpion toxin bound to the S3–S4 loop in domain II. *Neuron* **21**, 919–31.

[91] Ma, Z., Tang, L., Lu, S., Kong, J., Gordon, D. and Kallen, R.G. (2000) The domain 4 S3–S4 extracellular loop provides molecular determinants for binding of α-scorpion toxins (Lqh II, Lqh III and LqhαIT) to the voltage-gated rat skeletal muscle Na^+ channel (rSkM1). *Biophys. J.* **78**, A1011.

[92] Sato, C., Ueno, Y., Asal, K., Takahashi, K., Sato, M., Engel, A. and Fujiyoshi, Y. (2001) The voltage-sensitive sodium channel is a bell-shaped molecule with several cavities. *Nature* **409**, 1047–51.

[93] Lee, D., Park, Y., Brown, T.N. and Adams, M.E. (1999) Altered properties of neuronal sodium channels associated with genetic resistance to pyrethroids. *Mol. Pharmacol.* **55**, 584–93.

[94] Gilles, N., Krimm, I., Bouet, F., Froy, O., Gurevitz, M., Lancelin, J-M. and Gordon, D. (2000) Structural implications on the interaction of scorpion α-like toxins with the sodium channel receptor site inferred from toxin iodination and pH-dependent binding. *J. Neurochem.* **75**, 1735–45.

[95] Krimm, I., Gilles, N., Sautiere, P., Stankiewicz, M., Pelhate, M., Gordon, D., Lancelin, J-M. (1999) NMR structures and activity of a novel α-like toxin from the scorpion *Leiurus quinquestriatus hebraeus*. *J. Mol. Biol.* **285**, 1749–63.

14 Methodological Approaches to the Study of Ion Channels Using Peptide Toxins: Proposed Comprehensive Guidelines

MICHEL DE WAARD, JEAN-MARC SABATIER and HERVÉ ROCHAT*

In two decades, ion channels have gone through the stage of simple hypothetical structures to the description of real molecular entities. The first effort that has been made to classify these extremely diverse structures has been through functional and pharmacological criteria. The discovery of new natural compounds active on ion channels helped the classification efforts and served to validate new hypotheses on the existence of still more diverse structures. These molecular tools have generally been useful to the purification of ion channels on which they are active. For instance, the use of ω-conotoxin GVIA, a toxin isolated from the venom of a marine snail cone (*Conus geographus*), has been instrumental to the purification of voltage-dependent N-type calcium channels, a presynaptic oligomeric complex implicated in neurotransmitter release [1]. Purification of ion channels often leads the way to the molecular cloning of the underlying structures. Cloning of the α subunit of the voltage-gated sodium channel [2] was possible only after the purification of this channel from *Electrophorus electricus* electroplax membranes with the aid of tetrodotoxin [3], a toxin isolated from the puffer fish fugu. Toxins were then used in a post-cloning period to validate the cloning results. The cDNAs obtained have often been expressed in various recipient cell lines and the resulting functional ion channels were probed pharmacologically with the same molecular tools that served to trace these ion channels during their purification. These simple considerations illustrate the complex and permanent interplay that has arisen over time between the ligand (frequently, but not exclusively, a toxin) and the receptor (ion channel). This interplay is still pertinent and the methods used to study the relationship between these two partners are still evolving. This review focuses on the methodological approaches used to study such a relationship between these two interacting molecules or molecular complexes.

---

* Corresponding author

Perspectives in Molecular Toxinology
Edited by A Ménez © 2002 John Wiley & Sons, Ltd

14.1 ION CHANNELS ARE IMPORTANT TARGETS FOR PHARMACOLOGICAL AND THERAPEUTIC INTERVENTIONS

Most cells are excitable entities that require fast and controlled movement of ions from one side of the plasma membrane to the other. Each ion channel can be likened to an excitable macromolecular structure that is generally regulated by voltage and/or ligand binding and that reacts through conformational modifications leading to either opening or closing of its ionic pore element. The functional specificity of each ion channel is partly inferred by the ionic selectivity of the pore. Though some ion channels have a limited selectivity in that they are able to conduct multiple ion types (e.g. nicotinic receptors), most channels are extremely discriminative with regard to ionic permeability. For voltage-dependent ion channels, one can distinguish channels that conduct almost exclusively Na^+, K^+ or Ca^{2+} ions in physiological conditions. Such a selectivity is required to allow the desired changes in membrane potential, a key cellular event for cell-to-cell communication. However, several other parameters contribute to the required specificity of action of ion channels and they are no less important to the occurrence of an electrical modification of the cell. These parameters include ion channel localisation, cell surface density, biophysical properties and cellular regulation. Action potential generation is possible only because voltage-gated Na^+ channels open faster than K^+ channels in response to a depolarisation stimulus. Another important make-up of ion channel function is the ability of some of the transferred ions to behave as second messengers. This is clearly evidenced for Ca^{2+} which has several intracellular roles, the foremost of which is the control of the activity of other ion channels (mostly K^+ and Cl^- channels). A recent area of interest in the field of ion channels is based on the ability of these structures to interact with several intracellular components under direct control of their activity. For instance, skeletal muscle contraction takes place only because the voltage-gated Ca^{2+} channel of the plasma membrane controls the activity of another calcium channel present in the sarcoplasmic reticulum through direct molecular interaction. A similar, no less interesting, functional relationship seems to emerge from the interaction of neuronal voltage-gated Ca^{2+} channels with components of the SNARE complex, which is involved in neurotransmitter release [4]. These basic considerations highlight the importance of ion channels in cell functioning. Though this is not the topic of this review, there is growing evidence that ion channels are involved in multiple genetic and acquired human pathologies. Human diseases associated with ion channel dysfunction have even been grouped under the generic term of channelopathies. For example, amongst the acquired diseases known to affect Ca^{2+} channels and human health, the Lambert–Eaton Myasthenic Syndrome has been attributed to autoimmunity against voltage-gated N- and/or P/Q Ca^{2+} channels [5]. With regard to genetic deficiencies, several mutations have been identified in the gene coding for the α_1 subunit of P/Q Ca^{2+} channels. These genetic mutations can produce cerebellar degeneration accompanied by ataxia and/or epilepsy. Several cases of familial hemiplegic migraine have also been reported. These examples are far from being exhaustive and similar cases have been reported for many other ion channels (for review see [6]). They demonstrate the need to develop chemotherapeutic drugs capable of counteracting the negative side effects of these channelopathies.

Pharmacological intervention aimed at developing drugs active on ion channels may be an adequate strategy to treat several human diseases that do not directly involve ion channels. For instance, recent pharmacological efforts aim to develop novel peptides derived from toxin amino acid sequences able to treat pain [7], the consequences of ischemia [8] or to suppress immunological responses (e.g. kaliotoxin, margatoxin – [9, 10]). Another potential therapeutic value of this type of research is the development of drugs able to treat various neurological disorders (epilepsy, schizophrenia, Alzheimer's disease, multiple sclerosis, etc. [9, 11]). Also, potassium channel modulators could be used to treat cardiac ischemia and arrhythmia, type II diabetes, asthma, hypertension, alopecia, urinary incontinence and cerebral ischemia [12]. These examples are far from being exhaustive and illustrate the importance of ion channels as targets for the development of molecules of therapeutic interest.

14.2   TOXINS AS MOLECULAR TRACERS FOR THE PURIFICATION OF CLONED ION CHANNELS

With the completion of the Human Genome Sequencing Project, many new cDNA sequences are or will be available to the investigator interested in the cloning of ion channels. This *in silico* cloning strategy that is based on the search for homologous cDNA sequences has proven tremendously helpful in identifying many novel ion channels of interest. Such an approach shortcuts the classical efforts that were used hitherto in ion channel cloning, i.e. protein purification, peptide sequencing, antibody production and experimental cDNA screening. However, though classical cloning strategies often relied on the use of a molecular tracer during the purification procedure, such as a radio-labelled toxin, this novel cloning strategy bypasses the use of an ion channel ligand. More rarely, some investigators also used the strategy of expression cloning by cDNA fractioning/isolation and by screening the resultantion currents with the help of a toxin ligand. Successful cloning of a novel ion channel by a computerised approach requires the subsequent expression of the cDNA into a recipient cell line and a careful and thorough examination of its ionic properties. Such a cloning strategy is, however, flawed for two reasons: first, a cDNA cloned this way may not be functional in the absence of a required auxiliary subunit, and second, this strategy does not provide information on the real subunit composition of multimeric ion channels. This information will thus need to be completed by more classical biochemical approaches that may themselves require the use of a molecular tracer such as a toxin. It is therefore not unusual for toxins to be screened for their ability to modulate *in silico* cloned ion channels expressed in an appropriate cell line. The process of selecting adequate potential blockers is greatly facilitated by the analysis of the functional properties of the cloned ion channel that may be reminiscent of the properties observed for native ion channels. Selected toxins may therefore be used in a second step to purify the multimeric complex composing the ion channel of interest; a prelude to the characterisation of the auxiliary subunit(s). It was quite unusual to purify the α-dendrotoxin-sensitive voltage-gated K^+ channel from bovine brain [13], whereas the sequence of the main pore-forming subunit was long since known. An important contribution to this work was the discovery of a β subunit associated with the α ionic pore. This information proved extremely important, such that this initial study started the trend to the identification of novel β subunits involved in a molecular complex with other voltage-gated K^+ channels. A similar approach is envisaged for the purification of T-type calcium channels with the initiation of research programs looking for T-type-selective toxins that could be used as molecular tracers [14].

14.3 ON THE SEARCH FOR NEW NATURAL LIGANDS ACTING ON ION CHANNELS

There are two case scenarios that require the search for new ligands acting on ion channels. On the one hand, it may be necessary to look for a novel ligand that is more specific and/or has a higher affinity for its receptor. Also, it may be interesting to search for a ligand that regulates ion channel activity differently to ligands already available. On the other hand, one may wish to identify and characterise ligands that act on recently cloned ion channels of unknown function. We shall address existing strategies in the screening for novel peptide ligands acting on ion channels.

14.3.1 BIOLOGICAL SOURCES FOR NOVEL COMPOUNDS ACTING ON ION CHANNELS

Venoms of various biological origins (e.g. venomous animals, plants, micro-organisms) represent almost unlimited sources of pharmacologically active molecules. For instance, in a scorpion venom, several hundreds of molecules can be detected by mass spectrometry analysis; a number of these are active on a variety of ion channel types [15]. Though there is a huge diversity of compounds of interest present in a single venom, there is great difficulty in characterising them all. First, the quantity of a given toxin may vary greatly from one venom to the other, and the maximum quantities of any given toxin in a venom rarely exceed a few percent of the total protein content. Second, toxin isoforms

have been found in species and species-related venoms that differ only by one or
a few amino acid residues, thereby rendering the purification and characterisation tasks still more complex.

With these considerations in mind, the identification of novel toxins of
biological interest proceeds in several steps. First, one should select a biological
source that is most likely to provide the toxins of interest. For instance, spider
venoms appear to be rich in toxins active on calcium channels, whereas scorpion venoms are used as privileged sources for toxins active on sodium and
potassium channels. However, these statements should be viewed with caution
since in the venom of any given animal species, only a very limited number of
compounds have hitherto been really characterised. A practical approach to the
selection of a venom is an initial global screening relying on several *in vivo* and
in vitro tests, the nature of which depends on the target receptor of interest. We
shall succinctly detail some of these screening tests in the next section. Second,
once the biological source has been selected, sufficient amounts of venom are
collected, pooled and lyophilised. The venom is then divided into its various
fractions by using a variety of techniques such as size exclusion, reversed-phase
and ion exchange high-pressure liquid chromatography (HPLC). The homogeneity of the collected fractions is verified by analytical reversed-phase HPLC,
mass spectrometry analysis, and eventually Edman sequencing in the case
where several peptide molecules coelute in a given fraction. Fractions of interest
(e.g. peptides) can then be selected with the appropriate screening test and
undergo more complete physico-chemical and pharmacological characterisa-
tions. An alternative strategy to obtain the primary structure of a biologically
active molecule is based on the use of mass spectrometry. First, mass spectrometry can be used in conjunction with exopeptidase treatment for the direct
determination of the peptide amino acid sequence. Second, a rapid method
for the identification of a novel compound relies on the comparison of proteolytic peptide fragment profiles with theoretical profiles deduced from protein
databases, along with other physico-chemical characteristics. Such a strategy
supposes that extensive genomic data are available on the tissue and/or organism from which the peptide arises. This procedure will be invaluable to the
identification of novel compounds and should circumvent the laborious task of
peptide recovery and sequencing. Once natural compounds have been identified
and characterised, they can ultimately be synthesised by a chemical or molecular biology approach depending on their sizes. Before entering into some details
of peptide synthesis, we shall first introduce a number of tests that are useful to
the screening of novel compounds.

14.3.2   TOXIN SCREENING TESTS

One of the most crude screening tests that is used to assess the functionality of
venom fractions or purified toxins is based on *in vivo* toxicity evaluation.

Compounds are usually injected sub-cutaneously, intravenously or intra-cerebroventricularly into mice to assess the 50% lethal doses. The signs induced in the animal model are informative of the molecular targets of the toxins injected. For instance, toxins active on sodium channels often produce similar toxic signs (e.g. tremors, convulsion followed by death shortly after injection). Also, toxins active on potassium channels show high potency in mice only when injected by the intra-cerebroventricular route whereas toxins active on sodium channels produce lethal effects whatever the administration method. However, the LD_{50} values will vary depending on the mode of injection used, reflecting differences in bioavailability and pharmacokinetics of the injected toxin(s), but also the localisation of the most sensitive target organs.

Screening techniques most commonly used include (i) binding methods in which the ability of a peptide (or peptide fraction) to compete with the binding of radio-labelled ion channel-specific toxins to cell membranes (such as rat brain synaptosomes) is studied, and (ii) cell electrophysiology or other studies based on ion permeability through the lipid membrane ($^{45}Ca^{2+}$ fluxes, ion channel reconstitution, Ca^{2+} imaging), in which the ability of a toxin to inhibit ion fluxes is investigated. In the former case, binding methods require that one screens for a molecule able to compete with a known receptor ligand. This approach is limited by the need to radio-label the ligand that serves as a reference compound for the screening. It is also of limited usefulness for receptors for which (i) there are no ligands yet identified or (ii) the available ligands cannot be radio-labelled. For screening methods such as those based on electrophysiology, there is no need for a reference compound but the receptor targeted in screening is a well-identified molecule. One of the most convenient techniques for exogenous ion channel expression relies on the use of *Xenopus laevis* oocytes. The system appears to be adequate for electrophysiological and pharmacological studies because these cells express very low levels of endogenous ion channels or receptors. However, the overall inconvenience of the methods described herein is that they are material-intensive (cell membranes, cell culture, host expression, ion flux imaging, etc.) and time-consuming and therefore can only efficiently screen for a few compounds simultaneously. Several other novel approaches, currently under development, are designed to overcome some of these drawbacks. One of these is based on the use of surface plasmon resonance (SPR) biosensors which are electro-optical instruments used for analysing real-time protein–protein interactions. This technique has recently been used for the study of toxin–receptor interactions [16, 17]. Molecules can therefore be screened for their ability to bind either to the receptor or compete for a ligand to receptor interaction. There are several advantages to an SPR technique in the screening of novel compounds. Obviously, this technique is more appropriate for high-throughput screening of novel ligands. It can be declined in several ways, either by immobilising a competitive ligand or the receptor of interest itself. Interestingly, there is no need for a reference compound in this screening assay since crude or purified receptors bound onto the sensor chip are sufficient to screen for novel molecules. Also, screening of non-purified compounds is also possible. It should be mentioned that the use of SPR for screening new compounds active on ion channels or receptors should exponentially increase with the ability to produce recombinant proteins that structurally and functionally mimic the corresponding native proteins. For example, an active receptor site for kaliotoxin was obtained following the purification of chimeric recombinant KcsA/Kv1.3 ion channels produced in bacteria [18]. Such an approach should overcome the technical limitations inherent to protein quantities and the difficulties associated with the purification of native proteins. A second promising approach that might be useful in the large-scale screening of molecules is derived from a recent observation that chlorella virus PBCV-1 encodes a 94-mer functional Kcv1 potassium channel required for viral replication. Compounds, such as Ba^{2+}, able to inhibit the activity of Kcv1, also block viral replication [19]. Blockage of viral replication can be assessed visually by the resulting inhibition of viral plaque formation. This interesting implication of Kcv1 in viral replication could be used to develop a research programme for the screening of compounds able to block viral replication. By extension, Kcv1 could possibly be replaced by equivalent potassium channel sequences in the PBCV-1 genome for the screening of compounds active on the ionic pores of other channels.

Finally, it should be emphasised that all screening tests have their own limitations and it will undoubtedly be necessary to identify new methodologies for the screening of active compounds. For instance, one of the limitations often encountered is the apparent experimental discrepancies between K_d values obtained in binding experiments and the IC_{50} values observed in ionic current blockage, illustrating the difficulties of determining an *in vivo* active dose based on *in vitro* data. Therefore, the combination of several research strategies appears to represent a powerful addition to the screening approach.

14.3.3   PRODUCTION OF TOXINS

Screening of molecules isolated from biological sources will ultimately lead to the identification of a number of novel compounds of interest depending on the screening test used. Though biological sources such as venoms are of great interest to identify lead compounds, the major inconvenience faced by the investigator is that only tiny amounts of product can be purified and used for a complete characterisation of the product. To solve this problem, one has to produce the molecule in larger quantities by either chemical synthesis or molecular biology techniques. Being able to master the large-scale production of a toxin presents the additional advantage that the molecule can be modified for structure–function analysis aimed at increasing the selectivity and/or affinity of the molecule. There are now two routes commonly used for the production of toxins and their structural analogues. First, the solid-phase peptide synthesis can routinely be applied to the production of short- and medium-chain toxins (fewer than 60 amino acid residues. For an accurate description of the experimental procedures used, see [20]). Longer toxins could theoretically be synthesised but specific problems are often encountered during peptide-chain assembly and the final oxidation/folding process. These problems include incomplete incorporation of some amino acid residues during peptide assembly, side-reactions during final cleavage of the crude peptide from the peptidyl resin, and insolubility of either the reduced form of the toxin or its oxidation intermediates during folding/oxidation. Also, problems related to peptide aggregation and formation of stable but inactive oxidised species may occur. Besides, toxins are often highly cross-linked molecules that require the establishment of tricky experimental conditions for the formation of the correct half-cystine pairings. For instance, scorpion toxins active on voltage-gated sodium channels still remain difficult to obtain by the chemical synthesis route, whereas shorter toxins active on potassium channels are more easily synthesised. To circumvent the size limitation, molecular biology has recently been introduced to study various aspects of toxins such as toxin precursors and post-translational processing. For instance, a few cDNA libraries are now available for scorpion toxins [21–24]. Several recombinant scorpion toxins active on potassium channels have now been produced in bacteria; these include charybdotoxin [25], margatoxin [26] and agitoxin 2 [27]. Common yields of peptide recovery are generally in the mg/l range of bacterial culture. These toxins are often produced as fusion proteins with a cleavable tag that is required for the purification step. For instance, kaliotoxin 2 was produced as a fusion protein with the maltose-binding protein [28]. Following purification, this tag was removed by enterokinase treatment prior to final oxidation/folding to produce a fully functional toxin. However, tags may sometimes interfere with the correct toxin folding, and it appears that disulphide bond formation mostly occurs *in vitro* after purification and tag removal. Also, toxins produced in bacteria or yeast may reveal themselves toxic for their host cells. In spite of these inherent limitations, production of recombinant toxins through molecular biology presents the clear advantage that long-chain toxins can be produced. For several reasons, and whenever possible, it is generally preferable to synthesise toxins through a chemical approach. It is possible (i) to obtain peptides with either L- or D-amino acid residues, (ii) to incorporate unnatural amino acid residues, and (iii) to replace the peptide amide bonds (CO–NH) by a variety of other covalent linkages (e.g. CH_2-NH, CH_2-CH_2, NH–CO, etc.). In this way it may be possible to increase the *in vivo* half-life of therapeutically valuable compounds and/or their pharmacological characteristics. One should mention that although recombinant toxins can also be modified following purification, the chemical synthesis approach offers the definite advantage that structural modifications can be selectively introduced in various parts of the peptide backbone during the course of its assembly.

14.4 STRATEGIES IN THE DESIGN OF THERAPEUTICALLY ACTIVE MOLECULES

The identification, characterisation and synthesis of biologically active molecules issued from various screening tests usually represent the first steps to the design of structural analogues that possess the desired properties in terms of pharmacology, pharmacokinetics and therapeutic potentials. Once molecules have been identified, they can structurally be altered to increase their selectivity and/or affinity. Some other physico-chemical properties that need to be addressed include (i) peptide half-life, (ii) bio-availability, and (iii) its ability to cross the blood–brain barrier. Structural modifications of various natural toxins have been selected as a strategy for the *de novo* design of therapeutically useful molecules.

Most experimental strategies aiming at developing molecules of therapeutic interest are based on tests that resemble those used for screening the compounds. However, a refined structure–function analysis is required to yield the expected properties of the structural analogues synthesised. Much information can be collected for molecules that have undergone a thorough structural and pharmacological characterisation. Knowledge of the 3D-structure and of the key residues that affect peptide pharmacology contributes to the design of novel analogues.

14.5 STRUCTURE–FUNCTION RELATIONSHIP APPROACH FOR THE DESIGN OF NOVEL THERAPEUTIC AGENTS

14.5.1 TOXIN AND ION CHANNEL STRUCTURES

In order adequately to alter the structure of a molecule of interest (e.g. toxin), several crucial structural and functional details first need to be elucidated from the lead compound. Structural data can be obtained through molecular modelling; circular dichroism, ^{1}H-NMR, and/or crystallography. Molecular modelling may be a useful approach: however, its usefulness is often limited to the structural description of molecules that share over 30% sequence identity with other compounds of known 3-D structures. Therefore, when dealing with molecules not closely related to any other highly characterised product, this approach remains hazardous and of limited value. Other methods should then be preferred that will provide reliable experimental data. Circular dichroism is a rapid method of assessing peptide secondary structures with several advantages, including the capacity (i) to evaluate peptide conformation at different pH values, (ii) to assess real-time structural changes, and (iii) to recover and reuse the peptide sample. However, the technique is not relevant enough to account for slight structural variations. In contrast, finer structural information will be gained by the use of ^{1}H-NMR or crystallography techniques.

Though these techniques are extremely useful, there are some limitations in their application. First, ^{1}H-NMR structural determination is performed at acidic pH which might alter the physiologically-relevant structure of the peptide/toxin. For this reason, it is recommended that a ^{1}H-NMR structure obtained for a given product be validated by a circular dichroism analysis that addresses the pH-independence of its overall conformation (in the pH range of 2.5 to 7.4). This precaution is indicated for peptides whose degree of ionisation changes in this pH range: generally peptides that contain one or more His residue(s). Second, due to the difficulty of crystallising even highly cross-linked short-chain toxins, the crystallography approach has hitherto been successfully applied to long-chain toxins only (over 60 amino acid residues). The 3-D structures obtained by either one of these methods should be viewed with caution since they may not reflect structural adaptation that may occur upon interaction of this ligand with its receptor site. This consideration emphasises the need for co-crystallisation of the complex formed by both interacting molecules. In the present case, what would limit such a study is the difficulty often associated with the purification and crystallisation of membrane-inserted receptor (e.g. ion channels, G protein-coupled receptors, etc.). So far, the only ion channels for which structural data are available are the KcsA potassium channel [29], a two-transmembrane domain protein, and the voltage-dependent sodium channel [30]. Unfortunately, even for these two channels, no structural data are yet available with regard to ligand binding. Co-crystallisation of KcsA with an interacting ligand might represent the next step in the definition of the active binding sites. To overcome some of the difficulties inherent to the study of other native receptors, several experimental strategies have been developed based on the ability of receptor-derived fusion proteins to mimic the active binding site of the ligand. For instance, the binding of kaliotoxin onto Kv1.3 has been effectively maintained in purified KcsA/Kv1.3 chimera fusion proteins [18], thereby potentially simplifying the structural study of the interacting complex. Since such an approach is feasible, one might consider extending it to the study of other ionic pores through the production of KcsA chimera proteins in which the ionic pores of more complex channels (e.g. calcium channels) would be incorporated.

14.5.2   DESIGN OF NOVEL COMPOUNDS

Once structural data on the ligand and/or ligand–receptor complex have been gathered, several routes of investigation lead to the design of compounds with improved properties. These routes are based on a thorough knowledge of the interaction surface shared by the ligand and the receptor. This interaction surface represents the key structural element on which one may model or act in order to improve ligand-binding affinity and/or functional effects. Another essential feature regarding the interaction surface is that it is responsible for ligand selectivity. A unique interaction surface between any given ligand and receptor warrants a high ligand selectivity for that receptor. In contrast, the same ligand may interact with distinct receptors because they share structural homologies. Hence, the aim of a structure–function analysis is either to improve or, conversely, to decrease selectivity in order to obtain ligands that cover different spectra of pharmacological activities. Since no formal data on the topology of the ligand–receptor interaction surface are generally available, alanine scanning is widely used by investigators to probe the interaction surface. Complementary mutagenesis can also be applied to the receptor to investigate whether certain toxin residue(s) is (are) indeed interacting with specific residue(s) of the receptor. Such a study is complicated by the fact that toxin binding to its receptor frequently relies on several non-contiguous structural elements (multi-point interaction) that are spatially close but dispersed throughout the receptor primary structure. The concept of multi-point interaction explains why strict selectivity is difficult to attain for any given peptide toxin. With a clear picture in hand of the interacting surface of the toxin, it is then possible to design novel analogues aimed at altering this surface. The main strategy generally followed is based on the mutation or chemical modification of specific residues: such changes in structure can alter the 3-D presentation of key residues that are involved in peptide pharmacology. Alternatively, these modifications can directly target the key residues themselves without any significant variation in toxin conformation. Thus, in the first case, the spatial distribution of key residues may be affected by modifications in the toxin amino acid sequence outside the active interaction surface. These modifications include amino acid replacement, chemical modification and/or half-cystine pairings rearrangement in the case of cross-linked peptides. As an example, replacement of a pair of half-cystine residues by isosteric α-aminobutyrate derivatives in 34-mer scorpion toxin maurotoxin (MTX) leads to a three disulphide-bridged MTX analogue exhibiting a reorientation of the α-helix with respect to the β-sheet structure and presenting an increased pharmacological selectivity towards Kv1.2 channels [31]. In the second case, modifications in the toxin sequence can introduced directly in the active interaction surface. With such an approach, caution should be taken that amino acid substitution or chemical modification does not result in a global structural change of the toxin in which the spatial presentation of other key residues would be affected. For instance, substitution of a glycyl by an alanyl in position 33 of MTX alters its disulphide bridge pattern and pharmacology [32]. Notably, systematic alanine substitution, such as used in alanine scanning, is not always adequate since alanine residues do not necessarily take into account the nature of the substituted residue or the potential impact of the mutation on local secondary structures that may indirectly affect peptide pharmacology. Also, alanine residues themselves, present in the sequence, cannot be explored. Fine structural variations can indeed produce drastic changes in affinity and/or activity. For example, the carboxy-terminal amidation of P05 scorpion toxin results in increased affinity for apamin-sensitive SK channels [33].

One interesting feature of the structure–function approach is that the knowledge accumulated concerning the domains of the receptor that are involved in ligand recognition can serve as a molecular basis for the design of novel simplified screening tests. For instance, the ability of KcsA/Kv1.3 chimera fusion proteins to be recognised by kaliotoxin [18] could serve as a basis for the design of a new screening test in which kaliotoxin-like molecules would be evaluated for binding to fusion proteins immobilised on SPR biosensor chips. Such a high-throughput technique would therefore greatly improve and simplify the screening of molecules active on any given binding site.

Finally, it should be emphasised that rationale design of novel compounds should take into account the intimate molecular mechanisms of drug action. Not all toxins are ion channel blockers, and binding of a ligand onto an ion channel receptor may also result in channel gating modulation. In that case, the parameter of importance to screen for novel compounds is not the efficacy of ion flux blockage but instead variation in normal channel functionality. It is worth noting that binding of a ligand onto a defined receptor site does not necessarily allow one to predict the resulting behaviour of the receptor channel. For instance, ion channel pore blockers do not systematically inhibit 100% of the channel conductance, suggesting that incomplete occlusion of the ionic pore can also take place. Also, a ligand can bind onto a site distinct and remote from the ionic pore and still affect ion permeation. For instance, scorpion toxin maurocalcine partially diminishes ryanodine receptor (RyR) conductance in spite of the fact that, like imperatoxin A, a related natural toxin, it binds onto a cytoplasmic domain of RyR located far from the ionic pore [34, 35]. Obviously, information gained on the structure of ion channels and their biophysical functioning will be invaluable to the understanding of the mechanisms of action of toxins. The recent demonstration that voltage-dependent sodium channels are unexpectedly constituted of several cavities [30] highlights the intrinsic complexity of the molecular processes involved in toxin recognition and effects. These recent data weaken several computer-assisted and experimental database molecular modelling studies focusing on the docking of toxins onto the ionic pore. They certainly call into question the concept that the binding of a single toxin is generally necessary and sufficient for a full blockage of any given ionic conductance.

14.6   CONCLUDING REMARKS

The future in toxinology appears to be the continuation of some of the daunting and exploratory tasks initiated so far. First, novel ligands, extracted from numerous biological sources, should be identified and fully characterised, both structurally and functionally, in order to complement the growing list of molecules of potential therapeutic value. Second, a better understanding of the structure and functioning of ion channel receptors is also required. In particular, there are crucial needs for (i) novel 3-D structures of ion channel ligands, (ii) an increased definition of the intimate topology of toxin to receptor binding surfaces, and (iii) the assessment of the sequential molecular events leading to changes in ion channel behaviour following the interaction of the ligand with its receptor site. These tasks are inherently of high complexity and often require a serious simplification of the experimental procedures used to study toxin binding and resulting pharmacological effects. One of the simplification procedures in development relies on the isolation of active-toxin binding domains from the receptor. Basically, receptor oversimplification remains a hazardous practice since they are generally multi-protein complexes under strong regulatory influence that may structurally complicate the binding environment of these molecules. In the future, toxins identified and selected through simplified screening tests should be further characterised in more complex molecular and cellular environments.

REFERENCES

[1] Witcher, D.R., De Waard, M., Sakamoto, J., Franzini-Armstrong, C., Pragnell, M., Kahl, S.D. and Campbell, K.P. (1993) Subunit identification and reconstitution of the N-type Ca^{2+} channel complex purified from brain. *Science* **261**, 486–9.

[2] Noda, M., Shimizu, S., Tanabe, T., Takai, T., Kayano, T., Ikeda, T., Takahashi, H., Nakayama, H., Kanaoka, Y., Minamino, N., Kangawa, K., Matsuo, H., Raftery, M., Hirose, T., Inayama, S., Hayashida, H., Miyata, T. and Numa, S. (1984) Primary structure of Electrophorus electricus sodium channel deduced from cDNA sequence. *Nature* **312**, 121–7.

[3] Agnew, W.S., Levinson, S.R., Brabson, J.S. and Raftery, M.A. (1978) Purification of the tetrodotoxin-binding component associated with the voltage-sensitive sodium channel from Electrophorus electricus electroplax membranes. *Proc. Natl. Acad. Sci. USA* **75**, 2606–10.

[4] Fernandez-Chacón, R., Königstorfer, A., Gerber, S.H., García, J., Matos, M.F., Stevens, C.F., Brose, N., Rizo, J., Rosenmund, C. and Südhof, T.C. (2001) Synaptotagmin I functions as a calcium regulator of release probability. *Nature* **410**, 41–9.

[5] Vincent, A., Lang, B. and Newsom-Davis, J. (1989) Autoimmunity to the Voltage-Gated Calcium Channel Underlies the Lambert–Eaton Myasthenic Syndrome, a Paraneoplastic Disorder. *Trends Neurosci.* **12**, 496–502.

[6] Lorenzon, N.M. and Beam, K.G. (2000) Calcium channelopathies. *Kidney Int.* **57**, 794–802.

[7] Cox, B. and Denyer, J.C. (1998) N-type calcium channel blockers in pain and stroke. *Expert. Opin. Ther. Patents* **8**, 1237–50.

[8] Bowersox, S.S., Singh, T. and Luther, R.R. (1997) Selective blockade of N-type voltage-sensitive calcium channels protects against brain injury after transient focal cerebral ischemia in rats. *Brain Res.* **747**, 343–7.

[9] Beeton, C., Barbaria, J., Giraud, P., Devaux, J., Benoliel, A.M., Gola, M., Sabatier, J.M., Bernard, D., Crest, M. and Beraud, E. (2001) Selective blocking of voltage-gated K^+ channels improves experimental auto-immune encephalomyelitis and inhibits T-cell activation. *J. Immunol.* **166**, 936–44.

[10] Koo, G.C., Blake, J.T., Talento, A., Nguyen, M., Lin, S., Sirotina, A., Shah, N., Kulvany, K., Hora, D., Cunningham, P., Wunderler, D.L., McManus, O.E.,

Slaughter, R., Bugianesi, R., Felix, J., Garcia, M., Williamson, J., Kaczorowski, G., Sigal, N.H., Springer, M.S. and Feeny, W. (1997) Blockade of the voltage-gated potassium channel Kv1.3 responses in vivo. *J. Immunol.* **158**, 5120–8.

[11] Crest, M., Beraud-Juven, E. and Gola, M. (1999) Towards therapeutic applications of Kv1 channel blockers in neurological diseases. In *Animal Toxins and Potassium channels*, Darbon H. and Sabatier J-M., eds, *Perspectives in Drug Discovery and Design* vols 15/16, Kluwer / ESCOM publishers, pp. 333–42.

[12] Soria, B. (1998) Potential role of the pharmacology of K⁺ channels in therapeutics. In *Ion Channel Pharmacology*, B. Soria and V. Ceña, eds., pp. 229–45, Oxford University Press Inc., New York.

[13] Parcej, D.N., Scott, V.E. and Dolly, J.O. (1992) Oligomeric properties of alpha-dendrotoxin-sensitive potassium ion channels purified from bovine brain. *Biochemistry* **31**, 11084–8.

[14] Chuang, R.S., Jaffe, H., Cribbs, L., Perez-Reyes, E. and Swartz, K.J. (1998) Inhibition of T-type voltage-gated calcium channels by a new scorpion toxin. *Nat. Neurosci.* **1**, 668–74.

[15] Martin-Eauclaire, M-F. and Couraud, F. (1995) Scorpion neurotoxins: effects and mechanisms. in: *Handbook of Neurotoxicology*, Chang. LW and Dyer RS, eds, Marcel Dekker, Inc., New York, Basel, Hong Kong publishers, pp. 683–716.

[16] Faure, G., Villela, C., Perales, J. and Bon, C. (2000) Interaction of the neurotoxic and non-toxic secretory phospholipases A2 with the crotoxin inhibitor from Crotalus serum. *Eur. J. Biochem.* **267**, 4799–808.

[17] MacKensie, C.R., Hiriama, T. and Buckley, J.T. (1999) Analysis of receptor binding by the channel-forming toxin aerolysin using surface plasmon resonance. *J. Biol. Chem.* **274**, 22604–9.

[18] Legros, C., Pollmann, V., Knaus, H.G., Farrell, A.M., Darbon, H., Bougis, P.E., Martin-Eauclaire, M-F. and Pongs, O. (2000) Generating a high affinity scorpion toxin receptor in KcsA-Kv1.3 chimeric potassium channels. *J. Biol. Chem.* **275**, 16918–24.

[19] Plugge, B., Gazzarrini, S., Nelson, M., Cerana, R., Van Etten, J.L., Derst, C., DiFrancesco, D., Moroni, A. and Thiel, G. (2000) A potassium channel protein encoded by chlorella virus PBCV-1. *Science* **287**, 1641–4.

[20] Sabatier, J-M. (2000) Chemical synthesis and characterization of small proteins: example of scorpion toxins. In *Animal toxins*, H. Rochat and M-F. Martin-Eauclaire, eds., pp. 196–216, Birkhäuser Verlag.

[21] Bouhaouala-Zahar, B., Ducancel, F., Zenouaki, I., Ben Khalifa, R., Borchani, L., Pelhate, M., Boulain, J.C., El Ayeb, M., Ménez, A., and Karoui, H. (1996) A recombinant insect-specific alpha-toxin of Buthus occitanus tunecanus scorpion confers protection against homologous mammal toxins. *Eur. J. Biochem.* **238**, 653–60.

[22] Dimarq, J.L., Hoffmann, D., Meister, M., Bulet, P., Lanot, R., Reichhart, J.M. and Hoffmann, J.A. (1994) Characterization and transcriptional profiles of a Drosophila gene encoding an insect defensin. A study in insect immunity. *Eur. J. Biochem.* **221**, 201–9.

[23] Zilberberg, N. and Gurevitz, M. (1993) Rapid isolation of full length cDNA clones by 'inverse PCR': purification of a scorpion cDNA family encoding alpha-neurotoxins. *Ann. Biochem.* **209**, 203–5.

[24] Zilberberg, N., Gordon, D., Pelhate, M., Adams, M.E., Norris, T.N., Zlotkin, E., Gurevitz, M. (1996) Functional expression and genetic alteration of an alpha-scorpion neurotoxin. *Biochemistry* **35**, 10215–22.

[25] Park, C.S., Hausdorff, S.F. and Miller, C. (1991) Design, synthesis, and functional expression of a gene for charybdotoxin, a peptide blocker of K⁺ channels. *Proc. Natl. Acad. Sci. USA* **88**, 2046–50.

[26] Garcia-Caldo, M., Leonard, R.J., Novick, J., Stevens, S.P., Schmalhofer, W., Kaczorowski, G.J. and Garcia, M.L. (1993) Purification, characterization, and biosynthesis of margatoxin, a component of Centruroides margaritatus venom that selectively inhibits voltage-dependent potassium channels. *J. Biol. Chem.* **268**, 18866–74.

[27] Garcia, M.L., Garcia-Caldo, M., Hidalgo, P., Lee, A. and MacKinnon, R. (1994) Purification and characterization of three inhibitors of voltage-dependent K^+ channels from Leiurus quinquestriatus var. hebraeus venom. *Biochemitry* **33**, 6834–9.

[28] Legros, C., Feyfant, E., Sampieri, F., Rochat, H., Bougis, P.E. and Martin-Eauclaire, M-F. (1997) Influence of a NH_2-terminal extension on the activity of KTX2, a K^+ channel blocker purified from Androctonus australis scorpion venom. *FEBS Lett.* **417**, 123–9.

[29] Doyle, D.A., Morais-Cabral, J., Pfuetzner, R.A., Kuo, A., Guilbis, J.M., Cohen, S.L., Chait, B.T. and MacKinnon, R. (1998) The structure of the potassium channel: molecular basis of K^+ conduction and selectivity. *Science* **280**, 69–77.

[30] Sato, C., Ueno, Y., Asai, K., Takahashi, K., Sato, M., Engel, A. and Fujiyoshi, Y. (2001) The voltage-sensitive sodium channel is a bell-shaped molecule with several cavities. *Nature* **409**, 1047–51.

[31] Fajloun, Z., Ferrat, G., Carlier, E., Fathallah, M., Lecomte, C., Sandoz, G., di Luccio, E., Mabrouk, K., Legros, C., Darbon, H., Rochat, H., Sabatier, J-M. and De Waard, M. (2000a) Synthesis, 1H NMR structure, and activity of a three disulfide-bridged maurotoxin analog designed to restore the consensus motif of scorpion toxins. *J. Biol. Chem.* **275**, 13605–12.

[32] Fajloun, Z., Mosbah, A., Carlier, E., Mansuelle, P., Sandoz, G., Fathallah, M., di Luccio, E., Devaux, C., Rochat, H., Darbon, H., De Waard, M. and Sabatier, J-M. (2000b) Maurotoxin versus Pi1/HsTx1 scorpion toxins. Toward new insights in the understanding of their distinct disulfide bridge patterns. *J. Biol. Chem.* **275**, 39394–402.

[33] Sabatier, J-M., Zerrouk, H., Darbon, H., Mabrouk, K., Benslimane, A., Rochat, H., Martin-Eauclaire, M-F. and Van Rietschoten, J. (1993) P05, a new leiurotoxin I-like scorpion toxin: synthesis and structure activity relationships of the alpha-amidated analog, a ligand of Ca^{2+}-activated K^+ channels with increased affinity. *Biochemistry* **32**, 2763–70.

[34] Fajloun, Z., Kharrat, R., Chen, L., Lecomte, C., di Luccio, E., Bichet, D., El Ayeb, M., Rochat, H., Allen, P.D., Pessah, I.N., De Waard, M. and Sabatier, J-M. (2000c) Chemical synthesis and characterization of maurocalcine, a scorpion toxin that activates Ca^{2+} release channel/ryanodine receptors. *FEBS Lett.* **469**, 179–85.

[35] Samso, M., Trujillo, R., Gurrola, G.B., Valdivia, H.H. and Wagenknecht, T. (1999) Three-dimensional location of the imperatoxin A binding site on the ryanodine receptor. *J. Cell Biol.* **146**, 493–9.

15 Toxins as Probes for Structure and Specificity of Synaptic Target Proteins

PALMER TAYLOR*, BRIAN MOLLES, SIOBHAN MALANY and HITOSHI OSAKA

15.1 INTRODUCTION

Although the selective advantages in predation and protection from predation conferred by toxins to animal and plant species have been recognised for centuries, and extracts of the curare toxins were mentioned in 17th century literature and studied systematically in the mid-19th century [1], the potential of toxins in defining the molecular structure of the target site did not become fully evident until C.Y. Lee and his colleagues demonstrated the apparent irreversibility of action of α-bungarotoxin on the neuromuscular junction [2]. In 1970, Changeux and colleagues demonstrated that this toxin could be used to isolate and characterise the nicotinic receptor from the electric fish, *Torpedo* [3].

This toxin from the krait, *Bungarus multicinctus*, and orthologues from other species in the Elapidae family, have proven useful in isolation of nicotinic receptor from many sources, localisation of the receptor *in situ* and distinguishing subtypes of nicotinic receptors in a wide variety of muscle and neural tissues. Over this period, it also became evident that toxins from other species were effective blocking agents at the neuromuscular junction, many of which show selectivity for the nicotinic receptor subtype from muscle.

The pentameric muscle receptor, of $\alpha_2\beta\gamma\delta$ subunit composition in embryonic tissue and $\alpha_2\beta\epsilon\delta$ in adult muscle, contains two binding sites found at the $\alpha\delta$ and $\alpha\gamma(\epsilon)$ subunit interfaces. Simultaneous occupation of both sites by the agonist is required to elicit the conformational change that opens the gated channel. Thus, occupation by an antagonist toxin of a single site results in a nonfunctional receptor. If the two sites show nonequivalent affinities, the higher affinity site for the antagonist or toxin will govern functional antagonism.

* Corresponding author
Supported in part by USPHS Grant GM 18360 to PT

Perspectives in Molecular Toxinology
Edited by A Ménez © 2002 John Wiley & Sons, Ltd

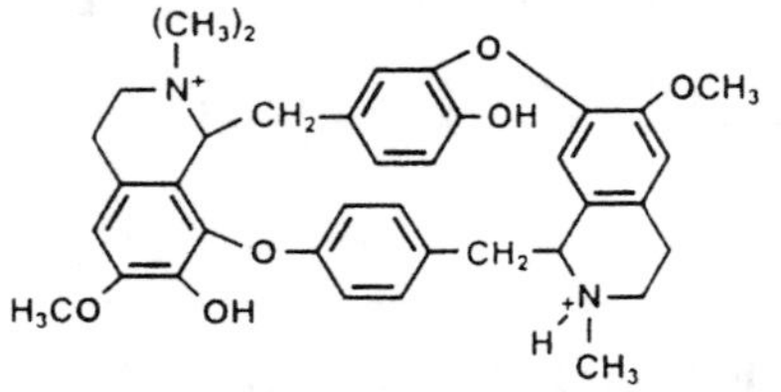

d-Tubocurarine 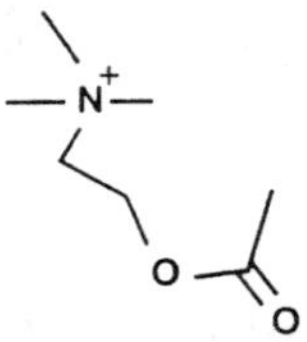   Acetylcholine 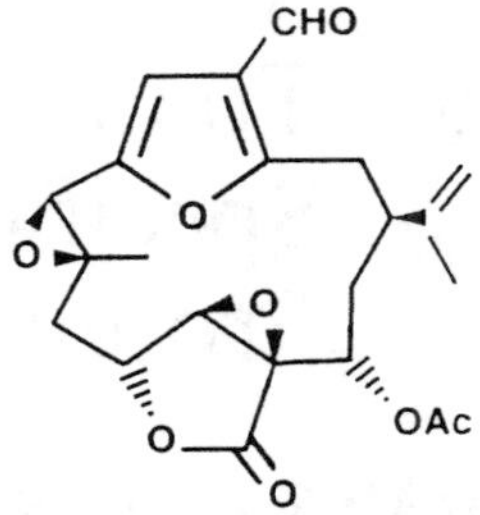   Lophotoxin

3-fingered snake toxins
   Short toxins (57-65 amino acids)
   Long toxins (70-76 amino acids)
   4 or 5 disulfides
MW=6000-8000 daltons

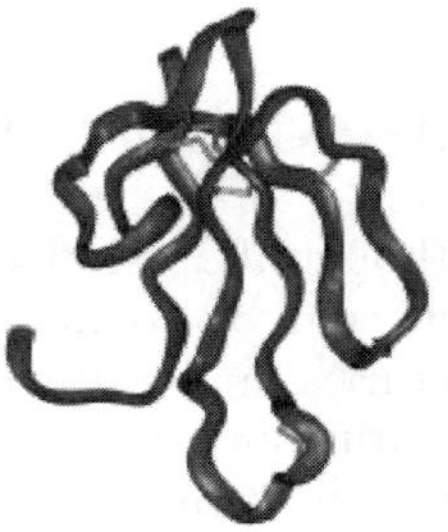

ECCNPACGRHYSC NH2   (GI)

ECCNPACGRHYSCGK NH2   (GIA)

ECCHPACGKHFSC NH2   (G2)

GRCCHPACGKNYSC NH2   (MI)

α–conotoxins

GGKPDLRPC**H**PPCHYIPRPKPR   (Wtx-1)

SLGGKPDLRPC**H**PPCHYIPRPKPR (Wtx-2)

GGKPDLRPC**Y**PPCHYIPRPKPR   (Wtx-3)

SLGGKPDLRPC**Y**PPCHYIPRPKPR (Wtx-4)

Waglerins

Figure 15.1 Structures of acetylcholine and various antagonists of the nicotinic receptor. *d*-Tubocurarine is an alkaloid from certain plants; lophotoxin is found in coral of the genus *Lophogorgia;* α-conotoxins are representative sequences from a larger family from fish-hunting cone snails, *Conus sp.*, and the waglerins constitute the four structures known to date from the pit viper, *Tropidolaemus wagleri.* Shown without sequences are the three-fingered alpha-neurotoxins, which are found in many members of the Elapidae family. The peptide α-carbon backbones are reproduced from structures determined by NMR or crystallography

Figure 15.1 shows a listing of toxins from plant, invertebrate and vertebrate sources that inhibit neuromuscular transmission in mammals. The corresponding structures also reveal the chemical diversity inherent in these blocking agents. The curare alkaloids are also blockers of serotonin (5HT-3) receptors, raising the question whether their deterrent actions could be directed to both grazing animals and insects. Several animal species also use alkaloids for protection, and in many of these cases they accumulate from arthropods in their diet [4]. For example, epibatidines isolated from amphibians and discussed in a companion chapter, interact with nicotinic systems.

The lophotoxins, derived from species of soft coral of the genus, *Lophogorgia* [5–7], are cyclic diterpenoids with a structural mimic of acetylcholine inherent in their lactone and adjoining epoxide moieties. Upon exposure to nicotinic receptors, these agents react covalently with Tyrosine 190 in the receptor α subunit giving rise to irreversible antagonism [7]. Coral as a stationary animal with a budding exoskeleton is continually bathed in sea water. The bathing medium serves as a perfect sink and would result in dissolution of stored peptide or alkaloid toxins from an organism lacking specialised glands. By contrast, the hydrophobic terpenoids can be stored within the membrane lipids. As is evident throughout the animal kingdom, peptides and their precursor pro-peptides are often contained within vesicular structures and well suited for storage within glands.

15.2 SITE SELECTIVITY OF THE TOXINS

Within the listing of toxins in Figure 15.1, all categories show some subunit selectivity for the muscle receptor with preferences for the $\alpha\gamma$, $\alpha\varepsilon$ or $\alpha\delta$ subunit interfaces. For example, *d*-tubocurarine has a preference for $\alpha\gamma$ and $\alpha\varepsilon$ over $\alpha\delta$ subunits [8–10] and lophotoxin has a slight preference for $\alpha\delta$ over $\alpha\gamma$ subunits [11]. However, we became familiar with the remarkable selectivity of the peptide toxins (Figure 15.2) when studying the α-conotoxin MI interaction with the mouse receptor, where it showed a 10 000-fold preference for the $\alpha\delta$ over the $\alpha\gamma$ subunit interface [12]. Other α-conotoxins also reveal a binding site selectivity, although not always of the same magnitude [13–15]. An exhaustive analysis, through formation of chimeras and then site-specific residue replacements of γ residues in a δ subunit template and δ residues in a γ subunit template, has enabled us to ascertain the determinants responsible for α-conotoxin specificity at these two subunit interfaces. Subsequent studies through mutations of the common α subunit have identified critical determinants in this subunit and indicate that α-conotoxins do not bind in identical orientations at the two subunit interfaces [16]. Studies using thermodynamic mutant cycle analysis, where pairwise mutations on the toxin and receptor binding sites are analysed for energetic linkages, have allowed Sine and colleagues to begin to orient the α-conotoxins at the $\alpha\gamma$ interface [17].

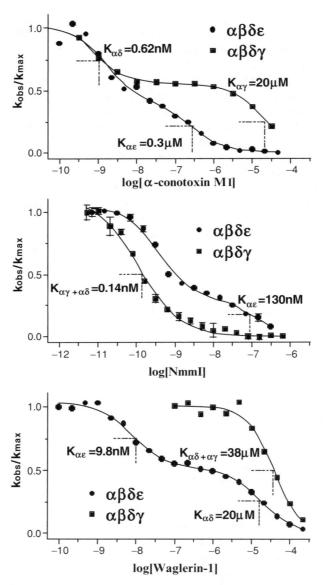

Figure 15.2 Measurements of peptide antagonist association with its two respective sites on mouse recombinant DNA derived nicotinic receptors determined by competition with the initial rate of α-bungarotoxin binding. **Top Panel** Association of α-conotoxin MI with the nicotinic receptor expressed as $\alpha_2\beta\gamma\delta$ and $\alpha_2\beta\epsilon\delta$. The K_D values are shown for the three subunit interface sites. **Middle Panel** Association of *Naja mossambica mossambica* (NmmI) α-neurotoxin with the same receptors. **Lower panel** Association of Waglerin-1 with the same nicotinic receptors as above. Dissociation constants are calculated from the competition curves for the two sites, assuming an equal population of sites

A second group of peptides from a genus of the Viperidae family, *Tropido-laemus wagleri*, have recently been characterised [18]. Only four naturally occurring peptides of this 22–24 amino acid family are known to date. These peptides have been shown to be selective for the $\alpha\varepsilon$ subunit interface over the $\alpha\gamma$ and $\alpha\delta$ interfaces [19]. Indications of both site and species selectivity emerged from toxicologic studies of the toxin that showed far greater toxicity in the mouse compared to the rat [20], and were remarkably selective for the adult mouse over the newborn animal [19]. The latter, in fact, suggested the ε subunit preference. Residues governing Waglerin-1 specificity have also been delineated by γ, δ and ε residue substitutions in the respective subunit templates [21, 22].

Finally, although most α-neurotoxins have not been found to exhibit a subunit interface preference, a short α-toxin from *Naja mossambica mossambica* (NmmI), closely related in structure to erabutoxin, shows a high affinity for the $\alpha\gamma$ and $\alpha\delta$ interfaces ($K_D = 140\,pM$), but a far lower affinity for the $\alpha\varepsilon$ interface ($K_D = 130\,nM$) [23].

Since the γ, δ and ε subunits show relatively close sequence identity, determinants of specificity were identified initially by making $\gamma\delta$, $\gamma\varepsilon$, and $\delta\varepsilon$ chimeras. This approach allows identification of regions where determinants reside. Systematically stepping across chimera boundaries often enables individual determinant identification that can be confirmed by specific residue substitution into the corresponding template. α Subunit determinants require direct substitutions to identify candidate residues. Even though a common face of the α subunit contributes to all binding sites, the contributions of individual α subunit residues to the binding affinity differ for the $\alpha\gamma$, $\alpha\delta$ and $\alpha\varepsilon$ subunit interface binding sites [16, 22–24].

These investigations have been extended to thermodynamic mutant cycle analyses, where the change in binding free energy associated with pairwise mutations on the toxin and receptor is compared with the change in energy associated with the individual mutations. Selective linkages can be identified and residue proximity ascertained from the linkage or interaction free energy. Since among the various molecular forces, electrostatic forces occur over the largest distance and are isotropic, analyses of charge reversal may be the most informative and straightforward strategy for altering structure.

Table 15.1 presents a compilation of the residues involved in the stabilisation of α-conotoxin MI, Waglerin1 and NmmI α-toxin at the agonist–antagonist binding site of the nicotinic receptor. It is not surprising that similar residues in common regions form the major determinants of binding energy for the three toxins. However, the critical determinants giving rise to the binding site selectivity differ for the three toxins. In the case of α-conotoxin, three residue positions, homologous with positions 34, 111 and 172 in the γ subunit, give rise to a subunit interface of selectivity $\alpha\delta > \alpha\varepsilon > \alpha\gamma$ for the toxin. In the case of waglerin, residues at 34, 57, 59, 115 and 173 in the ε-subunit are critical determinants for the selectivity of $\alpha\varepsilon > \alpha\delta = \alpha\gamma$. Although for NmmI α-toxin the specificity for subunit interfaces is inverted $\alpha\delta = \alpha\gamma > \alpha\varepsilon$, selectivity arises

Table 15.1 Subunit-selective toxins with residues identified as affinity determinants. Each of the residues indicated gives at least a 10-fold (1.36 kcal/mol) change in affinity when mutated.

toxin	selectivity	γ, δ, ε residues*	α-residues
Waglerin-1	αε > αδ = αγ	**34**, 57, 59, <u>115</u>, 172	**99**, <u>149</u>, 189, 190, 197, 198, 200
NmmI	αδ = αγ > αε	55, *119*, 175, 176	99, *149*, 188, 190, 198, *200*
α-conotoxin MI	αδ > αε > αγ	34, 111, 172	93, *188*, **190**, 198

italics = αγ site only; **bold** = αδ site only; <u>underlined</u> = αε site only

* Numbers indicate residues of sequence using γ subunit numbering scheme. Residues of ε and δ subunits as they align to the corresponding position of γ indicated as follows:

NmmI; *Naja mossambica mossambica* toxin I

γ mouse: ...^{34} K...^{55} WIEMQ...^{111} SPDGCIYWL...^{172} FIDPE...
ε mouse: ...^{34} K...^{55} WIGID...^{111} YEGGYVSWL...^{173} DIDTA...
δ mouse: ...^{36} S...^{57} WIDHA...^{113} YDSGYVTWL...^{178} IIDPE...

from two residues at the 176 and 177 positions in the γ subunit, a distinct set from those conferring selectivity to Waglerin1.

15.3   TOWARDS A STRUCTURE OF THE SUBUNIT BINDING FACE

These data on toxin selectivity may be combined with a large amount of affinity labelling data that have identified receptor subunit residues in proximity to reactive ligands (Figure 15.3). Analysis of these findings reveals several features of the binding site. Clearly there are 'hot spots' in the linear sequence where discrete regions govern binding of several ligands. In the α subunit, these are confined to two short regions that may form loops, in the vicinity of residue 93 and between residues 149 and 153. In addition, an extended region between residues 185 and 200 may form a cavity or pocket important for ligand binding. In the case of γ, δ and ε subunits, four specificity regions are in homologous positions, but differ from those found in the α subunit. Again, this is consistent with receptor structure, since all of the subunits in the pentamer should have the same handedness or chirality, and therefore the face of the α subunit would be opposite to that of γ, δ or ε subunits in forming a binding site.

15.4   DELINEATION OF INDIVIDUAL RESIDUE CONTRIBUTIONS TO THE BINDING ENERGY

The α-toxin-receptor interaction possesses several advantages for delineating the contributory amino acid residues. First, the high affinity or low K_D of the native complex allows one still to observe a specific association between toxin

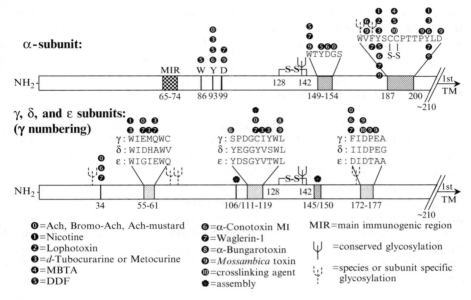

Figure 15.3 Maps of the amino-terminal domains of the α subunit and the γ, δ and ε subunits of the muscle nicotinic acetylcholine receptor showing positions of residues contributing to ligand recognition (agonist and competitive antagonist) sites, antigenic sites (main immunogenic region, MIR) and order of subunit assembly. The amino terminal domain, ~ 210 amino acids preceding the first trans-membrane span, constitutes virtually all of the extracellular portion of the homologous subunits. Only a short loop between transmembrane spans 2 and 3 and a short carboxyl-terminal sequence are also on the extracellular face. Residue assignments come from affinity labelling (dots 1–5), mutagenesis (dots 0, 1, 6–9) chemical cross-linking (dot 10), antibody reactivity (MIR), glycosylation (antennae) and subunit assembly patterns (petals). Experimental data are compiled in reference [27]

and binding site when the pairwise mutations on both molecules give rise to large energy reductions. Second, based on peptide–protein interactions of similar affinities and molecular weights of the associating molecules, a surface area of 1100–1500 Å is likely to be in van der Waals contact in forming the interaction between the toxin and its interfacial site on the receptor. This area is sufficient to encompass up to 30 amino acid residues on the two subunit interfaces. Third, comprehensive structure–activity studies have been conducted on a related short α-toxin, erabutoxin [25, 26].

Through initial studies with charge reversal, we have identified the critical residues involved in stabilising the NmmI α-toxin at the αγ site. In the case of the γ subunit, two pairwise residue positions, Arg33 with Leu119 and Lys27 with Glu176, show linkages upon charge reversal of 5.7 and 5.9 kcal, respectively [23]. Smaller contributions are found for Trp55 in NmmI toxin with these two receptor residues. In the case of the α subunit, a larger number of residue pairs, each with lower energy contributions, are involved [24]. Since Arg33 appears to contribute to stabilisation energy through residues on both the α subunit and the γ subunit, this toxin residue, located at the tip of the large second loop, is likely to be centred between the two subunits.

15.5 CONCLUDING REMARKS

Although analysis through site-specific mutagenesis and chemical modification is an arduous process for multi-subunit proteins and receptor binding sites that can accommodate a variety of ligands, it has yielded valuable information on the structure of the receptor. The high affinities, a variety of naturally occurring structures and the ease with which structures can be modified, either through peptide synthesis or recombinant DNA techniques, render naturally occurring peptide toxins ideally suited for such studies. Even with structural templates at atomic resolution soon becoming available for the receptor, establishing the energetics of the interaction at an individual residue level will be critical to understanding the precise structural details of the toxin-receptor complexes [24–26].

REFERENCES

[1] Bernard, C. (1856) Analyse physiologique des propriétés des systèmes musculaires et nerveux au moyen du curare. *C.R. Acad. Sci.* **43**, 825–9.
[2] Lee, C.Y. (1963) Isolation of neurotoxins from the venom of *Bungarus multicinctus* and their modes of neuromuscular blocking action. *Arch. Int. Pharmacodyn. Ther.* **144**, 241–57.
[3] Changeux, J-P., Kasai, M. and Lee, C.Y. (1970) Use of snake venom toxin to characterize the cholinergic receptor protein. *Proc. Natl. Acad. Sci. (USA)* **67**, 1241–5.
[4] Daly, J.W. (1998) Thirty years of discovery of arthropod alkaloids in amphibian skin. *J. Nat. Products.* **61**, 162–72.
[5] Fenical, W., Okuda, R.K., Bandurraga, M.M., Culver, P. and Jacobs, R.S. (1981) Lophotoxin, a novel neuromuscular toxin from sea whips of the genus *Lophogorgia*. *Science* **212**, 1512–4.
[6] Eterovic, V.A., Hann, R.M., Ferchmin, P.A., Rodriquez, A.D., Li, L., Lee, Y.H., and McNamee, M.G. (1993) Diterpenoids from caribbean gorgonians act as non-competitive inhibitors of the nicotinic acetylcholine receptor. *Cellular and Molecular Neurobiol.* **12**, 99–110.
[7] Abramson, S.N., Trischman, J.A., Tapiolas, D.M., Herold, E.E., Fenical., W. and Taylor, P. (1991) Structure/activity and molecular modeling studies of the lophotoxin family of irreversible nicotinic receptor antagonists. *J. Med. Chem.* **34**, 1798–1804.
[8] Chiara, D.C. and Cohen, J.B. (1997) Identification of amino acids contributing to high and low affinity *d*-tubocurarine sites in the *Torpedo* acetylcholine receptor. *J. Biol. Chem.* **272**, 37940–50.
[9] Sine, S. (1993) Molecular dissection of subunit interfaces in the acetylcholine receptor: Identification of residues that determine curare selectivity. *Proc. Natl. Acad. Sci.* **90**, 9436–40.

[10] Bren, N. and Sine, S. (1997) Identification of residues in the adult nicotinic acetylcholine receptor that confer selectivity for curariform antagonists. *J. Biol. Chem.* **272**, 30793–8.

[11] Culver, P., Fenical, W. and Taylor, P. (1984) Lophotoxin irreversibly inactivates the nicotinic acetylcholine receptor by preferential association at one of the two primary agonist sites. *J. Biol. Chem.* **259**, 3763–70.

[12] Kreienkamp, H-J., Sine, S.M., Maeda, R.K. and Taylor, P. (1995) Glycosylation sites selectively interfere with α-toxin binding to the nicotinic acetylcholine receptor. *J. Biol. Chem.* **269**, 8108–14.

[13] McIntosh, J.M., Santos, A.D., Olivera, B.M. (1999) *Conus* peptides targeted to specific nicotinic acetylcholine receptor subtypes. *Ann. Rev. Biochem.* **68**, 59–88.

[14] Hann, R.M., Pagan, O.R., Gregory, L.M., Jacome, J. and Eterovic, V.A. (1997) The 9-arginine residue of alpha-conotoxin G1 is responsible for its selective high affinity for the alpha gamma agonist site on the electric organ acetylcholine receptor. *Biochem.* **36**, 9051–6.

[15] Sine, S.M., Kreienkamp, H-J., Bren, N., Maeda, R.M. and Taylor, P. (1995) Molecular dissection of subunit interfaces in the acetylcholine receptor: Identification of residues that determine α conotoxin M1 selectivity. *Neuron* **15**, 205–11.

[16] Sugiyama, N., Marchot, P., Kawanishi, C., Osaka, H., Molles, B., Sine, S. and Taylor, P. (1998) Residues at the subunit interfaces of the nicotinic acetylcholine receptor that contribute to α-conotoxin M1 binding. *Mol. Pharmacol.* **53**, 787–94.

[17] Bren, N. and Sine, S. (2000) Hydrophobic pairwise interactions stablize α-conotoxin, MI in the muscle acetylcholine receptor binding site. *J. Biol. Chem.* **275**, 12692–700.

[18] Schmidt, J.J., Weinstein, S.A. and Smith L.A. (1992) Molecular properties and structure-function relationships of lethal peptides from the venom of Wagler's pit viper. *Toxicon* **30**, 1027–36.

[19] McArdle, J.J., Lentz, T.C., Witzewann, V., Schwarz, H., Weinstein, S.A. and Schmidt, J.J. (1999) Waglerin-1 selectively blocks the epsilon form of the muscle acetylcholine receptor. *J. Pharmacol. Exp. Ther.* **289**, 543–50.

[20] Lin, W.W., Smith, L.A. and Lee, C.Y. (1995) A study of the cause of death due to waglerinI, a toxin from *Trimeresurus wagleri. Toxicon* **33**, 111–14.

[21] Taylor, P., Osaka, H., Molles, B.E., Sugiyama, N., Marchot, P., Malany, S., McArdle, J.J., Sine, S.M. and Tsigelny, I. (1998) Toxins selective for subunit interfaces as probes of nicotinic acetylcholine receptor structure. *J. Physiol.* **92**, 79–83.

[22] Molles, B., Ph.D. Thesis (2001) Probing the muscle nicotinic receptor structure with Waglerin peptides. University of California, San Diego.

[23] Osaka, H., Malany, S., Molles, B.E., Sine, S.M. and Taylor, P. (2000) Pairwise electrostatic interactions between α-neurotoxins and γ, δ and ε subunits of the nicotinic acetylcholine receptor. *J. Biol. Chem.* **275**, 5478–84.

[24] Malany, S., Osaka, H., Sine, S.M. and Taylor, P. (2000) Orientation of α-neurotoxin at the subunit interfaces of the nicotinic acetylcholine receptor. *Biochem.* **39**, 15388–98.

[25] Michalet, S., Teixeira, F., Gilquin, B., Mourier, G., Servent, D., Drevet, P., Binder, P., Tzartos, S., Ménez, A. and Kessler, P. (2000) Relative spatial position of a snake neurotoxin and the reduced disulfide bond alpha (Cys 192 – Cys 193) at the alpha gamma interface of the nicotinic acetylcholine receptor. *J. Biol. Chem.* **275**, 25608–15.

[26] Tremeau, O., Lemaire, C., Drevet, P., Pinkasfeld, S., Ducancel, F., Boulain, J-C. and Ménez, A. (1995) Genetic engineering of snake toxins. The functional site of Erabutoxin, as delineated by site directed mutagenesis, includes variant residues. *J. Biol. Chem.* **270**, 9362–9.

[27] Taylor, P., Osaka, H., Molles, B., Keller, S.H, and Malany, S. (2000) Contributions of studies of the nicotinic receptor from muscle to defining structural and functional properties of ligand-gated ion channels, Chapter 5, In: *Neuronal Nicotinic Receptors*. F. Clementi, C. Gotti and D. Formaseri, eds. Handbook for Experimental Pharmacology.

16 Allosteric and Steric Interactions of Polyamines and Polyamine-containing Toxins with Nicotinic Acetylcholine Receptors

T.J. BRIER, I.R. MELLOR and P.N.R. USHERWOOD*

16.1 INTRODUCTION

Some arthropods produce polyamine-containing neurotoxins that paralyse prey by inhibiting ionotropic receptors (for glutamate and acetylcholine). These toxins, first discovered in a parasitic digger wasp (*Philanthus triangulum*) and orb-web spiders (*Argiope trifasciata* and *Joro spp*), have structural elements in common, i.e. a head-group containing an aromatic moiety, a polyamine and a terminal positive charge. The major paralytic agent in the venom of the wasp is philanthotoxin-433 (butanoyl-tyrosyl-thermospermine, PhTX-433, where the numerals denote the number of methylenes between the amino groups of the thermospermine moiety (Figure 16.1)). A comparable agent in the venom of *Argiope trifasciata* is argiotoxin-636 (2,4-dihydroxyphenylacetyl-asparaginyl-polyamine(533)-arginine, ArgTX-636, where the numerals refer to the molecular weight of the toxin (Figure 16.1)), although other equally powerful polyamine-containing toxins (hereafter called polyamine amides) are present in this venom [1].

Numerous polyamine amides have subsequently been discovered in arthropod venoms and angiosperm plants, e.g. the leaves of *Oncinotis tenuiloba* [2], and many neurobiologically active analogues have been synthesised [3, 4]. An early review of the properties of natural polyamine amides by Jackson and Usherwood [5] hinted at their possible multiple actions on excitable cells and this has been borne out by subsequent studies [6–9]. Their effects on nicotinic acetylcholine receptors (nAChR), ionotropic glutamate receptors and Ca^{2+} and K^+ channels have been well-documented, although some studies have focused on one mode of action to the exclusion of other important effects and have led to blinkered views about their neurobiological properties. Here we review the interactions of polyamine amides with vertebrate nAChR and provide explanations for their diverse effects on this class of ionotropic receptor.

---

* Corresponding author

Perspectives in Molecular Toxicology
Edited by A Ménez © 2002 John Wiley & Sons, Ltd

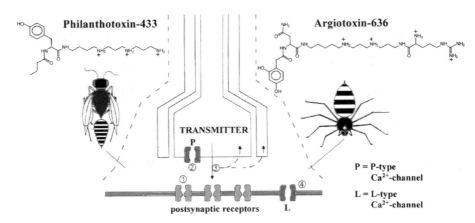

Figure 16.1 Structures of philanthotoxin-433 from the wasp *Philanthus triangulum* and argiotoxin-636 from the spider *Argiope trifasciata*. The actions of these and similar polyamine-containing toxins include: (1) inhibition of postsynaptic ionotropic receptors, (2) block of presynaptic P-type Ca^{2+}-channels [50], (3) inhibition of transmitter reuptake [51], and (4) block of muscle L-type Ca^{2+}-channels [50]

The nAChR are ligand-gated ion channels that include neuronal and muscle representatives and that undergo interconversion among several conformational states. They comprise five homologous subunits (~ 50 kDa) associated around a central axis, the ion channel. Skeletal muscle nAChR comprise $\alpha(\times 2)$, β, γ (e in adult muscle) and δ subunits. Each subunit has four putative membrane-spanning segments (M1–M4) with the M2 segments contributing to the wall of the ion channel pore and the M3 and M4 segments contributing to the protein–lipid interface. The nAChR of the *Torpedo* electroplax that have been used extensively for biochemical studies of polyamine amides are similar to nAChR of (immature) vertebrate skeletal muscle. Although neuronal nAChR are also pentamers, they usually have only two subunit types (α and β) and some are homo-oligomeric ($\alpha7$ nAChR). Each nAChR has two agonist-binding sites, which in the nAChR of *Torpedo* electroplax are located in the extracellular domain at the α–γ (low-affinity site) and α–δ subunit (high-affinity site) interfaces, in pockets about 30–35 Å deep. Electron microscopy and mutagenesis studies suggest an all-beta folding of the N-terminal extracellular domain of nAChR, with the connecting loops contributing to the ACh-binding pocket and to the subunit interfaces that mediate the allosteric interconversions between resting, open and desensitised conformational states [10]. The ion channel is funnel-shaped with extracellular and intracellular vestibules and a ~30 Å long pore, lined by M2 residues (Plate XI), which crosses the membrane bilayer. The extracellular vestibule is considered to extend ~60 Å above the outer surface of the membrane bilayer, whereas the intracellular vestibule projects ~12 Å beyond the cytoplasmic face of the bilayer. In its open state, the diameter of the nAChR channel ranges between 25 Å and 6–8 Å. It is narrowest at the so-called channel gate or ion filter which is located ~8 Å from the cytoplasmic entrance of the pore. In an α7 homo-pentameric nAChR model, the filter has an estimated radius of less than 2 Å when the channel is closed [11]. Current models of the nAChR pore contain two structural domains, an α-helical upper domain and a loop component domain in which the ion filter is located [10]. The pore contains eight rings of amino acids that either determine the kinetic and conductance properties of the channel and/or are binding sites for some non-competitive antagonists (Plate XI). The extracellular vestibule narrows to ~11 Å at the pore entrance where the wall is lined by a ring of three negatively-charged M2 residues which, presumably, attract cations to the pore. The intermediate ring, a ring of four negatively charged amino acids, lines the internal entrance of the pore. This is thought to be the ion filter. A second ring of four negatively charged amino acids, the internal ring, is located just below the filter. Immediately above the filter there are two hydrophilic rings containing predominantly threonine or serine residues, followed by leucine, valine and leucine rings. These rings of amino acids seemingly provide many possible binding sites for polyamine amides through hydrogen bonding, electrostatic interactions and hydrophobic interactions.

In a seminal publication, Heidmann et al. [12] identified two main categories of binding sites for noncompetitive inhibitors (NCIs) of vertebrate nAChR, a high-affinity site and a population of low-affinity sites. This has led to a generally held view of a unique NCI binding site [13]. However, according to Arias [14], there is more than one high-affinity site, with nAChR function being modulated either allosterically or sterically by high-affinity NCIs. Allosteric NCIs are said to promote a conformational change in nAChR that reduces the probability of channel opening. Such inhibitors stabilise either the resting or the desensitised states of nAChR by binding to sites either within the pore [15] or in the extracellular vestibule. In contrast, steric NCIs are said to block the open channel gated by nAChR physically by plugging the pore. Both allosteric and steric NCIs are considered to bind to sites on nAChR in regions usually distinct from agonist and competitive antagonist sites [14]. There is strong evidence from labelling and mutagenesis studies to support the contention that binding sites for some NCIs are located on the α-helical component of the pore.

Evidence that polyamine amide toxins are noncompetitive antagonists of ionotropic receptors was obtained before these natural products were chemically characterised [16]. Subsequently, most pharmacological studies have focused on their interactions with invertebrate and vertebrate glutamate receptors, although, recently, there has been growing interest in their noncompetitive inhibition of vertebrate nAChR. These studies have shown that, like ionotropic glutamate receptors [7, 17–19], nAChR have a complex pharmacological relationship with polyamine amides. Here, we review the electrophysiological and biochemical evidence that polyamine amides have multiple binding sites on these receptors. We shall also show that these compounds exhibit somewhat similar properties at nAChR to those of polyamines such as spermine, i.e. they are allosteric and steric NCIs of neuronal and muscle nAChR, but that they also bind to internal sites on neuronal nAChR to cause inward rectification. In addition to their antagonism of nAChR, polyamine amides and polyamines also potentiate responses of these receptors by inhibiting desensitisation.

16.2 ELECTROPHYSIOLOGICAL STUDIES

16.2.1 POTENTIATION

Hsu [20] and Shao *et al.* [21] used whole-cell and single-channel recording techniques to study the interactions of spermine with the muscle-type nAChR expressed in *Xenopus laevis* muscle and TE671 cells, respectively. These interactions included a potentiation of ACh-induced currents by 10 μM spermine. A similar phenomenon was previously described by Brackley *et al.* [17] for neuronal nAChR expressed in *Xenopus laevis* oocytes from rat brain RNA. The potentiation of TE671 nAChR results from inhibition of desensitisation and can be antagonised competitively by arcaine (1,4-diguanidinobutane). The onset of ACh-induced desensitisation of these nAChR is biphasic and spermine reduces both rates. It has not yet been established whether spermine affects the rate of recovery from desensitisation. Whole-cell recordings from TE671 cells have also revealed potentiation of responses to ACh at μM concentrations of PhTX-343 and some analogues of this polyamine amide. However, this phenomenon has not been consistently observed, possibly because it is masked by noncompetitive inhibition, which is obtained with lower concentrations of PhTX-343 than with spermine.

16.2.2 INHIBITION

It has been frequently demonstrated electrophysiologically that inhibition of an ionotropic receptor by an NCI is either voltage-dependent or voltage-independent, although inhibition by some NCIs contains voltage-dependent and voltage-independent components. Voltage-independent actions are thought to involve binding of NCIs to sites on receptors that are outside the membrane electric field, whereas voltage-dependent inhibition is considered to result from binding of these antagonists to sites within the membrane electric field, i.e. in the channel. In the former case the inhibition is allosteric, whereas in the latter case it may result in open channel block (steric inhibition).

Allosteric interactions

Most studies of polyamine amides have focused on their ability to block the open channels gated by cation-selective ionotropic receptors. As a result, there is a general assumption that multiple charged amine groups in the polyamine chain of these compounds are essential for inhibition of these receptors. Conversely, it has been assumed that the presence of these positively-charged moieties shows that the paralytic action of natural polyamine amides is due to their open channel block of cation-selective channels. Neither assumption is correct. Although the presence of multiple charged amine groups is a requirement for inhibition of ionotropic glutamate receptors [3, 22–24], this inhibition contains voltage-independent and voltage-dependent components [25]. In other words, the charged groups are required for binding to allosteric *and* steric sites on these receptors. In the case of nAChR, it seems that multiple charged groups are not essential for allosteric inhibition: indeed, they are detrimental. Inhibition during extracellular application of PhTX-343 to muscle and neuronal nAChR contains voltage-dependent and voltage-independent components, suggesting at least two sites of action for this polyamine amide. In a study performed to assess the importance of secondary amines in the spermine moiety of PhTX-343 for inhibition of the muscle-type nAChR of TE671 cells [26, 27], the amines were substituted either singly or as a pair by methylene or ether groups (Figure 16.2). None of the methylene or ether analogues was a less potent antagonist than PhTX-343. Indeed, the methylene analogues, especially PhTX-(12), were significantly more potent and although their inhibition contained a voltage-dependent component, this was smaller than that for PhTX-343. Consequently, PhTX-(12) is a potent inhibitor of nAChR of TE671 cells even at positive membrane potentials when PhTX-343 is only weakly inhibitory (Figure 16.3). It seems that PhTX-(12) promotes a closed channel conformation of muscle-type nAChR. This is not through competitive antagonism, because

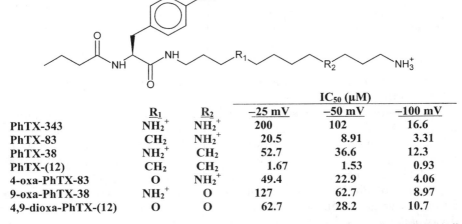

	R_1	R_2	IC$_{50}$ (µM)		
			−25 mV	−50 mV	−100 mV
PhTX-343	NH$_2^+$	NH$_2^+$	200	102	16.6
PhTX-83	CH$_2$	NH$_2^+$	20.5	8.91	3.31
PhTX-38	NH$_2^+$	CH$_2$	52.7	36.6	12.3
PhTX-(12)	CH$_2$	CH$_2$	1.67	1.53	0.93
4-oxa-PhTX-83	O	NH$_2^+$	49.4	22.9	4.06
9-oxa-PhTX-38	NH$_2^+$	O	127	62.7	8.97
4,9-dioxa-PhTX-(12)	O	O	62.7	28.2	10.7

Figure 16.2 Abasic analogues of PhTX-343. IC$_{50}$ values are for inhibition of TE671 cell nAChR by PhTX-343 analogues when co-applied with 10 µM ACh at holding potentials (V$_H$) of −25 mV, −50 mV and −100 mV

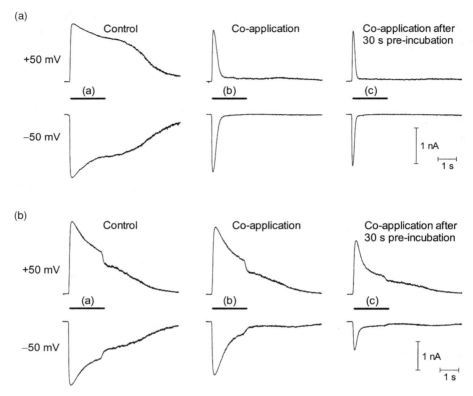

Figure 16.3 Whole-cell responses of TE671 cells to 2 s applications of 10 μM ACh in the absence and presence of PhTX-(12) (A) or PhTX-343 (B) at V_H of +50 mV and −50 mV. Horizontal bars indicate drug applications as follows: (a) 10 μM ACh alone (control); (b) co-application of 10 μM ACh with 10 μM PhTX-(12) (A) or 100 μM PhTX-343 (B); (c) same as (b) but immediately preceded by preincubation with 10 μM PhTX-(12) (A) or 100 μM PhTX-343 (B) for 30 s

this synthetic polyamine amide analogue has only a slight effect on the EC_{50} for ACh (∼ a three-fold increase).

Insight into the mechanism responsible for the voltage-independent inhibition of nAChR by PhTX-(12) has been obtained by comparing the effects of two application protocols. Responses of TE671 cells to a pulse of ACh comprise an inward current that reaches a peak before decaying rapidly to a low plateau current due to desensitisation of nAChR (Figure 16.3(a)). Co-application of PhTX-(12) with ACh results in a small reduction of the peak current, a greater reduction of the plateau current and an increased rate of decay of the whole-cell current (Figure 16.3A(b)), i.e. there is a time-dependence of antagonism. This is true at both negative and positive membrane potentials with only a minor difference between the two (Figure 16.3A(b)), suggesting that the major component of inhibition is not steric. When TE671 cells are pre-incubated in PhTX-(12) for up to 30 s prior to its co-application with ACh, the reductions of peak and plateau currents are greater than for co-application without pre-incubation, but again the effect of the polyamine amide on the peak response is small compared to that on the plateau response (Figure 16.3A(c)), i.e. the time-dependence is still present. Also, the rate of decay of the whole-cell current is greater with the pre-incubation protocol (Figure 16.3A(c)). Again, this is true for both positive and negative membrane potentials. In the pre-incubation experiment, this time-dependency would not be expected if PhTX-(12) interacts allosterically with the resting channel conformation of nAChR to inhibit channel gating by agonist. Rather, it suggests that PhTX-(12) binds to the resting channel conformation and enhances the rate of onset of agonist-induced desensitisation. Nevertheless, in both the presence and absence of pre-incubation there is a voltage-independent decline in peak response amplitude which does not readily lend itself to such an explanation. However, if the nAChR can transit directly from its closed state to a desensitised state, and the polyamine amide allosterically affects the equilibrium between these states, then desensitisation enhancement both in the presence and absence of ACh would account fully for the voltage-independent action of PhTX-(12). Electrophysiological evidence that an NCI enhances desensitisation was obtained by Ashford et al. [28] who showed that ketamine enhances desensitisation of locust muscle ionotropic glutamate receptors and that when desensitisation is inhibited by concanavalin A [29] this action of ketamine is abolished. Does PhTX-343 also enhance the rate of desensitisation onset of muscle-type nAChR? When this polyamine amide is briefly co-applied with ACh at negative membrane potentials it is an inhibitor (IC_{50} at -50 mV, $\sim 100\,\mu$M) of nAChR of TE671 cells, but only slight inhibition is observed at positive membrane potentials (Figure 16.3B(b)). However, following pre-incubation of these cells in PhTX-343, co-application with ACh now leads to significant inhibition at positive membrane potentials (Figure 16.3B(c)). It seems likely that, like PhTX-(12), PhTX-343 enhances the rate of onset of desensitisation allosterically by binding to the resting channel conformation of nAChR, but that its binding rate (affinity) at this allosteric site is lower than that of PhTX-(12).

Is their other evidence for allosteric effects of polyamine amides on nAChR? Jayaraman et al. [30], used whole-cell clamping and laser-pulse photolysis of caged carbamoylcholine to investigate inhibition of muscle-type nAChR in BC_3H1 cells. When the nAChR were activated by a high concentration (750 μM) of carbamoylcholine, the rate constant for nAChR channel opening was apparently reduced by the polyamine amide, thereby shifting the channel opening equilibrium toward a closed-channel state. This resulted in a decrease in the peak amplitude of the whole-cell current in the presence of the polyamine amide. Interestingly this inhibition was weakly voltage-dependent, increasing about two-fold as V_H was reduced from $+50$ mV to -60 mV, an affect attributed mainly to the action of PhTX-343 on the channel opening equilibrium constant. The possibility that PhTX-343 affects desensitisation was not considered.

Steric antagonism

Rozental *et al.* [31] were the first to demonstrate open-channel block of nAChR by a polyamine amide (PhTX-433), although their single-channel data from frog skeletal muscle indicated that the apparent channel opening rate was also lowered by this toxin. Jayaraman *et al.* [30] showed that PhTX-343 blocks the open channel of BC_3H1 nAChR, but its affinity for the open-channel conformation of this receptor is five times less than its affinity for the closed-channel conformation. PC12 cells express neuronal-type nAChR comprising mainly $\alpha3$ and $\beta4$ subunits [32]. PhTX-343 and JSTX-3 (a toxin from the Joro spider [33]) inhibit these receptors at sub-micromolar concentrations ($EC_{50} \sim 100\,nM$) in a voltage-dependent fashion [34]. Assuming that PhTX-343 is trivalent, Lui *et al.* [34] estimated that the binding site for JSTX-3 is located 16% within the membrane electric field, i.e. in the region of the outer leucine ring and the valine ring lining the nAChR pore. However, this estimate must be viewed with some caution because PhTX-343 may be at least 80% tetravalent at physiological pH (7.4) [35]. In TE671 cells, PhTX-343 inhibition of nAChR is strongly voltage-dependent (Figure 16.2 and 16.3A). Are the multiple charges on PhTX-343 essential for open channel block? It is clear that replacement of the secondary amines of PhTX-343 by methylenes, as in PhTX-(12), almost eliminates the voltage-dependent component of nAChR inhibition (Figures 16.2 and 16.3B). Inhibition by ether analogues of PhTX-343 (Figure 16.2) is strongly voltage-dependent and these compounds have qualitatively similar potencies to that of PhTX-343. It is likely, therefore, that the ether analogues interact with a steric inhibitory site on the nAChR. The possibility that the methylene analogues also bind to this site cannot be discounted, but the strong inhibition caused by their high-affinity binding to an allosteric site precludes us from investigating this possibility electrophysiologically until we have a polyamine amide competitor that has a high specific affinity for the allosteric site but does not inhibit the nAChR. At this time, our evidence suggests that polyamine amides, such as PhTX-343, with multiple charges sterically inhibit nAChR through open-channel block. Interestingly, inhibition by the singly-charged ether analogue of PhTX-343, 4,9-dioxa-PhTX-(12), is almost as voltage-dependent as that of PhTX-343.

Do polyamines also sterically inhibit nAChR? Extracellular application of spermine to neuronal-type nAChR causes noncompetitive antagonism, e.g. spermine inhibits recombinant $\alpha4\beta2$ receptors with an IC_{50} of $\sim 40\,\mu M$ [36]. Interestingly, spermine was equally potent as an inhibitor of $\alpha4\beta2$Glu-260 receptors and the wild-type $\alpha4\beta2$ receptors, which led Haghighi and Cooper [36] to conclude that the inhibition results from open-channel block, despite the fact that the inhibition was only weakly voltage-dependent.

Inward rectification

The idea that polyamines and polyamine amides might intracellularly modulate the functioning of an ionotropic receptor was first examined by Brundell *et al.* on quisqualate sensitive glutamate receptors (qGluR) of insect muscle ([37]; see also [38]). PhTX-343 and spermine were injected by micropipette into locust leg muscle fibres whilst recording either glutamate-induced postsynaptic currents or single extrajunctional qGluR channel openings. Both compounds elicited voltage-independent antagonism of qGluR, suggesting that they bind to a site on this receptor outside the membrane electric field. A common feature of neuronal nAChR in whole-cell recordings is that they conduct inward current at negative membrane potentials, but little outward current at positive membrane potentials. This inward rectification has recently been shown to result from channel block by intracellular spermine [36]; it is lost during the formation of outside-out patches for single nAChR channel recordings, but is retained if spermine (10 μM) is included in the patch pipette. Spermine antagonises neuronal nAChR when applied to the internal face of inside-out patches excised from superior cervical ganglion cell membrane and oocytes expressing recombinant $\alpha 3\beta 4$ and $\alpha 4\beta 2$ receptors [36], the proposed binding site for this inward rectification being about halfway into the pore of the nAChR channel. Using the Woodhull infinite barrier model for voltage-dependent ion channel block [39], and assuming a valency for spermine of 3.8 at physiological pH [40], it was estimated that for recombinant $\alpha 4\beta 2$ nAChR expressed in *Xenopus* oocytes $\sim 51\%$ of the membrane electrical field from the internal face of the membrane is sensed by the polyamine, whereas for nAChR of neurons of rat superior cervical ganglion $\sim 65\%$ of the field is sensed [36]. Although this difference may be significant, both estimates surprisingly suggest that the binding site for intracellular spermine is beyond the channel filter or gate. However, the use of the Woodhull model may be inappropriate for a multivalent ligand, such as spermine, which has a distributed charge. Also, this approach towards identifying binding sites in the channel is seriously flawed if the electric field distribution along the pore is not uniform. Haghighi and Cooper [41] later used site-directed mutagenesis and electrophysiology on $\alpha 4\beta 2$ and $\alpha 3\beta 4$ nAChR expressed in *Xenopus* oocytes to determine the binding site for spermine. Removal of the negative charges on the intermediate ring abolished inward rectification and diminished the high-affinity voltage-dependent interaction between intracellularly-applied spermine and the receptor. They concluded that the intermediate ring, which is only ~ 5 Å ($\sim 17\%$) inside the membrane electric field, forms a high-affinity binding site for intracellular polyamines, with the polyamines held at this site through electrostatic interactions with negatively-charged glutamic acid residues. In addition, the internal ring may concentrate spermine molecules in the intracellular vestibule, although mutations in this ring had little effect on inward rectification by spermine. Intracellularly-applied spermine [21] and PhTX-343 [30] do not antagonise muscle-type nAChR expressed in TE671 and BC$_3$H1 cells, respectively. The lower net negative charge on the intermediate ring of muscle nAChR compared with neuronal nAChR probably accounts for this difference [41].

16.2.3 LIGAND-BINDING STUDIES

The first biochemical studies of a polyamine-containing arthropod toxin were undertaken by Rozental et al. [31] who showed that PhTX-433 competitively inhibits the binding of [^{3}H]perhydrohistrionicotoxin ([^{3}H]H$_{12}$-HTX), an NCI of the resting and open states of nAChR, to membrane preparations of Torpedo electroplax with an IC$_{50}$ of $\sim$10 μM. Unsurprisingly, this inhibition was dependent on the presence of agonist. The authors proposed that PhTX-433 binds to a site on the open-channel conformation of nAChR. PhTX-433 also inhibits the binding of [^{125}I]α-bungarotoxin ([^{125}I]αBTX) and [^{3}H]ACh to nAChR, but at higher concentration (IC$_{50}$ $\sim$ 200 μM). Although 1 mM PhTX-433 reduced the initial rate of binding of [^{125}I]α-BTX by > 90 %, the initial rate of binding of [^{3}H]ACh was reduced by only 50 %, suggesting that this inhibition by PhTX-433 may not be competitive. It was concluded that PhTX-433 binds to an allosteric site on nAChR where it induces conformational changes affecting [^{3}H]ACh and [^{125}I]α-BTX binding. Rozental et al. [31] also found that concentrations (0.1 μM) of PhTX-433 lower than those that inhibited binding of [^{3}H]ACh potentiated the initial rate of [^{3}H]ACh and [^{3}H]H$_{12}$-HTX binding, by 20 % and 300 % respectively, although the equilibrium binding of [^{3}H]H$_{12}$-HTX was unaffected. Spermine and spermidine also inhibit [^{3}H]H$_{12}$-HTX and [^{125}I]α-BTx binding to Torpedo electroplax nAChR, although at higher concentrations (IC$_{50}$ 30–110 μM). Ethidium is a well-characterised NCI of nAChR, binding with high affinity in the pore region of the channel, and with lower affinity at the lipid–protein interface. Bixel et al. [42] examined the displacement of fluorescent ethidium from nAChR-containing membranes of Torpedo electroplax by polyamine-containing ligands, including some PhTX-433 analogues. The nAChR were cross-linked with glutardialdehyde to eliminate allosteric interactions between the fluorescent marker and the polyamine amides. The polyamine amides displaced ethidium bound to desensitised nAChR by a competitive non-allosteric mechanism. Bixel et al. [42] also showed that two molecules of the photolabile polyamine amide N$_3$-phenyl-^{125}I$_2$-PhTX-343-lysine reversibly bind with equal affinities to electroplax nAChR. This property differentiates this polyamine amide from the other NCIs of nAChR, which have a unitary stoichiometry.

16.2.4 PHOTOAFFINITY LABELLING EXPERIMENTS

These were first undertaken by Choi et al. [43] using N$_3$-phenyl-^{125}I$_2$-PhTX-343-lysine and nAChR purified from Torpedo electroplax and reconstituted into lipid vesicles. Cross-linking of this polyamine amide was to all subunits of nAChR in the absence of the 43 kDa protein associated with nAChR, but mainly to the α-subunit in the presence of this protein (also labelled). It was inhibited by chlorpromazine, which photo-cross-links to Ser-248 on the α-subunit, but not by α-BTX. Cross-linking of the polyamine amide was preferentially to a 20 kDa fragment of the α-subunit, viz. αSer-173 to αGlu-338, a fragment containing the agonist-binding site and the transmembrane segments M1–M3. Because α-BTx failed to inhibit binding of N_3-phenyl-$^{125}I_2$-PhTX-343-lysine, it was concluded that αSer-173 to αPro-211 and αVal-262 to αTyr-279, which are in the region of the α-BTX-binding site (αCys192,-193), do not bind the polyamine amide. Choi et al. [43] concluded that the aromatic head-group of N_3-phenyl-$^{125}I_2$-PhTX-343-lysine binds to a hydrophobic pocket formed by the hydrophobic residues intracellularly bordering the M2 segments of nAChR with the polyamine moiety interacting electrostatically with negatively charged residues of the intermediate ring. However, it is difficult to see how this could account for the inhibition of nAChR seen during extracellular application of polyamine amides to nAChR. Bixel et al. [42] revisited the labelling of Torpedo electroplax nAChR by N_3-phenyl-$^{125}I_2$-PhTX-343-lysine. They found that all five subunits of nAChR were labelled by this compound; carbachol increased this cross-linking, α-BTX did not affect the cross-linking, but triphenylmethylphosphonium ($TPMP^+$) reduced the cross-linking of the polyamine amide. They concluded that N_3-phenyl-$^{125}I_2$-PhTX-343-lysine binds to the closed- and open-channel conformations of nAChR. More recently, Bixel et al. [44] have studied the labelling of Torpedo electroplax nAChR by radio-iodinated MR44, a polyamine amide that carries a photolabile azido group on its aromatic moiety. ^{125}I-MR44 binds exclusively to the region αHis-186 to αLeu-199 of the α-subunits in the extracellular vestibule. This binding site is in the vicinity of a series of negatively charged amino acids that are known to bind Ca^{2+}. It is noteworthy that Ca^{2+} displaced reversibly-bound ^{125}I-MR44. The labelled sequence is also close to, but does not overlap with, the agonist-binding site, but cross-linking of ^{125}I-MR44 was unaffected by α-BTX. It has not yet been possible to identify the precise residues involved. Interestingly, the sequence αHis-186 to α-Leu-199 overlaps with the αSer-173 to αGlu-338 sequence labelled by N_3-phenyl-$^{125}I_2$-PhTX-343-lysine in the earlier studies of Choi et al. [43]. Because the photo-labile azido group of ^{125}I-MR44 is located on the aromatic moiety, the site of ^{125}I-MR44 cross-linking identifies the region to which this hydrophobic head-group binds and provides only limited information on the possible attachment sites for the polyamine moiety. It was concluded that the interaction of the head-group of ^{125}I-MR44 lies within the hydrophobic sequence α186-His to α189-Tyr. Electrophysiological data indicate that antagonism of nAChR by MR44 is largely voltage-independent, so presumably it also binds to a site(s) in the extracellular vestibule or at the entrance of the pore. The binding affinity of ^{125}I-MR44 increased when the agonist cabachol was present, as would be expected if antagonism resulted from enhancement of desensitisation. Cross-linking of ^{125}I-MR44 was displaced by the nAChR NCIs ethidium bromide and TPMP$^+$. The reason why labelling by N$_3$-phenyl-^{125}I$_2$-PhTX-343-lysine was more extensive in the studies of both Choi *et al.* [43] and Bixel *et al.* [42] than that by ^{125}I-MR44 can be accounted for by the greater propensity of the former to block the open channel gated by nAChR. However, Choi *et al.* [43] reported stronger labelling of the α-subunits than of the other subunits of *Torpedo* electroplax nAChR.

16.3 DISCUSSION

The concept of multiple binding sites for polyamine-containing ligands on ionotropic receptors is not new. Usherwood and Blagbrough [9] and Usherwood [7] reviewed evidence for four binding sites for PhTX-343 (an equipotent synthetic analogue of the wasp toxin PhTX-433) on locust muscle qGluR; an external potentiating site, external closed and open channel inhibitory sites and an internal rectifying site. Equivalent sites were later identified on mammalian ionotropic glutamate receptors [17, 19]. The discovery that polyamine amides such as N$_3$-phenyl-^{125}I^2-PhTX-343-lysine and MR44 have a binding stoichiometry of 2 at nAChR [44] places them in a different class from high-affinity NCIs that have a unitary stoichiometry [45]. Nevertheless, [^{125}I]MR44 can be competed off the polyamine amide sites by NCIs (chlorpromazine, ethidium bromide and TPMP$^+$) that have a unitary stoichiometry and which apparently bind to different hydrophobic regions of nAChR. These observations clearly need further thought. Labelling studies and mutagenesis experiments have disclosed high-affinity sites for chlorpromazine and TPMP$^+$ located in the nAChR pore. When these NCIs bind to such sites they are said to block the channel sterically. Chlorpromazine, which binds preferentially to the resting channel state of nAChR, has a large ($\sim$10 Å) aromatic head-group which should limit the depth to which it can descend into the pore. However, its quaternary ammonium, which is about $\sim$ 7 Å distant from the aromatic moiety, could reach the valine ring and, possibly, also the leucine ring. The reactive sites for photo-incorporation of chlorpromazine and TPMP$^+$ are contained within their aromatic moieties. In contrast, [^{3}H]meproadifen mustard, which contains a positively-charged photosensitive moiety (quarternary arizidium ion) and alkylates nAChR only when ACh is present, unsurprisingly does not alkylate α-subunit amino acids that are thought to be the sites for binding of [^{3}H]chlorpromazine [46, 47] and [^{3}H]TPMP$^+$[13]. Rather, it labels the extracellular ring at αGlu-262 [48]. Despite this, alkylation by meproadifen mustard prevents the binding of [^{3}H]HTX, an NCI which is considered to bind deep in the pore in the region of the serine and leucine rings [49] and which blocks the open channel in a highly voltage-dependent manner. Clearly, binding of an NCI to one site on nAChR may inhibit binding of NCIs to other sites on this receptor. Steric influences and/or the stabilisation of conformational states that select for a specific NCI or groups of NCIs may play a role in this phenomenon.

A mechanistic picture of the interaction of polyamines and polyamine amides with nAChR is now emerging from electrophysiological, ligand binding and cross-linking studies. Spermine, which lacks an aromatic head-group, has a maximum diameter of ~ 5 Å, a length of ~ 20 Å and four equidistant positive charges, could, in principle, bind to any complementary charged surface on nAChR. We suggest that the high-affinity site or sites responsible for potentiation (stabilisation of the open-state channel conformation) is located on the extracellular domain of nAChR. So far, we have no information on the binding stoichiometry for this potentiation. Spermine also causes voltage-independent and voltage-dependent inhibition of nAChR. We suggest that the former results from binding of the polyamine to the outer ring of the channel pore leading to enhancement of desensitisation. Notice that we do not envisage the hydrophobic sites on the α-subunits that were labelled by $[^{125}I]MR44$ being involved in either potentiation or inhibition of nAChR by spermine. Voltage-dependent inhibition possibly results from binding of spermine to one or more of the hydrophilic rings that are located in the pore above the filter, although the polyamine would need to pass deep in the channel to cause steric inhibition (Plate XI). Arcaine competes with spermine for the potentiation site and for the steric site, but not for the allosteric site responsible for voltage-independent inhibition. As expected, the picture emerging for polyamine amide action at nAChR is similar to that for polyamines, although the presence of aromatic head-groups on polyamine amides introduces additional properties and some limitations. There is evidence from ligand-binding studies that low concentrations of polyamine amides potentiate nAChR, but this phenomenon has not yet been seen routinely in electrophysiological experiments. The reason could be either that the head-group on the polyamine amide has a negative impact on binding affinity to the potentiation site or that potentiation is masked by inhibition. There is now good evidence to support previous suggestions that polyamine amides cause voltage-independent and voltage-dependent inhibition of nAChR, although the relative contribution of these two mechanisms to inhibition depends on the structure of the polyamine amide. For example, MR44 and PhTX-(12) are potent and predominantly voltage-independent antagonists. By allosterically enhancing desensitisation, they effectively inhibit channel openings and, thereby, exclude open channel block. In contrast, inhibition by more highly charged polyamine amides, such as PhTX-343 and ArgTX-636, is largely voltage-dependent, so one might envisage that these molecules bind in the pore and, thereby, sterically block the channel.

REFERENCES

[1] Budd, T., Clinton, P., Dell, A., Duce, I.R., Johnson, S.J., Quicke, D.L.J., Taylor, G.W., Usherwood, P.N.R. and Usoh, G. (1988). Isolation and characterization of glutamate receptor antagonists from venoms of orb-web spiders. *Brain Res.* **448**, 30–9.

[2] Doll, M.K-H., Guggisberg, A. and Hesse, M. (1994). N-4-benzoylspermidine from *Oncinotis tenuiloba* – analytical differentiation of the 3 isomeric N-benzoylspermidines. *Helv. Chim. Acta*, **77**, 1229–35.

[3] Bruce, M., Bukownik, R., Eldefrawi, A.T., Eldefrawi, M.E., Goodnow, R., Kallimopolous, T., Konno, K., Nakanishi, K., Niwa, M. and Usherwood, P.N.R. (1990). Structure-activity relationships of analogues of wasp toxin philanthotoxin: non-competitive antagonists of quisqualate receptors. *Toxicon* **38**, 1332–46.

[4] Nakajima, T., Toki, Y., Asami, T. and Kawai, N. (1992). Structure-activity relationship of spider toxin and utilization of synthetic analogs. In *Neuroreceptors, Ion Channels and the Brain*, eds: Kawai, N., Nakajima, T. and Barnard, E. Elsevier, Amsterdam, 22–33.

[5] Jackson, H. and Usherwood, P.N.R. (1988). Spider toxins as tools for dissecting elements of excitatory amino acid transmission. *Trends Neurosci.* **11**, 278–83.

[6] Usherwood, P.N.R. (1991). Natural toxins and glutamate transmission. In *'Excitatory Amino Acids'* eds: Meldrum, B.S., Moroni, F., Simon, R.P. and Woods, J.H. Raven Press, New York, pp. 379–95.

[7] Usherwood, P.N.R. (2000). Natural and synthetic polyamines: modulators of signalling proteins. Il *Pharmaco* **55**, 202–5.

[8] Usherwood, P.N.R. and Blagbrough, I.S. (1992). Spider toxins affecting glutamate receptors: polyamines in therapeutic neurochemistry. *Pharmac. Ther.* **52**, 245–68.

[9] Usherwood, P.N.R. and Blagbrough, I.S. (1994). Electrophysiology of polyamines and polyamine amides. In *'The Neuropharmacology of Polyamines'* ed. Carter, C., Acad. Press, London, pp. 185–204.

[10] Corringer, P.-J., Le Novere, N. and Changeux, J.-P. (2000). Nicotinic receptors at the amino acid level. *Annu. Rev. Pharmacol. Toxicol.* **40**, 431–58.

[11] Sankararamakrishnan, R., Adcock, C. and Sansom, M.S.P. (1996). The pore domain of the nicotinic acetylcholine receptor: molecular modeling, pore dimensions and electrostatics. *Biophys. J.* **71**, 1659–71.

[12] Heidmann, T., Oswald, R.E. and Changeux, J-P. (1983). Multiple sites of action for non-competitive blockers on acetylcholine receptor rich membrane fragments from *Torpedo marmorata. Biochem.* **22**, 3112–27.

[13] Hucho, F., Oberthur, W. and Lottspeich, F. (1986). The ion channel of the nicotinic acetylcholine receptor is formed by the homologous helices MII of the receptor subunit. *FEBS Lett.* **205**, 137–42.

[14] Arias, H.R. (1998). Binding sites for exogenous and endogenous non-competitive inhibitors of the nicotinic acetylcholine receptor. *Biochem. Biophys. Acta.* **1376**, 173–220.

[15] Utkin, Y.N., Tsetlin, V.I. and Hucho, F. (2000). Structural organization of nicotinic acetylcholine receptors. *Membr. Cell. Biol.* **13**, 143–64.

[16] Clark, R.B., Donaldson, P.L., Gration, K.A.F., Lambert, J.J., Piek, T., Ramsey, R.L., Spanjer, W. and Usherwood, P.N.R. (1982). Block of locust muscle glutamate receptors by δ-philanthotoxin occurs after receptor activations. *Brain Res.* **241**, 105–14.

[17] Brackley, P., Goodnow, R., Nakanishi, K., Sudan, H.L. and Usherwood, P.N.R. (1990). Spermine and philanthotoxin potentiate excitatory amino acid responses of *Xenopus* oocytes injected with rat and chick brain mRNA. *Neurosci. Lett.* **131**, 196–200.

[18] Brackley, P.T.H., Bell, D.R., Choi, S-K., Nakanishi, K. and Usherwood, P.N.R. (1993). Selective antagonism of native and cloned kainate and NMDA receptors by polyamine containing toxins. *J. Pharmacol. Exp. Ther.* **266**, 1573–80.

[19] Donevan, S.D. and Rogawski, M.A. (1996). Multiple actions of arylalkylamine arthropod toxins on the *N*-methyl-D-aspartate receptor. *Neurosci.* **70**, 361–75.

[20] Hsu, K.S. (1994). Modulation of the nicotinic acetylcholine receptor channels by spermine in *Xenopus* muscle cell culture. *Neurosci. Lett.* **182**, 99–103.

[21] Shao, Z., Mellor, I.R., Brierley, M.J., Harris, J. and Usherwood, P.N.R. (1998). Potentiation and inhibition of nicotinic acetylcholine receptors by spermine in the TE671 human muscle cell line. *J. Pharmacol. Exp. Therap.* **286**, 1269–76.

[22] Anis, N., Sherby, S., Goodnow, R. Jr., Niwa, M., Konno, K., Kallimopoulos, T., Bukownik, R., Nakanishi, K., Usherwood, P., Eldefrawi, A. and Eldefrawi, M. (1990). Structure–activity relationships of philanthotoxin analogs and polyamines on N-methyl-D-aspartate and nitotinic acetylcholine receptors. *J. Pharmacol. Exp. Therap.* **245**, 764–73.

[23] Karst, H., Piek, T., Van Marle, J., Lind, A. and Van Weeren-Kramer, J. (1991). Structure activity relationship of philanthotoxins. 1. Presynaptic and postsynaptic inhibition of the locust neuromuscular-transmission. *Comp. Biochem. Physiol.* **C98**, 471–7.

[24] Benson, J.A., Schürmann, F., Kaufmann, L., Gsell, L. and Piek, T. (1992). Inhibition of dipteran larval neuromuscular synaptic transmission by analogues of philanthotoxin-4.3.3.: a structure–activity. *Comp. Biochem. Physiol.* **C 102**, 267–72.

[25] Sudan, H.L., Kerry, C.J., Mellor, I.R., Choi, S.-K., Huang, D., Nakanishi, K. and Usherwood, P.N.R. (1995). The action of philanthotoxin-343 and photolabile analogues on locust (*Schistocerca gregaria*) muscle. *Invert. Neurosci.* **1**, 159–72.

[26] Strømgaard, K., Brierley, M.J., Andersen, K., Slok, F.A., Mellor, I.R., Usherwood, P.N.R., Krogsgaard-Larsen, P. and Jaroszewski, J.W. (1999). Analogues of neuroactive polyamine wasp toxins that lack inner basic sites exhibit enhanced antagonism toward a muscle-type mammalian nicotinic acetylcholine receptor. *J. Med. Chem.* **42**, 5224–34.

[27] Strømgaard, K., Bjornsdottir, I., Andersen, K., Brierley, M.J., Rizoli, S., Eldursi, N., Mellor, I.R., Usherwood, P.N.R., Hansen, S.H., Krogsgaard-Larsen, P. and Jaroszewski, J.W. (2000). Solid phase synthesis and biological evaluation of enantiometrically pure wasp toxin analogues PhTX-343 and PhTX-12. *Chirality* **12**, 93–102.

[28] Ashford, M.L.J., Boden, P., Ramsey, R.L. and Usherwood, P.N.R. (1989). Enhancement of desensitization of quisqualate-type glutamate receptor by the dissociative anaesthetic ketamine. *J. Exp. Biol.* **141**, 73–86.

[29] Mathers, D.A. and Usherwood, P.N.R. (1976). Concanavalin A blocks desensitization of glutamate receptors of insect muscle fibres. *Nature* (Lond.) **259**, 409–11.

[30] Jayaraman, V., Usherwood, P.N.R. and Hess, G.P. (1999). Inhibition of nicotinic acetylcholine receptor by philanthotoxin-343: kinetic investigations in the microsecond time region using a laser-pulse photolysis technique. *Biochem.* **38**, 11406–14.

[31] Rozental, R., Scobile, G.T., Albuquerque, E.X., Idriss, M., Sherby, S., Sattelle, D.B., Nakanishi, K., Konno, K., Eldefrawi, A.T. and Eldefrawi, M.E. (1989). Allosteric inhibition of nicotinic acetylcholine receptors of vertebrates and insects by philanthotoxin. *J. Pharmacol. Exp. Therap.* **249**, 123–30.

[32] Rogers, S.W., Mandelzys, A., Deneris, E.S., Cooper, E. and Heinemann, S. (1992). The expression of nicotinic acetylcholine receptors by PC12 cells treated with NGF. *J. Neurosci.* **12**, 4611–23.

[33] Arakami, Y., Yasukura, T., Higashijima, T., Yosioka, M., Miwa, A., Kawai, N. and Nakajima, T. (1986). Chemical characterization of spider toxin, JSTX and NSTX. *Proc. Japan. Acad. Sci.* **2(B)**, 359–62.

[34] Lui, M., Nakazawa, K., Inoue, K. and Ohno, Y. (1997). Potent and voltage-dependent block by philanthotoxin-343 of neuronal nicotinic receptor/channels in PC12 cells. *Brit. J. Pharmacol.* **122**, 379–85.

[35] Jaroszewski, J.W., Matzen, L., Frolund, B. and Krogsgard-Larsen, P. (1996). Neuroactive polyamine wasp toxins: nuclear magnetic resonance spectroscopic analysis of the proteolytic properties of philanthotoxin-343. *J. Med. Chem.* **39**, 515–21.

[36] Haghighi, A.P. and Cooper, E. (1998). Neuronal nicotinic acetylcholine receptors are blocked by intracellular spermine in a voltage-dependent manner. *J. Neurosci.* **18**, 4050–62.

[37] Brundell, P., Goodnow, R. Jr., Kerry, C.J., Nakanishi, K., Sudan, H.L. and Usherwood, P.N.R. (1991). Quisqualate-sensitive glutamate receptors of the locust *Schistocerca gregaria* are antagonised by intracellularly-applied philanthotoxin and spermine. *Neurosci. Lett.* **131**, 196–200.

[38] Kerry, C.J., Sudan, H.L., Nakanishi, K. and Usherwood, P.N.R. (1997). Intracellular application of polyamine and polyamine amide inhibits the quisqualate-sensitive ionotropic glutamate receptor of locust (*Schistocerca gregaria*) muscle. *Invert. Neurosci.* **2**, 223–33.

[39] Woodhull, A.M. (1973). Ionic blockade of sodium channels in nerve. *J. Gen. Physiol.* **61**, 687–708.

[40] Bowie, D. and Mayer, M.L. (1995). Inward rectification of both AMPA and kainate subtype glutamate receptors generated by polyamine-mediated ion channel block. *Neuron* **15**, 453–62.

[41] Haghighi, A.P. and Cooper, E. (2000). A molecular link between inward rectification and calcium permeability of neuronal nicotinic acetylcholine $\alpha 3\beta 4$ and $\alpha 4\beta 2$ receptors. *J. Neurosci.* **20**, 529–41.

[42] Bixel, G.M., Krauss, M., Liu, Y., Bolognesi, M.L., Rosini, M., Mellor, I.R., Usherwood, P.N.R., Melchiorre, C., Nakanishi, K. and Hucho, F. (2000). Structure–activity relationship and site of binding of polyamine derivatives at the nicotinic acetylcholine receptor. *Eur. J. Biochem.* **267**, 110–20.

[43] Choi, S-K., Kalivretenos, A.G., Usherwood, P.N.R. and Nakanishi, K. (1995). Labelling studies of photolabile philanthotoxins with nicotinic acetylcholine receptors: mode of interaction between toxin and receptor. *Chem. Biol.* **2**, 23–32.

[44] Bixel, G.M., Weise, C., Bolognesi, M.L., Rosini, M., Brierly, M.J., Mellor, I.R., Usherwood, P.N.R., Melchiorre, C. and Hucho, F. (2001). Location of the polyamine binding site in the vestibule of the nicotinic acetylcholine receptor ion channel. *J. Biol. Chem.* **276**, 6151–60.

[45] Arias, H.R. (1996). Luminal and non-luminal non-competitive inhibitor binding sites on the nicotinic acetylcholine receptor. *Mol. Memb. Biol.* **13**, 1–17.

[46] Giraudat, J., Dennis, M., Heidmann, T., Chang, J.Y. and Changeux, J.-P. (1986). Structure of the high affinity binding site for non-competitive blockers of the acetylcholine receptor: serine 262 of the delta subunit is labeled by [^{3}H]chlorpromazine. *Proc. Natl. Acad. Sci. USA* **83**, 2719–23.

[47] Giraudat, J., Dennis, M., Heidmann, T., Haumont, P.-Y., Lederer, F. and Changeux, J.-P. (1987). Structure of the high affinity binding site for non-competitive blockers of the acetylcholine receptor: [^{3}H]chlorpromazine labels homologous residues in the β and δ-chains. *Biochem.* **26**, 2410–18.

[48] Pedersen, S.E., Sharp, S.D., Liu, W.-S. and Cohen, J.B. (1992). Structure and non-competitive antagonist-binding site of the *Torpedo* nicotinic acetylcholine receptor. *J. Biol. Chem.* **267**, 10489–99.

[49] Dreyer, E.B., Hasan, F., Cohen, S.G. and Cohen, J.B. (1986). Reaction of [^{3}H]meproadifen mustard with membrane-bound *Torpedo* acetylcholine receptor. *J. Biol. Chem.* **261**, 13727–34.

[50] Norris, T.M., Moya, E., Blagbrough, I.S. and Adams, M.E. (1996). Block of high-threshold calcium channels by the synthetic polyamines sFTX-3.3 and FTX-3.3. *Mol. Pharmacol.* **50**, 939–46.

[51] Piek, T. (1982). Delta-philanthotoxin, a semi-irreversible blocker of ion-channels. *Comp. Biochem. Physiol.* **C72**, 311–15.

17 Anabaseine as a Molecular Model for Design of α7 Nicotinic Receptor Agonist Drugs

WILLIAM R. KEM

17.1 INTRODUCTION

Naturally occurring toxins, besides serving as chemical tools for biomedical research, can also serve as lead compounds for drug development. Each toxin has already been optimised over millions of years of evolution to interfere with some critical process essential for normal physiological function. Amongst toxins which cause rapid paralysis of prey or predator, those affecting nicotinic acetylcholine receptors at the neuromuscular junction are a rather large and diverse group. The most notable plant toxins stimulating these receptors are nicotine (Figure 17.1), cytisine and anatoxin (algal), while antagonists of plant origin include the *Erythrina* and *Delphiniium* alkaloids. Animal toxins include: (1) agonists such as leptodactyline, anabaseine (Figure 17.1) and epibatidine; (2) competitive antagonists such as elapid snake (example: α-bungarotoxin) and cone shell (α-conotoxin) peptides; and (3) noncompetitive antagonists, mainly frog and ant toxins (e.g. histrionicotoxin).

While revolutionary advances in protein chemistry and molecular biology have made the manipulation of peptide structures much easier, exogenous peptides are still largely perceived as unfavourable models for the design of drugs. A major problem is their greater complexity of interactions with receptor sites: the development of potent peptide-mimetics remains a formidable challenge. Additionally, foreign peptides are potentially antigenic and generally possess unfavourable pharmacokinetic problems relative to small molecules. Thus, they are unlikely to be acceptable drugs themselves. In the near future some of these apparent difficulties are likely to be surmounted, as new methods for administration are developed. Small molecules continue to be especially attractive lead compounds for drug development.

The first nicotinic receptor toxin (other than nicotine) to be used as a drug was tubocurarine, which was first used as a muscle relaxant sixty years ago. Since that time, considerable effort has been directed towards identifying safer muscle relaxants. A large variety of alkaloids were isolated from tropical plants and

Perspectives in Molecular Toxinology
Edited by A Ménez © 2002 John Wiley & Sons, Ltd

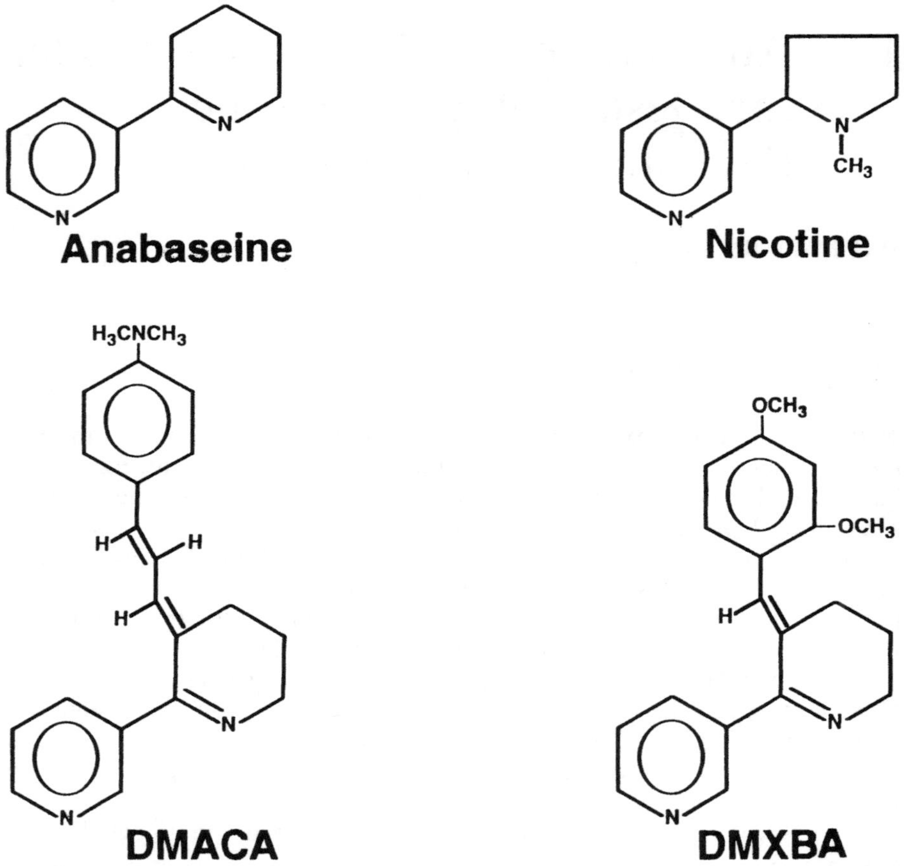

Figure 17.1 Chemical structures of anabaseine, nicotine, DMXBA (3-[2,4-dimethoxy-benzylidene]-anabaseine or GTS-21) and DMACA (3-[4-dimethylaminocinnamylidene]-anabaseine). Anabaseine differs from the tobacco alkaloid in possessing a tetrahydropyridyl rather than piperidyl ring. The compounds are synthesized as hydrochloride salts: anabaseine as the open-chain ammonium-ketone dihydrochloride (MW 251), DMXBA as the dihydrochloride salt (MW 381) and DMACA as the trihydrochloride (MW 427)

tested on neuromuscular preparations. For instance, the semi-synthetic muscle relaxant dihydro-B-erythroidine resulted from a Merck project which investigated compounds from over 50 species of *Erythrina* plants mainly collected in tropical regions of North and South America [1]. It is unfortunate that these laboriously collected and processed extracts are no longer available, as interest in finding new nicotinic compounds, especially agonists for central receptors, has lately become intense. There are many potent nicotinic compounds with known pharmacological as well as chemical properties, which can be used as molecular models for the design of new drug candidates. Nemertines, a phylum of carnivorous marine worms, produce a plethora of nicotinic alkaloids, including anabaseine [2–5]. After briefly considering the different nicotinic receptor subtypes, our efforts at drug design based upon anabaseine will be presented.

17.2 NEURONAL NICOTINIC RECEPTORS AS DRUG TARGETS

Like the neuromuscular receptor, neuronal nicotinic receptors are pentameric non-covalently bonded complexes of five polypeptide subunits. Each receptor contains at least two α subunits, which contain most of the binding determinants for ligands binding to the ACh recognition site. At least 10 different α subunits, each recognised by the presence of a characteristic disulphide bond between adjacent residues near the ACh-binding site, are expressed in the rat nervous system. In heteromeric neuronal receptors the other subunits are designated as β subunits. Brain receptors displaying high (nM) affinity for nicotine contain β2 subunits in combination with either α3 or α4 subunits; the α4-β2 combination accounts for at least 90% of these receptors in the rat brain. The most abundant subtype displaying lower (μM) affinity for nicotinic is a homomer composed only of α7 subunits. Stimulation of either the high or low affinity nicotinic receptors enhances learning and memory, which implicates α4–β2 and α7 subtypes, respectively [6]. An important physiological difference between neuronal and neuromuscular receptors is that the neuronal receptor channels are 20–fold more permeable to calcium ions than to sodium. The calcium influx through these receptor channels can serve as a second messenger inside the cell as well as cause membrane depolarisation. Central nicotinic receptors seem to tonically influence neuronal communication in this manner.

Cholinergic neurons in the cerebral cortex and hippocampus are particularly likely to degenerate in Alzheimer's disease. A drastic (>50%) decrease in high-affinity nicotinic receptor concentration is observed in Alzheimer's disease post-morten brains [7, 8]. Similar decreases in choline acetyltransferase and choline transporter sites were found in these brains. These findings, along with the demonstration of cognitive behavioural deficits in animals treated with cholinergic antagonists or whose projecting cholinergic neurons were chemically lesioned, led to the formulation of the 'cholinergic hypothesis' for the development of age-related dementias including Alzheimer's disease [9]. A therapeutic corollary of this hypothesis is that substances which reduce this cholinergic degeneration or counteract the resulting cholinergic hypofunction may be useful in treating Alzheimer's disease. Thus, CNS nicotinic receptors as well as the more traditional targets, cholinesterases and muscarinic receptors, are

Table 17.1   Some CNS effects mediated by stimulating the two major brain nicotinic receptors (actions are based on many studies, so they are not cited).

Action	Nicotinic receptor subtype	
	$\alpha7$	$\alpha4$–$\beta2$
Analgesia	?	+
Anxiolysis	?	+
Cognition	+	+
Hyperlocomotor	0	+
Hypothermia	?	+
Neuroprotection	+	0
Self-administration and cue	0	+
Neurotransmitter levels (PFC)		
ACh	0	+
Dopamine	+	+
Norepinephrine	+	+
Serotonin	0	+

attractive drug targets for treatment of Alzheimer's disease and a variety of other disorders. Since nicotinic agonists enhance attention and cognition, while antagonists have the opposite effects, the pharmaceutical focus is upon drug development of nicotinic agonists which are selective for the receptor subtypes that enhance cognition and are neuroprotective.

Which nicotinic receptor subtypes are the most attractive targets? Many of the high-affinity receptors are apparently localised on vulnerable cholinergic neurons projecting to the cortex and hippocampus, and their disappearance is related to the death of these neurons. In contrast, the $\alpha7$-type receptor appears to be found on noncholinergic neurons and its expression is not greatly altered in Alzheimer's disease [10]. This makes $\alpha7$ a particularly attractive CNS drug target for treating this neurodegenerative disease. Another reason why this receptor subtype is of great pharmaceutical interest is that most, if not all, neuroprotective actions of nicotinic agonists are mediated by this receptor [11–16]. Various effects resulting from selective stimulation of these two major CNS type receptors are summarised in Table 17.1. Another potential disadvantage of the $\beta2$-containing receptors, besides their inability to mediate neuroprotective responses, is their involvement in nicotine addiction. However, this may not be a problem in the treatment of neurodegenerative diseases, when $\beta2$-selective agonists are administered orally or transdermally [17].

17.3   ANABASEINE, A NATURAL TOXIN

This alkaloid was first isolated from a hoplonemertine worm, *Paranemertes peregrina*, which uses it as a chemical defence against predators and to paralyse its prey [2, 3, 5]. Subsequently it was also isolated from certain species of ants. While anabaseine, like nicotine, rather indiscriminately stimulates all nicotinic cholinergic receptors, it displays a particularly high potency upon two subtypes of nicotinic receptors which are most resistant to nicotine, namely the α-bungarotoxin sensitive neuromuscular and neuronal (α7) subtypes (Table 17.2). Since both compounds contain 3-substituted pyridyl rings, their different receptor subtype preferences depend upon the other ring, which is a 2-substituted tetrahydropyridyl in anabaseine. Because the imine bond in this ring is electronically conjugated with the pyridyl ring, the two rings of anabaseine are approximately co-planar, whereas in nicotine they are approximately orthogonal. This chemical difference probably determines the relatively high potency of anabaseine at α7 receptors and its relatively poor efficacy for stimulating α4-β2 receptors [18]. There are also important differences in apparent efficacy at these two receptors. Anabaseine is nearly a full agonist at the α7 receptor, while nicotine is a partial agonist (Figure 17.2A). Conversely, at the α4-β2 receptor, nicotine has high efficacy but anabaseine is only a weak partial agonist (Figure 17.2B). Both substances are essentially full agonists on the PC12 cell α3-β4 subtype receptor [18].

Table 17.2 Relative potencies and affinities of nicotinic agonists on various receptor subtypes. All estimates are in nM concentrations.

Cell type	Muscle EC$_{50}$		Neuronal K$_I$	
Compound	Frog α1, etc^a	Human α7^b	Rat Brain α4-β2^c	Rat α3-β4^d
Epitbatidine	18	21	—	0.38
(+)-Anatoxin-a	67	63	—	53
Leptodactyline	120	—	—	—
Anabaseine	740	60	75	—
DMXB–anabaseine	>10 000	650	85	—
DMAC–anabaseine	>10 000	110	350	—
Acetylcholine	530	9900	15	880
(S)-Nicotine	1960	16 100	3	480
Cytisine	6700	3900	0.5	195
Carbamylcholine	7400	22 000	—	3800
(S)-Anabasine	7000	—	—	—

^{a} Sources of all EC50 estimates for contracting frog muscle are found in [18]
^{b} Data obtained by displacement of iodinated α-bungarotoxin to HEK293 cells stably expressing human α7 receptors [52]
^{c} Data obtained by displacement of tritiated cytisine from rat brain membranes [24]
^{d} Data obtained by displacements of tritiated epibatidine binding to HEK293 cells stably expressing rat α3-β4 sympathetic ganglion type receptor [53]

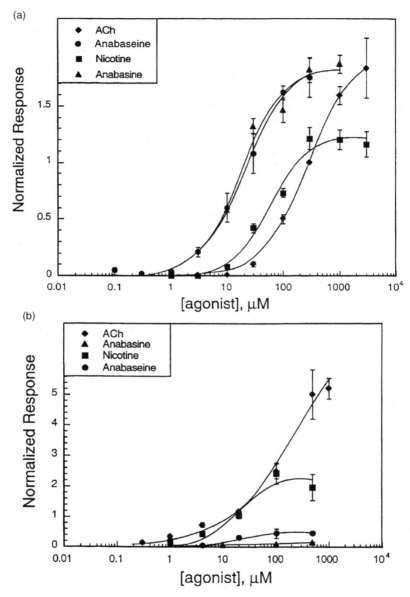

Figure 17.2   Comparison of the action of acetylcholine (ACh), anabaseine, (S)-nicotine and (S)-anabasine on the two major rat CNS nicotinic receptors. The peak current response of *Xenopus* oocytes is plotted against concentration. Anabasine is a tobacco alkaloid chemically related to anabaseine; it differs only in lacking the tetrahydropyridyl imine bond. (a) Full agonist actions of anabaseine and anabasine on the $\alpha7$ receptor; (b) Weak partial agonist actions of anabaseine and anabasine upon the $\alpha4$-$\beta2$ receptor. ACh is a full agonist and nicotine is a partial agonist at both of these receptors [18]

17.4 BENZYLIDENE AND CINNAMYLIDENE DERIVATIVES OF ANABASEINE

In the course of studying the distribution of anabaseine in nemertine tissues, the author prepared a benzylidene-adduct of anabaseine [2]. A decade later it was noticed that this compound lacked neuromuscular and ganglionic stimulatory effects. This led to the synthesis and testing of a wide variety of anabaseine derivatives to obtain more selective and less toxic nicotinic agonists with therapeutic potential in the late 1980s. From the initial group of compounds a disubstituted benzylidene–anabaseine derivative, 3-(2,4-dimethoxybenzylidene)-anabaseine (DMXBA) was selected for further development based upon a variety of preclinical tests. This compound was given the code name GTS-21 because it was only the twenty-first compound prepared in the author's lab as part of a joint project between University of Florida scientists in Gainesville and Taiho Pharmaceutical Company scientists in Tokushima, Japan. DMXBA is currently a drug candidate for treatment of Alzheimer's disease and schizophrenia.

DMXBA and other aryl-anabaseines display several interesting chemical properties. In contrast with anabaseine, DMXBA does not hydrolyse to an open-chain form in aqueous solution, even at relatively acidic pH [19, 20]. It is highly lipophilic, displaying an octanol:water partition coefficient of 1400 at pH 7.4 [21]. The anabaseine part of DMXBA provides the basic Beers and Reich [22] minimal nicotinic receptor pharmacore, namely a cationic N group separated from an H-bond acceptor group spaced about 5–6 Å apart. It is unlikely that the two anabaseine rings of DMXBA retain the same coplanar configuration as in the natural toxin, since there is considerable steric hindrance in the region where the pyridyl and benzylidene ring portions of the DMXBA molecule are connected to the tetrahydropyridyl ring [20, 21]. The pKa of the imine nitrogen is 7.6. Thus, at normal physiological pH approximately 60% of the DMXBA molecules in solution will be monocations and the remaining molecules will be unionised.

The cinnamylidene–anabaseines have very similar pharmacological as well as chemical properties to the benzylidene derivatives. Some are even more potent α7 agonists than DMXBA.

17.5 *IN VITRO* STUDIES OF DMXBA INTERACTION WITH NICOTINIC RECEPTORS

The *Xenopus* oocyte expression system has been very useful in elucidating the action of DMXBA upon various nicotinic receptor subtypes. α7 receptor stimulation is relatively difficult to measure because this receptor subtype displays such rapid desensitisation. Thus, measured concentration–response curves are shifted to higher apparent concentrations [23]. Both DMXBA and

DMACA behaved as strong partial agonists on the rat $\alpha7$ receptor expressed in the *Xenopus* oocyte [24]. DMACA displayed higher affinity and efficacy at this receptor. A large variety of benzylidene– and cinnamylidene–anabaseines have been made in this laboratory over the past 13 years. Several of these compounds possess pharmacokinetic and pharmacodynamic advantages over DMXBA and DMACA and thus have become 'lead' compounds for the development of even more optimised drug candidates.

In the oocyte system both DMXBA and DMAC–anabaseine generate currents that, relative to ACh, slowly develop and decay with continued bath application [24]. Thus, while they produce a smaller maximal peak current response than ACh, the total current generated over a longer period of time is similar [25]. Measurements of the residual inhibition of the response to a second ACh application 5 min after washing away the compound also indicated that the inhibitory effects of these two compounds also wash away slowly. This may be due to use-dependent inhibition caused by ion channel blockade or by a prolonged recovery from a desensitised state. Further experiments are required to understand these kinetic effects.

Two groups have found that DMXBA is a less efficaceous and potent agonist on the human form of the $\alpha7$ receptor relative to the rat receptor [26, 27]. However, the major phase I metabolite, 3-(4-hydroxy-2-methoxybenzylidene)-anabaseine, is just as good on the human receptor as DMXBA is on the rat receptor. The reason for this difference is not yet clear, but it is known that the human receptor possesses a tyrosyl side chain in its ACh recognition site where the rat form possesses a phenylalanyl side chain. Regardless of the molecular mechanism, it clearly becomes of the utmost importance to test new compounds on the human form of the $\alpha7$ receptor before considering them as serious drug candidates. So far, the $\alpha4$-$\beta2$ receptor has not yet exhibited important species differences in its responsiveness to various ligands.

One of the most tantalising questions in the central nicotinic receptor field is how relatively constant agonist concentrations are capable of enhancing cognition. Until recently a peripheral nervous system paradigm of nicotinic receptor function, emphasising phasic responses to pulses of neurotransmitter or agonist, was usually applied to explain the functional roles of central nicotinic receptors. Indeed, at both skeletal muscle and ganglionic synapses nicotine produces phasic depolarisation of the postjunctional membrane which are sufficient to trigger the electrical activation of the muscle or postganglionic neuron. Application of this 'neuromuscular' paradigm to explain the functional roles of CNS nicotinic receptors is probably inappropriate since the latter are less focally distributed and may also may receive more stimulation from endogenous choline than from ACh. Electrophysiological studies of neuronal receptors usually monitor peak responses to rapid applications of the nicotinic agonist, as in our *Xenopus* oocyte experiments. However, in an Alzheimer's patient, nicotinic agonist drug levels would be maintained relatively constant over periods of hours or days. At extremely low DMXBA concentrations there may be a relatively small, nondesensitising Ca influx generated by a small number of $\alpha 7$ receptor channels which is sufficient to enhance synaptic function and cognition.

Radcliffe and Dani [28] have shown that nicotine produces long-lasting as well as almost instantaneous enhancement of glutamatergic transmission. The long-lasting effect was suggested to be the result of a second messenger action of calcium ions entering the neuron through the nicotinic receptor channel. It has been demonstrated that nicotinic stimulation causes the release of nitric oxide, a gas which can readily affect a larger volume of neuronal tissue than would conventional, water soluble neurotransmitters. Recently, $\alpha 7$ receptors in the hippocampus were shown to co-localise with neurons possessing nitric oxide synthase; the auditory gating effects of DMXBA were also inhibited by NO synthase inhibitors [29].

17.6 DMXBA INTERACTION WITH OTHER RECEPTORS

A number of neurotransmitter receptors (5HT1A, 5HT2A, adenosine 1, adenosine 2A, NMDA-glutamate, and total glutamate receptors) and some voltage-gated ion channels (brain Kv1 channel, L-type Ca) have been screened by radioligand binding methods to assess the selectivity of DMXBA. Binding to some of these receptors was inhibited, but only at very high concentrations ($> 50\,\mu M$), which were at least one hundred times higher than the peak plasma concentrations which would be attained in the behavioural tests (Table 17.3).

A potentially significant interaction of DMXBA occurs at 5-HT$_3$ receptors, as observed with electrophysiological recordings from the *Xenopus* oocyte expression system [30]. DMXBA acts as a weak antagonist on the homodimeric mouse 5-HT$_{3A}$ receptor, but surprisingly the 2-hydroxy metabolite of DMXBA is almost as potent an agonist as 5-HT on this particular receptor. However, the importance of 5-HT$_3$ receptors for cognition and other mental functions has not yet been established.

Table 17.3 Effects of DMXBA upon behavioral tasks measuring cognition.

Behavioral assay	Effect (P $<$ 0.05)	Reference
Passive avoidance (lesioned rat)	++	48
Passive avoidance (adult mouse)	0	26
Active avoidance (aged rat)	++	49
Lashley III maze (aged rat)	++	49
Morris water maze (aged rat)	++	50, 51
11-Arm radial maze (aged rat)	++	49
Eyeblink conditioning (aged rabbit)	++	51
Delayed matching task (adult monkey)	++	26

17.7   EFFECTS OF DMXBA ON COGNITION AND SENSORY GATING

The effects of 1–10 mg/kg doses (s.c, or i.p.) of DMXBA upon a variety of cognitive behaviours have been investigated in four mammalian species at several laboratories around the world. Considering that there were differences in the behavioural protocols, the consistency of the results was rather impressive (Table 17.3). The lack of effect on passive avoidance in mice may have been partially related to the use of young adults rather than aging or lesioned animals. Considering that transgenic mice models of Alzeimer's disease are becoming popular, it is desirable that DMXBA should be tested further with this type of behavioural experiment.

DMXBA also improves defective sensory filtering or gating in a strain of mouse [29, 31, 32]. Freedman's lab has demonstrated that auditory gating is modulated by GABAergic interneurons in the hippocampus which possess α7-type nicotinic receptors. This mouse model may have applicability to the condition of schizophrenic individuals, who also filter repetitive auditory and visual stimuli poorly. Considerable evidence has been reported that α7 receptor gene expression is subnormal in schizophrenics [33]. If the human sensory deficiency can be improved by α7 agonists, then DMXBA or related compounds may prove useful in treating the negative symptoms of schizophrenia.

17.8   EFFECTS OF DMXBA UPON OTHER NEUROTRANSMITTER SYSTEMS

Microdialysis measurements of neurotransmitter level changes resulting from administration of nicotinic agonists to whole animals have revealed interesting differences between compounds preferentially stimulating either of the two major subtypes of brain receptor (Table 17.1). Nicotine, ABT-418 and other agonists which preferentially bind to the β2 subunit-containing receptors significantly increase ACh levels in the prefrontal cortex and hippocampus, while DMXBA does not increase the levels of this neurotransmitter, when administered alone. DMXBA elevated the two catecholamines, but not as greatly as did an equal dose of anabaseine or nicotine. The simplest interpretation for the different effects on ACh would be that β subunit-containing receptors modulate cholinergic neurons while α7 receptors are not a significant population on such neurons. Much more needs to be known about the distribution of nicotinic receptors on projecting neurons secreting the various neurotransmitters before the multiple actions of nicotinic agonists on brain neurotransmitters can be understood [34]. We plan further microdialysis experiments to extend these initial observations.

17.9 NEUROPROTECTIVE ACTIONS

Ionised calcium serves as a second messenger under normal physiological conditions. Elevating the free intracellular concentration can stimulate cellular functions at lower concentrations but cause irreversible damage at excessively high concentrations. In recent years several laboratories have shown that nicotine and DMXBA are neuroprotective against a variety of chemical stresses, including aspartate, glutamate, β-amyloid peptide and withdrawal of nerve growth factor. Also, it has recently been reported that nicotine protects cultured spinal motoneurons from programmed cell death resulting from trophic factor deprivation [35]. In many cases, where these studies were carried out on cultured cell lines, it has been demonstrated that protective actions of the nicotinic agonists can be inhibited by the co-administration of either the non-specific nicotinic antagonist mecamylamine or α7 receptor specific antagonists. The neuroprotective effects of ABT-418 and nicotine, which preferentially stimulate high-affinity receptors, were blocked by α7 receptor antagonists but not α4-β2 antagonists, leading to the conclusion that they primarily act through α7 receptor stimulation [16]. The nicotinic agonists were only active if they were administered prior to the damaging stimulus, which suggests that they act by allowing some Ca ions to enter the cell and trigger cellular processes which can then counteract a subsequent Ca overload. The possibility of actually slowing the progressive neurodegeneration of Alzheimer's disease with α7 nicotinic agonists is exciting, but further *in vivo* tests are required to determine whether this can occur in patients.

17.10 PHARMACOKINETICS AND BIOTRANSFORMATION OF DMXBA

This compound is very rapidly absorbed after oral administration and also readily enters the brain [36]. Both of these findings are understandable based upon its very high octanol:water partition coefficient [21]. Bioavailability was less than 25%, indicating that 'first-pass' intestinal and/or hepatic biotransformation was extensive. Consistent with this interpretation, less than 1% of the initially administered dose was recovered in the urine [36]. After intravenous administration, the plasma concentration–time curve displayed biphasic kinetics with plasma half-lives of 0.7 h and 3.7 h, respectively in lightly anaesthetised Sprague–Dawley rats. The apparent volume of distribution of DMXBA is moderately large, indicating that the compound probably enters all aqueous body compartments and binds to peripheral tissues.

Initial hepatic microsome incubation experiments indicated that O-demethylation of both ortho- and para-methoxy substituents occurred readily. The monohydroxy and dihydroxy metabolites were identified by HPLC-MS comparison with the synthetic compounds. The 4-OH monohydroxy metabolite was the predominant phase I biotransformation product *in vivo* in man as well as rat [37–39]. The two monohydroxy metabolites are significantly more potent agonists on rat $\alpha 7$ receptors relative to DMXBA (21). The apparent efficacy of acute DMXBA stimulation of the human $\alpha 7$ was only about half the apparent efficacies of the three hydroxy metabolites [21, 27; unpublished results].

17.11 EFFECTS OF CHRONIC NICOTINIC AGONIST ADMINISTRATION

It is well known that tobacco smokers increase their daily self-administration of nicotine as their dependence develops over time. This probably results primarily from the development of cellular type rather than metabolic tolerance. In a somewhat paradoxical fashion, the brain concentration of high-affinity receptors also increases considerably [40]. One would expect that this up-regulation would be accompanied by an increase in sensitivity to nicotinic agonists, but this probably is not the case, at least *in vivo*. The causal relationship between nicotine tolerance, and receptor up-regulation is still unclear. However, from the standpoint of using chronic nicotinic agonist administration for treating CNS disorders, development of tolerance would be undesirable. Recent reports suggest that both of the major CNS receptors are up-regulated functionally as   well as in number *in vitro*, by chronic exposure, although $\alpha 7$ receptor levels are much less sensitive to chronic administration. The implied functional up-regulation in nicotine responsiveness needs to be corroborated *in vivo* before we can assume that chronic agonist administration can actually increase nicotinic receptor function in Alzheimer's patients.

17.12 NICOTINE DEPENDENCE

A major concern related to chronic administration of nicotinic agonists is their potential for causing dependence or addiction. While studies [41] on $\beta 2$-subunit knockout mice have provided strong evidence for the major involvement of $\alpha 4$- and $\alpha 3$-$\beta 2$ combinations in nicotine self-administration, involvement of $\alpha 7$ receptors has been more controversial. Clearly this subtype is present in the ventral tegmentum and nucleus accumbens pathway implicated in generating drug reward. However, the only evidence for these receptors being functionally significant is based upon the ability of rather high doses (which could affect other, high-affinity nicotinic receptors) of methyllycaconitine (MLA) to inhibit self-administration of nicotine [42]. Recently, it was reported that two of the most selective $\alpha 7$ agonists available, DMAC-anabaseine and the Astra compound AR-R17779, failed to affect self-administration of nicotine [43]. Our results with DMXBA [44] are also consistent with the prediction that $\alpha 7$ agonists are less liable to cause dependence relative to agonists acting on the high-affinity receptors.

17.13   OTHER COMPOUNDS ACTING ON α7 RECEPTORS

Several compounds have been reported to be α7 receptor-selective. Of these, the Astra compound AR-R 17779, has attracted the most attention. This drug candidate was originally synthesized as a muscarinic receptor agonist, but turned out to be quite selective for the α7 type receptor. Within its structure one can see a restrained molecular conformation of ACh itself. Its pharmacological properties have not been reported as extensively as for DMXBA, but it appears to have higher binding selectivity than DMXBA, since its binding affinity for α4-β2 type receptors is much less than for α7 receptors. The published data for AR-R17779 and DMAC–anabaseine indicate that the former compound possesses about twice the efficacy of the latter compound on the human type α7 receptor expressed in the *Xenopus* oocyte [43]. The Astra compound also improves learning and memory and memory in rats [45]. This compound, besides being a promising drug candidate, should also become a useful neurobiological tool for selectivity stimulating α7 receptors in whole animal and isolated tissue slice experiments.

17.14   POSSIBLE TOXIC CONSEQUENCES OF EXCESSIVE α7 RECEPTOR STIMULATION

Since α7 receptors are located in many tissues besides the brain and are able to increase calcium permeability even under normal resting potential conditions, adverse effects may occur during administration, at least when these receptors are excessively stimulated. Li *et al.* [15] have shown that DMXBA at $30\,\mu M$ is toxic to differentiated PC12 cells expressing α7 receptors and that MLA blocked this action. However, this concentration of DMXBA would never be attained when doses optimal for cognition enhancement are used in whole animals. In fact, oral administration of a 20 mg/kg dose only produced peak plasma concentrations of about $0.5\,\mu M$. We have found that in plasma, about 95% of the DMXBA is bound to plasma proteins. Thus the PC12 data were obtained at free DMXBA concentrations which are nearly one to two hundred times higher than occur in the whole animal experiments summarised in Table 17.3.

It would be of considerable interest to quantitatively compare *in vitro* neuroprotective and neurotoxic concentrations of several α7 agonists such as DMXBA, DMACA and AR-R 17779, which differ in their efficacy of channel activation. This might enable prediction of an optimal efficacy which allows for sufficient drug safety as well as therapeutic potency.

17.15 CONCLUDING COMMENTS

Manipulating the structure of anabaseine to create 3-aryl derivatives has yielded α7-selective agonists. However, the resulting compounds also inhibit high-affinity receptors to varying degrees. Ultimately, it would be desirable to eliminate this antagonist activity and even possibly replace it with partial agonistic activity, thus being able to stimulate both nicotinic receptor subtypes which can enhance cognition. It would be helpful to be able to manipulate agonist structure using knowledge of the receptor's surface. Unfortunately, there is not yet a high-resolution structure for any nicotinic receptor. Several labs are currently attempting to obtain a high-resolution structure of the extracellular portion of the α subunit. Such a structure would be extremely useful for directing drug design. Currently, drug design for nicotinic receptors remains an interative process in which single structural modifications are made and tested. Recent research indicates that the cationic N group of ACh and of nicotinoid compounds nestles within a space surrounded by electronegative aromatic side-chains [46]. Theoretical computations indicate that the cationic groups of epibatidine and some other potent compounds should interact very strongly with a 'pocket' composed of aromatic side chains [47].

The pharmacokinetic properties of a drug candidate are just as important as its receptor affinity and efficacy. They should be considered as early in the drug development process as possible, after lead compounds with adequate potency at the receptor target are identified. The pharmacokinetic profile of DMXBA is not optimal. Although it is rapidly absorbed and distributed into the brain, it has a relatively modest bioavailability. Fortunately, the dose fraction which enters the systematic circulation is rapidly distributed to various tissues including the brain. While the monohydroxy metabolites of DMXBA are even more potent than DMXBA on human α7 receptors, they are also rapidly extracted from the circulation by the liver. Also, they do not readily enter the brain. Thus our present efforts are focused upon the design of new anabaseine analogues possessing greater bioavailability and brain penetration, as well as higher α7 receptor potency and selectivity.

ACKNOWLEDGEMENTS

I wish to thank the following colleagues for their important contributions to the DMXBA-Alzheimer's project, which already has now spanned 13 years. Chemists include Drs Linda Bloom, Katalin Prokai-Tatrai, Ferenc Soti and John Zoltewicz. Neuroscientists include Drs Catherine Adams, Gary Arendash, Kimberly Bjugstad, Robert Freedman, Ezio Giacobini, Tina Machu, Vladimir Mahnir, Ed Meyer, Roger Papke, Karen Stevens, Johanna Simosky, Kathleen Summers, Frans van Haaren and Diana Woodruff-Pak. Dr Laszlo Prokai and Mr XueFang Cao contributed greatly to understanding the metabolites of

DMXBA. The project has been largely supported by the Taiho Pharmaceutical Company, Ltd (Tokushima, Japan). The many contributions of the Taiho scientists, under leadership of Dr J. Yamamoto, are greatly appreciated. Finally, I thank Ms Judy Adams for assistance in preparing the manuscript.

REFERENCES

[1] Folkers, K. and Unna, K. (1939) Erythrina alkaloids. V. Comparative curare-like potencies of species of the genus Erythrina. J. Amer. Pharm. Assoc. 28, 1019–28.
[2] Kem, W.R. (1971) A study of the occurrence of anabaseine in Paranemertes and other nemertines. Toxicon 9, 23–32.
[3] Kem, W.R., Abbott, B.C. and Coates, R.M. (1971) Isolation and structure of a hoplonemertine toxin. Toxicon 9, 15–22.
[4] Kem, W.R., Scott, K.N. and Duncan, J.H. (1976) Hoplonemertine worms – a new source of pyridine neurotoxins. Experientia 32, 684–6.
[5] Kem, W.R. and Soti, F. (2001) Amphiporus alkaloid multiplicity implies functional diversity: Initial studies on crustacean pyridyl receptors. Hydrobiologia 456, 221–231.
[6] Felix, R. and Levin, E.D. (1997) Nicotinic antagonist administration into the ventral hippocampus and spatial working memory in rats. Neurosci. 81, 1009–17.
[7] Nordbeg, A. and Winblad, B. (1986) Reduced number of [^{3}H]nicotine and [^{3}H]acetylcholine binding sites in the frontal cortex of Alzheimer brains. Neurosci. Lett. 72, 115–21.
[8] Whitehouse, P.J., Martino, A.M., Antuono, P.G., Lowenstein, P.R., Coyle, J.T., Price, D.L. and Kellar, K.J. (1986) Nicotinic acetylcholine binding in Alzheimer's disease. Brain Res. 371, 146–51.
[9] Bartus, R.T. (2000) On neurodegenerative diseases, models, and treatment strategies: Lessons learned and lessons forgotten a generation following the cholinergic hypothesis. Exper. Neurol. 163, 495–529.
[10] Davies, P. and Feisullin, S. (1981) Postmorten stability of α-bungarotoxin binding sites in mouse and human brain. Brain Res. 216, 449–54.
[11] Martin, E.J., Panickar, K.S., King, M.A., Deyrup, M., Hunter, B.E., Wang, G. and Meyer, E.M. (1994) Cytoprotective actions of 2,4-dimethoxybenzylidene anabaseine in differentiated PC12 cells and septal cholinergic neurons. Drug Dev. Res. 31, 135–41.
[12] Kihara, T., Shimohama, S., Sawada, H., Kimura, J., Kume, T., Kochiyama, H., Maeda, T. and Akaike, A. (1997) Nicotinic receptor stimulation protects neurons against B-amyloid toxicity. Ann. Neurol. 42, 159–63.
[13] Meyer, E.M., Tay, E.T., Zoltewicz, J.A., Meyers, C., King, M.A., Papke, R.L. and de Fiebre, C.M. (1998) Neuroprotective and memory-related action of novel α-7 nicotinic agents with different mixed agonist/antagonist properties. J. Pharmacol. Exper. Therap. 284, 1026–32.
[14] Shimohama, S., Greenwald, D.L., Shafron, D.H., Akaika, A., Maeda, T., Kaneko, S., Kimura, J., Simpkins, C.E., Day, A.L. and Meyer, E.M. (1998) Nicotinic α7 receptors protect against glutamate neurotoxicity and neuronal ischemic damage. Brain Res. 779, 359–63.
[15] Li, Y., Papke, R.L., He, Y-J., Millard, W.J. and Meyer, E.M. (1997) Characterization of the neuroprotective and toxic effects of α7 nicotinic receptor activation in PC12 cells. Brain Res. 830, 218–25.

[16] Dajas-Bailador, F.A., Lima, P.A. and Wonnacott, S. (2000) The α7 nicotinic acetylcholine receptor subtype mediates nicotine protection against NMDA excitotoxicity in primary hippocampal cultures through a Ca^{2+} dependent mechanism. *Neuropharmacol.* **39**, 2799–807.

[17] Karan, L.D. and Rosecrans, J.A. (2000) Addictive capacity of nicotine. In *Nicotine in Psychiatry* (M. Piasecki and P.A. Newhouse, eds), Amer. Psychiat. Press, Washington, D.C., Ch. 4, pp. 83–107.

[18] Kem, W.R., Mahnir, V.M., Papke, R. and Lingle, C. (1997) Anabaseine is a potent agonist upon muscle and neuronal α-bungarotoxin sensitive nicotinic receptors. *J. Pharmacol. Exper. Therap.* **283**, 979–92.

[19] Zoltewicz, J.A., Bloom, L.B. and Kem, W.R. (1989) Quantitative determination of the ring-chain hydrolysis equilibrium constant for anabaseine and related tobacco alkaloids. *J. Org. Chem.* **54**, 4462–8.

[20] Zoltewicz, J.A., Prokai-Tatrai, K., Bloom, L.B. and Kem, W.R. (1993) Long range transmission of polar effects of cholinergic 3-arylideneanabaseines. Conformations calculated by molecular modelling. *Heterocycles* **35**, 171–9.

[21] Kem, W.R., Mahnir, V.M., Lin, B. and Prokai-Tatrai, K. (1996) Two primary GTS-21 metabolites are potent partial agonists at alpha7 nicotinic receptors expressed in the *Xenopus* oocyte. Soc. Neurosci. **22**, Abstr. 110.9.

[22] Beers, W.H. and Reich, E. (1970) Structure and activity of acetylcholine. *Nature* **228**, 917–22.

[23] Papke, R.L. and Thinschmidt, J.S. (1998) The correction of α7 nicotinic acetylcholine receptor concentration-response relationships in *Xenopus* oocytes. *Neurosci. Lett.* **256**, 163–6.

[24] DeFiebre, C.M., Meyer, E.M., Henry, J.C., Muraskin, S.I., Kem, W.R. and Papke, R.L. (1995) Characterization of a series of anabaseine-derived compound reveals that the 3-(4)-Dimethylaminocinnamylidine derivative (DMAC) is a selective agonist at neuronal nicotinic α7/[^{125}I] α-bungarotoxin receptor subtypes. *Mol. Pharmacol.* **47**, 164–71.

[25] Papke, R.L., Meyer, E., Nutter, T. and Uteshev, V.V. (2000) α7 receptor-selective agonists and modes of α7 receptor activation. *Eur. J. Pharmacol.* **393**, 179–95.

[26] Briggs, C.A., Anderson, D.J., Brioni, J.D., Buccafusco, J.J., Buckley, M.J., Campbell, J.E., Decker, M.W., Donnelly-Roberts, D., Elliott, R.L., Gopalakrishnan, M., Holladay, M.W., Hui, Y.-H., Kim, D.J.B., Marsh, K.C., O'Neill, A., Prendergast, M.A., Ryther, K.B., Sullivan, J.P. and Arneric, S.P. (1997) Functional characterization of the novel neuronal acetylcholine receptor ligand GTS-21 *In vitro* and *In vivo*. *Pharmacol. Biochem. Behav.* **57**, 231–41.

[27] Meyer, E.M., Kuryatov, A., Gerzanich, V., Lindstrom, J. and Papke, R.L. (1998) Analysis of 3-(4-hydroxy, 2-methoxybenzylidene)anabaseine selectivity and activity at human and rat α7 nicotinic receptors. *J. Pharmacol. Exper. Ther.* **287**, 918–25.

[28] Radcliffe, K.A. and Dani, J.A. (1998) Nicotinic stimulation produces multiple forms of increased glutamatergic synaptic transmission. *J. Neurosci.* **18**, 7075–83.

[29] Adams, C.E., Stevens, K.E., Kem, W.R. and Freedman, R. (2001) Inhibition of nitric oxide synthase prevents α7 nicotinic receptor-mediated restoration of inhibitory auditory gating in rat hippocampus. *Brain Res.* **877**, 235–44.

[30] Machu, T.K., Hamilton, M.E., Field, T.R., Shankin, C.L., Harris, M.C., Sun, H., Tenner, T.E., Jr., Soti, F. and Kem, W.R. 3-Benzylidene-anabaseine analogs display partial agonist and antagonist properties at the mouse 5-Hydroxytriptamine3A receptor. *J. Pharmacol. Exper. Therap.* **299**, 1112–1119.

[31] Stevens, K.E., Kem, W.R., Mahnir, V.M. and Freedman, R. (1998) Selective alpha7-nicotinic agonists normalize inhibition of auditory response in DBA mice. *Psychopharm.* **136**, 320–7.

[32] Simosky, J.K., Stevens, K.E., Kem, W.R. and Freedman, R. (2001) Intragastric DMXB-A, an α7 nicotinic agonist, improves deficient sensory inhibition in DBA/2 mice. *Biol. Psychiat.* **50**, 493–500.

[33] Leonard, S., Bresse, C., Adams, C., Benhammou, K., Gault, J., Stevens, K., Lee, M., Adler, L., Olincy, A., Ross, R. and Freedman, R. (2000) Smoking and schizophrenia: abnormal nicotinic receptor expression. *Eur. J. Pharmacol.* **393**, 237–42.

[34] Summers, K., Kem, W.R. and Giacobini, E. (1997) Nicotinic agonist modulation of neurotransmitter levels in the rat frontoparietal cortex. *Jap. J. Pharmacol.* **74**, 139–46.

[35] Messi, M.L., Renganathan, M., Grigorenko, E. and Delbono, O. (1997) Activation of α7 nicotinic acetylcholine receptor promotes survival of spinal cord motoneurons. *FEBS Lett.* **411**, 32–8.

[36] Mahnir, V.M., Lin, B., Prokai-Tatrai, K. and Kem, W.R. (1998) Pharmacokinetics and urinary excretion of DMXBA (GTS21), a compound enhancing cognition. *Biopharm. Drug Disp.* **19**, 147–51.

[37] Kem, W.R. (2000) The brain α7 nicotinic receptor may be an important therapeutic target for the treatment of Alzheimer's disease: Studies with DMXBA (GTS-21). *Behav. Brain Res.* **113**, 169–83.

[38] Azuma, R., Minami, Y. and Satoh, T. (1996) Simultaneous determination of GTS-21 and its metabolite in rat plasma by high-performance liquid chromatography using solid-phase extraction. *J. Chromatog.* **686**, 229–34.

[39] Kitagawa, H., Takenouchi, T., Wesnes, K., Kramer, W. and Clody, D.E. (1998) Phase 1 studies of GTS-21 to assess the safety, tolerability, PK and effects on measures of cognitive function in normal volunteers. *Neurobiol. Aging* **19**, S182 (abstract).

[40] Sparks, J.A. and Pauly, J.R. (1999) Effects of continuous oral nicotine administration on brain nicotinic receptors and responsiveness to nicotine in C57Bl/6 mice. *Psychopharmacol.* **141**, 145–53.

[41] Picciotto, M.R., Zoli, M., Rimondini, R., Lena, C., Marubio, L.M., Pich, M.E., Fuxe, K. and Changeux, J.-P. (1998) Acetylcholine receptors containing the B2 subunit are involved in the reinforcing properties of nicotine. *Nature* **391**, 173–7.

[42] Nomikos, G.G., Hildebrand, B.E., Panagis, G. and Svensson, T.H. (1999) Nicotine withdrawal in the rat: role of α7nicotinic receptors in the ventral tegmental area. *Neuroreport* **10**, 697–702.

[43] Grottick, A.J., Trube, G., Corrigall, W.A., Huwyler, J., Malherbe, P., Wyler, R. and Higgins, G.A. (2000) Evidence that nicotinic α7 receptors are not involved in the hyperlocomotor and rewarding effects of nicotine. *J. Pharmacol. Exper. Therap.* **294**, 1112–19.

[44] Van Haaren, F., Anderson, K.G., Haworth, S. and Kem, W.R. (1999) GTS-21, a mixed nicotinic receptor agonist/antagonist, does not affect the nicotine cue. *Pharmacol. Biochem. Behav.* **64**, 439–44.

[45] Levin, E.D., Bettegowda, C., Blosser, J. and Gordon, J. (1999) AR-R17779, an α7 nicotinic agonist, improves learning and memory in rats. *Behav. Pharmacol.* **10**, 675–80.

[46] Zhong, W., Gallivan, J.P., Zhang, Y., Li, L.T., Lester, H.A. and Dougherty, D.A. (1998) From *ab initio* quantum mechanics to molecular neurobiology: a cation–Pi binding site in the nicotinic receptor. *Proc. Natl. Acad. Sci. USA* **21**, 12088–93.

[47] Schmitt, J.D., Sharples, C.G.V. and Caldwell, W.S. (1999) Molecular recognition in nicotinic acetylcholine receptors: The importance of pi–cation interactions. *J. Med. Chem.* **42**, 3066–74.

[48] Meyer, E.M., de Fiebre, C.M., Hunter, B.E., Simpkins, C.E., Frauworth, N. and de Fiebre, N.E.C. (1994) Effects of anabaseine-related analogs on rat brain nicotinic receptor binding and on avoidance behaviors. *Drug Dev. Res.* **31**, 127–34.

[49] Arendash, G.W., Sengstock, G.J., Sanberg, R. and Kem, W.R. (1995) Improved learning and memory in aged rats with chronic administration of the nicotinic receptor agonist GTS-21. *Brain Res.* **674**, 252–9.

[50] Bjugstad, K.B., Mahnir, V.M., Kem, W.R. and Arendash, G.W. (1996) Long-term treatment with GTS-21 or nicotine enhances water maze performance in aged rats without affecting the density of nicotinic receptor subtypes in neocortex. *Drug Develop. Res.* **39**, 19–28.

[51] Woodruff-Pak, D.S., Li, Y-T and Kem, W.R. (1994) A nicotinic receptor agonist (GTS-21), eyeblink classical conditioning, and nicotinic receptor binding in rabbit brain. *Brain Res.* **645**, 309–17.

[52] Gopalakrishnan, M., Buisoon, B., Touma, E., Giordano, T., Campbell, J.E., Hu, I.C., Donnelly-Roberts, D., Arneric, S.P., Bertrand, D. and J.P. Sullivan (1995) Stable expression and pharmacological properties of the human alpha 7 nicotinic receptor. *Eur. J. Pharmacol.* **290**, 237–246.

[53] Xiao, Y., Meyer, E.L., Thompson, J.M., Surin, A., Wroblewski, J. and K.J. Kellar (1998) Rat alpha3-beta4 subtype of neuronal nicotinic acetylcholine receptor stably expressed in a transfected cell line: Pharmacology of ligand binding and function. *Mol. Pharmacol.* **54**, 322–333.

18 Understanding the Structure–Function Relationship of Snake Venom Cardiotoxins

T.K.S. KUMAR, S. SRISAILAM, R.R. VETHANAYAGAM and C. YU*

18.1 INTRODUCTION

Snakes belonging to the Elapidae family are known to contain two major classes of toxic proteins: cardiotoxins and neurotoxins [1–3]. Both these groups of toxins (cardiotoxins and neurotoxins) are small molecular weight proteins ($\sim$7–9 kDa) and share about 45–50% homology in their primary amino acid sequence [4–7]. Despite this homology, these two toxins exhibit drastically different biological activities [8]. Neurotoxins block nerve transmission by binding strongly to the acetylcholine receptor [7]. Cardiotoxins, on the other hand, exhibit diverse biological activities, which include depolarisation and contraction of muscular cells, prevention of platelet aggregation, lysis of erythrocytes, epithelial cells, and foetal lung cells [9, 10]. Interestingly, cardiotoxins have also been demonstrated to kill selectively certain types of tumour cells, such as the Yoshida sarcoma cells [11–13]. In addition, cardiotoxins are shown to inhibit the action of key cellular enzymes such as $Na^+ - K^+$-ATPase and protein kinase C [5, 7].

18.2 THE CHEMISTRY OF CARDIOTOXINS

Snake venoms are single chain highly basic (pI > 10.0) proteins, cross-linked by four disulphide bonds [14, 15]. The amino acid sequence of cardiotoxins are characterised by the presence of a signature tripeptide sequence, –I–D–V–, extending between the residues 39–41 [6, 16]. Further, cardiotoxins show the conspicuous absence of glutamic acid [7, 16]. Recently, based on interactions with zwitterionic phospholipid dispersions, two distinct classes of cardiotoxins, the P- and S-type, have been identified [17–19]. The S-type cardiotoxins are reported to be bestowed with one lipid-binding site whereas the P-type

---

* Corresponding author

Perspectives in Molecular Toxinology
Edited by A Ménez © 2002 John Wiley & Sons, Ltd

cardiotoxins are puntuated by the presence of proline-31 within a putative phospholipid binding site [18]. While the S-type cardiotoxins are characterised by the consistent presence of a serine residue at position 29, the P-type cardiotoxins are shown to exhibit higher membrane fusion activity than their S-type counterparts. Members belonging to the P-type show significant differences in the muscle depolarisation and haemolytic activities [18, 19]. P-type cardiotoxins exhibit higher muscle depolarisation activity, whereas the cardiotoxin members belonging to the S-type are found to display higher haemolytic activity [18].

18.3   STRUCTURE OF CARDIOTOXINS

To date, the three-dimensional structures of nine cardiotoxin isoforms have been elucidated ([20–27], Table 18.1). Barring subtle variations in the orientation of some side chain groups, the overall backbone folding of most of the cardiotoxin isoforms appears to be similar [28]. Cardiotoxins are popularly referred to as the 'the three-fingered' proteins [5]. The backbones of this class of toxins folds into three major loops emerging from a globular head (presenting a synonymous appearance of three fingers emerging from the palm of a hand). Cardiotoxins belong to the class of all β-sheet proteins (Figure 18.1). The secondary structural elements in these toxins include a large triple-stranded, anti-parallel β-sheet and a double stranded β-sheet (Figure 18.1 and [20, 21]). The β-strand domains (double and triple stranded β-sheets) are tethered together by four disulphide bonds [5, 15]. The disulphide bridges force the five β-strands in cardiotoxins to align themselves anti-parallel into double and triple stranded β-sheet domains [29, 30]. The disulphide bonds in the cardiotoxin molecules contribute significantly to their conformational stability (29, 30). In general, cardiotoxins resist denaturation even at high temperatures ($< 80\,^{\circ}\mathrm{C}$) and extreme pH conditions [30].

Table 18.1   List of cardiotoxin isoforms whose three-dimensional structures have been solved

Toxin	Source	PDB code	Reference
CTX I	*Naja naja atra*	2CDX	Jahnke[a] *et al.* [22]
CTX IB	*Naja mossambica mossambica*	2CCX	O'Connell[a] *et al.*[25]
CTX V	*Naja pallida*	1CXO	Bilwes[b] *et al.* [14]
CTX II	*Naja naja atra*	1CRF	Bhaskaran[a] *et al.* [20]
CTX V[II]	*Naja mossambica mossambica*	—	Otting[a] *et al.* [26]
CTX III	*Naja naja atra*	2CRS, 2CRT	Bhaskaran[a] *et al.* [20]
CTX IV	*Naja naja atra*	1KBS	Jang [a] *et al.* [23]
CTX V	*Naja naja atra*	1CHV	Jayaraman[a] *et al.* [40]

The superscripts 'a' and 'b' refer to the structures solved by NMR and X-ray crystallography, respectively. The numbers in parentheses refer to the reference of the original paper describing the three-dimensional structure of the cardiotoxin isoform.

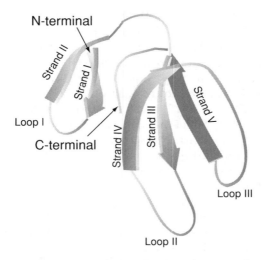

Figure 18.1 MOLSCRIPT representation of the general structural architecture of snake venom cardiotoxins. The arrow heads indicate the various β-strands in the toxin molecule

Loop-I extending between residues 2 to 14 is believed to play an important role in the haemolytic activity of cardiotoxins [7]. This loop lodges the double stranded β-sheet domain. It displays an asymmetric distribution of charged and nonpolar residues [20, 21]. The hydrophobic stretch of residues (residues 6 to 11) distributed at the extremity of loop-I is flanked on either side with positively charged residues [4, 5]. The asymmetric arrangement of the charged and hydrophobic residues in this loop is believed to provide a molecular architecture conducive to the erythrocytolytic action of cardiotoxins [23]. Lysine-12 located in loop I is a well-conserved residue and nullification of the positive charge on its (Lys12) side chain is shown to affect significantly the erythrocytolytic activity of cardiotoxins [31–33]. The loop II in the cardiotoxin molecule is comprised of residues 20 to 39. This loop lodges two of the three β-strands comprising the triple-stranded β-sheet domain [20]. β-strand III extends between residues 20 to 25 and β-strand IV spans residues 35 to 39 [5, 20, 21]. One of the characteristic features of the three-dimensional structures of cardiotoxins is the presence of a β-bulge [23, 25]. This β-bulge is formed by the well-conserved proline and valine at positions 33 and 34, respectively. According to the classification of Richardson, this β-bulge could be categorised as a G1 type [23]. The occurrence of the β-bulge is believed to compensate for the disruption of the regularity of the β-sheet due to the presence of a proline residue at position 33 [23]. Several residues in loop II are believed to have a functional role. These residues include Tyr22, Lys23, Met24, Met26, and Lys35 [8, 34]. The tyrosine residues, Tyr22 along with Tyr51 (located in loop III), are well conserved in the cardiotoxin family [34]. Two hydrophobic patches are organised (on the convex side of the molecule) around the two tyrosine (Tyr22 and Tyr51) residues. Met24, Gly37, Ile39, Pro43 and Cys53 (patch I) organise themselves around Tyr22. Tyr51 is encircled by Met24, Met26, Gly37, Val32, Val34 and Pro49 (patch II). These two nonpolar clusters contribute to the hydrophobic core of the cardiotoxin molecule [4]. In addition, the aromatic rings of Tyr22 and Tyr51 lie very close in space [34]. The aromatic rings are titled to about 80 °C with respect to each other [23]. Interestingly, chemical modification of Tyr22 and Tyr51 leads to substantial loss in toxicity of the cardiotoxin molecule [34]. Similarly, Met24 and Met26 are well conserved in all the snake venom cardiotoxins. Nonselective chemical modification of these methionine residues revealed that they are important for the lethal activity of cardiotoxins [35]. Due to the nonselective nature of the modification reagent used, it is not clear whether one or both of these methionine residues are involved in the toxicity of cardiotoxins [5]. Carlsson and Louw, studying the role of methionine residues in the activity of cardiotoxins V[11] 1 from *Naja melanoleuca*, predicted that the more solvent-accessible and flexible methionine residue could be responsible for the lethal action [35]. Recently, backbone dynamics experiments on cardiotoxin analogue II (CTX II) from *Naja naja atra* (at natural abundance of the ^{13}C nuclei) revealed that Met26 was more flexible than Met24 [36]. Thus, it appears that Met26 has a greater role in the lethal activity of cardiotoxins than Met24.

Loop III stretches between residues 50 and 55 [3]. This loop harbours β-strand V of the triple-stranded β-sheet domain. There is a short C-terminal loop held together in position by the disulphide bond between Cys55 and Cys59. One of the typical features of the cardiotoxin structures is the proximity of the N- and C-terminal ends. Both these ends are bridged by hydrogen bonds between Arg58NH and Lys2CO as well as between Asn4NH and Asn60COOH [4, 5, and 7]. These residues are well-conserved among cardiotoxins [7].

Cardiotoxins are highly basic molecules and are rich in cationic residues such as lysine [33]. There are three invariant lysine residues (Lys12, Lys18 and Lys35) in the cardiotoxin molecule [31, 33]. Elegant chemical modification studies by Menez and co-workers unambiguously revealed that modification of these invariant lysine residues nullifies the cell lysis activity [31,32]. Cardiotoxins are known to interact strongly with negatively charged phospholipid vesicles but not with neutral and positively charged vesicles [37–39]. The cationic centres on the cardiotoxin molecules are believed to facilitate binding of the toxin molecules to the erythrocyte membrane through charge interactions [33]. Thus, neutralisation of the positive charges (by chemical modification) appears to hinder the binding of the toxin molecules to the erythrocyte membrane and consequently inhibit their cell lysis activity [31, 33].

Multiple isoforms of cardiotoxins commonly occur in a single venom source [15, 7]. However, the rationale of the occurrence of the various cardiotoxin analogues in a single snake species remains elusive. In a study, Jang *et al.* [23] compared the erythrocytolytic activities of five cardiotoxin analogues isolated from the Taiwan cobra (*Naja naja atra*) venom. Interestingly, these cardiotoxin analogues showed significant differences in their erythrocytolytic ability [23]. For example, the erythrocytolytic activity of cardiotoxin analogue IV (CTX IV) was twice that of cardiotoxin analogue II even though these two CTX isoforms differ only in their N-terminal amino acid [23]. CTX IV is a unique cardiotoxin. The N-terminal amino acid in CTX IV is arginine, while in all other cardiotoxin isoforms isolated from various snake venom sources it is either leucine or isoleucine [7, 18, 23]. The solution structure of these two toxins (CTX II and CTX IV) revealed that the residues comprising the β-sheet segments are exactly the same [23]. However, comparison of the spatial distribution of the side chain groups in CTX II and CTX IV revealed that the clustering of the cationic residues is different in these two toxin analogues [23]. The charged guanido group of Arg1 in CTX IV is salt bridged (<4.0 Å) to the carbonyl group contributed by the side chain of Asp57 [23]. In addition the carbonyl group of Lys2 and the backbone amide group of Cys59 (located at the C-terminal) are found to be hydrogen bonded in CTX IV. As a consequence of these structural interactions, a 'dense' cationic cluster is formed at the N-terminal end comprising of Arg1, Lys2, Lys5, Lys23, Arg36 and Arg58 [23]. The presence of this cluster in the concave side of the structure of CTX IV is thought to be responsible for the enhanced erythrocytolytic activity. In contrast to CTX IV, the carbonyl group of Lys2 in CTX II is not hydrogen bonded to the imino (–NH) group of Cys59 and also the N-terminal amino acid (leucine) is not charged. The absence of these interactions at the N-terminal end of CTX II does not favour the development of the cationic cluster [23]. Hence, the erthyrocytolytic activity of CTX II is significantly lower than that of CTX IV [23].

The lethal potency of the various cardiotoxin (CTX) isoforms isolated from the Taiwan cobra was recently compared [40]. The LD_{50} values of CTX I and CTX V were about half those of the other cardiotoxin isoforms isolated from the Taiwan Cobra [40]. CTX IV was found to be the least potent among the five CTX isoforms from the Taiwan Cobra. CTX IV isolated from the Taiwan Cobra differs from the amino acid sequence of CTX V at five positions [18]. Leu1, Lys30, Met31, Ser45 and Leu47 in CTX V are replaced by Arg, Leu, Thr, Asn and Ala, respectively, in the corresponding positions in the amino acid sequence of CTX IV. Comparison of the solution structures of CTX IV and CTX V revealed that the two toxin analogues possess grossly similar 3D-structure [40]. Interestingly, the residues constituting the double- and triple-stranded β-sheet segments are identical. In addition, the structural interactions present on the concave surface in both these toxins showed no prominent difference [40]. However, significant differences could be discerned in the orientation of the residues located at the tip of loop III in CTX IV and CTX V [40]. A type I β-turn between residues 46 and 49 is found to occur consistently in the three-dimensional structures of all the snake cardiotoxins [6, 7]. The presence of this type I β-turn in CTX V is shown to be responsible for the orientation of the nonpolar side chains of Leu47 and Leu48 as a 'finger shaped' projection at the tip of loop III on the convex side of the molecule [40]. This nonpolar finger-shaped projection is believed to constitute a portion of the putative receptor-binding site. Interestingly, despite the presence of the conserved type I β-turn between Ser46 and Val49 in CTX IV, the non-polar residues (Ala47 and Leu48) at the tip of loop III do not clearly emerge out as a finger-shaped projection on the convex side [40]. Careful scrutiny of the interactions among the residues located at the tip of loop III in CTX IV revealed that Leu48 NH is hydrogen bonded to Val27CO (Figure 18.2). The solution structure of CTX V lacks this hydrogen bond (Figure 18.2). The presence of the hydrogen bond (between Val27CO and Leu48NH) between residues located at the tip of loops II and III was found to distort the backbone conformation at the tip of loop III and consequently positions this loop (loop III) closer to the tip of loop II. The net consequence of these structural interactions is the inability of the side chains of nonpolar residues (Ala47 and Leu48) to loop out as a finger-shaped projection (at the tip of loop III) in the solution structure of CTX IV [40]. In addition, the relatively smaller side chain of Ala47 also appears to be responsible for the lack of a prominent projection on the convex side of the CTX IV molecule [40]. The hydrogen bond between the Val27CO and Leu48NH in CTX IV also appears to affect the hydrophobic cluster at the tips of loops II and III. The solution structures of CTX IV and CTX V also revealed significant difference(s) in the degree of solvent exposure of Met26 [40]. Owing to the orientation of the nonpolar side chains of Leu47 and Leu48, a finger-shaped projection is formed between the residues located at the tip of loops II and III in the CTX V molecule. This structural feature is believed to increase the solvent exposure of Met26, and is crucial for the lethal activity of snake venom cardiotoxins [35]. The increased solvent accessibility of Met26 is thought to enhance the affinity

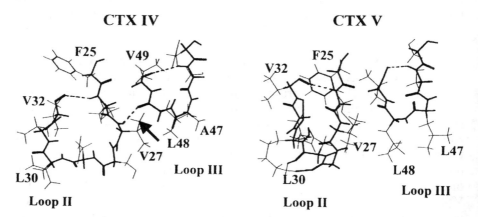

Figure 18.2 Depiction of the structural interactions among the residues at the tips of loops II and III. Hydrogen bonds among residues located at the distal regions of loops II and III are indicated by dotted lines. The arrow in the structure of CTX IV indicates the hydrogen bond between Val27CO and Leu48NH. This hydrogen bond is absent in CTX

of the toxin (CTX V) for its putative receptor [35]. However, in CTX IV the hydrogen bond between Val27CO and Leu48NH is shown to force the residues located at the tip of loops II and III to move spatially closer and reduce the size of the cleft formed between the loops II and III. The net consequence of this conformational topology is the decreased solvent exposure of Met26 in CTX IV. It is suggested that this decreased solvent exposure of Met26 results in weaker toxin-receptor interaction(s) accounting for the lower lethal potency of CTX IV.

The differences in the lethal potencies of CTX IV and CTX V are also attributed to the differential orientation of Lys44 [40]. Chemical modification studies revealed that Lys44 plays a crucial role in the lethal activity of cardio-toxins [32, 33]. Comparison of the solution structures of CTX IV and CTX V revealed significant differences in the orientation of the side chain of Lys44 in these toxins [40]. Most portions of the positively charged side chain of Lys44 are found to be projected on the convex side of the CTX IV molecule. However, in CTX IV the side chain of Lys44 is found to be projected on the convex side. This reversal of the direction of orientation of the positively charged side chain of Lys44 is suggested to influence the lethal activity of snake venom cardiotox-ins [40].

Critical comparison of the lethal potency and three-dimensional structures of the five cardiotoxin analogues from the Taiwan Cobra (*Naja naja atra*) revealed common structural features which could be crucial for the lethality of snake venom cardiotoxins [40]. They are, (1) the presence of a nonpolar finger-shaped projection at the tip of loop III, (2) occurrence of a prominent cleft between the residues at the tip of loops II and III and, (3) the degree of exposure of the cationic charge contributed by Lys44 on the convex surface of the CTX mol-ecule.

18.4 MODE OF ACTION OF CARDIOTOXINS

Although cardiotoxins exhibit a wide array of biological properties, the mechan-ism(s) underlying their multitude of activities is still obscure. It is generally believed that cardiotoxins exert their action(s) by perturbing the cell membrane. Using lipid binding studies monitored by intrinsic tryptophan fluorescence, Dufourcq and Faucon reported that cardiotoxins do not bind to neutral and zwitterionic phospholipids but interact specifically with negatively charged phos-pholipids such as phosphatidylserine, phosphatidylinositol and phosphatidic acid [37]. In studying the interaction of cardiotoxin II of *Naja mossambica mossambica* with cardiolipin model membranes, Batenburg proposed that cardi-otoxins initially electrostatically bind to the negatively charged cell membrane and subsequently penetrate and disrupt the cell bilayer [41]. Surewicz *et al.* studied the effect of lipid binding on the secondary structure of three cardiotoxin isoforms using Fourier transform infrared spectroscopy [42]. The binding of cardiotoxins to bilayers of dimyristol phosphatidyl glycerol results in an increased content of a β-structure at the expense of unordered conformation. Dauplais *et al.* investigated the interaction of toxin of a cardiotoxin from *Naja naja nigricollis* with perdeuterated dodecylphosphocholine micelles using two-dimensional proton NMR technique [43]. It was found that binding of the cardiotoxin to the lipid micelle was accompanied by stabilisation of the triple stranded β-sheet [43]. However, lipid binding was not found to involve any major conformational change(s) in the toxin. Further, based on the chemical shift perturbation data, the lipid interaction site was ascribed to the hydrophobic face of the toxin [43]. Carbone and Macdonald recently studied the interaction of a cardiotoxin analogue from the *Naja mossambica mossambica* venom, with mixtures of zwitterionic phopholipids using phosphorus (^{31}P) and deuterium (^{2}H) NMR spectroscopy [44]. The results of this study indicate that CTX II binds preferentially to anionic 1-palmitoyl-2-oleoyl-2*sn*-glycero-3-phosphoglycerol (POPG) and is able to laterally segregate POPG and 1-palmitoyl-2-olelyl-*sn*-glydero-3-phosphocholine. The phase composition was described in statistical-thermodynamics terms using a model of the cardiotoxin–phopholipid interactions involving two classes of phospholipid binding sites, one highly specific for phosphatidylglycerol and the other nonspecific and able to accommodate either phosphatidylglycerol or phosphadidylcholine [44]. Sue *et al.* recently studied the interaction of the cardiotoxin analogue III (CTX III) from the Taiwan Cobra with zwitterionic dodpalmitoyl phosphatidylcholine (DPPC) using ^{13}P and ^{2}H NMR [45]. This study for the first time showed that an all β-sheet protein (such as CTX III) behaves as an α-helix polypeptide mellitin in interacting with phospholipid bilayers. It is found that CTX III could lyse lipid bilayers into small aggregates. Elevation of the temperature to that of the liquid crystalline phase transition temperature is shown to result in the fusion of small aggregates into extended bilayers. Further, CTX is suggested to undergo a structural transition between the in-plane and transmembrane forms depending on the physical state of the lipid bilayers [45]. Based on the lipid transitions occurring upon binding toxin molecules, it is shown that the binding mode of P-type of cardiotoxins with zwitterionic phosphatidylcholine is different from that of the S-type cardiotoxins with negatively charged lipids. In a similar study, Dubovskii *et al.* studied the interaction of a P-type cardiotoxin from *Naja oxiana* with perdeuterated dodecylphosphocholine using ^{1}H-NMR and diffusion measurements [3]. Except for subtle changes in the tips of all three loops, no major conformational changes were detected upon the toxin binding to the lipid surface. Based on the results obtained in this study, a peripheral mode of binding of the toxin molecule to the micelle is suggested [3]. The micelle interaction domain includes a hydrophobic region within the extremities of loops I and III (residues 5–11 and 46–50), the basement of loop II (residues 24–29 and 31–37) and a stretch of polar residues encompassing these loops (lysines 4, 5, 12, 23, 50, serines 11, 46, histidine 31 and arginine 36).

There is much debate on the extent to which the cardiotoxins penetrate the membrane surface. Dufourcq *et al.* suggested that only the β-pleated loops of the CTX molecules interact with the hydrophobic interior of the membrane [39]. In contrast, Lauterwein *et al.* postulate that all three loops of the toxin molecule are inserted into the membrane [46]. They propose that the middle loop traverses the lipid bilayer and thereby allows the charged cationic groups at the tip of the middle loop to forge salt-bridge(s) with the polar region of the lipid head groups [46]. Although this model accounts for the high affinity of the cardiotoxin molecules to lipid surface, it cannot rationalise the apparent disparity in terms of the length of the hydrophobic core of a lipid bilayer. However, irrespective of the pros and cons of the proposals, the available experimental evidence suggests that cardiotoxins penetrate the lipid bilayer upon initial binding. The initial binding of the toxin is obviously favoured by the electrostatic interaction step, and the first hydrophobic loop of the CTX molecule is believed to penetrate the lipid phase of the membrane with the molecule in an 'edgewise' orientation [39, 47]. The resulting disorganisation of the membrane is proposed to lower the surface pressure, which consequently is believed to induce a rapid transition from the 'edgewise' to flat orientation of the CTX molecule [47]. Such a flopping of the toxin molecule is believed to amplify the structural perturbation of the membrane resulting in the disorganisation of the cell.

18.5 IS THERE A RECEPTOR FOR CARDIOTOXINS?

It is a common belief that cardiotoxins elicit their biological activities by a nonspecific, lipid phase transition effect on cell membranes. However, there are examples in the literature which hint at the possibility of cardiotoxins exerting their biological activities by binding to specific cell surface receptors. Thelestam and Molby investigating the binding of cardiotoxins to various cell types demonstrated the existence of receptor proteins on human fibroblast cell surface that bind specifically to snake venom cardiotoxins [48]. Similarly, Takechi [12] and Condrea [49] independently demonstrated the existence of specific high-affinity receptors on the surface of erythrocytes and certain muscle preparations. Using radio-labelled cardiotoxin analogues from *Naja naja atra*, Shiau Lin *et al.* unambiguously demonstrated the existence of high-affinity receptor binding sites on chick biventer cervicis muscle [50]. It is of interest that Rees and Bilwes, attempting to rationalise the multitude of biological activities through a common mechanism, suggested the possibility of definite cellular receptors for snake venom cardiotoxins [51]. Recently, Jayaraman *et al.* elucidated the three-dimensional structures of five cardiotoxin analogues from the Taiwan Cobra, and proposed that cardiotoxin molecules exert their defined action(s) on the target tissue by binding to the putative cell surface receptor in a lock- and-key mechanism [40]. Thus, in the context of the increasing evidence for the existence of specific cardiotoxin receptors, the mode of action of this group of toxin still is an open question.

ACKNOWLEDGEMENTS

We would like to thank all our collaborators and co-workers who have significantly contributed to research on snake venom cardiotoxins. We would also like to thank the National Science Council and Dr C.S. Tsong Memorial Medical Research Foundation for financially supporting research projects pertaining to snake venom cardiotoxins.

REFERENCES

[1] Kumar, T.K.S., Pandian, S.K., Srisailam, S. and Yu, C. (1998) *J. Toxicol.Toxin Rev.* **17**, 183–212.
[2] Fletcher, J.E., Hubert, M., Wieland, S.J., Gong, Q.H. and Jiang, M.S. (1996) *Toxicon* **34**, 1301–11.
[3] Dubovskii, P.V., Dementieva, D.V., Bochavlov, E.V., Utkin, Y.N. and Arseniev, A.S. (2001) *J. Mol. Biol.* **305**, 137–49.
[4] Kumar, T.K.S., Sivaraman, T. and Yu, C (1999) in *Natural and Selected Synthetic Toxins: Biological Implications* (Tu, A.T. and Gaffield, W., eds) Oxford University Press, New York, 222–48.
[5] Kumar, T.K.S., Jayaraman, G., Lee, C.S., Arun Kumar, A.I., Sivaraman, T., Samuel, D and Yu, C. (1997) *J. Biomolec. Struct. Dyn.* **15**, 431–63.
[6] Dufton, M.J. and Hider, R.C. (1991) in Harvey, A.L., ed. *Snake toxins*; New York, Pergamon Press, pp. 259–302.
[7] Dufton, M.J. and Hider, R.C. (1983) *CRC Crit. Rev. Biochem.* **114**, 113–71.
[8] Harvey, A.L. (1985) *J. Toxicol. Toxin Rev.* **4**, 41–69.
[9] Harvey, A.L. (1991) *Cardiotoxins from snake venoms*. In Tu, A.T., ed. *Handbook of Natural toxins*, Marcel Dekker, New York, pp. 85–106.
[10] Gilquin, B., Roumestand, C. Zinn-Justin, S., Ménez, A. and Toma, F. (1993) *Biopolymers*, **33**, 1659–75.
[11] Kumar, T.K.S. and Yu, C. (2000) In *The Encyclopedia of Molecular Medicine* [Creighton, T.E., ed.]., John Wiley and Sons, Inc., NewYork.
[12] Takechi, M., Tanaka, Y., Hayashi, K. (1985) *Biochem Int.* **11**, 795–800.
[13] Takechi, M., Tanaka, Y., Hayashi, K. (1986) *FEBS Lett.* **205**, 143–6.
[14] Bilwes, A., Rees, B., Moras, D., Ménez, R. and Ménez, A. (1994) *J Mol. Biol.* **239**, 122–36.
[15] Chang, J.Y., Kumar, T.K.S. and Yu, C. (1998) *Biochemistry* **37**, 6745–51.
[16] Sivaraman, T., Kumar, T.K.S., Yang, P.W. and Yu, C. (1997) *Toxicon* **35**, 1367–71.
[17] Chien, K.Y., Huang, W.N., Jean, J.H., Wu, W.G. (1991) *J. Biol. Chem.* **266**, 3252–9.
[18] Chien, K.Y., Chiang, C.M., Hseu, Y.C., Vyas, A.A, Rule, G.S. and Wu, W. (1994) *J. Biol.Chem.* **269**, 14473–83.
[19] Chiang, C.M., Chien, K.Y., Lin, H.J., Yen, H.C., Ho, P.L. and Wu, W.G. (1996) *Biochemistry* **35**, 9167–76.
[20] Bhaskaran, R., Huang, C.C., Chang, D.K. and Yu, C. (1994) *J. Mol. Biol.* **235**, 1291–301.
[21] Bhaskaran, R., Huang, C.C., Tsai, Y.C., Jayaraman, G., Chang, D.K. and Yu, C. (1994) *J. Biol. Chem.* **269**, 23500–8.
[22] Jahnke, W., Mierke, D.F., Bevers, L. and Kessler, H. (1994) *J. Mol. Biol.* **240**, 445–58.
[23] Jang, J.Y., Kumar, T.K.S., Jayaraman, G. and Yu, C. (1997) *Biochemistry* **36**, 14635–41.

[24] Dementieva, D.V., Bocharov, E.V. and Arseniev, A.S. (1999) *Eur J Biochem.* **263**, 152–62.
[25] O'Connell, J.F., Bougis, P.E and Wuthrich, K. (1993) *Eur. J. Biochem.* **213**, 891–900.
[26] Otting, G., Steinmetz. W.E., Bougis, P.E., Rochat, H. and Wuthrich, K. (1987) *Eur. J. Biochem.* **168**, 609–20.
[27] Sun, Y.J., Wu, W.G., Chiang, C.M., Hsin, A.Y., and Hsiao, C.D. (1997) *Biochemistry* **36**, 2403–13.
[28] Kumar, T.K.S., Lee, C.S. and Yu, C. (1996) *In Natural Toxins II* (Singh, B.R., Tu, A.T., eds) New York, Plenum Press, pp. 115–29.
[29] Sivaraman, T., Kumar, T.K.S. and Yu, C. (2000) *Biochemistry* **39**, 8705–10.
[30] Sivaraman, T., Kumar, T.K.S. and Yu, C. (1999) *Biochemistry* **38**, 9899–905.
[31] Ménez, A., Gatineau, F., Roumestand, C., Harvey, A.L., Mauawad, L., Gilquin, B. and Toma, F. (1990) *Biochimie* **72**, 575–88.
[32] Gatineau, E., Takechi, M., Bouet, F., Mansuelle, P., Rochat, H., Harvey, A.L., Montenay-Garestier, T. and Ménez, A. (1990) *Biochemistry* **29**, 6480–9.
[33] Kini, R.M. and Evans, H.J. (1989) *Biochemistry* **28**, 9209–215.
[34] Gatineau, E., Toma, F., Montenay-Garestier, T., Takechi, M., Fromageot, F., Ménez, A. (1987) *Biochemisty* **26**, 8046–55.
[35] Carlsson, F.H.H., Louw, A.I. (1978) *Biochem. Biophys. Acta.* **534**, 325–9.
[36] Lee, C.S., Kumar, T.K.S., Lian, L.Y., Cheng. J.W. and Yu, C. (1998) *Biochemistry* **37**, 155–64.
[37] Dufourcq. J. and Faucon, J.F. (1978) *Biochemistry* **17**, 1170–6.
[38] Picard, F., Pezolet, M., Bougis, P.E. and Auger, M. (1996) *Biophys. J.* **70**, 1737–44.
[39] Dufourcq, J., Faucon, J.F., Bernard, E., Pezolet, M., Tessier, M. and Bougis, P.E., Van Rietchotenim, J., Delori, P. and Rochat, H. (1982) *Toxicon* **20**, 165–74.
[40] Jayaraman, G., Kumar, T.K.S., Tsai, C.C., Sailam, S., Chou, S.H., Ho, C.L. and Yu, C. (2000) *Protein Sci.* **9**, 637–46.
[41] Batenburg, A.M., Bougis, P.E., Rochat, H., Verkleij, A.J. and Kruijff de, B. (1985) *Biochemistry* **24**, 7101–10.
[42] Surewicz, W.K., Stepanik, T.M., Szabo, A.G. and Mantsch, H.H. (1988) *J. Biol. Chem.* **263**, 786–90.
[43] Dauplais, M., Neumann, J.M., Pinkasfeld, S., Ménez, A. and Roumestand, C. (1995) *Eur. J. Biochem.* **230**, 213–20.
[44] Carbone, M.A. and Macdonald, P.M. (1976) *Biochemistry* **35**, 3368–78.
[45] Sue, S.C., Rajan, P.K., Chen, T.S., Hsieh, C.H. and Wu, W. (1997) *Biochemistry* **36**, 9826–36.
[46] Lauterwein, J. and Wuthrich, K. (1978) *FEBS Lett.* **93**, 181–5.
[47] Bougis, P., Tessier, M., Van Rietschoten, J., Rochat, H., Faucon, J.F. and Dufourcq, J. (1983) *Mol. Cell. Biochem.* **55**, 49–64.
[48] Thelestam, M. and Mollby, R. (1979) *Biochim.Biophys.Acta* **557**, 156–69.
[49] Condrea, E. (1974) *Experientia* **30**, 121–9.
[50] Shiau Lin, S.Y., Huang, M.C. and Lee, C.Y. (1976) *J. Pharmacol. Exp. Ther.* **196**, 758–70.
[51] Rees, B. and Bilwes, A. (1993) *Chem. Res. Toxicol.* **6**, 385–406.

19 Structure and Function of Disintegrins and C-lectins: Viper Venom Proteins Modulating Cell Adhesion

STEFAN NIEWIAROWSKI[§], CEZARY MARCINKIEWICZ,
IWONA WIERZBICKA-PATYNOWSKI, MARY ANN McLANE[*],
and JUAN J. CALVETE

19.1 INTRODUCTION

In this chapter we review four classes of viper venom proteins that modulate cell adhesion by selective recognition of integrins. These proteins include monomeric disintegrins, dimeric disintegrins, metalloproteinase–disintegrins, and C-lectins. Interaction of venom proteins with integrins and other adhesive receptors results in an impairment of platelet aggregation, defective haemostasis, tissue repair, and inflammatory responses. Monomeric disintegrins express RGD, KGD or MVD motifs in a hairpin loop and inhibit αIIbβ3, αvβ3, and α5β1 with different selectivity. Short monomeric disintegrins (echistatin and eristostatin) contain 48–49 amino acids and are linked by four intramolecular S–S bridges, whereas medium-size disintegrins (trigramin, flavoridin) containing about 70 amino acids are linked by six S–S bridges and long disintegrins (bitistatin) by seven intramolecular S–S bridges. Dimeric disintegrins are composed of two identical or homologous subunits, each containing 68–70 amino acids and 10 cysteines. These subunits form two intermolecular and four intramolecular S–S bonds. Each dimeric disintegrin contains two hairpin loops essential for integrin recognition. Heterodimeric disintegrins express in their hairpin loops other motifs than RDG that alter selective recognition of integrins. Substitution of the RGDD motif with MLDG in EC3 and EC6 disintegrins from *Echis carinatus* is responsible for the ability of these disintegrins to bind integrins α4β1, α4β7, and α9β1 and for the significant loss of their affinity for αvβ3. Selective recognition of integrins by disintegrins is also determined by motifs other than RGD or MLD. For instance, M28 in the

---

[*] Corresponding author
[§] Deceased; this paper is dedicated to the memory of Dr S. Niewiarowski

hairpin loop and HKGPAT in the C-terminus of echistatin are critical for the ability of this disintegrin to recognise α5β1. Viper venom metalloproteinase–disintegrins (haemorrhagins) may bind to integrins and degrade ligands and integrin receptors. Jararhagin from *Bothrops jararaca* specifically destroys collagen-receptor α2β1. This receptor is also targeted by EMS16, a C-lectin isolated from *Echis multisquamatus*, which at picomolar concentration inhibits adhesion of cells expressing α2β1. In our laboratory, we studied a number of C-lectins including echicetin, from *Echis carinatus*, and alboaggregins A and B, from *Tr. Albolabris*, interacting with the GPIb/IX complex. The activity of C-lectins depends on their secondary structure.

19.2   VIPER VENOM ANTI-ADHESIVE MOLECULES

Venoms of different vipers contain a number of different proteins that interfere with normal haemostasis and tissue repair by blocking integrin function. These proteins can be divided into four categories : (1) metalloproteinase–disintegrins (hemorrhagins or reprolysins), (2) monomeric disintegrins, (3) dimeric and heterodimeric disintegrins, and (4) lectin-like proteins, also called CDR-lectins or C-lectins. Metalloproteinase–disintegrins represent a subgroup of reprolysins, potent haemorrhagic toxins that contain the disintegrin domain. They occur both in *Crotalidae* and in *Viperinae*. Jararhagin, isolated and cloned from the venom of *Bothrops jararaca*, contains metalloproteinase, a disintegrin-like domain, and a cysteine-rich domain. The RGD motif of the disintegrin domain of jararhagin is substituted with the MSEC motif. Jararhagin binds to the I domain of the α2β1 integrin and degrades this receptor [1] but it does not interact with α1β1 [2]. In this article, we will focus on monomeric and dimeric disintegrins and on CDR lectins occurring in viper venoms.

19.3   MONOMERIC DISINTEGRINS

Monomeric disintegrins were first described as unusually potent inhibitors of platelet fibrinogen receptors associated with integrin αIIbβ3 [3–8]. They occur in vipers living in different geographical locations. Monomeric disintegrins contain 48 to 84 amino acids, 8, 12, or 14 highly conserved cysteines, and a hairpin loop with RGD or KGD motifs maintained in an appropriate conformation by the disulphide bridge. To date, more than 30 monomeric disintegrins have been described. They are divided into short disintegrins containing 48–49 amino acids (echistatin, eristostatin, acutin), medium-size disintegrins containing approximately 70 amino acids (trigramin, flavoridin, albolabrin, kistrin, and botroxostatin), and long disintegrins (bitistatin, 84 amino acids) (Figure 19.1). The RGD motif is critical for the expression of biological activity of disintegrins since the substitution of Arg24 in echistatin from *Echis carinatus* [9]

```
Echistatin                                         ECESGPCCRNCKFLKEGTICKRARGDDMDDYCNGKTCDCPRNPHKGPAT
Eristostatin                                       QEEPCATGPCCRRCKFKRAGKVCRVARGDWNNDYCTGKSCDCPRNPWNG
Trigramin         EAGEDCDCGSP----ANPCCDAATCKLIPGAQCGEGLCCDQCSFIEEGTVCRIARGDDLDDYCNGRSAGCPRNPFH
Albolabrin:       EAGEDCDCGSP----ANPCCDAATCKLLPGAQCGEGLCCDQCSFMKKGTICRRARGDDLDDTCNGISAGCPRNPLHA
Flavoridin:          GEECDCGSP----SNPCCDAATCKLRPGAQCADGLCCDQCRFKKKTGICRIARGDFPDDRCTGLSNDCPRWNDL
Rhodostomin          GKECDCSSP----ENPCCDAATCKLRPGAQCGEGLCCEQCKFSRAGKICRIPRGDMPDDRCTGQSADCPRYH
Barbourin         EAGEECDCGSP----ENPCCDAATCKLRPGAQCADGLCCDQCRFMKKGTVCRVAKGDWNDDTCTGQSADCPRNGLYG
Bitistat3  VSPPVCGNKILEQGEDCDCGSPANC-QDQCCNAATCKLTPGSQCNHGECCDQCKFKKARTVCRIARGDWNDDYCTGKSSDCPWNH
```

Figure 19.1 Amino acid sequences of monomeric disintegrins: echistatin, eristostatin, trigramin, albolabrin, flavoridin, kistrin (rhodostomin), barbourin, and bitistatin. Summarised following the review by Gould *et al.* [4] and Niewiarowski *et al.* [7]

and Arg49 and Asp51 in kistrin from *Agkistrodon rhodostoma* [10] with alanine results in an almost complete loss of disintegrin activity. The RGD sequence is substituted with KGD in barbourin, resulting in an ability to selectively recognise the αIIbβ3 integrin [11]. More recently, the MVD sequence has been found to replace RGD in atrolysin E, a disintegrin that potently inhibits platelet aggregation [12]. Medium-size disintegrins occur exclusively in *Crotalidae*, short and long disintegrins predominate in *Viperinae*.

Reduction and alkylation also result in an almost complete loss of disintegrin biological activity. The disulphide bond pattern of disintegrins has been determined by chemical methods [13–15] and by NMR spectroscopy [16–17]. NMR studies on the disintegrins kistrin, flavoridin (from *Tr. flavoviridis*), and echistatin revealed that an RGD motif is on the apex of a mobile loop between two short β strands of the protein protruding 14–17 Å from the protein core [16–17]. Almost all monomeric disintegrins are potent antagonists of integrins αvβ3 and αIIbβ3 integrins; they inhibit platelet aggregation and endothelial cell adhesion to vitronectin at 30 nM to 300 nM. Some monomeric disintegrins, such as echistatin and flavoridin, also inhibit α5β1 integrin interaction with fibronectin. On the other hand, the disintegrin eristostatin (from *Eristocophis macmahoni*), which is highly homologous with echistatin, is quite a selective inhibitor of the αIIbβ3 integrin. Lu *et al.* [18] and McLane *et al.* [19] provided evidence that amino acids flanking the RGD motif in the hairpin loop of the disintegrin contribute to the selective recognition of the integrin.

In order to identify the echistatin motifs required for selective inhibition of αvβ3 and α5β1, Wierzbicka-Patynowski *et al.* [20] expressed in *E. coli* recombinant echistatin, eristostatin and 15 hybrid molecules of these disintegrins (Figure 19.2). Despite the presence of four S–S bonds, both recombinant molecules appeared to fold properly. They eluted at the same acetonitrile gradient concentration from an HPLC column as native molecules. They had the same molecular mass as determined by mass spectrometry, the same biological activity, and the same amino acid sequences. The ability of recombinant echistatin and eristostatin to inhibit adhesion of different cell lines to fibronectin and von Willebrand Factor and to express the ligand-induced binding site epitope has been tested. The results showed that Asp[27] and Met[28] support recognition of both αvβ3 and α5β1. Replacement of Met[28] with Asn completely abolished echistatin's ability to recognise each of the integrins, while replacement of Met[28]

Dimeric Disintegrins

rEchistatin (wildtype)		^{20}CKRARGDDMDDYC^{32}.........^{40}PRNPHKGPAT^{49}	
rEchis M28L		L	
rEchis M28N		N	
rEchis R22V/D27W/M28L	V	WL	
rEchis M28L/WNG		L	WNG
rEchis 1-43/M28L		L	~~HKGPAT~~
rEchis 1-40/M28L		L	~~RNPHKGPAT~~
rEchis 1-40/R22V/D27W/M28L	V	WL	~~RNPHKGPAT~~
rEristostatin (wildtype)		^{23}CKVARGDWNDDYC^{35}.........^{43}PRNPWNG^{49}	
rEristo W30D		D	
rEristo V25R	R		
rEristo V25R/W30D	R	D	
rEristo N31M		M	
rEristo W30D/N31M		DM	
rEristo 1-43			~~WNG~~
rEristo 1-46			~~RNPWNG~~
rEristo 1-43/HKGPAT			HKGPAT

Figure 19.2 Alignment of eristostatin and echistatin amino acid sequences

with Leu selectively decreased echistatin's ability to recognise $\alpha5\beta1$ only. Eristostatin, in which the C-terminal WNG sequence was substituted with HKGPAT, exhibited new activity with $\alpha5\beta1$, equal to about 50% of recombinant echistatin activity. On the basis of this data, Wierzbicka-Patynowski *et al.* [20] proposed a hypothesis that the hairpin loop and C-terminus of echistatin interact with separate sites on the $\beta1$ or $\beta3$ integrin.

19.4 DIMERIC DISINTEGRINS

A homodimeric disintegrin, contortrostatin, has been isolated from *Agkistrodon contortrix* (family *Crotalidae*) by Markland and his colleagues [21, 22]. Extensively characterised contortrostatin contains two identical hairpin loops expressing RGD motifs. Another homodimeric disintegrin, recently isolated from the venom of *Agkistrodon ussuriensis*, is ussuristatin-2. It contains two identical hairpin loops with a KGD motif [23]. Studies in our laboratory identified a new subfamily of viper venom disintegrins, heterodimeric disintegrins. This subfamily includes the heterodimeric disintegrins EC3 and EC6 from *Echis carinatus* [24, 25], and EMF10 from *Eristocophis macmahoni* [26, 27], and is shown in Figure 19.3. Newly identified CC8 from *Cerastes cerastes* and VLO5 isolated from *Vipera labetina* belong to the same subfamily (Marcinkiewicz *et al.*, unpublished data).

Identification and isolation of heterodimeric disintegrins was accomplished by a combination of reverse-phase HPLC and cell adhesion assay [24, 25]. In brief, aliquots of viper venoms were dissolved in 1% trifluoroacetic acid and

EC3A	NSVHPCCDPVKCEPREGEHCISGPCCRNCKFLRAGTVCKRAVGDDVDDYCSGITPDCPRNRYKGKED
EC3B	NSVHPCCDPVKCEPREGEHCISGPCCRNCKFLNAGTICKRAMLDGLNDYCTGKSSDCPRNRYKGKED
EMF10A	MNSANPCCDPITCKPKKGEHCVSGPCCRNCKFLNPGTICKKGRGDNLNDYCTGVSSDCPRNPWKSEEED
EMF10B	ELLQNSGNPCCDPVTCKPRRGEHCVSGPCCDNCKKLNAGTVCWPAMGDWNDDYCTGISSDCPRNPVFK
EC6A	NSVHPCCDPVTCEPREGEHCISGPCCRNCKFLNAGTICKKAMLDGLNDYCTGISSDCPRNRYKGKEDD
EC6B	NSVHPCCDPVTCKPKRGKHCASGPCCENCYIVGVGTVCNPARGDWNDDNCTGVSSDCPPNPWNGKPSDN

Figure 19.3 Amino acid sequences of heterodimeric disintegrins: EC3A and EC3B [23], EMF10A and EMF10B [25, 26], and EC6A and EC6B [24]

applied on a C-16 column (Vydac). A number of fractions were eluted using acetonitrile gradient. Eluted fractions were lyophilised, dissolved in buffer, and immobilised on 96-well ELISA plates. Cells expressing various integrins were labelled with 5'-chloromethylfluorescein diacetate and added to the plates. Fractions that bound cells were further assayed for inhibition of the adhesion of the same cells to immobilised ligands. Recombinant VCAM-1 was used as a ligand for Jurkat or Ramos cells expressing α4β1, and fibronectin was used as a ligand for K562 cells expressing α5β1. Three heterodimeric disintegrins, EC3 and EC6 in *Echis carinatus* and EMF10 in *Eristocophis macmahoni*, were identified by this method. It should be mentioned that heterodimeric disintegrins are weak inhibitors of platelet aggregation, and screening their activity by inhibition of platelet aggregation assay is not adequate. The molecular weight of heterodimeric or dimeric disintegrins is about 14.5 kDa; after reduction and pyridylethylation, they yield two subunits of approximate MW 7.2 kDa. The dimeric disintegrin, contortrostatin [21, 22], represents a disulphide-linked dimer with an RGD motif in each subunit. It inhibits melanoma cell and breast cancer cell metastases by blocking the α5β1 integrin. It also blocks αvβ3, αIIbβ3, and αvβ5 integrins, whereas monomeric disintegrins block platelet aggregation and inhibit phosphorylation of platelet proteins on tyrosine. Contortrostatin triggers tyrosine phosphorylation on the tyrosine kinase pp72^{syk} and inhibits platelet aggregation [27].

In contrast to monomeric disintegrins, heterodimeric disintegrins EC3 and EMF10 are very weak antagonists of β3 integrins, but they significantly inhibit α4β1 and α5β1 integrins. EC3 (M_r 14 762) is composed of two subunits, EC3A and EC3B, linked covalently with S–S bonds. Each subunit contains 67 amino acids. EC3 inhibited adhesion of cells that express the α4β1 and α4β7 integrins to the natural ligand's vascular cell adhesion molecule 1 (VCAM-1) and to the mucosal cell adhesion molecule 1 (MadCAM-1), with $IC_{50} = 6$–30 nM, and adhesion of K562 cells expressing α5β1 to fibronectin, with $IC_{50} = 150$ nM. It also inhibited adhesion of cells expressing α9β1 to immobilised VCAM-1. Anti-α4 activity resided in subunit B of EC3 that expressed the MLDG motif in the hairpin loop. Linear MLDG peptide inhibited adhesion of Jurkat cells to VCAM-1 in a dose-dependent manner, whereas RGDS peptide was not active

[23]. Recently we identified and isolated from *Echis carinatus suchoreki* another heterodimeric disintegrin, EC6, which inhibits α4β1 and α9β1 with the same potency as EC3 [24]. EC6 contains the MLDG motif in the loop of the A subunit and the RGDW motif in the loop of the B subunit. This RGDW motif appears to be responsible for the strong inhibitory effect of EC6 on the adhesion of K562 cells to fibronectin. Most recently, another heterodimeric disintegrin, VLO5, has been isolated in our laboratory from *Vipera labetina* venom (Marcinkiewicz *et al.*, unpublished observation). VLO5 shows a high degree of amino acid sequence similarity with EC3, including the presence of the MLDG motif in one of its hairpin loops. It is known that the ILDV motif present on the CS1 fragment of fibronectin inhibits binding of cells expressing α4β1 to immobilised VCAM-1. Tselepis *et al.* [28] found that the substitution of the PRGD motif in kistrin with the ILDV motif resulted in the loss of the mutant ability to inhibit β3 integrins. On the other hand, ILDV-kistrin inhibited binding of the CS1 fragment of fibronectin (containing the LDV motif) to immobilised α4β1 integrin. There is evidence that the fibronectin CS1 fragment inhibits α4β1, but it does not inhibit the α9β1 integrin (Sheppard, D., personal communication).

EMF10 (M_r 14949) is composed of two subunits, A containing 68 amino acids, and B containing 69 amino acids. The pattern of intermolecular and intramolecular bridges in the EMF10 molecule has been established [25, 26] (Figure 19.4). Four N-terminal cysteines of this molecule are involved in forming two intermolecular S–S bonds; the remaining cysteines form intramolecular bonds. EMF10 inhibits adhesion of cells expressing α5β1 to fibronectin ($IC_{50} = 1-4$ nM), and it causes expression of the ligand-induced binding site (LIBS epitope) on the β1 subunit of the α5β1 integrin. It inhibits adhesion of cells expressing α4β1 to VCAM1 at a concentration higher than 500 nM, and it inhibits adhesion of CHO cells expressing αvβ3 to vitronectin at a concentration higher than 5000 nM. It does not induce the LIBS epitope in the β3 integrin.

In putative hairpin loops, EMF10A and EMF10B expressed CKKGRG DNLNDYC and CWPAMGDWNDDYC motifs, respectively. On the basis

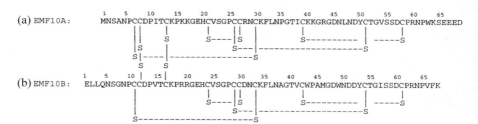

Figure 19.4 (a) Inter- and intra-molecular S–S bridges in EMF10. (b) Molecular model of EMF10. Reprinted from Calvete J.J., *et al.*, *Biochemical J.* **345**, 573–81, 2000, with the permission of the publisher

of the study with synthetic peptides and reduced and alkylated subunits, we proposed that the α5β1 recognition site of EMF10 is associated with the MGDW motif located in a putative hairpin loop of the β subunit. The expression of activity may also depend on the RGDN motif in subunit A and on the C-termini of both subunits [25]. In a computer model of EMF10, the N-terminal polypeptide of EMF10B is close to the hairpin loop of EMF10A subunit, suggesting that it may potentially interfere with the function of the RGDN motif of EMF10A [26].

The common feature of dimeric disintegrins is that they all contain 10 cysteines per subunit and 20 cysteines per molecule. It can be speculated that the two N-terminal cysteines in the dimeric disintegrins create favourable conditions for dimerisation.

Figure 19.5 compares selective recognition of α4β1, α5β1 and αIIbβ3 integrins by EC3, EC6, EMF10, echistatin, flavoridin, eristostatin, and kistrin. It can be concluded that EC3 and EC6 are most selective in inhibiting α4β1 and α9β1 and EMF10 is most specific for α5β1, whereas echistatin, eristostatin, kistrin, and flavoridin are potent inhibitors of β3 integrins.

19.5   C-LECTIN-LIKE PROTEINS

The mannose binding protein is a prototype of C-lectin-like proteins (CLP) or CDR-lectins. In contrast to typical lectins, which recognise carbohydrate moieties, CLP also recognises protein motifs. Snake venom CLPs have several biological activities. Echicetin [29], agkicetin, flavocetins, and tokarecetin [30] bind to the GPIb/IX complex and inhibit vWF binding to this protein, thus blocking platelet agglutination. Injection of echicetin to mice results in a significant prolongation of bleeding time [29]. This haemostatic defect is probably due to the inhibition of both platelet and endothelial GPIb/IX complexes [31]. Studies on the structure–function relationship of C-lectins are much less advanced than the studies on the structure–function of disintegrins.

Both of the biological activites of echicetin [32] and of Jararaca GPIb binding protein [33] are associated with the β subunits of these C-lectins. A high degree of amino acid sequence homology between the β chains of echicetin and Jararaca GPIb binding protein suggests the involvement of a similar motif in binding to the GPIb/IX complex. It appears that the function of C-lectins depends on their degree of polymerisation. Alboaggregin A and alboaggregin B isolated from *Trimeresurus albolabris* by Peng et al. [34, 35] represent a unique group of C-lectins. They inhibit the binding of von Willebrand Factor to the GPIb complex and activate this complex without any cofactor. We reported the primary structure of both alboaggregins [36]. Four chains of alboaggregin A and two chains of alboaggregin B share a high degree of homology, and all cysteines in both alboaggregins are highly conserved. Both alboaggregins showed a similar effect on agglutination of fixed platelets and stimulated

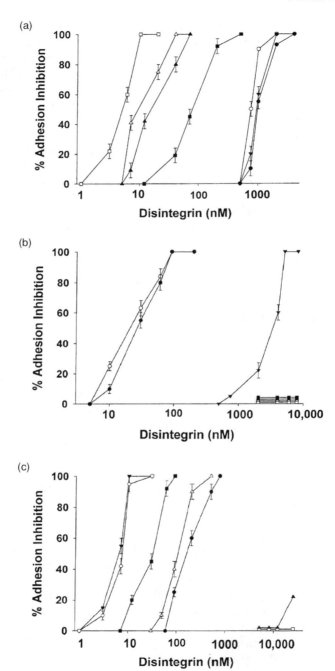

Figure 19.5 Effects of various concentrations of monomeric and heterodimeric disintegrins on the adhesion of Jurkat cells to immobilised VCAM-1, K562 cells to immobilised

human endothelial cells to a similar extent [31]. Alboaggregin A induced the platelet aggregation and release reaction with an IC_{50} about 20 times lower than that of alboaggregin B. These data suggest that the dimeric structure of alboaggregin B is sufficient to activate the GPIb complex on fixed platelets and on endothelial cells, whereas the platelet aggregation and release reaction requires the tetrameric structure of alboaggregin A.

Other CLPs include botrocetin, which binds to the A domain of von Willebrand Factor, Factor IX/X binding protein [30], and convulxin, an activator of the platelet collagen receptor [37]. Most recently we have isolated and characterised EMS16, a potent and selective inhibitor of the $\alpha 2 \beta 1$ integrin from *Echis multisquamatus* venom [38]. Its amino acid sequence is homologous with other CLPs. EMS16 (Mr = 33 kDa) is a heterodimer composed of two subunits linked by S–S bonds. K562 cells, transfected with the $\alpha 2$ integrin, selectively adhered to immobilised EMS16, but not to two other snake venom-derived CLPs, echicetin and alboaggregin B. EMS16 inhibited adhesion of $\alpha 2$-expressing cells to immobilised collagen type I at picomolar concentrations, and the platelet–collagen type I interaction at nanomolar concentration. EMS16 inhibited binding of isolated, recombinant I domain of $\alpha 2$ integrin to collagen in ELISA but not the interaction of collagen type IV of the isolated recombinant domain of $\alpha 1$ integrin. EMS16 inhibited collagen-induced platelet aggregation but had no effect on platelet aggregation induced by other agonists, such as ADP, SFLLRN, thromboxan analogue, and convulxin. EMS16 also inhibited TNSα-stimulated HUVEC migration in a collagen type I gel. Our data, obtained with a number of cells transfected with various integrins, indicate that EMS16 is a highly selective inhibitor of $\alpha 2 \beta 1$. No integrin other than $\alpha 2 \beta 1$ interacting with EMS16 has been identified [38]. Independent of our research, a potent inhibitor of $\alpha 2 \beta 1$ integrin named rhodocetin has been identified in *Agkistrodon rhodostoma* venom [39, 40]. C-lectins occur in a variety of species of *Viperidae*.

19.6 IMPLICATION IN BIOMEDICAL RESEARCH

Disintegrins have found numerous applications in studies on platelet thrombosis, angiogenesis, cancer, bone destruction, and inflammation. They have been used to prevent experimental arterial thrombosis in animal models. A synthetic derivative of the disintegrins barbourin, integrilin, has been developed as a drug that prevents arterial thrombosis after angioplasty (see [7] and [8] for

---

fibronectin (b), and A5 CHO cells stably transfected with $\alpha IIb\beta 3$ to immobilised fibrinogen (c). Fluorescently labelled cells (10^5 cells/well) were mixed with disintegrins, added to 96-well plates coated overnight with relevant ligand, incubated at 37 °C for 30 min, and washed. Bound cells were lysed in 0.5% Triton X100, fluorescence was measured and percent inhibition of cell adhesion was calculated in comparison to fluorescence of adherent cells in the absence of disintegrins. Data are mean +/− SE from at least three experiments. Reprinted from Marcinkiewicz C. *et al.*, *J. Biol. Chem.*, **275**, 31930–7, 2000, with permission of the publisher

references). An interesting application of disintegrins is visualisation of thrombi and emboli. Knight *et al.* [41] found that the radiolabelled disintegrin bitistatin is quite a selective agent in visualising venous thrombi and pulmonary emboli in animal models, and is quite superior to other disintegrins and synthetic peptides that block fibrinogen receptors.

Several observations indicate that disintegrins may cause endothelial cell apoptosis, inhibition of migration, and angiogenesis. We observed that HUVEC adhering to immobilised kistrin undergoes apoptosis in contrast to HUVEC adhering to immobilised vitronectin [42]. Both kistrin and vitronectin recognize αvβ3, but only vitronectin can initiate signals which promote cell growth and protect against apoptosis. Recently, the effect of monomeric disintegrins on angiogenesis has been extensively studied by a Taiwanese group. Yeh *et al.* [43] demonstrated that accutin, a short disintegrin isolated from *Agkistrodon acutus*, blocks the αvβ3 integrin expressed on endothelial cells, induces apoptosis of these cells, and inhibits angiogenesis *in vitro* and *in vivo* in chick chorioallantoic membrane assay. Triflavin (flavoridin) also inhibited angiogenesis with a much higher potency that the monoclonal anti-αvβ3 antibody [44]. It is possible that this potent inhibitory effect results from the ability of triflavin to inhibit both the αvβ3 and α5β1 integrin. It has recently been found that EMF10 is a potent inhibitor of angiogenesis in quail chorioallantoic membrane system assay (Varner, personal communication). EMF10 has very little effect on the adhesion of αvβ3 transfected CHO cells to vitronectin and von Willebrand Factor. However, it appears to inhibit αvβ3 on endothelial cells (Varner, personal communication). Therefore, the anti-angiogenic effect of EMF10 may reflect the inhibition of both α5β1 and αvβ3 on endothelial cells. However, the effect of EMF10 on αvβ3 may be secondary, followed by the inhibition of α5β1. In a recent study, Kim *et al.* [45] presented evidence that integrin α5β1 regulates the function of αvβ3 on endothelial cells. Endothelial cell attachment to vitronectin suppresses protein kinase A activity, and monoclonal anti-α5β1 antibodies block endothelial cell migration on vitronectin and fibronectin.

In numerous studies, monomeric disintegrins have been used to inhibit cancer cell metastases in mouse models (see [7] and [8] for review). The mechanism of the anti-metastatic effect may be due to different factors including inhibition of the formation of cancer cell-platelet aggregates, apoptosis, inhibition of invasion of endothelia by malignant cells, and inhibition of angiogenesis. Staiano *et al.* [46] suggested that that anti-metastatic effect of echistatin may be related to the decrease in pp125[FAK] phosphorylation and increase in actin disassembly.

Echistatin binds to αvβ3 receptor expressed on osteoclasts that are involved in bone destruction. Masarachia *et al.* [47] reported prevention of bone destruction by echistatin *in vivo* in an animal model. Synthetic compounds blocking αvβ3 destruction on osteoclasts are currently being used in clinical trials to prevent bone destruction.

In recent last years, a number of short peptides and peptidomimetics inhibiting α4β1 and α4β7 have been developed for application to the treatment of autoimmune diseases [48, 49]. Our preliminary experiments have shown that EC3 administered to nonobese diabetic (NOD) mice attenuated infiltration of Langerhans islets and salivary ducts by lymphocytes [50]. Application of compounds containing MLDG in some inflammatory and autoimmune diseases may be of special interest since these compounds inhibit both the invasion of lymphocytes expressing $\alpha4\beta1$ and $\alpha4\beta7$ and the invasion of neutrophils expressing predominantly $\alpha9\beta1$.

In conclusion, research on disintegrins and C-lectins is important in understanding the biology of viper venom toxins, and also provides information of the new structures and amino acid motifs involved in integrin recognition that may be useful in basic and clinical research.

REFERENCES

[1] Kamiguti, A.S., Markland, F.S., Zhou, Q., Laing, G.D., Theakston, R.D.G. and Zuzel, M. (1997) Proteolytic cleavage of the $\beta1$ subunit of platelet $\alpha2\beta1$ integrin by the metalloproteinase Jararhagin compromises collagen-stimulated phosphorylation of pp72^{SYK}. *J. Biol. Chem.* **272**, 32599–605.

[2] Moura-da-Silva, A.M., Marcinkiewicz, C.M.M. and Niewiarowski, S. Selective recognition of $\alpha2\beta1$ integrin by jararhagin, a metalloproteinase/disintegrin from *Bothrops jararaca* venom. *Thromb Res.*, in press.

[3] Huang, T.-F., Holt, J.C., Lukasiewicz, H. and Niewiarowski, S. (1987) Trigramin: a low molecular weight peptide inhibiting fibrinogen interaction with platelet receptors expressed on glycoprotein IIb-IIIa complex. *J. Biol. Chem.* **262**, 16157–63.

[4] Huang, T.-F., Holt, J.C., Kirby, K.P. and Niewiarowski, S. (1989) Trigramin: primary structure of trigramin and its inhibition of von Willebrand Factor binding to glycoprotein IIb/IIIa complex on human platelets. *Biochemistry* **28**, 661–6.

[5] Gould, R.J., Polokoff, M.A., Friedman, P.A., Huang, T.-F., Holt, J.C., Cook, J.J. and Niewiarowski, S. (1990) Disintegrins: a family of integrin inhibitor proteins from viper venoms. *Proc. Soc. Exp. Biol. Med.* **195**, 168–71.

[6] Dennis, M.S., Henzel, W.J., Pitti, R.M., Lipari, M.T., Napier, M.A., Deisher, T.A., Bunting, S. and Lazarus, R.A. (1990) Platelet glycoprotein IIb/IIIa protein antagonists from snake venoms: evidence for a family of platelet-aggregation inhibitors. *Proc. Natl. Acad. Sci. USA* **187**, 2471–5.

[7] Niewiarowski, S., McLane, M.A., Kloczewiak, M. and Stewart, G.J. (1994) Disintegrins and other naturally occurring antagonists of platelet fibrinogen receptor. *Semin. Hematol.* **31**, 289–300.

[8] McLane, M.A., Marcinkiewicz, C., Vijay-Kumar, S., Wierzbicka-Patynowski, I. and Niewiarowski, S. (1998) Viper venom disintegrins and related molecules. *Proc. Soc. Exp. Biol. Med.* **219**, 109–19.

[9] Garsky, V.M., Lumma, P.K., Freidinger, R.M., Pitzenberger, S.M., Randall, W.C., Veber, D.F., Gould, R.J. and Friedman, P.A. (1989) Chemical synthesis of echistatin, a potent inhibitor of platelet aggregation from *Echis carinatus*: synthesis and biological activity of selected analogs. *Proc. Natl. Acad. Sci. USA* **86**, 4022–6.

[10] Dennis, M.S., Carter, P. and Lazarus, R.A. (1993) Binding interaction of kistrin with platelet glycoprotein IIb/IIIa. Active site by directed mutagenesis. *Proteins* **15**, 312–21.

[11] Scarborough, R.M., Rose, J.W., Naughton, M.A., Phillips, D.R., Nannizzi, L., Arfsten, A., Campbell, A.M. and Charo, I.F. (1993) Characterization of the integrin specificities of disintegrins isolated from American pit viper venoms. *J. Biol. Chem.* **268**, 1058–65.

[12] Shimokawa, K., Jia, L.G., Shannon, J.D. and Fox, J.W. (1998) Isolation, sequence analysis, and biological activity of atrolysin E/D, the non-RGD disintegrin domain from *Crotalus atrox* venom. *Arch. Biochem. Biophys.* **354**, 239–46.

[13] Calvete, J.J., Shafer, W., Soszka, T., Lu, W., Cook, J.J., Jameson, B. and Niewiarowski, S. (1991) Identification of the disulfide pattern in albolabrin, an RGD-containing peptide from the venom of *Tr. albolabris* : significance for the expression on platelet aggregation inhibitory activity. *Biochemistry* **30**, 5225–9.

[14] Calvete, J.J., Wang, Y., Mann, K., Shafer, W., Niewiarowski, S. and Stewart, G.J. (1992) The disulfide bridge pattern of snake venom disintegrins, flavoridin and echistatin. *FEBS Lett.* **309**, 316–20.

[15] Calvete, J.J., Schrader, K., Raida, M., McLane, M.A., Romero, A. and Niewiarowski, S. (1997) The disulfide bond pattern of bitistatin, a disintegrin isolated from the venom of the viper *Bitis arietans. FEBS Lett.* **416**, 197–202.

[16] Adler, M., Lazarus, R.A., Dennis, M.S. and Wagner, G. (1991) Solution structure of kistrin, a potent platelet aggregation inhibitor and GPIIb/IIIa antagonist. *Science* **253**, 445–448.

[17] Senn, H. and Klaus, W. (1993) The nuclear magnetic resonance solution structure of flavoridin, an antagonist of the platelet GPIIb/IIIa receptor. *J. Mol. Biol.* **232**, 907–25.

[18] Lu, X., Rahman, S., Williams, J.A., Kakkar, V. and Authi, K.S. (1996) Substitutions of proline 42 to alanine and methionine 46 to asparagine around the RGD domain of the neurotoxin dendroaspin alter its preferential antagonism to that resembling the disintegrin elegantin. *J. Biol. Chem.* **271**, 289–94.

[19] McLane, M.A., Vijay-Kumar, S., Calvete, J.J. and Niewiarowski, S. (1996) Importance of the structure of the RGD containing-loop in disintegrins echistatin and eristostatin for recognition of αIIbβ3 and αvβ3 integrins. *FEBS Lett.* **391**, 139–43.

[20] Wierzbicka-Patynowski, I., Niewiarowski, S., Marcinkiewicz, C.M.M., Calvete, J.J. and McLane, M.A. (1999) Structural requirements of echistatin for the recognition of αvβ3 and α5β1 integrin. *J. Biol. Chem.* **274**, 37809–14.

[21] Markland, F.S. (1998) Snake venoms and the hemostatic system. *Toxicon* **36**, 1749–800.

[22] Zhou, Q., Hu, P., Ritter, M.R., Swenson, S.D., Argounova, S., Epstein, A.L. and Markland, F.S. (2000) Molecular cloning and functional expression of contortrostatin, a homodimeric disintegrin from southern copperhead snake venom. *Arch. Biochem. Biophys.* **375**, 278–88.

[23] Oshikawa, R. and Terada, S. (1999) Ussuristatin 2, a novel KGD-bearing disintegrin from *Agkistrodon ussuriensis* venom. *J. Biochem.* **125**, 31–5.

[24] Marcinkiewicz, C., Calvete, J.J., Marcinkiewicz, M.M., Raida, M., Vijay-Kumar, S., Huang, Z., Lobb, R.R. and Niewiarowski, S. (1999a) EC3, a novel heterodimeric disintegrin from *Echis catina* venom, inhibits α4β1 and α5β1 integrins in an RGD-independent manner. *J. Biol. Chem.* **274**, 12468–73.

[25] Marcinkiewicz, C., Taooka, Y., Yokosaki, Y., Calvete, J.J., Marcinkiewicz, M.M., Lobb, R.R., Niewiarowski, S. and Sheppard, D. (2000) Inhibitory effects of MLDG-containing heterodimeric disintegrins reveal distinct structural requirements for interaction of the integrin α9β1 with VCAM-1, tenascin-C and osteopontin. *J. Biol. Chem.* **275**, 31930–7.

[26] Marcinkiewicz, C., Calvete, J.J., Vijay-Kumar, S., Marcinkiewicz, M.M., Raida, M., Schick, P., Lobb, R.R. and Niewiarowski, S. (1999b) Structural and functional characterization of EMF10, a heterodimeric disintegrin from *Eristocophis macmahoni* venom that selectively inhibits α5β1 integrin. *Biochemistry* **38**, 13302–9.

[27] Calvete, J.J., Jurgens, M., Marcinkiewicz, C., Romero, A., Schrader, M. and Niewiarowski, S. (2000) Disulfide bond pattern and molecular modeling of the dimeric disintegrin EMF10, a potent and selective integrin α5β1 antagonist from *Eristocophis macmahoni* venom. *Biochemical J.* **345**, 573–81.

[28] Clark, E.A., Trikha, M., Markland, F.S. and Brugge, J.S. (1994) Structurally distinct disintegrins contortrostatin and multisquamatin differentially regulate platelet tyrosine phosphorylation. *J. Biol. Chem.* **269**, 21940–3.

[29] Tselepis, V.H., Green, L.J. and Humphries, M.J. (1997) An RGD to LDV Motif conversion within the disintegrin kistrin generates an integrin antagonist that retains potency but exhibits altered receptor specificity. *J. Biol. Chem.* **272**, 21341–8.

[30] Peng, M., Lu, W., Beviglia, L., Niewiarowski, S. and Kirby, E.P. (1993) Echicetin: a snake venom protein that inhibits binding of von Willebrand Factor and alboaggregins to platelet glycoprotein Ib. *Blood* **81**, 2321–8.

[31] Fujimura, Y., Kawasaki, T. and Titani, K. (1996) Snake venom proteins modulating the interaction between von Willebrand Factor and platelet glycoprotein Ib. *Thromb. Haemost.* **76**, 633–9.

[32] Tan, L., Kowalska, M.A., Romo, G.M., Lopez, J.A., Darzynkiewicz, Z. and Niewiarowski, S. (1999) Identification and characterization of endothelial glycoprotein Ib using viper venom proteins modulating cell adhesion. *Blood* **93**, 2605–16.

[33] Peng, M., Holt, J.C. and Niewiarowski, S. (1994) Isolation, characterization and amino acid sequence of echicetin β subunit, a specific inhibitor of von Willebrand factor and thrombin interaction with glycoprotein Ib. *Biochem. Biophys. Res. Comm.* **205**, 8–72.

[34] Kawasaki, T., Fujimura, F., Usami, Y., Suzuki, M., Miura, S., Sakurai, Y., Makita, K., Taniuchi, Y., Hirano, K. and Titani, K. (1996) Complete amino acid sequence and identification of the platelet glycoprotein Ib-binding site of jararaca GPIb-BP, a snake venom protein isolated from *Bothrops jararaca*. *J. Biol. Chem.* **271**, 10635–9.

[35] Peng, M., Lu, W. and Kirby, E.P. (1991) Alboaggregin-B : a new platelet agonist that binds to platelet membrane glycoprotein Ib. *Biochemistry* **30**, 11529–632.

[36] Peng, M, Lu, W. and Kirby, E.P. (1992) Characterization of three alboaggregins purified from *Tr. albolabris* venom. *Thromb. Haemost.* **67**, 702–7.

[37] Kowalska, M.A., Tan, L., Holt, J.C., Peng, M., Karczewski, J., Calvete, J.J. and Niewiarowski, S. (1998) Alboaggregins A and B. Structure and interaction with human platelets. *Thromb. Haemost.* **79**, 609–13.

[38] Leduc, M. and Bon, C. (1998) Cloning of subunits of convulxin, a collagen-like platelet-aggregating protein from *Crotalus durissus terrificus* venom. *Biochem. J.* **333**, 389–93.

[39] Marcinkiewicz, C., Lobb, R.R., Marcinkiewicz, M.M., Daniel, J.L., Smith, J.B., Dangelmaier, C., Weinrab, P.H., Beacham, D.A. and Niewiarowski, S. (2000) Isolation and characterization of EMS16, a C-lectin type protein from *Echis multisquamatus* venom, a potent and selective inhibitor of the α2β1 integrin. *Biochemistry* **39**, 9859–67.

[40] Wang, R., Kini, R.M. and Chung, M.C. (1999) Rhodocetin, a novel platelet aggregation inhibitor from the venom of *Calloselasama rhodostoma* (Malayan pit viper): synergistic and noncovalent interaction between its subunits. *Biochemistry* **38**, 7584–93.

[41] Eble, J.A., Beermann, B., Hinz, H-J. and Schmidt-Hederich, A. α2β1 integrin is not recognized by rhodocytin, but is the specific, high-affinity target of rhodocetin, an RGD-independent disintegrin and potent inhibitor of cell adhesion to collagen. *J. Biol. Chem.*, in press (published 12/19/00).

[42] Knight, L.C., Baidoo, K.E., Romano, J.E., Gabriel, J.L. and Maurer, A.H. (2000) Imaging pulmonary emboli and deep venous thrombi with Tc-99m-bitistatin, a platelet-binding polypeptide from viper venom. *J. Nucl. Med.* **41**, 7056–64.

[43] Juliano, D., Wang, Y., Marcinkiewicz, C., Rosenthal, L.A., Stewart, G.J. and Niewiarowski, S. (1996) Disintegrin interaction with αvβ3 integrin on human umbilical vein endothelial cells: expression of ligand-induced binding site on β3 subunit. *Exp. Cell. Res.* **225**, 132–42.

[44] Yeh, C.H., Peng, H.C. and Huang, T-F. (1998) Accutin, a new disintegrin inhibits angiogenesis *in vitro* and *in vivo* by acting as integrin αvβ3 antagonist and inducing apoptosis. *Blood* **92**, 3268–76.

[45] Kim, S., Harris, M. and Varner, J.A. (2000) Regulation of integrin αvβ3-mediated endothelial cell migration and angiogenesis by integrin α5β1 and protein kinase A. *J. Biol. Chem.* **275**, 33920–8.

[46] Staiano, N., Garbi, C., Squillacioti, C., Esposito, S., De Martino, E., Belisaro, M.A., Nitch, L. and Di Natale, P. (1997) Echistatin induces decrease of pp125[FAK] phosphorylation, disassembly of actin cytoskeleton and focal adhesion and detachment of fibronectin adherent melanoma cells. *Eur. J. Cell. Biol.* **73**, 298–305.

[47] Masarachia, P., Yamamoto, M., Leu, C., Rodan, G. and Duong, L. (1998) Histomorphometric evidence for echistatin inhibition of bone resorptions in mice with secondary hyperparathyroidism. *Endocrinology* **139**, 1401–10.

[48] Jackson, D.Y., Quan, C., Artis, D.R., Rawson, T., Blackburn, B., Struble, M., Fitzgerald, G., Chan, K., Mullins, S., Bumier, F.P., Fairbrother, W.J., Clark, K., Berisini, M., Chui, H., Renz, M., Jones, S. and Fong, S. (1997) Potent α4β1 peptide antagonists as potential anti-inflammatory agents. *J. Med. Chem.* **40**, 3359–68.

[49] Vanderslice, P., Ren, K.I., Revelle, J.K., Kim, D.C., Scott, D., Bjorcke, R.J., Yeh, E.T.H., Beck, P.I. and Koaan, T.P. (1997) A cyclic hexapeptide is a potent antagonist of α4 integrins. *J. Immunol.* **158**, 1710–19.

[50] Brando, C., Marcinkiewicz, C., Goldman, B., McLane, M.A. and Niewiarowski, S. (2000) EC3, a heterodimeric disintegrin from *Echis carinatus* inhibits human and murine α4 integrins and attenuates lymphocyte infiltration of Langerhans islets in pancreas and salivary glands in non-obese diabetic mice. *Biochem. Biophys. Res. Comm.* **267**, 413–17.

20 Prothrombin Activators from Snake Venoms

R. MANJUNATHA KINI*, JEREMIAH S. JOSEPH and VEENA S. RAO

20.1 INTRODUCTION

The circulation of blood is essential for life and the integrity of the process is crucial for the survival of the organism. It occurs in a closed system in which the volume of circulating fluid is maintained fairly constant. Haemostasis, defined (by *Stedman's Medical Dictionary*) as 'the arrest of bleeding' following vascular injury, is strictly regulated by the interplay of three major processes: (a) platelet aggregation, a cellular event of activation, adhesion and aggregation of circulating platelets, at the site of injury to form a physical plug to stop the leak; (b) blood coagulation, a series of proteolytic activations culminating in the conversion of soluble fibrinogen into an insoluble clot to cover the ruptured area; and (c) vasoconstriction by which the lumen of the blood vessel that is affected is narrowed to reduce blood flow through the area. These processes are interrelated and synergistic, and they occur with such rapidity that they can be considered more or less simultaneous and prevent blood loss.

Venom from several snakes interferes in thrombosis and haemostasis in addition to affecting several other physiological processes. A large number of toxins, which affect blood coagulation and platelet aggregation, have been purified and characterised. Most of these toxins interfere in either platelet aggregation or blood coagulation. However, some toxins, particularly phospholipases A2 and proteinases, interfere in both these processes. Several reviews are available on the toxins that interfere in platelet aggregation (for inventories, see [1–4]; for reviews, see [5–7]), which are beyond the scope of this review. Among toxins that interfere in the coagulation cascade, some are anticoagulants and they prolong coagulation times, whereas others exhibit procoagulant effects and shorten coagulation times. All the snake venom procoagulants characterised to date are proteinases that activate factor X or prothrombin, or convert fibrinogen to fibrin. One of the prothrombin activators, oscutarin (*Oxyuranus scutellatus*), also activates factor VII [8]. Two factor V-activating

* Corresponding author

Perspectives in Molecular Toxinology
Edited by A Ménez © 2002 John Wiley & Sons, Ltd

proteinases, RVV-V (Russell's viper venom, *Daboia russelli*) and thrombocytin (*Bothrops atrox*), have also been characterised [9]. A large number of snake species contain prothrombin activators in their venoms (for an inventory, see [10] and for reviews, see [11–13]. Based on their structural properties, functional characteristics and cofactor requirements, they have been categorised into four groups (Plate XII) [14, 15, 11]. In this review, we will describe structural and functional properties of all these procoagulant toxins.

20.2 GROUP A PROTHROMBIN ACTIVATORS

These are proteinases that efficiently activate prothrombin without the requirement for any cofactors, such as Ca^{2+} ions, phospholipids or factor Va [10]. They are widely distributed in many kinds of viper venoms. They are presumably toxic, since they are resistant to the natural coagulation inhibitors (serpins), such as antithrombin III [15]. The best-known example is ecarin from the venom of the saw-scale viper *Echis carinatus* [16]. The sequence of this protein was deduced from the sequences of cDNA clones [17]. The mature protein is a metalloproteinase with 426 amino acids and shares 64% identity with the heavy chain of RVV-X (Plate XII and Figure 20.1). It has three domains, a metalloproteinase, disintegrin and a Cys-rich domain. The metalloproteinase domain contains the consensus sequence HEXXHXXGXXH, corresponding to the zinc-chelating active site. In its disintegrin-like domain, the RGD sequence (found in disintegrins) is replaced by an RDD tripeptide sequence; consequently, it has no inhibitory effect on platelet aggregation (Figure 20.1). The function of the Cys-rich domain in ecarin and other similar metalloproteinases is unknown. Ecarin is a highly efficient enzyme with a low Km for prothrombin and a high kcat [13]. It cleaves the $Arg_{320}-Ile_{321}$ bond in prothrombin and produces meizothrombin, which is ultimately converted to α-thrombin by autolysis. Ecarin can also activate descarboxyprothrombin which accumulates in plasma during warfarin therapy. Other prothrombin activators in this class [10, 11], for example, those isolated from the *Bothrops* species, also have similar properties [13].

20.3 GROUP B PROTHROMBIN ACTIVATORS

Yamada *et al.* [15] characterised Ca^{2+}-dependent prothrombin activator, carinactivase-1 from *Echis carinatus* venom. This activator consisted of two subunits noncovalently held together: a 62 kD metalloproteinase and a 25 kD C-type lectin-like disulphide-linked dimer (Plate XII). The C-type lectin-related subunit is homologous to the factor IX/X-binding protein from *Trimeresurus flavoviridis* venom [18]. In contrast to the Group A activators, carinactivase-1 required millimolar concentrations of Ca^{2+} for activity and it had virtually no

(a) Metalloproteinase Domain

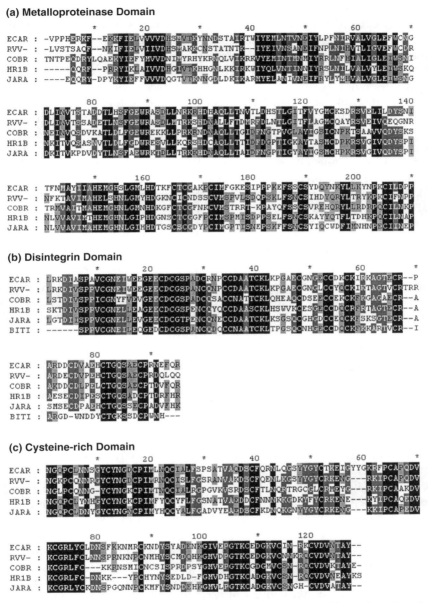

Figure 20.1 Amino acid sequence of ecarin (Group A prothrombin activator) and its sequence homology with snake venom metalloproteinases and bitistatin, a disintegrin. The references for the sequences are ecarin (ECAR) from *Echis carinatus* [17], heavy chain of factor X activator (RVV-) from *Vipera russelli* [65], Cobrin (COBR) from *Naja naja* (Gen-Bank: AF063190), HR1B from *Trimeresurus flavoviridis* [64], jararhagin (JARA) from *Bothrops jararaca* [66] and bitistatin (BITI) from *Bitis arietans* [76] venoms

activity in the absence of Ca^{2+}. Also, unlike ecarin, it does not activate pro-thrombin derivatives, such as prethrombin-1 and descarboxyprothrombin, in which Ca^{2+}-binding has been perturbed. Based on this property, Yamada and Morita [19] developed a chromogenic assay for normal prothrombin in the plasma of warfarin-treated individuals. The metalloproteinase catalytic subunit taken in isolation is similar to ecarin: it no longer requires Ca^{2+} for activity. Reconstitution of the C-type lectin-related subunit restores Ca^{2+} dependence. Prothrombin activation by carinactivase-1 is inhibited by prothrombin frag-ment 1, and the isolated regulatory subunit is capable of binding fragment 1 in the presence of Ca^{2+} ions. Hence this protein recognises the Ca^{2+}-bound conformation of the Gla domain in prothrombin via the 25 kD regulatory subunit, and the subsequent conversion of prothrombin is catalysed by the 62 kD catalytic subunit. Multactivase from *Echis multisquamatus* venom also had similar properties [20].

20.4   GROUP C PROTHROMBIN ACTIVATORS

These are serine proteinases found exclusively in the venoms of Australian elapids [10, 11], and they require only Ca^{2+} ions and negatively charged phos-pholipids, but not factor Va, for maximal activity. They are high molecular weight (>250 kD) prothrombin activators and have multiple subunit enzymes. They have been purified and characterised from *Oxyuranus scutellatus* [21, 22, 23] and *Pseudonaja textilis* venoms [24, Rao, V.S. and Kini, R.M., manuscript in preparation].

The native activators have a molecular mass of ~ 300 kD and consist of a factor Xa-like catalytic subunit (60 kD) and a nonenzymatic, factor Va-like subunit (~ 200 kD) (Plate XII). The catalytic subunit is composed of two disulphide-linked chains, whereas the two chains of the factor Va-like cofactor are held together by noncovalent interactions. The catalytic subunit, similar to factor Xa, has weak catalytic activity in isolation that is greatly stimulated by the factor Va-like subunit [23]. Bovine factor Va can substitute for the cofactor subunit of the enzyme [23, Rao, V.S. and Kini, R.M., manuscript in prepar-ation]. Analysis of prothrombin activation products by these activators indi-cates that they cleave at both $Arg_{271}-Thr_{272}$ and $Arg_{320}-Ile_{321}$ bonds of prothrombin [23], converting it to mature thrombin, in contrast to Group A and B activators, which only convert prothrombin to meizothrombin.

We recently initiated structural studies of pseutarin from *Pseudonaja textilis* venom. The factor Xa-like subunit of pseutarin is homologous to the light and heavy chains of mammalian factor Xa and other group D activators (see below). The internal peptides of the factor Va-like subunit are homologous to the light and heavy chains of mammalian factor Va. Pseutarin thus consists of a trocarin (or factor Xa) homologue noncovalently complexed to a factor Va homologue (Rao and Kini, manuscript in preparation). Thus group C prothrombin activators structurally and functionally resemble the mammalian factor Xa-factor Va complex.

Stocker et al. [25] suggested that the Pseudonaja textilis venom might contain two different prothrombin activators: one each of Group C and D activators. They purified textarin, a $\sim 50\,kD$ serine proteinase, from this venom. The prothrombin activation properties of this protein are similar to those of other group D prothrombin activators [25]. Since pseutarin, a group C activator, has also been purified from the same venom [24, Rao, V.S. and Kini, R.M., manuscript in preparation], it is possible that textarin could just be the factor Xa-like subunit of pseutarin.

20.5   GROUP D PROTHROMBIN ACTIVATORS

These serine proteinases are found exclusively in the venoms of Australian elapids [10, 11] and their activities are strongly stimulated by Ca^{2+} ions, coagulation factor Va and negatively charged phospholipid vesicles [26, 27, 28]. Several prothrombin activators of this group have been purified and characterised [29, 24, 30, 31, 25, 32, 33]. Their molecular mass ranges from 45 000 to 47 000 (based on mass spectrometry and primary structure [33, 34], or from 52 000 to 58 000 (based on SDS-PAGE [29, 24, 30, 35, 25, 32]. They have two chains held together by disulphide bonds (Plate XII). The serine proteinase active site is located in the heavy chain [29, 33]. Unlike other snake venom proteins, which are found in multiple forms, generally a single isoform with varying degrees of glycosylation (most likely a single gene product) has been found so far. However, we found two isoforms of this enzyme with heterogeneity in their amino acid sequence in Notechis scutatus scutatus venom [34]. Currently, we are not certain whether they are due to geographic variation or difficulties in the identification of snakes in closely related species [36].

Their procoagulant effects and the mechanism of prothrombin activation are comparable to those of mammalian factor Xa [29, 33]. Like factor Xa, they cleave prothrombin at two sites, $Arg_{274} - Thr_{275}$ and $Arg_{323} - Ile_{324}$. The electrophoretic pattern of prothrombin cleavage products by human factor Xa and this group of prothrombin activators was identical [29, 33]. The site of cleavage was confirmed by retention times of the cleavage products on reverse phase HPLC, mass spectrometry and amino terminal sequencing [33]. Similar to factor Xa, their poor activator activity was stimulated more than one million-fold by negatively charged phospholipids, factor Va and Ca^{2+} ions [29, 33]. They hydrolyse factor Xa-specific synthetic chromogenic substrates, such as CBS 31.39, S-2222 and S-2765. However, the Vmax for synthetic substrate hydrolysis is two to three orders of magnitude slower [29, 33]. Interestingly, in the absence of Ca^{2+}, they cleave macromolecular substrate prothrombin faster than factor Xa [33]. Similar to their prothrombin activation, the amidolytic activity on chromogenic substrates also increases ($\sim$ 7–14-fold) upon their interaction with factor Va [29, 33]. Factor Xa, on the other hand, does not show such increase in its amidolytic activity. Overall, the prothrombin activators of this group are 'remarkably similar' to factor Xa.

Recently, we determined the complete amino acid sequences of trocarin [33] and hopsarin (Rao *et al.*, unpublished observations) (Figure 20.2). Trocarin shares high identity (53–60%) and homology (62–70%) with factor Xa [33].

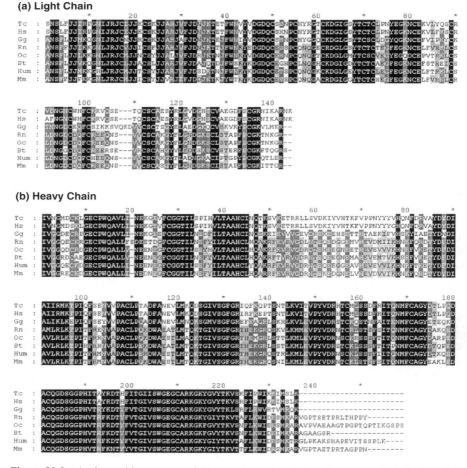

Figure 20.2 Amino acid sequence of Group C prothrombin activators and their sequence homology to factor Xa. The references for the sequences of Group C prothrombin activators are trocarin (Tc) from *Tropidechis carinatus* [33] and hopsarin (Hs) from *Hoplocephalus stephensi* (Rao *et al.*, manuscript in preparation) venoms. Factor Xa sequences are from Swiss-Prot or Genbank databases: Gg, *Gallus gallus* (Chicken); Rn, *Rattus norvegicus* (Rat); Oc, *Oryctolagus cuniculus* (Rabbit); Bt, *Bos taurus* (Bovine); Hum, *Homo sapiens* (Human); and Mm, *Mus musculus* (Mouse)

Based on homology, similar domain architecture is proposed for trocarin. The light chain is homologous to the light chains of vitamin K-dependent coagulation factors, especially that of factor Xa (53–59% identity, 60–67% homology with factor Xa [33]. It consists of an N-terminal Gla domain (residues 1–39), followed by two epidermal growth factor (EGF)-like domains, EGF-I (residues 50–81) and EGF-II (residues 89–124) (Plate XII and Figure 20.2). Cysteine residues are completely conserved.

The Gla domain of trocarin has eleven Gla (γ-carboxy glutamic acid) residues [33]. The first ten lie in positions conserved in the Gla domains of vitamin K-dependent coagulation factors. The eleventh Gla residue in position 35 of trocarin is conserved in other factor Xa's but is replaced by Asp in human factor Xa. On the other hand, Gla 39 found in other factor Xa's is missing in trocarin [33]. These Gla-rich domains of prothrombin activators and other vitamin K-dependent coagulation factors mediate their Ca^{2+}-dependent binding to negatively charged phospholipids. The first EGF domain of factor Xa is involved in calcium binding. This domain of trocarin shares 63–81% identity and 66–84% homology with those of other factor Xa's [33]. However, the second EGF domain of trocarin has only a 38–56% identity and a 44–61% homology with factor Xa EGF-II domains [33].

The heavy chain is the catalytic subunit and is homologous to serine proteinases. It contains all functional residues, including the triad of the charge-relay system: His 42, Asp 88 and Ser 185. Comparison with other factor Xa's show identities and homologies ranging from 52–61% and 61–71%, respectively. Cysteine residues are completely conserved. The residues in factor Xa forming the substrate-binding pocket (Asp 179, Ala 180, Cys 181, Gln 182) and those involved in forming the hydrogen-bonded structure anti-parallel to the substrate (Ser 204, Trp 205, Gly 206, Gly 216) are identical to those in trocarin [33].

In addition to γ-carboxylation of glutamic acid residues, trocarin is also glycosylated. In the light chain Ser_{52} is O-glycosylated, whereas Asn_{45} in the heavy chain has an N-linked glycosylation [33]. Although the zymogen factor X is a glycoprotein, all carbohydrate-bearing peptides are removed during activation [37, 38]. These carbohydrates are important for the activation of zymogen factor X by both intrinsic and extrinsic tenase complexes [39, 38]. In general, glycosylation has been thought to play a role in the stabilisation of protein conformation and protection from proteolysis [40–43] as well as cell-surface recognition phenomena in multicellular organisms (for a review, see [44]). The importance of glycosylation in trocarin has not yet been determined. The glycosylated Asn 45 in trocarin is just three residues removed from the active site His 42. This ~ 3 kD mass carbohydrate is located on the lip to the entrance of the active site (Plate XIII) and its role in the activity of trocarin would be of interest.

Overall, group D venom prothrombin activators are the true structural and functional homologues of blood coagulation factors. Their study should contribute significantly to our understanding of the molecular details of prothrombinase complex formation and activation of prothrombin (discussed below).

20.6 IMPLICATIONS OF STRUCTURAL STUDIES OF PROTHROMBIN ACTIVATORS

Group A and B prothrombin activators are metalloproteinases or their complexes with C-type lectin-related proteins. Structurally, they are not similar to mammalian coagulation factors, which are serine proteinases. Thus these metalloproteinases appear to have evolved with substrate specificity towards prothrombin. Structure–function relationships of these prothrombin activators should contribute to our understanding of substrate recognition and the segments of the metalloproteinase that interact with prothrombin. In the case of group B activators, the C-type lectin-related proteins act as regulatory subunit [15], and in their absence the metalloproteinase subunit is a weak activator of prothrombin. Understanding of structure–function relationships of this group of prothrombin activators should contribute to the delineation of the prothrombin recognition site of the C-type lectin-related protein subunit. This should help in identifying characteristic features that are involved in the recognition of various protein targets, such as prothrombin, factor X and/or IX, glycoprotein Ib, von Willebrand factor and thrombin, recognised by the versatile group of C-type lectin-related proteins. The characterisation of the interaction sites between the C-type lectin-related protein subunit and the metalloproteinase subunit should help us design novel metalloproteinases with specific protein targets. A good example for this scenario is the factor X activator isolated from *Vipera russelli* snake venom [31]. In this case, the metalloproteinase is targeted to factor X, probably through its C-type lectin-related protein subunit [45].

Group C and D prothrombin activators, on the other hand, are serine proteinases and are the first procoagulant factors that are structurally homologous to mammalian blood coagulation activators. Other snake venom serine proteinase-procoagulants activate factor X [45] or act directly on soluble fibrinogen to form a fibrin clot [46]. The structures of the former group have not yet been investigated, but several members of the latter group, named as thrombin-like enzymes (TLEs), have been thoroughly characterised [46]. Their description as 'TLEs' is a misnomer, since they have several distinct structural and functional differences compared with mammalian thrombin [47–50]. TLEs are single chain molecules; thrombin has two chains. They show $< 32\%$ identity ($< 43\%$ homology) with the human thrombin heavy chain and $< 28\%$ identity/ $< 38\%$ homology with intact thrombin. Comparable homology exists between any serine proteinase and thrombin (for example, human trypsin shares 34% identity/41% homology with the human thrombin heavy chain). Only six of the 12 cysteines in TLEs are in identical positions to the seven on the thrombin heavy chain. Unlike the thrombin zymogen, prothrombin, there are no Gla and kringle domains in the precursor of TLEs. Functionally, there are differences between TLEs and thrombin. TLEs cleave fibrinogen and preferentially release either fibrinopeptide A or B, whereas thrombin does not show any preference

[46, 49]. These differences combined with their inability to coagulate snake plasmas [51, 52 – Joseph *et al.*, unpublished observations] indicate that venom TLEs are most likely different from snake thrombin. Thrombin activates factor XIIIa, whereas none of the TLEs so far are known to activate factor XIIIa [49]. Although thrombin is a potent inducer of platelet aggregation, only some TLEs aggregate platelets, but with much lower potency [48]. Thus, while TLEs are serine proteinases that have substrate specificity towards fibrinogen, their dissimilarity to thrombin suggests that determination of their structure–function relationships would contribute only to our understanding of fibrinogen substrate recognition.

In contrast, studies on the structure–function relationships of group C and D prothrombin activators should contribute to our understanding of the molecular details of the prothrombin activation in the mammalian coagulation cascade and the formation of the prothrombinase complex. To this end, so far, the three-dimensional structure of factor Xa has been determined by NMR and X-ray diffraction techniques [53, 54]. But, all the three-dimensional structural information available on factor V/Va has been obtained only by electron microscopy [55]. Recently, Macedo-Ribeiro *et al.* [56] determined the crystal structure of C2 domain of factor V. Using this structure and that of ceruloplasmin [57], Pellequer *et al.* [58] modelled the entire structure of factor Va (1fv4, PDB). The interaction sites between factor Xa and factor Va and the molecular mechanisms of the formation of the prothrombinase complex are not yet identified. Furthermore, molecular details of the intrinsic tenase complex, another coagulation complex formed by factor IXa and factor VIIIa, are also not known. These coagulation factors are structurally closely related to factor Xa and factor Va, respectively, and hence the prothrombinase and intrinsic complexes are deemed to share similar characteristics. A clear understanding of the formation of coagulation complexes should be helpful in developing a novel class of anticoagulant agents that interfere in the complex formation. These anticoagulants will differ from current lead compounds that are targeted to block the active site of specific coagulation proteinases and are useful in the development of therapeutic agents in the treatment of formation of unwanted clots as in, for example, cardiovascular and cerebrovascular diseases. The phylogenetic distance between snakes and mammals will provide the added advantage, since the venom components efficiently recognise and interact with the mammalian counterparts and hence the interaction sites of these proteins have to be conserved. Thus structure–function relationships of group C and D prothrombin activators should provide exciting insights.

20.7 PHYSIOLOGICAL ROLE OF PROTHROMBIN ACTIVATORS

In general, viperid snakes contain group A and B prothrombin activators, whereas group C and D activators have been found only in Australian elapid

venoms [11]. Recently, we purified and characterised the first group A prothrombin activator, Mikarin from the venom of *Micropechis ikaheka*, an elapid
snake [59]. It is interesting to note that two distinct classes of protein toxins,
metalloproteinases in viperid snakes and serine proteinases in elapid snakes,
target this key step in the blood coagulation cascade. In some cases, for example
Hoplocephalus stephensi venom, prothrombin activators of both serine proteinase and metalloproteinase classes are present. A similar situation also exists in
platelet aggregation inhibitors. The final step in platelet aggregation, the interaction between fibrinogen and its receptor glycoprotein IIb/IIIa complex is
inhibited by disintegrins derived by proteolysis of metalloproteinase precursor
in viperid and crotalid venoms [60–66 and others], whereas mambin (or dendroaspin) from mamba (an elapid) venom belongs to the three-fingered toxin
family [67]. Thus distinct molecular molds or templates are utilised to target
critical physiological steps.

    In snake venoms, prothrombin activators are found in large quantities. Sev-
eral Australian elapids are particularly rich sources of prothrombin activators.
In *Pseudonaja textilis* venom, the amount of group C prothrombin activator is as
high as 20–30% of the venom [24]. *Hoplocephalus stephensi* venom contains large
amounts of group D (∼5–6%) and yet to be characterised metalloproteinase
(∼ 8–10%) (Joseph, J.S. and Kini, R.M., unpublished observations). Group D
activators are generally about 3–6% of the total venom protein. Thus snake
venom contains up to 11.5 mg/ml of Group D activators (structurally similar to
factor Xa), whereas human blood contains only 8 µg/ml of factor X. Hence these
factor Xa homologues are present in the venom at a concentration ∼ 1500-fold
that in blood plasma [33]. Thus snake venom is the richest known source of factor
Xa-like proteins. In fact, there is very likely more factor Xa in the venom of an
individual snake than the factor X contained in its entire blood.

    All the Australian elapid venoms, which contain Group C and D prothrom-
bin activators, cause disseminated intravascular coagulopathy (DIC) in victims,
among other symptoms. This condition is characterised by defibrination, and *in
vitro* is reflected by incoagulable plasma, undetectable fibrinogen and a massive
elevation of fibrin(ogen) degradation products [28, 68, 49, 69]. The procoagulants in these venoms clot the victim's blood in the region of the bite. Thrombi
thus formed are potentially fatal, as they may migrate and block vital blood
vessels elsewhere. Furthermore, there is a massive depletion in plasma levels of
coagulation factors, leading to an imbalance in haemostasis in the rest of the
body, resulting in haemorrhage.

    Purified prothrombin activators are highly toxic and they induce cyanosis,
DIC and death [24, 33]. The prothrombin activator from *Pseudonaja textilis*
venom (pseutarin) caused death in rats within several minutes of intravenous
injection at doses as low as 23 µg/kg body weight [24]. Pre-injection of heparin
(intravenous) neutralised the lethal effects of the procoagulant [24], indicating
that lethality was due to disseminated intravascular coagulopathy. Trocarin
induces cyanosis and death in mice within about 320 and 150 minutes at 1 and

10 mg/kg (i.p.), respectively [33]. The appearance of cyanosis in the extremities seems to indicate that trocarin-induced thrombi occluded blood vessels, impeding blood circulation. Thus prothrombin activators play the role of toxins in these venoms.

As described earlier, group C and D prothrombin activators are structurally and functionally similar to mammalian factor Xa–Va complex and coagulation factor Xa, respectively. Since a similar extrinsic coagulation cascade exists in snake blood [51], it is logical to propose that similar, if not identical, coagulation factors are found in the snake blood and they would play a critical role in the haemostasis. However, blood coagulation factors X will be in the zymogen form, unlike the active venom prothrombin activators. Thus, in some snakes, two closely related (if not identical) proteins play distinctly different physiological roles. It is also important to note that coagulation proteins are synthesised primarily in the liver, whereas prothrombin activators are expressed in a nonhepatic tissue, the venom gland. In addition to their extremely high expression, the venom prothrombin activators are expressed inducibly, as are all venom gland proteins [70–73]. The synthesis of blood coagulation factor X in the liver, in contrast, is essentially constitutive, to keep plasma levels constant [74, 75]. Thus, structure–function analysis of the genes for these hepatic and venom proteins would contribute not only to the tissue-specific expression, but also to the transcription control of these closely related proteins.

ACKNOWLEDGMENTS

This work was supported by a financial grant from Academic Research Grants of the National University of Singapore.

REFERENCES

[1] Kini, R.M. and Chow, G. (2001) Exogenous inhibitors of platelet aggregation from animal sources. *Thromb. Haemost.* **85**, 179–81.
[2] Chow, G. and Kini, R.M. (2001) Exogenous factors from animal sources that induce platelet aggregation. *Thromb. Haemost.* **85**, 177–8.
[3] Teng, C.M. and Huang, T.F. (1991) Inventory of exogenous inhibitors of platelet aggregation. *Thromb. Haemost.* **65**, 624–6.
[4] Smith, S.C. and Brinkhous, K.M. (1991) Inventory of exogenous platelet-aggregating agents derived from venoms. *Thromb. Haemost.* **66**, 259–63.
[5] Kini, R.M. and Evans, H.J. (1997) Effects of phospholipase A$_2$ enzymes on platelet aggregation. In: *Venom phospholipase A2 enzymes: structure, Function and mechanism*, Kini, R.M. (ed.), John Wiley & Sons, Ltd, Chichester, pp. 369–87.
[6] Kini, R.M. and Evans, H.J. (1990) Effects of snake venom proteins on blood platelets. *Toxicon* **28**, 1387–422.
[7] Ouyang, C., Teng, C.M. and Huang, T.F. (1992) Characterization of snake venom components acting on blood coagulation and platelet function. *Toxicon* **30**, 945–66.

[8] Nakagaki, T., Lin, P. and Kisiel, W. (1992) Activation of human factor VII by the prothrombin activator from the venom of *Oxyuranus scutellatus* (Taipan snake). *Thromb. Res.* **65**, 105–16.

[9] Tokunaga, F. and Iwanaga, S. (1998) Proteases activating factor V. In: Bailey, G.S. (ed.), *Enzymes from snake venom*, Alaken, Fort Collins, CO, pp. 209–25.

[10] Rosing, J. and Tans, G. (1991) Inventory of exogenous prothrombin activators. *Thromb. Haemost.* **65**, 627–30.

[11] Rosing, J. and Tans, G. (1992) Structural and functional properties of snake venom prothrombin activators. *Toxicon* **30**, 1515–27.

[12] Tans, G. and Rosing J. (1993) Prothrombin activation by snake venom proteases. *J. Toxicol. Toxin Rev.* **12**, 155–73.

[13] Petrovan, R., Tans, G. and Rosing, J. (1998) Proteases activating prothrombin. In: Bailey, G.S. (ed.), *Enzymes from snake venom*, Alaken, Fort Collins, CO, pp. 227–52.

[14] Kini, R.M., Morita, T. and Rosing, J. (2001) Classification and nomenclature of prothrombin activators isolated from snake venoms. *Thromb. Haemost.* (in press).

[15] Yamada, D., Sekiya, F. and Morita, T. (1996) Isolation and characterization of carinactivase, a novel prothrombin activator in *Echis carinatus* venom with a unique catalytic mechanism. *J. Biol. Chem.* **271**, 5200–7.

[16] Kornalik, F. and Blomback, B. (1975) Prothrombin activation induced by ecarin – a prothrombin converting enzyme from *Echis carinatus* venom. *Thromb. Res.* **6**, 57–63.

[17] Nishida, S., Fujita, T., Kohno, N., Atoda, H., Morita, T., Takeya, H., Kido, I., Paine, M.J.I., Kawabata, S. and Iwanaga, S. (1995) cDNA cloning and deduced amino acid sequence of prothrombin activator (ecarin) from Kenyan *Echis carinatus* venom. *Biochemistry* **34**, 1771–8.

[18] Atoda, H., Hyuga, M. and Morita, T. (1991) The primary structure of coagulation factor IX/factor X-binding protein isolated from the venom of *Trimeresurus flavoviridis*: homology with asialoglycoprotein receptors, proteoglycan core protein, tetranectin, and lymphocyte Fcε receptor for immunoglobulin E. *J. Biol. Chem.* **266**, 14903–11.

[19] Yamada, D. and Morita, T. (1999) CA-1 method, a novel assay for quantification of normal prothrombin using a Ca^{2+}-dependent prothrombin activator, carinactivase-1. *Thromb. Res.* **94**, 221–6.

[20] Yamada, D. and Morita, T. (1997) Purification and characterization of a Ca^{2+}-dependent prothrombin activator, multactivase, from the venom of *Echis multisquamatus*. *J. Biochem.* **122**, 991–7.

[21] Owen, W.G. and Jackson, C.M. (1973) Activation of prothrombin with *Oxyuranus scutellatus scutellatus* (Taipan snake) venom. *Thromb. Res.* **3**, 705–14.

[22] Walker, F.J., Owen, W.G. and Esmon, C.T. (1980) Characterization of the prothrombin activator from the venom of *Oxyuranus scutellatus scutellatus* (Taipan venom). *Biochemistry* **19**, 1020–3.

[23] Speijer, H., Govers-Riemslag, J.W.P., Zwaal, R.F.A. and Rosing, J. (1986) Prothrombin activation by an activator from the venom of *Oxyuranus scutellatus* (Taipan snake). *J. Biol. Chem.* **261**, 13258–67.

[24] Masci, P.P., Whitaker, A.N. and de Jersey, J. (1987) Purification and characterization of the prothrombin activator of the venom of *Pseudonaja textiles*. In: Gopalakrishnakone P., Tan C.K. (eds), *Progress in Venom and Toxin Research*, National University of Singapore, Singapore, pp. 209–219.

[25] Stocker, K., Hauer, H., Muller, C. and Triplett, D.A. (1994) Isolation and characterization of Textarin, a prothrombin activator from eastern brown snake (*Pseudonaja textilis*) venom. *Toxicon* **32**, 1227–36.

[26] Jobin, F. and Esnouf, M.P. (1966) Coagulant activity of tiger snake (*Notechis scutatus scutatus*) venom. *Nature* 211, 873–5.

[27] Chester, A. and Crawford, G.P.M. (1982) *In vitro* coagulant properties of venoms from Australian snakes. *Toxicon* 20, 501–4.

[28] Marshall, L.R. and Herrmann, R.P. (1983) Coagulant and anticoagulant actions of Australian snake venoms. *Thromb. Haemost.* 50, 707–11.

[29] Tans, G., Govers-Riemslag, J.W.P., van Rihn, J.L.M.L. and Rosing, J. (1985) Purification and properties of a prothrombin activator from the venom of *Notechis scutatus scutatus*. *J. Biol. Chem.* 260, 9366–72.

[30] Morrison, J.J., Masci, P.P., Bennett, E.A., Gauci, M., Pearn, J., Whitaker, A.N. and Korschenko, L.P. (1987) Studies of the venom and clinical features of the Australian rough-scaled snake (*Tropidechis carinatus*). In: Gopalakrishnakone P., Tan C.K. (eds), *Progress in Venom and Toxin Research*, National University of Singapore, Singapore, pp. 220–33.

[31] Williams, W.J. and Esnouf, M.P. (1962) The fractionation of Russell's viper (*Vipera russelli*) venom with special reference to the coagulant protein. *Biochem. J.* 84, 52–62.

[32] Marsh, N.A., Fyffe, T.L. and Bennett, E.A. (1997) Isolation and partial characterization of a prothrombin-activating enzyme from the venom of the Australian rough scaled snake (*Tropidechis carinatus*). *Toxicon* 35, 563–71.

[33] Joseph, J.S., Chung, M.C., Jeyaseelan, K. and Kini, R.M. (1999) Amino acid sequence of trocarin, a prothrombin activator from *Tropidechis carinatus* venom: its structural similarity to coagulation factor Xa. *Blood* 94, 621–31.

[34] Joseph, J.S. (2001) Procoagulant snake venom protein similar to mammalian blood coagulation factor Xa: structural and functional characterization. Ph.D. Thesis, National University of Singapore.

[35] Williams, V. and White, J. (1989) Purification and properties of a procoagulant from Peninsula tiger snake (*Notechis ater niger*) venom. *Toxicon* 27, 773–9.

[36] Mirtschin, P.J., Crowe, G.R. and Davis, R. (1990) Dangerous snakes of Australia. In: Gopalakrishnakone P., Chou L.M. (eds), *Snakes of Medical Importance (Asia-Pacific region)*, National University of Singapore, Singapore, pp. 1–174.

[37] Fujikawa, K., Titani, K. and Davie, E.W. (1975) Activation of bovine X (Stuart factor): conversion of factor Xa_α to factor Xa_β. *Proc. Natl. Acad. Sci., USA* 72, 3359–63.

[38] Inoue, K. and Morita, T. (1993) Identification of O-linked oligosaccharide chains in the activation peptides of blood coagulation factor X. The role of the carbohydrate moieties in the activation of factor X. *Eur. J. Biochem.* 218, 153–63.

[39] Sinha, U. and Wolf, D.L. (1993) Carbohydrate residues modulate the activation of coagulation factor X. *J. Biol. Chem.* 268, 3048–51.

[40] Walsh, M.T., Watzlawick, H., Putnam, F.W., Schmid, K. and Brossmer, R. (1990) Effect of the carbohydrate moiety on the secondary structure of β 2-glycoprotein. I. Implications for the biosynthesis and folding of glycoproteins. *Biochemistry* 29, 6250–7.

[41] Bernard, E.R., Sheila, A.N. and Olden, K. (1983) Effect of size and location of the oligosaccharide chain on protease degradation of bovine pancreatic ribonuclease. *J. Biol. Chem.* 258, 12198–202.

[42] Rudd, P.M., Joao, H.C., Coghill, E., Fiten, P., Saunders, M.R., Opdenakker, G. and Dwek, R.A. (1994) Glycoforms modify the dynamic stability and functional activity of an enzyme. *Biochemistry* 33, 17–22.

[43] Wang, C., Eufemi, M., Turano, C. and Giartosio, A. (1996) Influence of the carbohydrate moiety on the stability of glycoproteins. *Biochemistry* 35. 7299–307.

[44] Kobata, A. (1992) Structures and functions of the sugar chains of glycoproteins. *Eur. J. Biochem.* 209, 483–501.

[45] Morita, T. (1998) Proteases which activate factor X. In: Bailey G.S. (ed.), *Enzymes from snake venom*, Alaken, Fort Collins, CO, pp. 179–208.

[46] Pirkle, H. and Stocker, K. (1991) Thrombin-like enzymes from snake venoms; an inventory. *Thromb. Haemost.* **65**, 444–50.

[47] Aronson, D.L. (1976) Comparison of the actions of thrombin and the thrombin-like venom enzymes ancrod and batroxobin. *Thromb. Haemost.* **36**, 9–13.

[48] Teng, C.M. and Ko, F.N. (1988) Comparison of the platelet aggregation induced by three thrombin-like enzymes of snake venoms and thrombin. *Thromb. Haemost.* **59**, 304–9.

[49] Hutton, R.A. and Warrell, D.A. (1993) Action of snake venom components on the haemostatic system. *Blood Rev.* **7**, 176–89.

[50] Bell, Jr., W.R. (1997) Defibrinogenating enzymes. *Drugs* **54**, 18–30.

[51] Denson, K.W.E. (1976) The clotting of a snake (*Crotalus viridis helleri*) plasma and its interaction with various snake venoms. *Thromb. Haemost.* **35**, 314–23.

[52] Nahas, L., Kamiguti, A.S., Sousa e Silva, M.C.C., Ribeiro de Barros, M.A.A. and Morena, P. (1983) The inactivating effect of *Bothrops jararaca* and *Waglerophis merremii* snake plasma on the coagulant activity of various snake venoms. *Toxicon* **21**, 239–46.

[53] Padmanabhan, K., Padmanabhan, K.P., Tulinsky, A., Park, C.H., Bode, W., Huber, R., Blankenship, D.T., Cardin, A.D. and Kisiel, W. (1993) Structure of human des(1–45) factor Xa at 2.2 Å resolution. *J. Mol. Biol.* **232**, 947–66.

[54] Kamata, K., Kawamoto, H., Honma, T., Iwama, T. and Kim, S.-H. (1998) Structural basis for the chemical inhibition of human blood coagulation factor Xa. *Proc. Natl. Acad. Sci., USA* **95**, 6630–5.

[55] Fowler, W.E., Fay, P.J., Arvan, D.S. and Marder, V.J. (1990) Electron microscopy of human factor V and factor VIII: correlation of morphology with domain structure and localization of factor V activation fragments. *Proc. Natl. Acad. Sci., USA* **87**, 7648–52.

[56] Macedo-Ribeiro, S., Bode, W., Huber, R., Quinn-Allen, M.A., Kim, S.W., Ortel, T.L., Bourenkov, G.P., Bartunik, H.D., Stubbs, M.T., Kane, W.H. and Fuentes-Prior, P. (1999) Crystal structures of the membrane-binding C2 domain of human coagulation factor V. *Nature* **402**, 434–9.

[57] Zaitseva, I., Zaitsev, V., Card, G., Moshkov, K., Bax, B., Ralph, A. and Lindley, P. (1996) The X-ray structure of human serum ceruloplasmin at 3.1 Å: nature of the copper centres. *J. Biol. Inorg. Chem.* **1**, 15–23.

[58] Pellequer, J.-L., Gale, A.J., Getzoff, E.D. and Griffin, J.H. (2000) Three-dimensional model of coagulation factor Va bound to activated protein C. *Thromb. Haemost.* **84**, 849–57.

[59] Gao, R., Kini, R.M. and Gopalakrishnakone, P. (2001) A novel prothrombin activator from the venom of *Micropechis ikaheka*: isolation and characterization. *FEBS Lett.* submitted.

[60] Kini, R.M. and Evans, H.J. (1992) Structural domains in venom proteins: evidence that metalloproteinases and nonenzymatic platelet aggregation inhibitors (disintegrins) from snake venoms are derived by proteolysis from a common precursor. *Toxicon* **30**, 265–93.

[61] Hite, L.A., Shannon, J.D., Bjarnason, J.B. and Fox, J.W. (1992) Sequence of a cDNA clone encoding the zinc metalloproteinase hemorrhagic toxin *e* from *Crotalus atrox*: evidence for signal, zymogen, and disintegrin-like structures. *Biochemistry* **31**, 6203–11.

[62] Hite, L.A., Bjarnason, J.B. and Fox, J.W. (1994) cDNA sequences for four snake venom metalloproteinases: structure, classification, and their relationship to mammalian reproductive proteins. *Arch. Biochem. Biophys.* **308**, 182–91.

[63] Au, L.C., Chou, J.S., Chang, K.J., Teh, G.W. and Lin, S.B. (1993) Nucleotide sequence of a full-length cDNA encoding a common precursor of platelet aggregation inhibitor and hemorrhagic protein from *Calloselasma rhodostoma* venom. *Biochim. Biophys. Acta* **1173**, 243–5.

[64] Takeya, H., Oda, K., Miyata, T., Omori-Satoh, T. and Iwanaga, S. (1990) The complete amino acid sequence of the high molecular mass hemorrhagic protein HR1B isolated from the venom of *Trimeresurus flavoviridis*. *J. Biol. Chem.* **265**, 16068–73.

[65] Takeya, H., Nishida, S., Miyata, T., Kawada, S.I., Saisaka, Y., Morita, T. and Iwanaga, S. (1992) Coagulation factor X activating enzyme from Russell's viper venom (RVV-X). A novel metalloproteinase with disintegrin (platelet aggregation inhibitor)-like and C-type lectin-like domains. *J. Biol. Chem.* **267**, 14109–17.

[66] Paine, M.J.I., Desmond, H.P., Theakston, R.D.G. and Crampton, J.M. (1992b) Purification, cloning, and molecular characterization of a high molecular weight hemorrhagic metalloproteinase, jararhagin, from *Bothrops jararaca* venom. Insights into the disintegrin gene family. *J. Biol. Chem.* **267**, 22869–76.

[67] McDowell, R.S., Dennis, M.S., Louie, A., Shuster, M., Mulkerrin, M.G. and Lazarus, R.A. (1992) Mambin, a potent glycoprotein IIb-IIIa antagonist and platelet aggregation inhibitor structurally related to the short neurotoxins. *Biochemistry* **31**, 4766–72.

[68] Marsh, N.A. (1994) Snake venoms affecting the hemostatic mechanism – a consideration of their mechanisms, practical applications and biological significance. *Blood Coag. Fibrinolysis* **5**, 399–410.

[69] White, J. (1998) Envenoming and antivenom use in Australia. *Toxicon* **36**, 1483–92.

[70] Rotenberg, D., Bamberger, E.S. and Kochva, E. (1971) Studies on ribonucleic acid synthesis in the venom glands of *Vipera palaestinae* (Ophidia, Reptilia). *Biochem. J.* **121**, 609–12.

[71] De Lucca, F.L., Imaizumi, M.T. and Haddad, A. (1974) Characterization of ribonucleic acids from the venom glands of *Crotalus durissus terrificus* (Ophidia, Reptilia) after manual extraction of the venom. Studies on template activity and base composition. *Biochem. J.* **139**, 151–6.

[72] Brown, R.S., Brown, M.B., Bdolah, A. and Kochva, E. (1975) Accumulation of some secretory enzymes in venom glands of *Vipera palaestinae*. *Am. J. Physiol.* **229**, 1675–9.

[73] Paine, M.J., Desmond, H.P., Theakston, R.D. and Crampton, J.M. (1992a) Gene expression in *Echis carinatus* (carpet viper) venom glands following milking. *Toxicon* **30**, 379–86.

[74] Bahnak, B.R., Howk, R., Morrissey, J.H., Ricca, G.A., Edgington, T.S., Jaye, M.C., Drohan, W.W. and Fair, D.S. (1987) Steady state levels of factor X mRNA in liver and HepG2 cells. *Blood* **69**, 224–30.

[75] Miao, C.H., Leytus, S.P., Chung, D.W. and Davie, E.W. (1992) Liver-specific expression of the gene coding for human factor X, a blood coagulation factor. *J. Biol. Chem.* **267**, 7395–401.

[76] Shebuski, R.J., Ramjit, D.R., Bencen, G.H. and Polokoff, M.A. (1989) Characterization and platelet aggregation inhibitory activity of bitistatin, a potent arginine-glycine-aspartic acid-containing peptide from the venom of the viper *Bitis arietans*. *J. Biol. Chem.* **264**, 21550–6.

21 C-type Lectins from Snake Venoms: New Tools for Research in Thrombosis and Haemostasis

ANNE WISNER, MIREILLE LEDUC and CASSIAN BON*

21.1 INTRODUCTION

Venoms from *Viperidae* and *Crotalidae* snakes contain a large variety of proteins and peptides affecting the haemostatic system of mammals [1]. Consequently, the envenomations by these snakes generally result in persistent bleeding because the venom causes considerable degradation of fibrinogen and other coagulation factors, preventing clot formation [2]. Many of the snake venom proteins that affect haemostasis are enzymes such as nucleotidases, phospholipases A$_2$ (PLA$_2$), metalloproteinases and serine proteinases, whereas others, such as disintegrins and C-type lectins, have no enzymatic activity [1, 3–5]. Snake venom metalloproteinases are often multidomain proteins composed of a catalytic domain and one or several noncatalytic domains structurally related to disintegrins and/or a cysteine-rich elements and/or C-type-lectin (CTL) domains [6, 7]. Thus snake venom components have been reported to interfere with various haemostatic processes. They usually have a selective action against a specific molecular target and they are resistant to physiological inhibitors. These snake venom components affecting thrombosis and haemostasis (SVC&TH) therefore appear to be potentially useful tools for investigating blood coagulation mechanisms and have been extensively used in the development of diagnostic tests [8]. In addition, SVC&TH are of value for structure–function studies since their structures are very similar while they possess different physiological activities. Furthermore, in many circumstances the various biological activities of a unique human coagulation factor are

Abbreviations: PLA$_2$, phospholipases A$_2$; CTL, C-type-lectin; SVC&TH, snake venom components affecting thrombosis and haemostasis; vWF, von Willebrand factor; vWD, von Willebrand disease; GpIIb/IIIa, glycoprotein IIb/integrin IIIa; GP-Ib-IX-V, glycoprotein complex Ib-IX-V; FXa, factor X; FII, prothrombin; FVa, activated factor V; HLIT, lithostatin; CRD, carbohydrate recognition domain; MBP-A, mannose-binding protein-A; RSL, rattlesnake lectin; RVV-X, factor X activator of Russell viper venom; Bc, botrocetin; IX/X-BP, factor IX /factor X binding protein; IX-BP, factor IX binding protein; CVX, Convulxin; FITC-CVX, fluorescein-labelled CVX; Bj, Bothrojaracin
* Corresponding author

Perspectives in Molecular Toxinology
Edited by A Ménez © 2002 John Wiley & Sons, Ltd

carried by different SVC&TH. For example, the venom serine proteinases that
have been characterised as thrombin-like enzymes possess only some of the
effects of thrombin on platelets [9–11], fibrinogen [12–14], factor V [15] or
protein C [16]. Thus the SVC&TH appear to be interesting molecular models
for the development of new therapeutic agents or drugs. This chapter of
Perspectives in Toxinology focuses on the most recent insights into the structure–activity relationships of snake venom CTLs that act on haemostasis, and
on the ways in which the SVC&TH contribute to the understanding of the
molecular mechanisms of haemostasis, in many unexpected ways.

Haemostasis involves not only clot formation, but also clot dissolution [2] as
shown in Figure 21.1. These two opposite processes are in equilibrium *in vivo*
and maintain constant the haemodynamic properties of the blood [17]. Under
physiological conditions, coagulation is initiated by a vascular injury, which
leads locally to a reflex vasoconstriction, changes in the vascular endothelial
surface and the release of procoagulant factors that are responsible for clot
formation. In plasma exposed to high hydrodynamic shear, as occurs in the case
of the rupture of an atherosclerotic plaque, platelet adhesion is initiated by the
binding of the von Willebrand factor (vWF) to its specific membrane receptor,
the glycoprotein GPIb-IX-V. At low shear, other receptors, such as the collagen
receptors, contribute to platelet adhesion. Platelet aggregation then continues
via platelet/platelet interactions involving the association of fibrinogen or vWF
with the glycoprotein IIb/integrin IIIa (GpIIb/IIIa) complex (Figure 21.1).
Finally, a fibrin network is formed, strengthening the plug, which is transformed into a solid clot. The clot is made of fibrin and results in the activation
of fibrinogen by thrombin. Thrombin is generated during a cascade of activation of serine proteinase zymogens in which functional enzymes activate each
other, amplifying the signal (Figure 21.1). This amplification is well illustrated
by the prothrombinase complex (leading to thrombin production), which involves three factors, activated factor X (FXa), prothrombin (FII) and activated
factor V (FVa), which bind via calcium bridges to the negatively charged
phospholipids exposed at the surface of platelets during their activation [18].
Thrombin formation is a critical step in coagulation because thrombin not only
initiates fibrin formation but also activates FV, factor VIII, factor XIII and
platelets. After clot formation, once bleeding has been stopped, the continuity
of the endothelium is restored. Then, the fibrinolytic system is activated and
gradually dissolves the clot (Figure 21.1).

Mammalian CTLs constitute a large superfamily of proteins which can be
subdivided into several classes. This family includes extracellular proteins such
as receptors (selectins and type II receptors) and soluble proteins (collectins and
proteoglycans) [19]. Most of them are multimeric proteins that contain one or
several 'carbohydrate recognition domains' (CRDs) [20]. With the exception of
lithostatin (HLIT), the CRDs of mammalians CTLs bind sugars in a calcium-
dependent manner [21–23]. Mammalian CRDs comprise about 120 amino acid
residues and are structurally conserved. As shown in Figure 21.2, CRDs possess

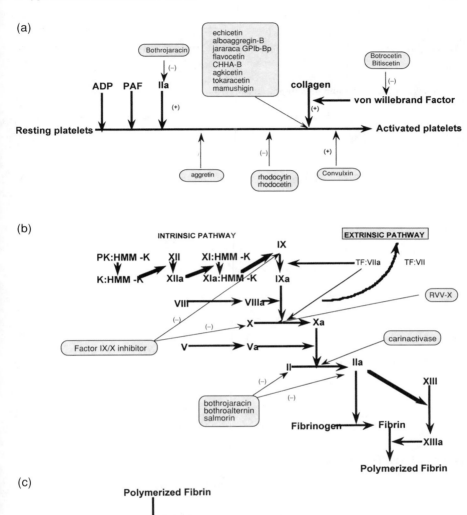

Figure 21.1 Physiological processes in haemostasis and site of action of various snake venom CTLs.
(a) Platelet aggregation. (b) Coagulation. (c) Fibrinolysis.
Heterodimeric and multimeric CTLs are indicated in rounded boxes; metalloproteinases associated with a CTL subunit are shown in right-angled boxes

six conserved cysteine residues involved in three S–S bridges which stabilise their scaffold [24], as well as one triplet EPN or QPD and three amino acid residues E, N and D which are implicated, respectively, in the binding of the

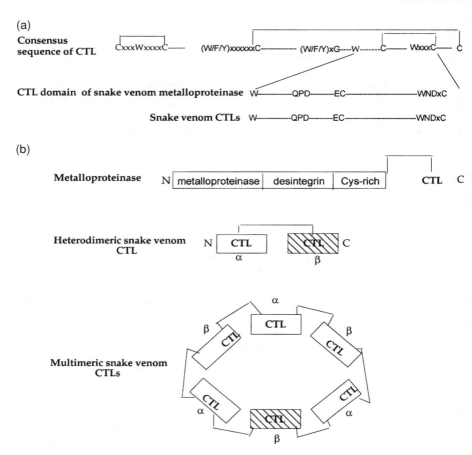

Figure 21.2   Snake venom CTLs.
(a) Representation of the various domains. (b) Conserved residues in the CRD of CTLs. The sites for sugar binding 'QPN' and the site 'END' for Ca²⁺ binding are indicated in boxes

sugar, mannose or galactose, and of Ca^{2+} [19]. HLIT is a soluble protein consisting of a single domain structurally homologous to CRD, which is unable to bind sugar and accordingly does not contain the conserved residues implicated in the binding of sugar and Ca^{2+} [25]. The crystallographic structures of the CRDs of three mammalian CTLs have been elucidated in the case of mannose-binding protein-A (MBP-A), selectin-E and HLIT [25–28]. The superposition of these three structures clearly shows the homology of their architecture, which comprises two orthogonal α-helices separated by a β-sheet and which is stabilised by two disulphide bridges (Plate XIV A). Variable and exposed loops, implicated in the interaction with sugar, extend from a second β-sheet placed above the first one.

A homogenous group of venom proteins from *Crotalidae* and *Viperidae* snakes is associated with the CTL superfamily because they possess most of the characteristics defined for the CRDs of CTLs, as reported by Drickamer [22] and Spiess [24]. The sequences of the CRD domain of snake venom CTLs present a large homology (Figure 21.3), although these snake venom proteins

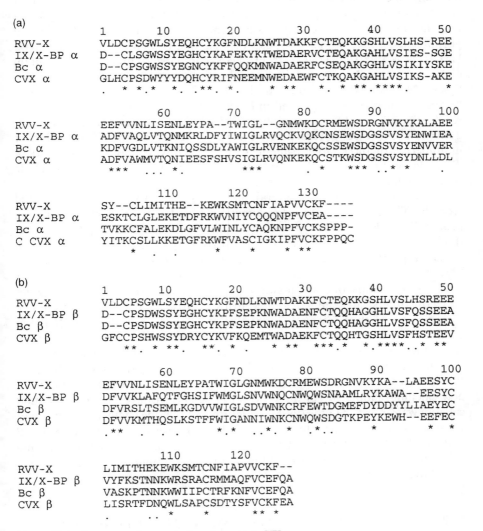

Figure 21.3 Sequence alignment of snake venom CTLs.
The residues that are identical in all sequences are indicated by asterisks

can be subdivided into true lectins, which bind galactose in a calcium-dependent manner, and homologous proteins, like HLIT, which do not possess these binding characteristics (Figure 21.2A). The first group includes homodimeric CTLs, like the 'rattlesnake lectin' (RSL), which have two CRDs joined by a disulphide bridge [29], and some metalloproteinases containing a catalytic domain linked by a disulphide bond to a single CRD motive, as in the factor X activator of Russell viper venom (RVV-X) and in other metalloproteinases that act on haemostasis or possess a haemorrhagic activity [30]. The second group includes heterodimeric CTLs which contain two different although similar CRDs, α and β. The CRDs from snake venom heterodimeric CTLs contain eight cysteine residues. According to the structural model of CRDs proposed by Spiess [24], six of these residues are involved in three intracatenary disulphide bridges, as chemically confirmed in botrocetin (Bc) [31]. The last two cysteine residues form an additional disulphide bridge linking the α and β chains (Figure 21.2B). In some cases, several heterodimeric CTLs might be associated in higher multimeric structures, either trimeric $(\alpha\beta)_3$, as in convulxin, or tetrameric $(\alpha\beta)_4$, as in flavocetin A [32–34]. In this case, additional intercatenary disulphide bridges stabilise the oligomeric structure [32–34].

The crystal structures of several snake venom CTLs have been reported recently. Three of them are heterodimeric CTLs: factor IX /factor X binding protein (IX/X-BP), factor IX binding protein (IX-BP) and Bc [35], the last one, flavocetin A, is a heterodimeric multimer $(\alpha\beta)_4$ [34]. In all these structures, each CRD adopts the canonical fold observed for the CRD of CTLs. However, the crystal structures of snake venom CTLs showed that the dimerisation of their CRDs is an example of domain-swapping [36] vice versa (Plate XIV B). Mizuno and co-workers suggested that new features arise in these CTLs from the dimerisation of their CRDs [37, 38]. In the case of IX/X-BP and IX-BP in particular, it has been suggested that the concave surface which appears between the two subunits is part of their binding site for the Gla domain of factor IX and factor X [37, 38]. The crystal structures of IX/X-BP and IX-BP confirmed that the canonical calcium-binding site was absent but revealed also a new calcium-binding site in the CTL fold [37, 38]. In flavocetin A, no bound calcium was observed in the crystal structure, but the side chain amino group of a lysine residue was shown to occupy the same place as the calcium in IX/X-BP and IX-BP [34]. Finally, in Bc, Sen et al. [35] observed a canonical Ca^{2+}-binding site in the β subunit, whereas the situation in the α subunit was much the same as in flavocetin A.

21.2 SNAKE VENOM CTL AFFECTING PLATELET FUNCTIONS

Among the snake venom proteins that affect platelet functions, either promoting or inhibiting platelet aggregation, a large number of snake venom CTLs have been recognised to interact selectively with various platelet receptors.

21.2.1 INTERACTION OF SNAKE VENOM CTLs WITH PLATELET GPIb

Von Willebrand factor is a heterogeneous highly multimeric glycoprotein found in plasma, platelet α granules and sub-endothelial matrix [39]. It is well established that its binding to its platelet membrane receptor, the glycoprotein complex Ib-IX-V (GP-Ib-IX-V), is the first contact adhesion of platelets to the exposed sub-endothelium and that this step plays a crucial role in the control of haemostasis [40]. The GPIb-IX-V complex is a multimeric molecule consisting of several components: GPIbα (Mr = 145 kDa), GPIbβ (Mr = 22 kDa), GPIX (Mr = 17 kDa) and GPV (Mr = 82 kDa). Among these components, the α-subunit of GPIb is the structural and functional binding site of vWF on platelets. Recently, a number of snake venom CTLs have been isolated and characterised to interact with the GPIb α subunit [41–43] (Table 21.1). Most of them are heterodimeric CTLs of 28 kDa, although two high molecular weight multimeric CTLs of 139–149 kDa, flavocetin A and flavocetin B isolated from *Trimeresurus flavoviridis* venom, were also reported to interact with GPIb α subunit [44]. The multimers have a higher affinity than heterodimers and it has been suggested that the cyclic tetrameric structure of flavocetin A explains their higher activity on human platelets [34, 44].

Heterodimer studies of the α and β subunits of echicetin [45] and jararaca GPIb binding protein [46] indicated that the β subunit alone has binding properties, while the role of the α subunit remains obscure [47]. It should, however, be noted that the αβ heterodimer has 100-fold higher affinity for GPIb than the β subunit alone. The α subunit therefore may also play a role in the interaction of the heterodimer with its target. On the other hand, it has been shown that the heterodimeric echicetin and the multimeric CTL alboaggregin B bind to the same site on platelet membranes but that alboaggregin B activates platelets while echicetin inhibits their aggregation, even when stimulated by alboaggregin B [48]. Furthermore, the fact that alboaggregin B loses its GPIb-binding ability after the reduction of its disulphide bridges suggests that it might be able to associate with several GPIb molecules and to cross-link them, in the same way as its natural ligand vWF [45].

Although the snake venom CTLs that interact with the vWF receptor show 45–55 % sequence identity, there is evidence that their binding sites on GPIb α are not identical. First, the platelet binding properties of alboaggregin-B and echicetin are inhibited in different ways by an anti-GPIbα monoclonal antibodies [49]. Second, although they all affect the platelet response to vWF, only echicetin and flavocetin also inhibit the platelet response to thrombin [50]. Thus these various snake venom-CTLs which bind to the GPIb-IX-V receptor interact with the vWF binding site slightly differently. The precise elucidation of their structure and function thus appears an interesting challenge in understanding the functional interaction between vWF and its GPIb-IX-V receptor and possibly in developing new agonists and antagonists with therapeutic potential.

Table 21.1 Snake venom C-type lectin functions in haemostasis

		Activate platelet function		
Botrocetin	*Bothrops jararaca*	Binds vWF A1 domain	No	[53]
Bitiscetin	*Bitis arietans*	Binds vWF A3 domain	No	[62]
Alboaggregin A	*Trimeresurus albolabris*	Binds GP Ib / Mimics vWF	No	[49] [101]
Convulxin	*Crotalus durissus terrificus*	Binds GP VI / Mimics collagen	No	[68]
Rhodocetin	*Calloselasma rhodostoma*	Unknown mechanism	No	[102]
Rhodocytin	*Calloselasma rhodostoma*	Unknown mechanism	No	[103]
Aggretin	*Calloselasma rhodostoma*	Binds GPIa/IIa	No	[104]

		Activate platelet function		
Echicetin	*Echis carinatus*	Binds to GPIb/blocks vWF	No	[45]
Alboaggregin B	*Trimeresurus albolabris*	Binds to GPIb/blocks vWF	No	[48]
Jararaca GP Ib-BP	*Bothrops jararaca*	Binds to GPIb/blocks vWF	No	[46]
Flavocetin	*Trimeresurus flavoviridis*	Binds to GPIb/blocks vWF	No	[44]
CHH-A and B	*Crotalus horridus horridus*	Binds to GPIb/blocks vWF	No	[49]
Agkicetin	*Agkistrodon acutus*	Binds to GPIb/blocks vWF	No	[105]
Tokaracetin	*Trimeresurus tokarensis*	Binds to GPIb/blocks vWF	No	[106]
Mamushigin	*Agkistrodon halys blomhoffi*	Binds to GPIb/blocks vWF	No	[107]

		Activate coagulation factors		
RVV-X	Russel's viper	Cleaves factor X	M	[92]
RVV-X	*Bothrops jararaca*	Cleaves factor X	M	[94]
Factor IX activator	Russel's viper	Cleaves factor IX	M	[99]
Prothrombin activator	*Echis carinatus*	Cleaves factor II	M	[100]

		Inhibit coagulation factors		
Bothrojaracin	*Bothrops jararaca*	Binds to FII-FIIa / Inhibits FIIa	No	[82]
Bothroalternin	*Bothrops alternatus*	Binds to FII / Inhibits FII a	No	[84]
Salmorin	*Calloselasma rhodostoma*	Binds to FII & FIIa	No	[85]
Triabin	*Triatoma pallidepennis*	Binds to FIIa	No	[108]
Factor IX/X inhibitor	*Trimeresurus flavoviridis*	Binds to FIX-FX / Inhibits FIX-FX	No	[87]
Factor IX inhibitor	*Trimeresurus flavoviridis*	Bind to FIX / Inhibit FIX	No	[109]

21.2.2 SNAKE VENOM CTLs INTERACTING WITH vWF

As mentioned above, vWF mediates platelet adhesion at sites of vascular injury through its binding to its platelet receptor – GPIBα. The adhesive properties of vWF are tightly regulated so that vWF circulating in plasma is unable to interact with GPIbα until the vessel wall has been damaged [39]. Unfortunately, we still lack knowledge of the molecular events underpinning the ability of vWF to induce platelet agglutination. Read and co-workers [51] demonstrated that the snake venom heterodimeric CTL Bc interacts with vWF and that the Bc-vWF complex then induces agglutination [41, 53, 54]. On the other hand, it has been suggested that Bc could mimic a protein from the vessel wall since a monoclonal antibody raised against Bc cross-reacts with fibronectin and an as yet unidentified 130 kDa protein present in plasma and in endothelial cells [52]. Recent progress made in elucidating vWF structure and function has contributed to the identification of the Bc binding site of vWF on its A1 domain [46, 55, 56]. The crystal structure of Bc has revealed a negatively charged surface on this snake venom-CTL that might play a major role in its association with the A1 domain of vWF [35]. Recently, a Bc-like snake venom CTL has been isolated from the venom of *Trimeresurus albolabris*, and has been used to measure the number of vWF binding sites on the GPIb molecule [57].

Platelet agglutination by Bc depends on the presence of high molecular weight vWF in plasma and this property has been use to assay vWF [51, 58]. The vWF assay with Bc is different from that with ristocetin, a low molecular weight molecule with antibiotic activity which binds to vWF and induces platelet activation, like Bc. However, in addition to low molecular weight vWF, ristocetin can also be used to quantify high molecular weight polymers of vWF, which promote platelet aggregation more potently than the low molecular forms. Bc can only be used to quantify high molecular weight vWF polymers. Thus the two agonists can be used to differentiate molecular weight variants of vWF. This is important in some circumstances. For example, Bc partially aggregates platelets from patients suffering from Bernard Soulier disease where GPIb is absent, while ristocetin does not induce platelet aggregation in these patients since its action is strictly dependent on GPIb [59]. The same difference is noticed in type Ia von Willebrand disease (vWD) characterised in particular by the absence of high molecular weight multimers of vWF [59]. Similarly, the combination of ristocetin and of Bc assays is valuable in the measure of functional vWF [60] in plasma, as shown for example in the detection of a missense mutation in type B vWF [61].

Another snake venom CTL, bitiscetin, binds to vWF and induces platelet agglutination like Bc [62]. However, it is a basic protein in contrast to Bc, which is rather acidic. Further analysis of the binding of bitiscetin to various proteolytic fragments of vWF, and of the effects of anti-vWF monoclonal antibodies on bitiscetin-induced platelet activation in the presence of vWF, showed that bitiscetin, unlike Bc, does not bind to the A1 domain of vWF but rather to its homologous A3 domain [63].

21.2.3 SNAKE VENOM CTL INTERACTING WITH THE COLLAGEN RECEPTOR GPVI

Convulxin (CVX), a tri-heterodimeric $(\alpha\beta)_3$ snake venom CTL isolated from *Crotalus durissus terrificus*, is a very potent platelet aggregating protein [32, 64–66]. Pharmacological investigations first demonstrated that CVX activates platelets by a mechanism very similar to that of collagen, and that a cross-desensitisation exists between these two platelet agonists [66]. Ligand blotting, performed with ^{125}I-CVX, then established that CVX binds to a protein of 62 kDa, the molecular weight of GPVI [67, 68, 69]. Recently, S. Watson *et al.* [69] showed that CVX is a more powerful GPVI agonist than collagen [70], and CVX proved to be an excellent tool in studying this collagen receptor. ^{125}I-CVX and fluorescein-labelled CVX (FITC-CVX) were used to quantify GPVI on platelet membranes and to study its interactions in the presence of collagen-related peptides (CRPs) [71–75]. More importantly, CVX proved to be an exceptionally valuable tool not only in identifying GP-VI but also in purifying and characterising it [76–78]. The binding of CVX or collagen to GPVI induces its cross-linking, and is associated with the tyrosine phosphorylation of a number of proteins such as FcRγ chain, Syk and phospholipase γ2 and with the subsequent increase in the level of second messengers and intracellular Ca^{2+} that leads to platelet shape change and aggregation.

21.3 SNAKE VENOM CTLs AFFECTING BLOOD CLOTTING

21.3.1 SNAKE VENOM CTLs WITH ANTI-COAGULANT ACTIVITY

In addition to proteinases that degrade fibrinogen and/or other coagulation factors and disintegrins that inhibit platelet aggregation, several snake venom CTLs are also able to prevent blood coagulation by binding to various factors, in particular thrombin, and by inhibiting their physiological function. Thrombin is a multifunctional serine proteinase which plays a major role in the maintenance and regulation of haemostasis. It stimulates platelet aggregation and plasmatic coagulation and thus possesses a pro-coagulant action. It also has anti-coagulant properties, since after its interaction with thrombomodulin it activates protein C which is a major anti-coagulant factor [79]. The functional residues involved in the pro-coagulant and anti-coagulant functions of thrombin are located on the same hemisphere of the molecule [80]. This hemisphere includes both the active site cleft and the highly basic exosite I [81].

Bothrojaracin (Bj) is a 27 kDa heterodimeric CTL which has been isolated from *Bothrops jararaca* venom [82]. The sequences of the α and β chains of Bj have been determined by molecular cloning, thus confirming the CTL structure of Bj (Plate XIV). Bj is a potent and selective thrombin inhibitor that forms a noncovalent 1:1 complex with thrombin. It interacts with the two thrombin exosites, and thus competes with the macromolecular ligands of thrombin, but it does not affect the catalytic site [83]. Nevertheless, the sequence of Bj does not show the linear arrangement of acidic residues that has been recognised in these ligands. Two possibilities may therefore explain the binding of Bj to thrombin: either the binding mechanism of Bj to exosite I is different from that of other inhibitors or the expected groups of acidic residues result from a conformational arrangement between the α and β chains of this heterodimeric snake venom CTL.

Other snake venom CTLs have been recently reported to possess properties more or less similar to that of Bj. Bothroalternin, from the venom of *Bothrops alternatus*, is a thrombin inhibitor somewhat less efficient than Bj since it does not completely inhibit the thrombin effects on platelet aggregation [84]. Salmorin purified from *Agkistrodon halys brevicaudus* inhibits not only thrombin-induced fibrinogen clotting like Bj but, unlike Bj, also inhibits factor Xa-induced prothrombin activation [85]. Like Bj, salmorin does not bind to the thrombin catalytic site but interacts with thrombin and prothrombin exosites. The inhibition by salmorin of prothrombin activation by factor X is Ca^{2+}-independent and is associated with its interaction with prothrombin and not factor Xa. It is worth remembering that Bj forms a similar complex with prothrombin but that Bj does not inhibit the prothrombin activation by factor Xa [86]. The identification of the structural determinants responsible for the functional differences between Bj, bothroalternin and salmorin will enhance understanding of the structure/function of prothrombin and thrombin, and in particular their interaction with their macromolecular ligands. In particular, these snake venom CTLs with thrombin and/or prothrombin inhibitory activities appear to be valuable tools for developing anti-thrombotic drugs.

Snake venom CTL which inhibits factor IX and factor X has been isolated from *Trimeresurus flavoviridis* [87]. The structure of this protein called IX/X Bp has been elucidated, demonstrating that it belongs to the group of heterodimeric CTLs [87]. IX/X bp does not bind to prothrombin but inhibits prothrombin activation, forming in a Ca^{2+}-dependent manner a complex with factors IX and X [88]. The IX/X binding protein binds to the γ-carboxy glutamic acid residues of factor IX and X [89] induces a conformational change in the anticoagulant protein [90]. A CTL factor X inhibitor was also isolated from *Deinagkistrodon acutus* venom. It binds to FXa in the presence of Ca^{2+} and prevents its involvement in the prothrombin activation complex by a mechanism similar to the one described for IX/X Bp [91].

21.3.2 SNAKE VENOM CTLs WITH PRO-COAGULANT ACTIVITY

Many serine proteinases and metalloproteinases from snake venoms possess pro-coagulant activity. Some of these metalloproteinases also contain a CRD characteristic of CTLs. In contrast to those of heterodimeric CTLs, the CRDs of these proteins exhibit the binding properties of lectins for sugars.

A potent human blood coagulation factor X activator (RVV-X) was the first characterised in Russel's viper venom [92, 93]. Similar factor X activators have been isolated from *Bothrops jararaca* venom [94] and from the venoms of several other snake species [95–97]. RVV-X is a disulphide-linked two-chain protein. The heavy chain is composed of a metalloproteinase domain, a disintegrin-like domain and cysteine-rich domain. The light chain is homologous to the CRD of CTLs (Figure 21.2). Interestingly, factor X activators have also been described in the venom of *Ophiophagus hannah* and *Bungarus fasciatus*, but, unlike RVV-X, these activators from *Elapidae* snakes are serine proteinases [95, 96]. RVV-X activates human factor X like the enzymes of the blood coagulation pathway, by cleaving a specific peptide bond in the amino-terminal region of factor X heavy chain [98]. This results in the formation of factor IXa without any change in its molecular weight [99].

Russel's viper venom also contains a factor IX activator whose structure is similar to that of RVV-X. Its activation mechanism implies the cleavage of a unique peptide bond in factor IX. This contrasts with the activation mechanism of factor IXa, the physiological activator factor IX, which involves the release of a peptide and the reduction in the molecular weight of activated factor IXa.

A very active and Ca^{2+}-dependent prothrombin activator, carinactivase, was recently described in *Echis carinatus* venom [100]. This activator is composed of two noncovalently associated subunits, a metalloproteinase of 62 kDa and a 25 kDa regulatory subunit. The lower molecular weight subunit, which consists of two polypeptides of 17 and 14 kDa linked by disulphide bonds, is similar to IX/X-BP which binds to the Gla domain of coagulation factors IX and X. The heavy chain is a metalloproteinase homologous to ecarin, another prothrombin activator previously detected in *Echis carinatus* venom, which is less potent than carinactivase and Ca^{2+}-independent. The authors hypothesise first that the CTL subunit of the enzyme complex carinactivase recognises the Gla domain of prothrombin in a Ca^{2+}-dependent manner, and second that the catalytic, ecarin-like, subunit activates prothrombin into thrombin in a more selective fashion [100].

21.4 CONCLUSION

Numerous CTL proteins have been purified from snake venoms and characterised. Recent investigations on their structure, function and mechanism of action indicated that these proteins are very similar in structure. In particular, their polypeptide sequences are remarkably similar, while they interact selectively with very different targets. Interestingly, the targets of these snake venom CTLs play crucial roles in blood coagulation mechanisms and the venom molecules impair their physiological functions, preventing their ligand-specific interactions. The snake venom CTLs are thus useful structural templates for the identification and characterisation of the interaction sites of these key molecules of blood coagulation. They also appear to be excellent tools for the development of new pharmacological agents, some of which may prove diagnostically and/or therapeutically valuable.

REFERENCES

[1] Markland, F.S. (1998) Snake venoms and the hemostatic system. *Toxicon* **36**, 1749–800.
[2] Samama, M. (1990) *Physiologie et exploration de l'hémostase*, Doin Editeurs, Paris.
[3] Iyaniwura, T.T. (1991) Snake venom constituents: biochemistry and toxicology (Part 2). *Veterinary & Human Toxicology* **33**, 475–80.
[4] Iyaniwura, T.T. (1991) Snake venom constituents: biochemistry and toxicology (Part 1). *Veterinary & Human Toxicology* **33**, 468–74.
[5] Marsh, N.A. (1994) Snake venoms affecting the haemostatic mechanism – a consideration of their mechanisms, practical applications and biological significance. *Blood Coagulation & Fibrinolysis* **5**, 399–410.
[6] Bjarnason, J.B., and Fox, J.W. (1994) Hemorrhagic metalloproteinases from snake venoms. *Pharmacology & Therapeutics* **62**, 325–72.
[7] Kamiguti, A.S., Zuzel, M., and Theakston, R.D. (1998) Snake venom metalloproteinases and disintegrins: interactions with cells. *Brazilian Journal of Medical & Biological Research* **31**, 853–62.
[8] Marsh, N.A., and Fyffe, T.L. (1996) Practical applications of snake venom toxins in haemostasis. *Bollettino Societa Italiana Biologia Sperimentale* **72**, 263–78.
[9] Serrano, S.M., Mentele, R., Sampaio, C.A., and Fink, E. (1995) Purification, characterization, and amino acid sequence of a serine proteinase, PA-BJ, with platelet-aggregating activity from the venom of *Bothrops jararaca*. *Biochemistry* **34**, 7186–93.
[10] Marrakchi, N., Zingali, R.B., Karoui, H., Bon, C., and el Ayeb, M. (1995) Cerastocytin, a new thrombin-like platelet activator from the venom of the Tunisian viper *Cerastes cerastes*. *Biochim. Biophys. Acta* **1244**, 147–56.
[11] Kirby, E.P., Niewiarowski, S., Stocker, K., Kettner, C., Shaw, E., and Brudzynski, T.M. (1979) Thrombocytin, a serine protease from *Bothrops atrox* venom. 1. Purification and characterization of the enzyme. *Biochemistry* **18**, 3564–70.
[12] Stocker, K., Fischer, H., and Meier, J. (1982) Thrombin-like snake venom proteinases. *Toxicon* **20**, 265–73.
[13] Pirkle, H., and Theodor, I. (1990) Thrombin-like venom enzymes: structure and function. *Advances in Experimental Medicine & Biology* **281**, 165–75.
[14] Pirkle, H. (1998) Thrombin-like enzymes from snake venoms: an updated inventory. Scientific and standardization committee's regristry of exogenous hemostatic factors. *Thromb. Haemost.* **79**, 675–83.
[15] Kisiel, W. (1979) Molecular properties of the factor V-activating enzyme from Russell's viper venom. *J. Biol. Chem.* **254**, 12230–4.
[16] Stocker, K., Fischer, H., Meier, J., Brogli, M., and Svendsen, L. (1987) Characterization of the protein C activator protac from the venom of the southern copperhead (*Agkistrodon contortrix*) snake. *Toxicon* **25**, 239–52.
[17] Davie, E.W., Fujikawa, K., and Kisiel, W. (1991) The coagulation cascade: initiation, maintenance, and regulation. *Biochemistry* **30**, 10363–70.
[18] Rosing, J., Tans, G., Govers-Riemslag, J.W., Zwaal, R.F., and Hemker, H.C. (1980) The role of phospholipidds and factor Va in the prothrombinase complex. *J. Biol. Chem.* **255**, 274–83.

[19] Hirabayashi, J. (1994) Two distinct families of animal lectins: speculations on their raisons d'être. *Lectins: biology, biochemistry, clinical biochemistry Vol.* **10**, 205–19.

[20] Taylor, M.E., Bezouska, K., and Drickamer, K. (1992) Contribution to ligand binding by multiple carbohydrate-recognition domains in the macrophage mannose receptor. *J. Biol. Chem.* **267**, 1719–26.

[21] Drickamer, K. (1988) Two distinct classes of carbohydrate-recognition domains in animal lectins. *J. Biol. Chem.* **263**, 9557–60.

[22] Drickamer, K., and Taylor, M.E. (1993) Biology of animal lectins. *Annual Review of Cell Biology* **9**, 237–64.

[23] Quesenberry, M.S., and Drickamer, K. (1991) Determination of the minimum carbohydrate-recognition domain in two C-type animal lectins. *Glycobiology* **1**, 615–21.

[24] Spiess, M. (1990) The asialoglycoprotein receptor: a model for endocytic transport receptors. *Biochemistry* **29**, 10009–18.

[25] Bertrand, J.A., Pignol, D., Bernard, J.P., Verdier, J.M., Dagorn, J.C., and Fontecilla-Camps, J.C. (1996) Crystal structure of human lithostathine, the pancreatic inhibitor of stone formation. *EMBO J.* **15**, 2678–84.

[26] Rini, J.M. (1995) Lectin structure. *Annual Review of Biophysics & Biomolecular Structure* **24**, 551–77.

[27] Weis, W.I. (1994) Lectins on a roll: the structure of E-selectin. *Structure* **2**, 147–50.

[28] Weis, W.I., Crichlow, G.V., Murthy, H.M., Hendrickson, W.A., and Drickamer, K. (1991) Physical characterization and crystallization of the carbohydrate-recognition domain of a mannose-binding protein from rat. *J. Biol. Chem.* **266**, 20678–86.

[29] Hirabayashi, J., Kusunoki, T., and Kasai, K. (1991) Complete primary structure of a galactose-specific lectin from the venom of the rattlesnake Crotalus atrox. Homologies with Ca2(+)-dependent-type lectins. *J. Biol. Chem.* **266**, 2320–6.

[30] Takeya, H., Nishida, S., Miyata, T., Kawada, S., Saisaka, Y., Morita, T., and Iwanaga, S. (1992) Coagulation factor X activating enzyme from Russell's viper venom (RVVXà. A novel metalloproteinase with disintegrin (platelet aggregation inhibitor)-like and C-type lectin-like domains. *J. Biol. Chem.* **267**, 14109–17.

[31] Usami, Y., Fujimura, Y., Suzuki, M., Ozeki, Y., Nishio, K., Fukui, H., and Titani, K. (1993) Primary structure of two-chain botrocetin, a von Willebrand factor modulator purified from the venom of *Bothrops jararaca*. *Proc. Nat. Acad. Sci. USA* **90**, 928–32.

[32] Marlas, G. (1985) Isolation and characterization of the alpha and beta subunits of the platelet-activating glycoprotein from the venom of *Crotalus durissus cascavella*. *Biochimie* **67**, 1231–9.

[33] Shin, Y., Okuyama, J., Hasegawa, J., and Morita, T. (2000) Molecular cloning of glycoprotein Ib-binding protein, flavocetin-A, which inhibits platelet aggregation. *Thrombosis Research* **99**, 239–47.

[34] Fukuda, K., Mizuno, H., Atoda, H., and Morita, T. (2000) Crystal structure of flavocetin-A, a platelet glycoprotein Ib-binding protein, reveals a novel cyclic tetramer of C-type lectin-like heterodimers. *Biochemistry* **39**, 1915–23.

[35] Sen, U., Vasudevan, S., Subbarao, G., McClintock, R.A., Celikel, R., Ruggeri, Z.M., and Varughese, K.I. (2001) Crystal structure of the von Willebrand factor modulator botrocetin. *Biochemistry* **40**, 345–52.

[36] Bennett, M.J., Schlunegger, M.P., and Eisenberg, D. (1995) 3D domain swapping: a mechanism for oligomer assembly. *Protein Sci.* **4**, 2455–68.

[37] Mizuno, H., Fujimoto, Z., Koizumi, M., Kano, H., Atoda, H., and Morita, T. (1999) Crystal structure of coagulation factor IX-binding protein from habu snake venom at 2.6 Å: implication of central loop swapping based on deletion in the linker region. *J. Mol. Biol.* **289**, 103–12.

[38] Mizuno, H., Fujimoto, Z., Koizumi, M., Kano, H., Atoda, H., and Morita, T. (1997) Structure of coagulation factors IX/X-binding protein, a heterodimer of C-type lectin domains. *Nature Struct. Biol.* **4**, 438–41.

[39] Ruggeri, Z.M. (1999) Structure and function of von Willebrand factor. *Thromb. Haemost.* **82**, 576–84.

[40] Sakariassen, K.S., Bolhuis, P.A., and Sixma, J.J. (1979) Human blood platelet adhesion to artery subendothelium is mediated by factor VIII-Von Willebrand factor bound to the subendothelium. *Nature* **279**, 636–8.

[41] Fujimura, Y., Kawasaki, T., and Titani, K. (1996) Snake venom proteins modulating the interaction between von Willebrand factor and platelet glycoprotin Ib. *Thromb. Haemost.* **76**, 633–9.

[42] Clemetson, K.J., Polgar, J.M., and Clemetson, J.M. (1998) Snake venom C-type lectins as tools in platelet research. *Platelets* **9**, 165–9.

[43] Andrews, R.K., and Berndt, M.C. (2000) Snake venom modulators of platelet adhesion receptors and their ligands. *Toxicon* **38**, 775–91.

[44] Taniuchi, Y., Kawasaki, T., Fujimura, Y., Suzuki, M., Titani, K., Sakai, Y., Kaku, S., Hisamichi, N., Satoh, N., Takenaka, T., *et al.* (1995) Flavocetin-A and -B, two high molecular mass glycoprotein Ib binding proteins with high affinity purified from *Trimeresurus flavoviridis* venom inhibit platelet aggregation at high shear stress. *Biochim. Biophys. Acta* **1244**, 331–8.

[45] Peng, M., Lu, W., Beviglia, L., Niewiarowski, S., and Kirby, E.P. (1993) Echicetin: a snake venom protein that inhibits binding of von Willebrand factor and alboaggregins to platelet glycoprotein Ib. *Blood* **81**, 2321–8.

[46] Fujimura, Y., Ikeda, Y., Miura, S., Yoshida, E., Shima, H., Nishida, S., Suzuki, M., Titani, K., Taniuchi, Y., and Kawasaki, T. (1995) Isolation and characterization of jararaca GPIb-BP, a snake venom antagonist specific to platelet glycoprotein. Ib. *Thromb. Haemost.* **74**, 743–50.

[47] Peng, M., Holt, J.C., and Niewiarowski, S. (1994) Isolation, characterization and amino acid sequence of echicetin beta subunit, a specific inhibitor of von Willebrand factor and thrombin interaction with glycoprotein Ib. *Biochem. Biophys. Res. Commun.* **205**, 68–72.

[48] Peng, M., Lu, W., and Kirby, E.P. (1992) Characterization of three alboaggregins purified from *Trimeresurus albolabris* venom. *Thromb. Haemost.* **67**, 702–7.

[49] Andrews, R.K., Kroll, M.H., Ward, C.M., Rose, J.W., Scarborough, R.M., Smith, A.I., Lopez, J.A., and Berndt, M.C. (1996) Binding of a novel 50-kilodalton alboaggregin from *Trimeresurus albolabris* and related viper venom proteins to the platelet membrane glycoprotein Ib-IX-V complex. Effect on platelet aggregation and glycoprotein Ib-mediated platelet activation. *Biochemistry* **35**, 12629–39.

[50] Peng, M., Emig, F.A., Mao, A., Lu, W., Kirby, E.P., Niewiarowski, S., and Kowalska, M.A. (1995) Interaction of echicetin with a high affinity thrombin binding site on platelet glycoprotein GPIb. *Thromb. Haemost.* **74**, 954–7.

[51] Read, M.S., Smith, S.V., Lamb, M.A., and Brinkhous, K.M. (1989) Role of botrocetin in platelet agglutination: formation of an activated complex of botrocetin and von Willebrand factor. *Blood* **74**, 1031–5.

[52] Katayama, M., Nagata, S., Hirai, S., Miura, S., Fujimura, Y., Matusi, T., Kato, I., and Titani, K. (1995) Fibronectin and 130-kDa molecule complex mimics snake venom botrocetin-like structure potentially modulating association between von Willebrand factor and vascular vessel wall. *J. Biochem. (Tokyo)* **117**, 331–8.

[53] Fujimura, Y., Titani, K., Usami, Y., Suzuki, M., Oyama, R., Matsui, T., Fukui, H., Sugimoto, M., and Ruggeri, Z.M. (1991) Isolation and chemical characterization of two structurally and functionally distinct forms of botrocetin, the platelet coagglutinin isolated from the venom of *Bothrops jararaca*. *Biochemistry* **30**, 1957–64.

[54] Andrews, R.K., Booth, W.J., Gorman, J.J., Castaldi, P.A., and Berndt, M.C. (1989) Purification of botrocetin from *Bothrops jararaca* venom. Analysis of the botrocetin-mediated interaction between von Willebrand factor and the human platelet membrane glycoprotein Ib-IX complex. *Biochemistry* **28**, 8317–26.

[55] Sugimoto, M., Mohri, H., McClintock, R.A., and Ruggeri, Z.M. (1991) Identification of discontinuous von Willebrand factor sequences involved in complex formation with botrocetin. A model for the regulation of von Willebrand factor binding to platelet glycoprotein Ib. *J. Biol. Chem.* **266**, 18172–8.

[56] Matsushita, T., and Sadler, J.E. (1995) Identification of amino acid residues essential for von Willebrand factor binding to platelet glycoprotein Ib. Charged-to-alanine scanning mutagenesis of the A1 domain of human von Willebrand factor. *J. Biol. Chem.* **270**, 13406–14.

[57] Yoshida, E., Fujimura, Y., Ikeda, Y., Takeda, I., Yamamoto, Y., Nishikawa, K., Miyataka, K., Oonuki, M., Kawasaki, T., Katayama, M., *et al.* (1995) Impaired high-shear-stress-induced platelet aggregation in patients with chronic renal failure undergoing haemodialysis. *Br. J. Haematol.* **89**, 861–7.

[58] Brinkhous, K.M., Read, M.S., Fricke, W.A., and Wagner, R.H. (1983) Botrocetin (venom coagglutinin): reaction with a broad spectrum of multimeric forms of factor VIII macromolecular complex. *Proc. Natl. Acad. Sci. USA* **80**, 1463–6.

[59] Bloom, A.L. (1991) Progress in the clinical management of haemophilia. *Thromb. Haemost.* **66**, 166–77.

[60] Williams, S.B., McKeown, L.P., Krutzsch, H., Hansmann, K., and Gralnick, H.R. (1994) Purification and characterization of human platelet von Willebrand factor. *Br. J. Haematol.* **88**, 582–91.

[61] Rabinowitz, I., Tuley, E.A., Mancuso, D.J., Randi, A.M., Firkin, B.G., Howard, M.A., and Sadler, J.E. (1992) von Willebrand disease type B: a missense mutation selectively abolishes ristocetin-induced von Willebrand factor binding to platelet glycoprotein Ib. *Proc. Natl. Acad. Sci. USA* **89**, 9846–9.

[62] Hamako, J., Matsui, T., Suzuki, M., Ito, M., Makita, K., Fujimura, Y., Ozeki, Y., and Titani, K. (1996) Purification and characterization of bitiscetin, a novel von Willebrand factor modulator protein from Bitis arietans snake venom. *Biochem. Biophys. Res. Commun.* **226**, 273–9.

[63] Obert, B., Houllier, A., Meyer, D., and Girma, J.P. (1999) Conformational changes in the A3 domain of von Willebrand factor modulate the interaction of the A1 domain with platelet glycoprotein Ib. *Blood* **93**, 1959–68.

[64] Marlas, G., Joseph, D., and Huet, C. (1983) Subunit structure of a potent platelet-activating glycoprotein isolated from the venom of *Crotalus durissus cascavella*. *Biochimie* **65**, 619–28.

[65] Francischetti, I.M., Saliou, B., Leduc, M., Carlini, C.R., Hatmi, M., Randon, J., Faili, A., and Bon, C. (1997) Convulxin, a potent platelet-aggregating protein from *Crotalus durissus terrificus* venom, specifically binds to platelets. *Toxicon* **35**, 1217–28.

[66] Vargaftig, B.B., Joseph, D., Wal, F., Marlas, G., Chignard, M., and Chevance, L.G. (1983) Convulxin-induced activation of intact and of thrombin-degranulated rabbit platelets: specific crossed desensitisation with collagen. *Eur. J. Pharmacol.* **92**, 57–68.

[67] Jandrot-Perrus, M., Lagrue, A.H., Okuma, M., and Bon, C. (1997) Adhesion and activation of human platelets induced by convulxin involve glycoprotein VI and integrin alpha2beta1. *J. Biol. Chem.* **272**, 27035–41.

[68] Polgar, J., Clemetson, J.M., Kehrel, B.E., Wiedemann, M., Magnenat, E.M., Wells, T.N.C., and Clemetson, K.J. (1997) Platelet activation and signal transduction by convulxin, a C-type lectin from *Crotalus durissus terrificus* (tropical rattlesnake) venom via the p62/GPVI collagen receptor. *J. Biol. Chem.* **272**, 13576–83.

[69] Watson, S.P. (1999) Collagen receptor signaling in platelets and megakaryocytes. *Thromb. Haemost.* **82**, 365–76.

[70] Asazuma, A., Wilde, J.I., Berlanga, O., Leduc, M., Leo, A., Scheighoffer, E., Tybulewicz, V., Bon, C., Liu, S.K., McGlade, C.J., Schraven, B., and Watson, S. (2000) Interaction of linker for activation of T cells with multiple adapter proteins in platelets activated by the glycoprotein VI-selective ligand, convulxin. *J. Biol. Chem.* **275**, 33427–34.

[71] Niedergang, F., Alcover, A., Knight, C.G., Farndale, R.W., Barnes, M.J., Francischetti, I.M., Bon, C., and Leduc, M. (2000) Convulxin binding to platelet receptor GPVI: competition with collagen related peptides. *Biochem. Biophys. Res. Commun.* **273**, 246–50.

[72] Kehrel, B., Wierwille, S., Clemetson, K.J., Anders, O., Steiner, M., Knight, C.G., Farndale, R.W., Okuma, M., and Barnes, M.J. (1998) Glycoprotein VI is a major collagen receptor for platelet activation: it recognizes the platelet-activating quaternary structure of collagen, whereas CD36, glycoprotein IIb/IIIa, and von Willebrand factor do not. *Blood* **91**, 491–9.

[73] Barnes, M.J., Knight, C.G., and Farndale, R.W. (1996) The use of collagen-based model peptides to investigate platelet-reactive sequences in collagen. *Biopolymers* **40**, 383–97.

[74] Morton, L.F., Hargreaves, P.G., Farndale, R.W., Young, R.D., and Barnes, M.J. (1995) Integrin alpha 2 beta 1-independent activation of platelets by simple collagen-like peptides: collagen tertiary (triple-helical) and quaternary (polymeric) structures are sufficient alone for alpha 2 beta 1-independent platelet reactivity. *Biochem. J.* **306**, 337–44.

[75] Knight, C.G., Morton, L.F., Onley, D.J., Peachey, A.R., Ichinohe, T., Okuma, M., Farndale, R.W., and Barnes, M.J. (1999) Collagen-platelet interaction: Gly-Pro-Hyp is uniquely specific for platelet Gp VI and mediates platelet activation by collagen. *Cardiovasc. Res.* **41**, 450–7.

[76] Clemetson, J.M., Polgar, J., Magnenat, E., Wells, T.N., and Clemetson, K.J. (1999) The platelet collagen receptor glycoprotein VI is a member of the immunoglobulin superfamily closely related to FcalphaR and the natural killer receptors. *J. Biol. Chem.* **274**, 29019–24.

[77] Ezumi, Y., Uchiyama, T., and Takayama, H. (1998) Thrombopoietin potentiates the protein-kinase-C-mediated activation of mitogen-activated protein kinase/ERK kinases and extracellular signal-regulated kinases in human platelets. *Eur. J. Biochem.* **258**, 976–85.

[78] Miura, Y., Ohnuma, M., Jung, S.M., and Moroi, M. (2000) Cloning and expression of the platelet-specific collagen receptor glycoprotein VI. *Thromb. Res.* **98**, 301–9.

[79] Fenton J.W., O.F.A., Moon D.G., Maraganore J.M. (1991) Thrombin structure and function: why thrombin is the primary target for antithrombotics. *Blood coagulation and fibrinolysis* **2**, 69–75.

[80] Guillin M.C., B.A., Bouton, M.C., Jandrot-Perrus M. (1995) Thrombin specificity. *Thromb. Haemost.* **74**, 129–33.

[81] Tsiang, M., Jain, A.K., Dunn, K.E., Rojas, M.E., Leung, L.L., and Gibbs, C.S. (1995) Functional mapping of the surface residues of human thrombin. *J. Biol. Chem.* **270**, 16854–63.

[82] Zingali, R.B., Jandrot-Perrus, M., Guillin, M.C., and Bon, C. (1993) Bothrojaracin, a new thrombin inhibitor isolated from *Bothrops jararaca* venom: characterization and mechanism of thrombin inhibition. *Biochemistry* **32**, 10794–802.

[83] Arocas, V., Castro, H.C., Zingali, R.B., Guillin, M.C., Jandrot-Perrus, M., Bon, C., and Wisner, A. (1997) Molecular cloning and expression of bothrojaracin, a potent thrombin inhibitor from snake venom. *Eur. J. Biochem.* **248**, 550–7.

[84] Castro, H.C., Dutra, D.L., Oliveira-Carvalho, A.L., and Zingali, R.B. (1998) Bothroalternin, a thrombin inhibitor from the venom of *Bothrops alternatus*. *Toxicon* **36**, 1903–12.

[85] Koh, Y.S., Chung, K.H., and Kim, D.S. (2000) Purification and cDNA cloning of salmorin that inhibits fibrinogen clotting. *Thromb. Res.* **99**, 389–98.

[86] Arocas, V., Lemaire, C., Bouton, M.C., Bezeaud, A., Bon, C., Guillin, M.C., and Jandrot-Perrus, M. (1998) Inhibition of thrombin-catalyzed factor V activation by bothrojaracin. *Thromb. Haemost.* **79**, 1157–61.

[87] Atoda, H., Hyuga, M., and Morita, T. (1991) The primary structure of coagulation factor IX/factor X-binding protein isolated from the venom of *Trimeresurus flavoviridis*. Homology with asialoglycoprotein receptors, proteoglycan core protein, tetranectin, and lymphocyte Fc epsilon receptor for immunoglobulin E. *J. Biol. Chem.* **266**, 14903–11.

[88] Atoda, H., and Morita, T. (1989) A novel blood coagulation factor IX/factor X-binding protein with anticoagulant activity from the venom of *Trimeresurus flavoviridis* (Habu snake): isolation and characterization. *J. Biochem. (Tokyo)* **106**, 808–13.

[89] Atoda, H., Yoshida, N., Ishikawa, M., and Morita, T. (1994) Binding properties of the coagulation factor IX/factor X-binding protein isolated from the venom of *Trimeresurus flavoviridi*. *Eur. J. Biochem.* **224**, 703–8.

[90] Sekiya, F., Yoshida, M., Yamashita, T., and Morita, T. (1996) Magnesium(II) is a crucial constituent of the blood coagulation cascade. Potentiation of coagulant activities of factor X by Mg2+ ions. *J. Biol. Chem.* **271**, 8541–4.

[91] Ouyang, C., and Teng, C.M. (1972) Purification and properties of the anticoagulant principle of *Agkistrodon acutus* venom. *Biochim. Biophys. Acta* **278**, 155–62.

[92] Kisiel, W., Hermodson, M.A., and Davie, E.W. (1976) Factor X activating enzyme from Russell's viper venom: isolation and characterization. *Biochemistry* **15**, 4901–6.

[93] Furie, B.C., and Furie, B. (1976) Coagulant protein of Russell's viper venom. *Methods Enzymol.* **45**, 191–205.

[94] Hofmann, H., and Bon, C. (1987) Blood coagulation induced by the venom of *Bothrops atrox*. 2. Identification, purification, and properties of two factor X activators. *Biochemistry* **26**, 780–7.

[95] Lee, W.H., Zhang, Y., Wang, W.Y., Xiong, Y.L., and Gao, R. (1995) Isolation and properties of a blood coagulation factor X activator from the venom of king cobra (*Ophiophagus hannah*). *Toxicon* **33**, 1263–76.

[96] Zhang, Y., Xiong, Y.L., and Bon, C. (1995) An activator of blood coagulation factor X from the venom of *Bungarus fasciatus*. *Toxicon* **33**, 1277–88.

[97] Stocker, K., Hauer, H., Muller, C., and Triplett, D.A. (1994) Isolation and characterization of Textarin, a prothrombin activator from eastern brown snake (*Pseudonaja textilis*) venom. *Toxicon* **32**, 1227–36.

[98] Di Scipio, R.G., Hermodson, M.A., and Davie, E.W. (1977) Activation of human factor X (Stuart factor) by a protease from Russell's viper venom. *Biochemistry* **16**, 5253–60.

[99] Lindquist, P.A., Fujikawa, K., and Davie, E.W. (1978) Activation of bovine factor IX (Christmas factor) by factor XIa (activated plama thromboplastin antecedent) and a protease from Russell's viper venom. *J. Biol. Chem.* **253**, 1902–9.

[100] Yamada, D., Sekiya, F., and Morita, T. (1996) Isolation and characterization of carinactivase, a novel prothrombin activator in *Echis carinatus* venom with a unique catalytic mechanism. *J. Biol. Chem.* **271**, 5200–7.

[101] Kowalska, M.A., Tan, L., Holt, J.C., Peng, M., Karczewski, J., Calvete, J.J., and Niewiarowski, S. (1998) Alboaggregins A and B. Structure and interaction with human platelets. *Thromb. Haemost.* **79**, 609–13.

[102] Wang, R., Kini, R.M., and Chung, M.C. (1999) Rhodocetin, a novel platelet aggregation inhibitor from the venom of *Calloselasma rhodostoma* (Malayan pit viper): synergistic and noncovalent interaction between its subunits. *Biochemistry* **38**, 7584–93.

[103] Shin, Y., and Morita, T. (1998) Rhodocytin, a functional novel platelet agonist belonging to the heterodimeric C-type lectin family, induces platelet aggregation independently of glycoprotein Ib. *Biochem. Biophys. Res. Commun.* **245**, 741–5.

[104] Huang, T.F., Liu, C.Z., and Yang, S.H. (1995) Aggretin, a novel platelet-aggregation inducer from snake (*Calloselasma rhodostoma*) venom, activates phospholipase C by acting as a glycoprotein Ia/IIa agonist. *Biochem. J.* **309**, 1021–7.

[105] Chen, Y.L., and Tsai, I.H. (1995) Functional and sequence characterization of agkicetin, a new glycoprotein Ib antagonist isolated from *Agkistrodon acutus* venom. offf2p4. *Biochem. Biophys. Res. Commun.* **210**, 472–7.

[106] Kawasaki, T., Taniuchi, Y., Hisamichi, N., Fujimura, Y., Suzuki, M., Titani, K., Sakai, Y., Kaku, S., Satoh, N., Takenaka, T., *et al.* (1995) Tokaracetin, a new platelet antagonist that binds to platelet glycoprotein ib and inhibits von Willebrand factor-dependent shear-induced platelet aggregation. *Biochem. J.* **308**, 947–53.

[107] Sakurai, Y., Fujimura, Y., Kokubo, T., Imamura, K., Kawasaki, T., Handa, M., Suzuki, M., Matsui, T., Titani, K., and Yoshioka, A. (1998) The cDNA cloning and molecular characterization of a snake venom platelet glycoprotein Ib-binding protein, mamushigin, from *Agkistrodon halys blomhoffii* venom. *Thromb. Haemost.* **79**, 1199–207.

[108] Noeske-Jungblut, C., Haendler, B., Donner, P., Alagon, A., Possani, L., and Schleuning, W.D. (1995) Triabin, a highly potent exosite inhibitor of thrombin. *J. Biol. Chem.* **270**, 28629–34.

[109] Atoda, H., Ishikawa, M., Mizuno, H., and Morita, T. (1998) Coagulation factor X-binding protein from *Deinagkistrodon acutus* venom is a Gla domain-binding protein. *Biochemistry* **37**, 17361–70.

22 Toxins Leading to Medicines

ALAN L. HARVEY

22.1 INTRODUCTION

As discussed elsewhere in this volume, venoms are highly complex mixtures of (most usually) peptides and proteins. The effects of venoms in an envenomed animal are complex because different components have different actions and because some of the components may act in concert with others. But the individual toxins often have highly specific and very potent effects at key biological targets within the body. Isolated toxins, with potent and specific actions, can, therefore, be useful experimental tools for studying physiological processes (see, e.g. [1]). Examples include toxins that can discriminate better than any other compounds between different subtypes of receptors or ion channels or between isoforms of enzymes. Specific examples are described in several chapters of this volume.

In addition, toxins are usually highly structured and very stable molecules. Understanding their three-dimensional structures can provide structural information to guide the design of smaller analogues. And on relatively rare occasions, toxins can themselves be used therapeutically [2].

This article will summarise why toxins may provide leads for drug discovery and highlight some of the recent developments in medicines originating from research into toxins.

22.2 NEUROMUSCULAR BLOCKING AGENTS

Perhaps the best known example of a toxin leading to a medicine is that of the muscle relaxant tubocurarine. This plant alkaloid is the active ingredient of some South American arrow poisons. It blocks nicotinic acetylcholine receptors at neuromuscular junctions and, hence, causes paralysis. The purified alkaloid was introduced into medical practice to reduce limb fractures during use of electroconvulsant shock therapy, and then it was more widely applied as a muscle relaxant in association with general anaesthesia in surgery. Tubocurarine and related naturally occurring alkaloids acted as the starting point for the design of synthetic analogues with fewer side effects.

Perspectives in Molecular Toxinology
Edited by A Ménez © 2002 John Wiley & Sons, Ltd

Another muscle paralysing type of toxin, botulinum toxin, has somewhat surprisingly found its way into medical practice. Botulinum toxins are a group of proteins produced from clostridial bacteria. They enter motor nerve terminals and cause an irreversible block of the release of the neurotransmitter acetylcholine. Because the blockade lasts until new components of the secretory mechanism are synthesised, the muscle paralysis can last for several weeks. However, if absorbed systemically, botulinum toxins cause a fatal respiratory paralysis. Despite this, purified botulinum toxins are being used to treat unwanted muscle spasms in conditions such as blepharospasm. In these conditions, very small amounts of toxin are injected locally to the hyperactive muscles. Preparations of botulinum A have been used in cerebral palsy, cervical dystonia (spastic torticollis), laryngeal dystonia, spastic paresis, and hyperhidrosis (excessive sweating) [3]. They are also being used to reduce facial wrinkles as an alternative to cosmetic surgery, and they have been claimed to be effective in the prophylaxis of migraine and pain, although the mechanism for this possible effect is not clear [4]. More recently, a preparation of botulinum B has been introduced for treatment of cervical dystonia.

Another class of nicotinic acetylcholine blocking toxin that attracted some interest as a potential starting point for a new type of muscle relaxant is the α-conotoxins from marine snails of the *Conus* genus. Although these small peptides are highly specific for the neuromuscular acetylcholine receptors, they have not given rise to medicinal products. However, other types of conotoxins may have more potential, as explained in the following sections. These include the calcium channel blocking ω-conotoxins and the NMDA receptor antagonist conantokins [5].

22.3 ANALGESIC EFFECTS OF ω-CONOTOXINS AND THEIR ANALOGUES

ω-Conotoxins from venoms of various species of *Conus* marine snails block voltage-activated Ca^{2+} channels in excitable cells. These toxins are highly homologous peptides containing 25–29 amino acid residues and three disulphide bonds. Individual toxins have different selectivities for the different subtypes of Ca^{2+} channel [6]. Of particular interest is ω-conotoxin MVIIA from *Conus magus* because it is specific for N-type Ca^{2+} channels. Synthetic ω-conotoxin MVIIA (named ziconotide) is at the registration stage for treatment of chronic intractable pain [7, 8]. When administered intrathecally, ziconotide acts on N-type Ca^{2+} channels in neurons in the spinal cord responsible for transmitting pain stimuli. Block of these channels reduces the amount of neurotransmitter released from nerve endings and therefore reduces the strength of the signal. Results of several clinical trials are encouraging. In a Phase III trial in nonmalignant neuropathic pain which was not improved by opioid drugs, ziconotide produced significant improvements in more than half of treated patients. In one published case study, for example, ziconotide gave complete pain relief to a patient who had suffered from intractable 'phantom limb' pain after an amputation 23 years previously. Side effects were minimal and were decreased by adjusting the dose of ziconotide [9]. Wider experience, however, has led to reports of side effects with ziconotide [10]. In addition to its uses in cases of chronic intractable pain, ziconotide may provide an alternative to morphine for control of acute postoperative pain [11].

Given the relatively small number of *Conus* species whose venom has been investigated, it will not be surprising if additional ω-conotoxins are discovered, and some of these may also have potentially useful therapeutic applications. For example, a toxin with greater selectivity than MVIIA for N-type over P/Q-type Ca^{2+} channels has been isolated from *Conus catus* [12].

Naturally occurring ω-conotoxins and synthetic variants with mutated sequences have been compared for effects on N-type and P/Q-type Ca^{2+} channels. Detailed structural analysis by NMR indicated the possible structure of the pharmacophore responsible for selective binding to N-type channels [13], leading to the possibility of designing small molecular weight mimics. Moreover, the knowledge of the effectiveness of ziconotide on various animal models provided the confirmation that spinal N-type Ca^{2+} channels were valid therapeutic targets for analgesic molecules, and this led to a search for low molecular weight compounds with selective actions on N-type Ca^{2+} channels. This approach is beginning to be successful. For example, one compound from a series of 4–benzyloxyanilines was particularly effective in a variety of functional models for pain and epilepsy [14–16]. Unlike ziconotide, this compound blocked Na^+ channels in superior cervical neurons at similar concentrations to those that blocked N-type Ca^{2+} channels. If the apparently greater efficacy of this compound holds up in further *in vivo* testing, this may prompt searches for compounds with a mixed blocking effect.

22.4   CONANTOKINS AS POTENTIAL ANTI-EPILEPTIC AGENTS

Another class of conotoxins is under investigation for anti-epileptic properties. This is the conantokins, which block the NMDA type of glutamate receptor [17]. The lead compound for development, conantokin G, is a 17-residue peptide originally isolated from *Conus geographus*. It produces hyperactivity when injected into the brains of adult mice, and it is a potent blocker of the effects of NMDA on rat cerebellar brain slices. The block is noncompetitive. Conantokin G has been shown to be neuroprotective in a rat model of transient brain ischaemia [18], and it is also active in a model of Parkinson's disease [19]. The peptide is unusual in having five γ-carboxyglutamic acid residues and an amidated C-terminal asparagine amide. Its three-dimensional structure has been solved by NMR investigation [20]. Conantokin G is being developed as an anti-epileptic compound and it is currently in phase I trials. Conantokin G may differ from known anti-epileptic agents because it blocks only certain subtypes of NMDA receptor, those containing an NR2B protein subunit [21].

Another peptide, conantokin-R from *Conus radiatus*, is also a subtype-selective blocker of NMDA receptors, and it has been shown to be a potent anticonvulsant in animal models of epilepsy [22]. Conantokin-R has 27 amino acid residues and one disulphide bond. Structure–activity studies are underway [23, 24].

22.5   FROM ARROW-POISON TO ANALGESIC

Plant-derived arrow-poisons gave rise to the tubocurarine type of skeletal muscle relaxant. Another potential therapeutic development from arrow-poisons relates to analgesic compounds. In particular, the alkaloid epibatidine isolated from the toxic skin secretions of the arrow-poison frog *Epipedobates tricolor* has led to a synthetic homologue ABT-594 that is in phase II clinical trials as an analgesic and which is also under investigation as a memory enhancer. This compound acts as an agonist at some neuronal nicotinic acetylcholine receptors [25, 26]. It is as least as active as morphine in several pain models, but it does not appear to cause opioid-like withdrawal effects [27, 28]. The structural information from epibatidine and related compounds has allowed the development of a model pharmacophore which should lead to the design and synthesis of new compounds [29].

22.6   LEADS FROM SCORPION TOXINS

Scorpion venoms are particularly rich sources of toxins that specifically block ion channels (see chapters by Gordon, Possani and Rochat in this volume). Some of these toxins have been central to current drug discovery programmes in two key areas: immunosuppressants and anti-cancers.

Potassium channels exist in a very wide variety of types and subtypes. Scorpion toxins have been useful pharmacological tools in studying the physiological functions of particular classes of K^+ channels. In particular, toxins that block the Kv1.3 type of voltage-activated K^+ channel have been useful in demonstrating that block of such channels prevents the proliferation of human T-lymphocytes. These toxins include charybdotoxin and margatoxin (see review [30]). This work has led to the discovery of small molecular weight inhibitors of Kv1.3 channels that might have useful immunosuppressant properties, although greater selectivity for the target ion channels is probably necessary.

Another scorpion venom polypeptide was isolated from *Leiurus quinquestriatus* because of its ability to block low conductance Cl^- channels [31]. It is structurally homologous to other short scorpion toxins [32]. Chlorotoxin is now

being examined for its anti-cancer potential. This is because glioma cells were
found to express a chloride ion channel that is sensitive to chlorotoxin, while
normal glial cells do not express the same type of channel [33, 34]. Radiolabelled chlorotoxin appears to bind selectively to tumour cells in mice that
have been implanted with gliomas [35]. Chlorotoxin was demonstrated to
reduce the proliferation of glioma cells in culture [36], and to block their ability
to migrate [37].

22.7   ANTI-PLATELET AGENTS FROM SNAKE VENOMS

As described elsewhere in this volume, many snake venoms affect blood coagulation and many active components have been isolated and studied for their
actions on different parts of the coagulation cascade. In particular, attention
has focused on the disintegrins that prevent platelet aggregation by binding to a
surface receptor on platelets (glycoprotein GPIIb/IIIa, or integrin $\alpha_{IIb}\beta_3$) to
block their binding to fibrinogen. Two disintegrins, barbourin (from the South
Eastern pygmy rattlesnake *Sistrurus miliarius barbouri*) and echistatin (from the
sawscaled viper, *Echis carinatus*), have been central to the design of new
therapeutic agents.

   Most of the disintegrins contain the amino acid sequence RGD (arginine-
glycine-aspartic acid) that is critical for the interaction between fibrinogen and
platelet GP IIb/IIIa. Barbourin is unusual in having lysine instead of arginine,
and this may lead to different specificity against different integrins. Scarborough and colleagues have examined many small cyclic analogues of barbourin
and discovered eptifibatide (Integrilin), which is a cyclic hexapeptide inhibitor
of GPIIb/IIIa binding to fibrinogen [38]. Nonpeptide analogues of echistatin
have also been developed. Active compounds require a positive charge at one
end of the molecule (from an amino group), a hydrophobic group in the middle,
and a negative charge (from a carboxylate group) at the other end. The final
compound was tirofiban (so called because of the use of tyrosine as the scaffold
for the functional groups).

   Eptifibatide and tirofiban represent a new class of drug (the 'fibans') and
they are under extensive clinical trials in patients (more than 18 000 enrolled)
thought to be at risk of a myocardial infarction. When used alone or in
combination with other agents, the fibans reduce the incidence of death or
serious coronary events [39].

22.8   HYPOGLYCAEMICS FROM GILA MONSTER VENOM

The salivary secretions of the Gila monster (*Heloderma suspectum*) have small
proteins (exendins) with structural similarities to glucagon-like peptide (for
review, see [40]). The exendins bind to the GLP-1 receptor and stimulate the

release of insulin from pancreatic beta cells [41]. Unlike glucagon, their hypogly-
caemic effects can last several hours. Exendin-4 is being studied extensively in
animal models of type II diabetes (i.e. noninsulin dependent) and obesity. Very
low doses (< 0.1 μg/kg) of exendin-4 have prolonged actions in diabetic rats and
mice, both acutely and chronically, leading to improved control of blood sugar
levels and to weight loss [42]. Exendin-4 stimulates beta cell proliferation and
differentiation in a model of pancreatic damage [43]. Exendin-4 has also been
demonstrated to be effective in diabetic Rhesus monkeys, suggesting that it should
also work in humans [42]. Intracerebroventricular infusion of exendin-4 by osmotic minipumps reduced body weight in normal and genetically obese rats [44].

REFERENCES

[1] Harvey, A.L. (2001) Twenty years of dendrotoxins. *Toxicon* **39**, 15–26.
[2] Harvey, A.L., Bradley, K.N., Cochran, S.A., Rowan, E.G., Pratt, J.A., Quillfeldt, J.A. and Jerusalinsky, D.A. (1998) What can toxins tell us for drug discovery? *Toxicon* **36**, 1635–40.
[3] Bakheit, A.M., Severa, S., Cosgrove, A., Morton, R., Rousso, S.H., Doderlein, L. and Lin, J.P. (2001) Safety profile and efficacy of botulinum toxin A (Dysport) in children with muscle spasticity. *Dev. Med. Child. Neurol.* **43**, 234–8.
[4] Schott, G.D. (2001) Botulinum toxin type A in pain management. *J Neurol Neurosurg Psychiatry* **70**, 709E.
[5] Shen, G.S., Layer, R.T. and McCabe, R.T. (2000) Conopeptides: from deadly venoms to novel therapeutics. *Drug Discovery Today* **5**, 98–106.
[6] Uchitel, O.D. (1997) Toxins affecting calcium channels in neurones. *Toxicon* **35**, 1161–91.
[7] Cox, B. (2000) Calcium channel blockers and pain therapy. *Curr Rev Pain* **4**, 488–98.
[8] Jain, K.K. (2000) An evaluation of intrathecal ziconotide for the treatment of chronic pain. *Expert. Opin. Investig. Drugs* **9**, 2403–10.
[9] Brose, W.G., Gutlove, D.P., Luther, R.R., Bowersox, S.S. and McGuire, D. (1997) Use of intrathecal SNX-111, a novel, N-type, voltage-sensitive, calcium channel blocker, in the management of intractable brachial plexus avulsion pain. *Clin. J. Pain* **13**, 256–9.
[10] Penn, R.D. and Paice, J.A. (2000) Adverse effects associated with intrathecal administration of ziconotide. Pain **85**, 291–6.
[11] Atanassoff, P.G., Hartmannsgruber, M.W., Thrasher, J., Wermeling, D., Longton, W., Gaeta, R., Singh, T., Mayo, M., McGuire, D. and Luther, R.R. (2000) Ziconotide, a new N-type calcium channel blocker, administered intrathecally for acute postoperative pain. *Reg. Anesth. Pain Med.* **25**, 274–8.
[12] Lewis, R.J, Nielsen, K.J, Craik, D.J, Loughnan, M.L, Adams, D.A, Sharpe, I.A, Luchian, T, Adams, D.J, Bond, T, Thomas, L, Jones, A, Matheson, J-L, Drinkwater, R, Andrews P.R and Alewood, P.F (2000) Novel ω-conotoxins from *Conus catus* discriminate among neuronal calcium channel subtypes. *J. Biol. Chem.* **275**, 35335–44.
[13] Nielsen, K.J., Adams, D., Thomas, L., Bond, T., Alewood, P.F., Craik, D.J. and Lewis, R.J. (1999) Structure–activity relationships of omega-conotoxins MVIIA, MVIIC and 14 loop splice hybrids at N and P/Q-type calcium channels. *J. Mol. Biol.* **289**, 1405–21.
[14] Hu, L.Y., Ryder, T.R., Rafferty, M.F., Feng, M.R., Lotarski, S.M., Rock, D.M., Sinz, M., Stoehr, S.J., Taylor, C.P., Weber, M.L., Bowersox, S.S., Miljanich, G.P.,

Millerman, E., Wang, Y.X. and Szoke, B.G. (1999) Synthesis of a series of 4-benzyloxyaniline analogues as neuronal N-type calcium channel blockers with improved anticonvulsant and analgesic properties. *J. Med. Chem.* **42**, 4239–49.

[15] Hu, L.Y., Ryder, T.R., Rafferty, M.F., Siebers, K.M., Malone, T., Chatterjee, A., Feng, M.R., Lotarski, S.M., Rock, D.M., Stoehr, S.J., Taylor, C.P., Weber, M.L., Miljanich, G.P., Millerman, E. and Szoke, B.G. (2000a) Neuronal N-type calcium channel blockers: a series of 4-piperidinylaniline analogs with analgesic activity. *Drug. Des. Discov.* **17**, 85–93.

[16] Hu, L.Y., Ryder, T.R., Rafferty, M.F., Taylor, C.P., Feng, M.R., Kuo, B.S., Lotarski, S.M., Miljanich, G.P., Millerman, E., Siebers, K.M. and Szoke, B.G. (2000b) The discovery of [1-(4-dimethylamino-benzyl)-piperidin-4-yl]-[4-(3,3-dimethylbutyl)-phenyl]-(3-methyl-but-2-enyl)-amine, an N-type Ca^{2+} channel blocker with oral activity for analgesia. *Bioorg. Med. Chem.* **8**, 1203–12.

[17] Mena, E.E., Gullak, M.F., Pagnozzi, M.J., Richter, K.E., Rivier, J., Cruz, L.J. and Olivera, B.M. (1990) Conantokin-G: a novel peptide antagonist to the N-methyl-D-aspartic acid (NMDA) receptor. *Neuroscience Letters* **118**, 241–4.

[18] Williams, A.J., Dave, J.R., Phillips, J.B., Lin, Y., McCabe, R.T. and Tortella, F.C. (2000) Neuroprotective efficacy and therapeutic window of the high-affinity N-methyl-D-aspartate antagonist conantokin-G: *in vitro* (primary cerebellar neurons) and *in vivo* (rat model of transient focal brain ischemia) studies. *J.Pharmacol. Exp. Ther.* **294**, 378–86.

[19] Adams, A.C., Layer, R.T., McCabe, R.T. and Keefe, K.A. (2000) Effects of con-antokins on L-3,4-dihydroxyphenylalanine-induced behavior and immediate early gene expression. *Eur. J. Pharmacol.* **404**, 303–13.

[20] Rigby, A.C., Baleja, J.D., Furie, B.C. and Furie, B. (1997) Three-dimensional structure of a gamma-carboxyglutamic acid-containing conotoxin, conantokin G, from the marine snail *Conus geographus*: the metal-free conformer. *Biochemistry* **36**, 6906–14.

[21] Donevan, S.D. and McCabe, R.T. (2000) Conantokin G is an NR2B-selective competitive antagonist of N-methyl-D-aspartate receptors. *Molecular Pharmacology* **58**, 614–23.

[22] White, H.S., McCabe, R.T., Armstrong, H., Donevan, S.D., Cruz, L.J., Abogadie, F.C., Torres, J., Rivier, J.E., Paarmann, I., Hollmann, M. and Olivera, B.M. (2000) *In vitro* and *in vivo* characterization of conantokin-R, a selective NMDA receptor antagonist isolated from the venom of the fish-hunting snail *Conus radiatus*. *Journal of Pharmacology and Experimental Therapeutics* **292**, 425–32.

[23] Blandl, T., Warder, S.E., Prorok, M. and Castellino, F.J. (2000a) Structure-function relationships of the NMDA receptor antagonist peptide, conantokin-R. *FEBS Lett.* **470**, 139–46.

[24] Blandl, T., Zajicek, J., Prorok, M. and Castellino, F.J. (2000b) Sequence requirements for the NMDA receptor antagonist properties of conantokin-R. *J Biol Chem.* November 28 epublication.

[25] Bannon, A.W., Decker, M.W., Holladay, M.W., Curzon, P., Donnelly-Roberts, D., Puttfarcken, P.S., Bitner, R.S., Diaz, A., Dickenson, A.H., Porsolt, R.D., Williams, M. and Arneric, S.P. (1998) Broad-spectrum, non-opioid analgesic activity by selective modulation of neuronal nicotinic acetylcholine receptors. *Science* **279**, 77–81.

[26] Boyce, S., Webb, J.K., Shepheard, S.L., Russell, M.G., Hill, R.G. and Rupniak, N.M. (2000) Analgesic and toxic effects of ABT-594 resemble epibatidine and nicotine in rats. *Pain* **85**, 443–50.

[27] Decker, M.W. and Meyer, M.D. (1999) Therapeutic potential of neuronal nicotinic acetylcholine receptor agonists as novel analgesics. *Biochemical Pharmacology* **58**, 917–23.

[28] Kesingland, A.C., Gentry, C.T., Panesar, M.S., Bowes, M.A., Vernier, J.M., Cube, R., Walker, K. and Urban, L. (2000) Analgesic profile of the nicotinic acetylcholine receptor agonists, (+)-epibatidine and ABT-594 in models of persistent inflammatory and neuropathic pain. *Pain* **86**, 113–18.

[29] Meyer, M.D., Decker, M.W., Rueter, L.E., Anderson, D.J., Dart, M.J., Kim, K.H., Sullivan, J.P. and Williams, M. (2000) The identification of novel structural compound classes exhibiting high affinity for neuronal nicotinic acetylcholine receptors and analgesic efficacy in preclinical models of pain. *Eur. J. Pharmacol.* **393**, 171–7.

[30] Chandy, K.G., Cahalan, M., Pennington, M., Norton, R.S., Wulff, H. and Gutman, G.A. (2001) Potassium channels in T lymphocytes: toxins to therapeutic immunosuppressants. *Toxicon* **39**, 1269–1276.

[31] DeBin, J.A., Maggio, J.E. and Strichartz, G.R. (1993) Purification and characterization of chlorotoxin, a chloride channel ligand from the venom of the scorpion. *Am. J. Physiol.* **264**, C361–9.

[32] Lippens, G., Najib, J., Wodak, S.J. and Tartar, A. (1995) NMR sequential assignments and solution structure of chlorotoxin, a small scorpion toxin that blocks chloride channels. *Biochemistry* **34**, 13–21.

[33] Ullrich, N., Gillespie, G.Y. and Sontheimer, H. (1996) Human astrocytoma cells express a unique chloride current. *Neuroreport* **7**, 1020–4.

[34] Ullrich, N., Bordey, A., Gillespie, G.Y. and Sontheimer, H. (1998) Expression of voltage-activated chloride currents in acute slices of human gliomas. *Neuroscience* **83**, 1161–73.

[35] Soroceanu, L., Gillespie, Y., Khazaeli, M.B. and Sontheimer, H. (1998) Use of chlorotoxin for targeting of primary brain tumors. *Cancer Research* **58**, 4871–9.

[36] Rouzaire-Dubois, B., Milandri, J.B., Bostel, S. and Dubois, J.M. (2000) Control of cell proliferation by cell volume alterations in rat C6 glioma cells. *Pflugers Archiv.* **440**, 881–8.

[37] Soroceanu, L., Manning, T.J. and Sontheimer, H. (1999). Modulation of glioma cell migration and invasion using Cl⁻ and K⁺ ion channel blockers. *Journal of Neuroscience* **19**, 5942–54.

[38] Scarborough, R.M. (1999) Development of eptifibatide. *Am. Heart J.*, 1999, **138**, 1093–104.

[39] Harrington, R.A. (1999) Overview of clinical trials of glycoprotein IIb-IIIa inhibitors in acute coronary syndromes. *Am. Heart J.* **138**, 276–86

[40] Doyle, M.E. and Egan, J.M. (2001) Glucagon-like peptide-1. *Recent Prog. Horm. Res.* **56**, 377–99.

[41] Parkes, D.G., Pittner, R., Jodka, C., Smith, P. and Young, A. (2001) Insulinotropic actions of exendin-4 and glucagon-like peptide-1 *in vivo* and *in vitro*. *Metabolism* **50**, 583–9.

[42] Young, A.A., Gedulin, B.R., Bhavsar, S., Bodkin, N., Jodka, C., Hansen, B. and Denaro, M. (1999) Glucose-lowering and insulin-sensitizing actions of exendin-4: studies in obese diabetic (ob/ob, db/db) mice, diabetic fatty Zucker rats, and diabetic rhesus monkeys (Macaca mulatta). *Diabetes* **48**, 1026–34.

[43] Xu, G., Stoffers, D.A., Habener, J.F. and Bonner-Weir, S. (1999) Exendin-4 stimulates both beta-cell replication and neogenesis, resulting in increased beta-cell mass and improved glucose tolerance in diabetic rats. *Diabetes* **48**, 2270–6.

[44] Al-Barazanji, K.A., Arch, J.R., Buckingham, R.E. and Tadayyon, M. (2000) Central exendin-4 infusion reduces body weight without altering plasma leptin in (fa/fa) Zucker rats. *Obesity Res.* **8**, 317–23.

Part IV Evolution of Animal Toxins

23 Accelerated and Regional Evolution of Snake Venom Gland Isozymes

MOTONORI OHNO, TOMOHISA OGAWA, NAOKO ODA-UEDA, TAKAHITO CHIJIWA and SHOSAKU HATTORI

After two decades of study, it became evident that snake venom isozymes with a variety of physiological activities have evolved in an accelerated manner. Such accelerated evolution has rarely been known for general (ordinary) isozymes. In addition, it was recently found that *Trimeresurus flavoviridis* species isolated in separate islands over a long period have undergone island-specific changes in the genes encoding venom phospholipase A_2 (PLA_2) isozymes. We will mostly focus on accelerated and regional evolution of snake venom-gland PLA_2 isozymes from Crotalinae snakes such as *T. flavoviridis*, *T. okinavensis* and *T. gramineus*, although a brief description is also given of the evolution of serum inhibitors against venom PLA_2 isozymes. Outstanding problems regarding the molecular evolution of snake venom isozymes will be discussed.

23.1 ASSOCATION OF ACCELERATED EVOLUTION WITH FUNCTIONAL DIVERSITIES IN SNAKE VENOM ISOZYMES

23.1.1 STRUCTURES OF PLA₂ ISOZYMES FROM THE VENOMS OF *TRIMERESURUS* SNAKES

PLA_2 [EC 3.1.1.4] catalyses the hydrolysis of the 2-acyl ester linkage of 3-*sn*-phosphoglycerides with the requirement of Ca^{2+} to produce 3-*sn*-lysophospho-glycerides and fatty acids [1–3]. Snake venom PLA_2s are classified into groups I and II based on the mode of disulphide pairings [4]. Group I PLA_2s are found in Elapidae (Elapinae and Hydrophiinae) snake venoms. By contrast, Viperidae (Viperinae and Crotalinae) snake venom PLA_2s belong to a group II PLA_2 family. Group II PLA_2s are divided into two subgroups, $[Asp^{49}]PLA_2$ forms

Perspectives in Molecular Toxinology
Edited by A Ménez © 2002 John Wiley & Sons, Ltd

and [Lys^{49}]PLA$_2$ forms [5, 6]. They are known to exhibit a variety of physiological functions [7–13].

   Trimeresurus snakes belong to a Crotalinae subfamily. *T. flavoviridis* and *T.
okinavensis* snakes inhabit the south-western islands of Japan: Amami-Oshima,
Tokunoshima and Okinawa. Unless otherwise specified, we dealt with *T.
flavoviridis* and *T. okinavensis* from Tokunoshima and Amami-Oshima islands,
respectively. Specimens of *T. gramineus* are from Taiwan.

   PLA$_2$ enzymes in the venoms of *Trimeresurus* snakes form an isozyme
family. For example, five PLA$_2$ isozymes were isolated from *T. flavoviridis*
(Tokunoshima) venom. They all consist of 122 amino acid residues and belong
to a group II PLA$_2$ family. They are [Asp^{49}]PLA$_2$ (pI 7.9) called PLA2 [14–16],
basic [Asp^{49}]PLA$_2$ (pI 8.6) called PLA-B [17], more basic [Asp^{49}]PLA$_2$ (pI 10.3)
called PLA-N (unpublished work), and two [Lys^{49}]PLA$_2$s (pI 10.1 and 10.2)
called BPI and BPII, respectively [18, 19]. Three cDNAs encoding PLA2, BPI
and BPII have been cloned together with two cDNAs encoding PL-X$'$ and
[Thr^{38}]PLA$_2$, both [Asp^{49}]PLA$_2$s, whose proteins have not been isolated
[16, 20]. Four [Asp^{49}]PLA$_2$s, named PLA$_2$-I, PLA$_2$-II, PLA$_2$-III and PLA$_2$-
IV [21–23], and one [Lys^{49}]PLA$_2$, named PLA$_2$-V [24], have been isolated
from *T. gramineus* (Taiwan) venom. Their amino acid sequences are shown
in Figure 23.1 together with that of bovine pancreatic [Asp^{49}]PLA$_2$ [25]
which belongs to a group I PLA$_2$ family. It should be noted that BPI and
BPII are identical in sequence except only for one amino acid substitution at
position 67 (the numbering according to Figure 23.1), Asp for BPI and Asn
for BPII [18–20]. It is well known that group II PLA$_2$s have the same scaffold
[26].

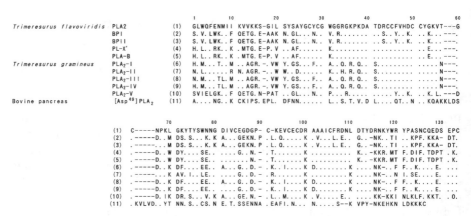

Figure 23.1 The aligned amino acid sequences of PLA$_2$s from *T. flavoviridis* (Tokunoshima) and *T. gramineus* (Taiwan) venoms together with that of bovine pancreatic
[Asp^{49}]PLA$_2$

23.1.2   DIVERSE PHYSIOLOGICAL FUNCTIONS OF PLA₂ ISOZYMES FROM *T. FLAVOVIRIDIS* VENOM

Actions on micelles and liposomes

When assayed with an egg-yolk emulsion, BPI and BPII showed only 1-2% lipolytic activity of PLA2 [18, 19]. However, when assayed with 1-stearoyl-2-arachidonoyl-*sn*-glycero-3-phosphocholine (SAPC) liposomes, BPI and BPII liberated arachidonic acid although they were 30–70% as active as PLA2 [27, 28], indicating that BPI and BPII can act on phospholipid membranes.

Hemolysis, necrosis and membrane depolarisation

PLA2 showed a strong haemolytic activity in the presence of phospholipids, whereas almost no haemolytic activity was detected for BPI and BPII [12]. When PLA2, BPI and BPII were injected intramuscularly into the hind thigh of mice, all isozymes induced the same histological disorder of myonecrosis in muscle fibres [12]. Creatine kinase (CK) was released into blood stream when these enzymes were injected intramuscularly into mice [12]. The CK levels produced by BPI and BPII were several times greater than that produced by PLA2, showing that the myolytic activity of BPI and BPII is stronger than that of PLA2. On the other hand, BPI and BPII can depolarise muscle membranes in a dose-dependent manner but PLA2 can not even at higher doses [12].

Muscle contraction, oedema-inducing activity and neurotoxicity

When guinea pig ileum or artery was incubated with BPI or BPII, BPII induced muscle contraction 10–100 times as strong as BPI [27]. Since Asn-67 of BPII is replaced by Asp in BPI, the absence of a charged residue at position 67 seems favourable for contraction. This is supported by the fact that PLA2 with Asn at position 67 (Figure 23.1) contracted the muscle as strongly as BPII [27, 29].

   PLA-B, an [Asp⁴⁹]PLA₂, which has 30% of the lipolytic activity of PLA2, can induce an oedema although PLA2 cannot [17]. PLA-N, an [Asp⁴⁹]PLA₂, which shows 10% of the lipolytic activity of PLA2, exhibited presynaptic neurotoxicity with $LD_{50} = 1.34\,\mu g/g$ for mice (unpublished work).

Apoptosis-inducing activity

It was found that BPI and BPII, but not PLA2, gave induced the apoptosis of HL-60 cells (unpublished work) which has not been observed before for any

PLA$_2$ isozymes. This was evidenced by the following observations: (a) morphological changes in light microscopy; (b) significant increase in hypodiploid cells detected by propidium iodide staining in flow cytometry; and (c) abnormal chromosome condensation found by Hoechst 33258 staining in fluorescence microscopy. Detailed experiments suggested that BPI-or BPII-induced apoptosis is not associated with the caspase 3 or 6 apoptotic cascade.

As mentioned above, PLA$_2$ isozymes from *T. flavoviridis* venom exhibit a variety of physiological activities. Since these PLA$_2$ isozymes have the same scaffold, it can be said that proper amino acid substitutions among them which have occurred in the evolutionary process produced their diverse physiological activities.

23.1.3 STRUCTURES OF THE cDNAs AND GENES ENCODING *TRIMERESURUS* VENOM-GLAND PLA$_2$ ISOZYMES

Five cDNAs encoding *T. flavoviridis* (Tokunoshima) venom-gland PLA$_2$s were cloned [16, 20]. Northern blot analysis with these PLA$_2$ cDNAs showed that the mRNAs coding for venom PLA$_2$s are expressed only in the venom gland but not in other tissues [16, 20]. All cDNAs contain an open reading frame of 414 bp coding for 138 amino acid residues. The comparison of the nucleotide sequences of the cDNAs indicated two remarkable characteristics. First, the nucleotide identities are 67% for the protein-coding region and 89 and 98% for the 5'- and 3'-untranslated regions (UTRs), respectively. This is the first time that much less sequence homology has been noted in the protein-coding region compared to the UTRs. The opposite pattern, that is greater sequence homology in the protein-coding region than in the UTRs, is usual for general isozymes as observed for various isoforms of a G-protein α-subunit family [30, 31] and a protein kinase C family [32, 33]. The signal peptide-coding domain coding for 16 amino acid residues is exceptionally conserved because of their common functions in membrane penetration and susceptibility to a common signal peptidase. Second, the nucleotide substitutions at the first, second and third positions of the triplet codons in the protein-coding region occurred at similar rates, 32.1, 30.0 and 28.6%, respectively, indicating that the substitution rates at the first and second positions are unusually high compared to the case of ordinary cDNAs. Since substitutions at the first and second positions in the codons tend to cause amino acid change, high substitution rates at the first and second positions cause the production of isozyme proteins with plenty of substituted amino acids, and thus with diverse physiological activities. Such observations in *T. flavoviridis* venom-gland PLA$_2$ isozyme cDNAs suggest that they have been much less constrained in their evolution.

For an in-depth search, six genes encoding *T. flavoviridis* venom-gland PLA$_2$ isozymes were isolated from its liver genomic DNA library [34]. Four genes

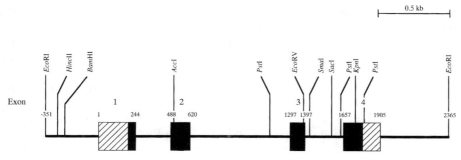

Figure 23.2 Schematic representation of a common structure of *T. flavoviridis* (Tokunoshima) and *T. gramineus* (Taiwan) venom PLA₂-encoding genes. Four exons are indicated by boxes and the UTRs are hatched. The nucleotide position numbers represent those of pgPLA 1a coding for *T. flavoviridis* PLA2

(pgPLA 1a, pgPLA 1b, pgPLA 2a and pgPLA 2b) encoded [Asp^{49}]PLA$_2$ isozymes and two genes (BP-I and BP-II) encoded [Lys^{49}]PLA$_2$ isozymes. Each gene spanned 1.9 kb, consisted of four exons and three introns and encoded 138 amino acid residues, including a signal peptide of 16 amino acids (Figure 23.2). In addition, four genes encoding *T. gramineus* venom-gland PLA$_2$ isozymes were also isolated [35]. Two genes (gPLA2-I and gPLA2-VI) encoded [Asp^{49}]PLA$_2$ isozymes and two genes (gPLA2-V and gPLA2-VII) encoded [Lys^{49}]PLA$_2$ isozymes. The constructs of these genes were exactly the same as those of *T. flavoviridis* PLA$_2$ isozyme genes.

When the nucleotide sequences of six *T. flavoviridis* and four *T. gramineus* PLA$_2$ isozyme genes were compared, those of introns were much less variable as in the case of the 5'- and 3'-UTRs of their cDNAs [34, 35]. For example, the sectional homologies between pgPLA 1a and BP-I are 99% for the 5'-UTR and 98% for the signal peptide-coding domain (1st exon), 94% (1st intron), 68% (2nd exon), 93% (2nd intron), 82% (3rd exon), 97% (3rd intron) and 75% for the protein-coding region and 92% for the 3'-UTR (4th exon). It has been claimed that the evolutionary rate of introns is much greater than that of the protein-coding regions of exons because introns experience fewer evolutionary constraints [36]. However, the structural characteristics of *T. flavoviridis* and *T. gramineus* PLA$_2$ isozyme genes are inconsistent with those expected from the so-called neutral theory [36].

23.1.4   ACCELERATED EVOLUTION OF VENOM-GLAND PLA$_2$ ISOZYMES OF *TRIMERESURUS* SNAKES

The evolutionary significance of *T. flavoviridis* and *T. gramineus* venom-gland PLA$_2$ isozymes became evident from the novel structures of their genes and from the manifestation of their diverse physiological activities. Mathematical computation was done for pairs of PLA_2 isozyme genes [37, 38]. The numbers of nucleotide substitutions per site (K_N) for the noncoding regions, the numbers of nucleotide substitutions per synonymous site (K_S), and the numbers of nucleotide substitutions per nonsynonymous site (K_A) for the protein-coding regions were computed with corrections for multiple substitutions [36]. Two features are noted. First, K_N values for introns are approximately one-fourth of K_S values for all the pairs, indicating that introns are much less variable than the protein-coding regions. Second, K_A/K_S values are close to or larger than unity: 0.752 for pgPLA 1a-pgPLA 1b, 1.17 for pgPLA 1a-BP-I and 1.75 for pgPLA 1b-BP-I. These values are much greater than those reported for other isoprotein genes which are 0.2 or so on average [39–42]. Frequency of non-synonymous substitution in the protein-coding region is much greater in *T. flavoviridis* PLA_2 isozyme genes than in ordinary isoprotein genes. It can be said that *T. flavoviridis* PLA_2 isozyme genes have been evolving in an accelerated manner.

K_N/K_S and K_A/K_S values of PLA_2 isozyme genes were also computed for all the pairs within species (*T. gramineus* vs *T. gramineus*) and between species (*T. gramineus* vs *T. flavoviridis*) [35]. The K_A/K_S values obtained, which are close to or larger than unity, also indicated accelerated evolution of these genes. It can be assumed that accelerated evolution has occurred in Crotalinae snake venom-gland PLA_2 isozyme genes. This conclusion is supported by the fact that in the genes encoding heterodimeric mojave toxin from *C. scutulatus scutulatus* (Crotalinae) venom [43] and ammodytoxin C from *Vipera ammodytes ammodytes* (Viperinae) venom [44], the protein-coding regions are much more variable than the noncoding regions. In addition, when the mean values of K_A and K_S for 13 Viperidae (Viperinae and Crotalinae) snake venom PLA_2 cDNAs and for 4 mammalian PLA_2 cDNAs were computed, the K_A/K_S value for snake venom PLA_2 cDNAs was 1.19, whereas that for mammalian PLA_2 cDNAs was 0.29, again indicating that the protein-coding regions of Viperidae snake venom PLA_2 isozyme genes have evolved via accelerated evolution [45].

Analysis of six cDNAs encoding serine protease isozymes of *T. flavoviridis* (Tokunoshima) and *T. gramineus* (Taiwan) venoms [46] and three cDNAs encoding metalloprotease isozymes of *T. flavoviridis* venom (unpublished work) also showed that accelerated evolution has occurred in these isozymes. It was recently found that cobra (Elapinae) venom-gland PLA_2 isozymes, which belong to a group I PLA_2 family, have also evolved in an accelerated manner [47].

23.1.5   MECHANISM OF ACCELERATED EVOLUTION IN SNAKE VENOM-GLAND PLA_2 ISOZYME GENES

The detailed analysis of the nucleotide sequences of snake venom isozyme genes and cDNAs indicates that accelerated evolution has occurred in these genes.

There are two possibilities for accelerated evolution. One is the rapid change in the protein-coding region while the noncoding region has changed at a rate similar to those of ordinary genes. The other is the slower change in the noncoding region while the protein-coding region has changed at a rate similar to those of ordinary genes. In order to examine which possibility is true, K_N values of introns for a pair of PLA$_2$ isozyme genes were compared with those of introns for a pair of TATA box-binding protein (TBP) genes, as the representatives of ordinary genes. Two TBP genes were isolated from *T. flavoviridis* and *T. gramineus* liver genomic DNA libraries [48]. Both TBP genes spanned about 19 kb and consisted of eight exons and seven introns. The K_N, K_S and K_A values were computed between gPLA2-VI and BP-I and between *T. flavoviridis* and *T. gramineus* TBP genes [35]. As shown in Figure 23.3, K_N values of introns for a

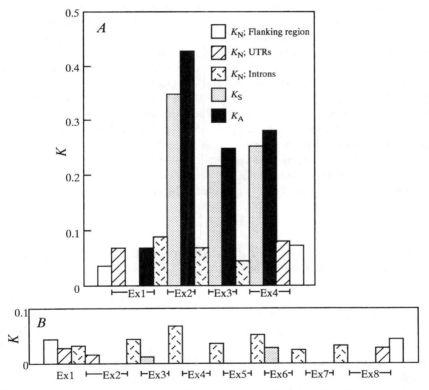

Figure 23.3 Schematic representation of sectional K_N, K_S and K_A values for a pair of gPLA2-VI and BP-I (*A*) and those for *T. gramineus* and *T. flavoviridis* TBP genes (*B*). Ex designates exons, and introns are portions sandwiched between exons. Each K is expressed by different patterns, as indicated in the key in *A*. In PLA$_2$ isozyme genes (*A*), K_S for exon 1 is zero. In TBP genes (*B*), exon 1 contains the UTR only. In TBP genes, K_A for exons 3 and 6 and K_S and K_A for exons 2, 4, 5, 7 and 8 are all zero. It is also evident from *A* that in PLA$_2$ isozyme genes, K_S and K_A values are much greater than K_N values

venom PLA_2 isozyme gene pair are judged to be comparable to those of introns of a TBP gene pair. This shows that introns of venom PLA_2 isozyme genes have evolved at rates similar to those of TBP genes. Since TBP genes have evolved in accordance with the neutral theory [36], it can be concluded that introns of venom PLA_2 isozyme genes have evolved at rates similar to those of ordinary genes. Thus, accelerated evolution of venom-gland PLA_2 isozyme genes can be attributed only to the rapid change in the protein-coding region.

The K_S and K_A values are greater than K_N values in *Trimeresurus* venom-gland PLA_2 isozyme genes. On the other hand, K_S values are smaller than K_N values in ordinary genes as in the case of TBP genes. These facts indicate that the protein-coding regions of PLA_2 isozyme genes have changed, after gene duplication, so as to code for PLA_2 proteins carrying new, different physiological activities. Gene conversion is thought to be the mechanism that produces diversity in genes, and this idea was examined. *T. flavoviridis* and *T. gramineus* PLA_2 isozyme genes form multigene families [34, 35]. By contrast, only three PLA_2 isozyme genes ($gPLA_2$-o1, $gPLA_2$-o2 and $gPLA_2$-o3) were detected from the *T. okinavensis* liver genomic DNA library [49]. They consisted of four exons and three introns and showed the same structural features as *T. flavoviridis* and *T. gramineus* PLA_2 isozyme genes. The K_A/K_S values (1.20–1.53) and K_N/K_S values (0.28–0.33) for three pairs of the genes indicated the occurrence of accelerated evolution. If gene conversion occurs within only three genes, it would be expected that all genes converge into similar ones. Thus, the genes with high K_A/K_S or low K_N/K_S values could not be formed. Gene conversion can not therefore be the mechanism that brings about the diversity in the genes. Hence, it can be proposed that accumulation of point mutations produced the diversity in venom PLA_2 isozyme genes.

23.1.6   EVOLUTION OF *T. FLAVOVIRIDIS* SERUM INHIBITORS AGAINST ITS VENOM PLA_2s

It is known that venomous snakes are resistant to their own venoms. This natural resistance to their own toxins is due to neutralising factors in their sera. These factors must protect against toxicity due to accidental bites by the snake itself or by fellow snakes. As mentioned, the genes encoding venom PLA_2 isozymes have evolved in an accelerated manner. As a consequence, it was thought that serum inhibitors have been evolving in an accelerated manner in accordance with venom PLA_2 isozymes. *T. flavoviridis* serum inhibitor capable of neutralising *T. flavoviridis* PLA2 actually consists of two subunits, PLA_2 inhibitor (PLI)-A and PLI-B, and their amino acid sequences of 129 amino acid residues have been determined [50]. They contain the sequences homologous to the carbohydrate recognition domain (CRD) noted for C-type lectins [51].

Their genes, gPLI-A and gPLI-B, have been cloned [52]. Both genes consist of four exons and three introns and encode 147 amino acid residues, including a signal peptide of 18 amino acid residues. The nucleotide sequences of gPLI-A and gPLI-B are highly homologous except for exon 3 where K_A/K_S was estimated to be 3.61. It is thus assumed that the variable regions of PLI-A and PLI-B encoded by exon 3 are crucial in determining the specificity of interaction with PLA2. The fragments containing these regions do in fact interact with PLA2 [53]. When *T. flavoviridis* serum was fractionated on four columns, each conjugated with one of four PLA$_2$s, PLA2, PLA-B, BPI and BPII, five binding proteins, named PLI-I-V, were obtained [54]. PLI-IV and PLI-V were identical to PLI-A and PLI-B, respectively. PLI-I was a major component of inhibitory proteins against three basic PLA$_2$ isozymes, PLA-B, BPI and BPII. Its cDNA encoded 200 amino acid residues, including a signal peptide of 19 amino acid residues, and its gene of 9.6 kb consisted of five exons and four introns [54]. From the comparison of the exon–intron structures, it was suggested that the PLI-I gene belongs to a urokinase-type plasminogen activator [55], Ly-6 [56], CD59 [57] and neurotoxin [58] gene family, whose protein products are assumed to have three-fingered motif(s) consisting of approximately 90 amino acid residues. Thus, PLI-I, which is assumed to have two three-fingered motifs, is structurally different from PLI-IV (PLI-A) and PLI-V (PLI-B), which have a CRD. The gene organisation of PLI-I is also different from those of PLI-IV and PLI-V. It can be said that the evolutionary origin of PLI-I differs from those of PLI-IV and PLI-V. Thus, co-evolution of serum PLA$_2$ inhibitors against venom PLA$_2$ isozymes is unlikely to occur. It is likely that some of the serum proteins that had some degree of spontaneous affinity for PLA$_2$ have mutated in an adaptive manner to the forms capable of binding tightly to and inhibiting particular PLA$_2$ species in the venom, as a requisite for self-protection.

23.2   REGIONAL EVOLUTION OF VENOM-GLAND PLA$_2$ ISOZYMES OF *T. FLAVOVIRIDIS* SNAKES

23.2.1   DISTRIBUTION OF *T. FLAVOVIRIDIS* SNAKES

T. flavoviridis snakes inhabit the south-western islands of Japan: Amami-Oshima, Tokunoshima and Okinawa. Amami-Oshima island is the northern-most and Tokunoshima island is 30 km south of Amami-Oshima. Okinawa island is located a further 120 km south of Tokunoshima island. These islands are thought to have been separated by eustacy (changes in sea level) in the orogenic stage one to two million years ago [59]. After this stage, ancestral *T. flavoviridis* species in the former Okinawa continent were scattered to these islands and have been kept isolated.

23.2.2   REGIONAL DIFFERENCES IN EVOLUTION OF VENOM PLA$_2$ ISOZYMES

Okinawa T. flavoviridis lacks BPI and BPII isozymes

It was recently found in chromatographic analysis that myotoxic BPI and BPII, which are expressed abundantly in Amami-Oshima and Tokunoshima *T. flavoviridis* venoms, are completely missing from Okinawa *T. flavoviridis* venom [60]. Northern blot analysis and single-stranded conformational polymorphism–polymerase chain reaction [61] for venom-gland mRNAs of *T. flavoviridis* from three islands showed that BPI and BPII mRNAs are not expressed in Okinawa *T. flavoviridis* venom gland. However, the PCR experiments for Okinawa *T. flavoviridis* genomic DNA species with a variety of primers indicated that, although the upstream and downstream regions of the genes for BPI and/or BPII exist, most of the second exon at its 3′ end and the first half of the second intron are lost. Analysis of the sequences containing polymorphisms (positions 236 and 1353) between the genes for BPI and BPII [34] confirmed that the upstream region of the gene for BPI down to the 5′ moiety of the second exon is followed, with a possible insertion, by the downstream region of the gene for BPII starting from the middle portion of the second intron in the Okinawa *T. flavoviridis* genome (Figure 23.4). It is considered that ancestral *T. flavoviridis* in the former Okinawa continent had genes for BPI and BPII that were arranged in tandem in this order on one chromosome. After separation into islands one to two million years ago, the genes for BPI and BPII were disrupted to form a pseudogene only in Okinawa *T. flavoviridis*.

Possible relationship between venom components and feeding habits

The loss of BPI and BPII greatly decreases the venom toxicity of Okinawa *T. flavoviridis* because their myotoxic activity is estimated to be several times

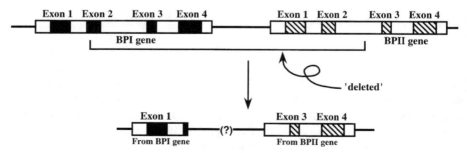

Figure 23.4   Conceptual model showing the pseudogene derived from the genes encoding BPI and BPII in Okinawa *T. flavoviridis* genome

stronger than that of PLA2 [12] and the combined quantity of BPI and BPII is comparable to that of PLA2 in the venom. The lack of strongly toxic BPI and BPII would seem to be of great disadvantage to Okinawa *T. flavoviridis*, although it seems to thrive well enough.

One possibility for the disappearance of strongly toxic BPI and BPII from Okinawa *T. flavoviridis* venom is that the prey species of Okinawa *T. flavoviridis* are partly or largely different from those of Amami-Oshima and Tokunoshima *T. flavoviridis*, such that strong venom toxicity might not be required for Okinawa *T. flavoviridis*. It is conceivable that the lack of necessity for a strongly toxic venom gave rise to the inactivation or disappearance of genes encoding such highly toxic proteins. According to a statistical investigation in Amami-Oshima, for example, *T. flavoviridis* snakes feed on rats (86%) and also birds, reptiles and amphibians (14% in total) in the farmlands and in the areas close to villages [62]. However, in Yambaru, a mountain area covered with forests in the northern part of Okinawa, *T. flavoviridis* snakes mostly (approximately 90%) prey on Holst's frogs (*Rana holsti*, a native animal designated by Okinawa Prefecture) inhabiting the streams in this area (K. Terada, personal communication). This natural, wild area seems to reflect the environment in Okinawa in which *T. flavoviridis* snakes lived in ancient times. It could be assumed that these feeding habits over a long period affected the venom components in Okinawa *T. flavoviridis*. This might be a sort of adaptation to the environment. In this connection, in the same snake species of pit viper *Calloselasma rhodostoma* (Viperinae) inhabiting the remote areas of Southeast Asia, changes in the compositions of venom proteins reflect adaptation to prey species [63].

23.3 PROBLEMS IN THE FUTURE

Why do snake venoms contain a variety of PLA_2 isozymes or other isozymes? It is thought that plural PLA_2 isozymes or other isozymes with diverse physiological activities act on target tissues or cells of prey animals in a synergistic manner and thus strengthen the venom toxicity. This is a great advantage in the survival of snakes. Now we understand that accelerated evolution is associated with production of isozymes with diverse physiological activities. However, the true mechanism of accelerated evolution in venom isozyme genes is unknown. As mentioned above for *T. okinavensis* venom PLA_2 isozymes [49], the diversity of the genes seems to have been acquired through accumulation of point mutations rather than gene conversion. It seems that diverse physiological activities in isozymes have been acquired by proper substitutions of particular amino acid residues. For instance, replacement of Asp (GAC) by Lys (AAA) at position 49 in PLA_2 molecules drastically changes their functions. Although there is only one amino acid substitution at position 67 between BPI and BPII in *T. flavoviridis* venom (Figure 23.1), Asp (GAC) for BPI and Asn (AAC) for BPII [18–20], BPII was 10–100 times more active than BPI in muscle contraction. Point mutation in the genes seems to have occurred for the purpose of producing new functions in isozymes. Thus, it is strongly felt that random point mutation can not be the mechanism for this purpose. The problem is to solve how point mutation in the genes took place at proper positions to produce the functionally well-directed isozymes.

Okinawa *T. flavoviridis* venom lost strongly myotoxic BPI and BPII which are abundant in Amami-Oshima and Tokunoshima *T. flavoviridis* venoms [60]. This is because the genes for BPI and BPII formed a pseudogene. It is thought that Okinawa *T. flavoviridis* need not preserve BPI and BPII because there is no need for a strong venom toxicity when feeding mostly on Holst's frogs. The thorny problem now facing us is how such information was transmitted to the genes to make them inactive. We plan to sequence the genome domain (possibly 100 kb) coding for a cluster of venom-gland PLA_2 isozymes of *T. flavoviridis* in the hope of finding clues to the mechanisms of accelerated evolution and regional changes of the genes.

REFERENCES

[1] Slotboom, A.J., Jansen, E.H.J.M., Vlijm, H., Pattus, F., Soares de Araujo, P. and de Haas, G.H. (1978) *Biochemistry* **17**, 4593–600.
[2] Dijkstra, B.W., Kalk, K.H., Hol, W.G.J. and Drenth, J. (1981) *J. Mol. Biol.* **147**, 97–123.
[3] Ohno, M., Honda, A., Tanaka, S., Mohri, N., Shieh, T.-C. and Kihara, H. (1984) *J. Biochem. (Tokyo)* **96**, 1183–91.
[4] Dufton, M.J. and Hider, R.C. (1983) *Eur. J. Biochem.* **137**, 545–51.
[5] Maraganore, J.M., Merutka, G., Cho, W., Welches, W., Kézdy, F.J. and Heinrikson, R. (1984) *J. Biol. Chem.* **259**, 13839–43.
[6] Maraganore, J.M. and Heinrikson, R.L. (1986) *J. Biol. Chem.* **261**, 4797–804.
[7] Fohlman, J., Eaker, D., Karlsson, E. and Thesleff, S. (1976) *Eur. J. Biochem.* **68**, 457–69.
[8] Verheij, H.M., Boffa, M.-C., Rothen, C., Bryckaert, M.-C., Verger, R. and de Haas, G.H. (1980) *Eur. J. Biochem.* **112**, 25–32.
[9] Condrea, E., Fletcher, J.E., Rapuano, B.E., Yang, C. and Rosenberg, P. (1981) *Toxicon* **19**, 705–20.
[10] Gopalakrishnakone, P., Dempster, D.W., Hawgood, B.J. and Elder, H.Y. (1984) *Toxicon* **22**, 85–98.
[11] Huang, H.-C. (1984) *Toxicon* **22**, 359–72.
[12] Kihara, H., Uchikawa, R., Hattori, S. and Ohno, M. (1992) *Biochem. Int.* **28**, 895–903.
[13] Ohno, M., Ménez, R., Ogawa, T., Danse, J.M., Shimohigashi, Y., Fromen, C., Ducancel, F., Zinn-Justin, S., Le Du, M.H., Boulain, J.-C., Tamiya, T. and Ménez, A. (1998) *Progress in Nucleic Acid Research and Molecular Biology* (Moldave, K., ed.), pp. 307–64, Academic Press, New York.
[14] Ishimaru, K., Kihara, H. and Ohno, M. (1980) *J. Biochem. (Tokyo)* **88**, 443–51.
[15] Tanaka, S., Mohri, N., Kihara, H. and Ohno, M. (1986) *J. Biochem. (Tokyo)* **99**, 281–9.
[16] Oda, N., Ogawa, T., Ohno, M., Sasaki, H., Sakaki, Y. and Kihara, H. (1990) *J. Biochem. (Tokyo)* **108**, 816–21.

[17] Yamaguchi, Y., Shimohigashi, Y., Chijiwa, T., Nakai, M., Ogawa, T., Hattori, S. and Ohno, M. (2001) *Toxicon*, **39**, 1069–76.

[18] Yoshizumi, K., Liu, S.-Y., Miyata, T., Saita, S., Ohno, M., Iwanaga, S. and Kihara, H. (1990) *Toxicon* **28**, 43–54.

[19] Liu, S.-Y., Yoshizumi, K., Oda, N., Ohno, M., Tokunaga, F., Iwanaga, S. and Kihara, H. (1990) *J. Biochem. (Tokyo)* **107**, 400–8.

[20] Ogawa, T., Oda, N., Nakashima, K., Sasaki, H., Hattori, M., Sakaki, Y., Kihara, H. and Ohno, M. (1992) *Proc. Natl. Acad. Sci. USA* **89**, 8557–61.

[21] Oda, N., Nakamura, H., Sakamoto, S., Liu, S.-Y., Kihara, H., Chang, C.-C. and Ohno, M. (1991) *Toxicon* **29**, 157–66.

[22] Fukagawa, T., Matsumoto, H., Shimohigashi, Y., Ogawa, T., Oda, N., Chang, C.-C. and Ohno, M. (1992) *Toxicon* **30**, 1331–41.

[23] Fukagawa, T., Nose, T., Shimohigashi, Y., Ogawa, T., Oda, N., Nakashima, K., Chang, C.-C. and Ohno, M. (1993) *Toxicon* **31**, 957–67.

[24] Nakai, M., Nakashima, K., Ogawa, T., Shimohigashi, Y., Hattori, S., Chang, C.-C. and Ohno, M. (1995) *Toxicon* **33**, 1469–78.

[25] Fleer, E.A.M., Verheij, H.M. and de Haas, G.H. (1978) *Eur. J. Biochem.* **82**, 261–9.

[26] Suzuki, A., Matsueda, E., Yamane, T., Ashida, T., Kihara, H. and Ohno, M. (1995) *J. Biochem. (Tokyo)* **117**, 730–40.

[27] Shimohigashi, Y., Tani, A., Matsumoto, H., Nakashima, K., Yamaguchi, Y., Oda, N., Takano, Y., Kamiya, H., Kishino, J., Arita, H. and Ohno, M. (1995) *J. Biochem. (Tokyo)* **118**, 1037–44.

[28] Shimohigashi, Y., Tani, A., Yamaguchi, Y., Ogawa, T. and Ohno, M. (1996) *J. Mol. Recogn.* **9**, 639–43.

[29] Matsumoto, H., Shimohigashi, Y., Nonaka, S., Saito, R., Takano, Y., Kamiya, H. and Ohno, M. (1991) *Biochem. Int.* **24**, 181–6.

[30] Matsuoka, M., Itoh, H., Kozasa, T. and Kajiro, Y. (1988) *Proc. Natl. Acad. Sci. USA* **85**, 5384–8.

[31] Strathmann, M. and Simon, M.I. (1990) *Proc. Natl. Acad. Sci. USA* **87**, 9113–17.

[32] Ohno, S., Kawasaki, H., Imajoh, S., Suzuki, K., Inagaki, M., Yokokura, H., Sakoh, T. and Hidaka, H. (1987) *Nature* **325**, 161–6.

[33] Kubo, K., Ohno, S. and Suzuki, K. (1987) *FEBS Lett.* **223**, 138–42.

[34] Nakashima, K., Ogawa, T., Oda, N., Hattori, M., Sakaki, Y., Kihara, H. and Ohno, M. (1993) *Proc. Natl. Acad. Sci. USA* **90**, 5964–8.

[35] Nakashima, K., Nobuhisa, I., Deshimaru, M., Nakai, M., Ogawa, T., Shimohigashi, Y., Fukumaki, Y., Hattori, M., Sakaki, Y., Hattori, S. and Ohno, M. (1995) *Proc. Natl. Acad. Sci. USA* **92**, 5605–9.

[36] Kimura, M. (1983) The Neutral Theory of Molecular Evolution, Cambridge Univ. Press, Cambridge, U. K.

[37] Miyata, T. and Yasunaga, T. (1980) *J. Mol. Evol.* **16**, 23–36.

[38] Nei, M. and Gojobori, T. (1986) *Mol. Biol. Evol.* **3**, 418–26.

[39] Miyata, T., Yasunaga, T. and Nishida, T. (1980) *Proc. Natl. Acad. Sci. USA* **77**, 7328–32.

[40] Miyata, T., Hayashida, H., Kuma, K. and Yasunaga, T. (1987) *Proc. Jpn. Acad.* **63B**, 327–31.

[41] Wu, C.-I. and Li, W.-H. (1985) *Proc. Natl. Acad. Sci. USA* **82**, 1741–5.

[42] Nei, M. (1987) Molecular Evolutionary Genetics, Columbia Univ. Press, Irvington-on-Hudson, NY.

[43] John, T.R., Smith, L.A. and Kaiser, I.I. (1994) *Gene* **139**, 229–34.

[44] Kordis, D. and Gubensek, F. (1996) *Eur. J. Biochem.* **240**, 83–90.

[45] Ogawa, T., Kitajima, M., Nakashima, K., Sakaki, Y. and Ohno, M. (1995) *J. Mol. Evol.* **41**, 867–77.

[46] Deshimaru, M., Ogawa, T., Nakashima, K., Nobuhisa, I., Chijiwa, T., Shimohiga-
     shi, Y., Fukumaki, Y., Niwa, M., Yamashina, I., Hattori, S. and Ohno, M. (1996)
     FEBS Lett. **397**, 83–8.
[47] Chuman, Y., Nobuhisa, I., Ogawa, T., Deshimaru, M., Chijiwa, T., Tan, N.-H.,
     Fukumaki, Y., Shimohigashi, Y., Ducancel, F., Boulain, J.-C., Ménez, A. and
     Ohno, M. (2000) *Toxicon* **38**, 449–62.
[48] Nakashima, K., Nobuhisa, I., Deshimaru, M., Ogawa, T., Shimohigashi, Y.,
     Hattori, M., Sakaki, Y., Hattori, S. and Ohno, M. (1995) *Gene* **152**, 209–13.
[49] Nobuhisa, I., Nakashima, K., Deshimaru, M., Ogawa, T., Shimohigashi, Y., Fuku-
     maki, Y., Hattori, S., Kihara, H. and Ohno, M. (1996) *Gene* **172**, 267–72.
[50] Inoue, S., Kogaki, H., Ikeda, K., Samejima, Y. and Ohmori-Satoh, T. (1991) *J.
     Biol. Chem.* **266**, 1001–7.
[51] Drickamer, K. (1988) *J. Biol. Chem.* **263**, 9557–60.
[52] Nobuhisa, I., Deshimaru, M., Chijiwa, T., Nakashima, K., Ogawa, T., Shimohiga-
     shi, Y., Fukumaki, Y., Hattori, S., Kihara, H. and Ohno, M. (1997) *Gene* **191**, 31–7.
[53] Nobuhisa, I., Chiwata, T., Fukumaki, Y., Hattori, S., Shimohigashi, Y. and Ohno,
     M. (1998) *FEBS Lett.* **429**, 385–9.
[54] Nobuhisa, I., Inamasu, S., Nakai, M., Tatsui, A., Mimori, T., Ogawa, T., Shimo-
     higashi, Y., Fukumaki, Y., Hattori, S., Kihara, H. and Ohno, M. (1997) *Eur. J.
     Biochem.* **249**, 838–45.
[55] Wang, Y., Dang, J., Johnson, L.K., Selhamer, J.J. and Doe, W.F. (1995) *Eur. J.
     Biochem.* **227**, 116–22.
[56] Fleming, T.J., O'hUigin, C. and Malek, T.R. (1993) *J. Immunol.* **150**, 5379–90.
[57] Petranka, J.G., Fleenor, D.E., Sykes, K., Kaufman, R.E. and Rosse, W.F. (1992)
     Proc. Natl. Acad. Sci. USA **89**, 7876–9.
[58] Fuse, N., Tsuchiya, T., Nonomura, Y., Ménez, A. and Tamiya, T. (1990) *Eur. J.
     Biochem.* **193**, 629–33.
[59] Hoshino, M. (1975) Eustacy in relation to orogenic stage, Tokai Univ. Press,
     Tokyo.
[60] Chijiwa, T., Deshimaru, M., Nobuhisa, I., Nakai, M., Ogawa, T., Oda, N., Naka-
     shima, K., Fukumaki, Y., Shimohigashi, Y., Hattori, S. and Ohno, M. (2000)
     Biochem. J. **347**, 491–9.
[61] Orita, M., Suzuki, Y., Sekiya, T. and Hayashi, K. (1989) *Genomics* **5**, 874–9.
[62] Mishima, S. (1966) *Sanit. Zool.* **17**, 1–21.
[63] Daltry, J.C., Wuster, W. and Thorpe, R.S. (1996) *Nature* **379**, 537–40.

24 Functional Diversification of Animal Toxins by Adaptive Evolution

DUŠAN KORDIŠ*, IGOR KRIŽAJ and FRANC GUBENŠEK

24.1 INTRODUCTION

Venomous animals are to be found in a variety of invertebrate phyla, including Cnidaria, Platyhelmintha, Annelida, Arthropoda, Mollusca and Echinodermata. In vertebrates, venomous representatives are known from fishes (Chondrichtyes and Osteichtyes), Amphibia and Reptilia (snakes and the venomous lizard (*Heloderma*)). These venomous animals have developed a number of different systems for the production and delivery of venom [1, 2]. Animal venoms contain numerous peptides and proteins with a variety of pharmacological functions. Many toxic proteins are small in size and possess a high density of disulphide bonds. The protein targets of animal toxins are receptors, ion channels, and enzymes, resulting in serious perturbation of physiological processes of the prey on toxin binding [3].

A diverse array of animal toxin multigene families (ATMFs) exists in the genomes of venomous animals and is responsible for a variety of biochemical activities and pharmacological functions. The molecular evolution of toxins [4] has been studied in a limited number of species, some invertebrate [5–8] and a few more vertebrate [9–23]. Among the former, extremely rapid evolution of conotoxin [5–7] and scorpion toxin genes [8] has recently been observed, a phenomenon that was recognised in the early 1990s in snake PLA$_2$ multigene families [9]. We shall review the current state of knowledge concerning the adaptive evolution of animal toxins.

24.2 BIOLOGICAL ROLES OF VENOMS

Venom is a mixture of different components with widely differing functions. Toxic venom proteins play a number of adaptive roles: immobilising, paralysing,

Abbreviations: PLA$_2$, phospholipase A$_2$; ATMF, animal toxin multigene family.
* Corresponding author

Perspectives in Molecular Toxinology
Edited by A Ménez © 2002 John Wiley & Sons, Ltd

killing, liquefying prey and deterring competitors. The synergistic action of other venom proteins may enhance the activity or spreading of toxins. Venoms can be part of the feeding system, may serve in digestion or can play a defensive role [24].

Different animal toxins are maximally effective on different prey species, and this is linked to toxin evolution. Each venomous animal is usually highly specialised with respect to its prey. Predatory *Conus* snails have specialised in feeding on hemichordates, echiuroids, molluscs, polychaete worms and, surprisingly, fish. Their venoms are extremely complex, and an individual *Conus* may have 50–200 distinct, biologically active components. The peptides found in *Conus* venom differ markedly between *Conus* species [25]. Scorpion toxins target ion channels with high specificity and selectivity, being able to discriminate between mammalian and insect sodium channels [26].

Highly specialised systems for production and delivery of venom have developed during evolution. Snake venom systems, which are complex and often quite different, originated in Colubroidea snakes. Evolution of the equipment for delivering oral secretions (e.g. jaws, fangs, muscles), the appropriate behavioural patterns [24], and the adaptive evolution of toxin multigene families has been occurring in concert. Advanced snakes (viperids and elapids) have the ability to kill prey quickly. The proteolytic enzymes accompanying the toxins in viperids accelerate the digestion of bulky prey. The presence of venom initiates and shortens the digestive process in the prey, even before it is swallowed. In snakes the evolution of oral secretions and the jaw apparatus that deploys them depends on the biological roles served. Quite different prey capture strategies and associated behaviours are known in advanced snakes; and these are reflected in the different compositions of venom components and in jaw design. Effective hunting performance has been an important factor bearing on the evolution of the prey capture strategies of advanced venomous snakes resulting in the induction of rapid prey death [24].

24.3   TOXINS BELONG TO DIFFERENT CLASSES OF PROTEINS

Through adaptive evolution, in each taxonomic group of venomous invertebrates (e.g. in scorpions and *Conus* snails), a huge number of toxins has evolved from a single protein scaffold. The invertebrate toxins have evolved highly refined specificity and selectivity in targeting different types of ion channels and receptors.

The toxins from *Conus* snails are unusually small peptides, the majority being in the range of 12–30 amino acids [25, 27–29]. Conotoxins are channel-blocking ligands that are smaller than the polypeptide toxins used by other venomous animals. The evolution of such ligands in *Conus* venoms may be a response to strong selection for rapid prey paralysis, best met by venom components that diffuse rapidly upon injection [27]. Conotoxins are invariably highly constrained conformationally and exhibit one of three characteristic arrangements of cysteine residues, the two-, three- and four-loop frameworks. Although there are only a few, though widely shared structural motifs, the sequences of the conotoxins diverge remarkably between different *Conus* species [25, 27, 29].

Scorpion toxins comprise 60–76 amino acid residues with a conformation stabilised by four disulphide bridges. Similarly to *Conus* snails, each species of scorpion possesses around 70 different toxins. While the primary structures of scorpion toxins can be quite different, a single scaffold is consistently conserved among these families of proteins [30]. All scorpion toxins have a highly conserved, dense core formed by an α-helix and two to three β-strands, stabilised by disulphide bridges. In evolutionarily conserved 'scorpion fold' insertions, deletions and mutations can appear, conferring diverse specificities on scorpion toxins [26].

In contrast to the invertebrates, venomous vertebrates, such as snakes, use several types of toxin scaffold from different protein classes that have evolved highly refined specificity and selectivity in targeting different types of ion channels and receptors. Venom components can act more or less specifically on different physiological systems of prey, such as on brain and neuromuscular junctions (neurotoxins) or on the haemostatic system (serine proteinases, metalloproteinases, disintegrins). Large numbers of pharmacological activities, such as pre- and post-synaptic neurotoxicity, myotoxicity, cardiotoxicity, anticoagulant effects, platelet aggregation (inhibition or initiation), internal haemorrhage, antihaemorrhagic activity, convulsant activity, hypotensive activity, oedema-inducing activity and organ or tissue damage are associated with snake venom phospholipases A_2 (PLA$_2$s) [31]. Similarly, diverse pharmacological activities are typical also for other toxins present in snake venoms.

Most of the proteins found in a single snake venom belong to a small number of protein superfamilies, such as the PLA$_2$s, three-fingered neurotoxins, proteinase inhibitors, serine proteinases and metalloproteinases. The members of a single family show remarkable similarity in their primary, secondary and tertiary structures but differ from each other in their biological targeting and hence pharmacological effects, thus posing exciting challenges in structure-function relationships [32].

A tremendous diversity of secretory PLA$_2$s exists in animal venoms; they have been classified into different groups: group I in Elapidae snakes, group II in Viperidae snakes and group III in *Heloderma* lizard, insects and hydrozoans. They have a catalytic mechanism similar to that of mammalian secretory PLA$_2$s and highly conserved three-dimensional structures. The structures share a core of three α-helices, a Ca^{2+}-binding loop and a hydrophobic channel which binds the substrate fatty-acyl chains. Most venom secretory PLA$_2$s play a role in prey digestion and are potent toxins exerting a wide spectrum of pharmacological effects [31].

Elapidae snakes possess three-fingered neurotoxins with 60–70 amino acid residues, whose three-dimensional structures are highly conserved. A wide range of activities has been found in three-fingered neurotoxins, including inhibition of acetylcholinesterase by fasciculin from mamba snake and blocking of nicotinic acetylcholine receptor by α- and κ-neurotoxins. Mamba snakes possess muscarinic toxins that are agonists of muscarinic acetylcholine receptor, whereas calciseptine and FS2 toxin are blockers of potassium and calcium channels. All three-fingered neurotoxins share a common structural framework secured by four disulphide bridges [3, 19, 33].

Dendrotoxins from mamba snakes are homologous to Kunitz/BPTI-type serine proteinase inhibitors. The dendrotoxins block particular subtypes of voltage-dependent potassium channels in neurons but exhibit little or no anti-proteinase activity. They consist of 57–60 amino acid residues in a single-polypeptide chain crosslinked by three disulphide bridges [34]. A highly conserved three-dimensional structure has been found in all Kunitz/BPTI homologues [35, 36].

24.4   TOXINS HAVE MANY MOLECULAR TARGETS

Animal toxins affect the vital functions of another organism through specific interaction with their target. The efficiency of venom is crucial for survival of the animals and toxins have adapted to their molecular targets during evolution. They interact with each other through rather small but specific portions of their surfaces, which confer selectivity and specificity between complementary proteins. A response to selective pressure during the evolution of animal toxins was their strict target-selectivity, discriminating between closely related receptor subtypes, and hence distinguishing between physiologically relevant and irrelevant targets [25, 27].

24.4.1   ION CHANNELS

Ion channels involved in the generation of electrical signals are either voltage-gated, in which case they open in response to a depolarisation from the resting membrane potential, or ligand-gated, their integral ion pores being opened by binding a particular ligand (e.g. a neurotransmitter). Ion channels constitute a physiologically versatile and highly important group of proteins and, as such, present a natural target for many animal toxins. Together, ion channels and animal toxins constitute a remarkable world of diverse molecular complementarity. Venomous animals feed on diverse prey, and individual toxins have developed during their evolution target selectivity against specific isoforms of ion channels and neuronal receptors of particular prey species [37, 5, 25, 26].

24.4.2   VOLTAGE-ACTIVATED ION CHANNELS

In many cell types voltage-sensitive ion channels play a pivotal role in complex cell signalling systems. These channels induce transmembrane ion flows in response to changes in transmembrane potential. Many animal toxins affect voltage-gated Na^+-, Ca^{2+}, K^+- and Cl^--channels.

Voltage-gated Na^+-channels are targets for δ- and μ-conotoxins. δ-conotoxins increase the conductance of Na^+-channels by inhibiting their inactivation. μ-conotoxins discriminate between muscle and neuronal Na^+-channels and block preferentially the skeletal muscle channel type, inducing paralysis in vertebrates [25, 38–40].

Sodium channel-specific scorpion toxins bind to the voltage-sensitive Na^+-channels of excitable cells and modify their gating properties. Structural differences in the C-terminal region of these toxins are responsible for their species specificity, α- and β-scorpion toxins bind to different regions on the Na^+-channels, receptor sites 3 or 4, respectively, and affect the inactivation or activation steps of the Na^+-channel [8, 26, 39, 41].

Voltage-activated Ca^{2+}-channels are the targets for the ω-conotoxins from the venom of fish-hunting *Conus* snails. These toxins are 24–30 amino acids long, with three disulphide bonds, and selectively block voltage-dependent Ca^{2+}-channels of N-, P- and Q-type in brain neurons [25, 40].

Voltage-gated Ca^{2+}-channel effectors are found also in the three-fingered toxin family. Calciseptine and FS2 toxin from mamba venoms selectively block the L-type Ca^{2+}-channels in a variety of cells with tissue-dependent sensitivity. Calcicludine, the Kunitz-type proteinase inhibitor homologue, potently inhibits all types of high voltage-activated Ca^{2+}-channels, that are species- and tissue-specifically susceptible to this toxin [3].

Voltage-dependent K^+-channels are affected by a number of structurally diverse toxins from different venomous animals. In *Conus* snails they have been found to be inhibited by κ-conotoxins [25, 40].

Scorpion venoms contain K^+-channel toxins of 29–41 amino acids, held in a tight conformation by three or four disulphide bonds [41]. These toxins vary considerably in their K^+-channel binding specificity [8, 42].

Dendrotoxins from mamba snake venoms belong structurally to Kunitz-type proteinase inhibitors. They block particular subtypes of voltage-dependent K^+-channels in neurons. In spite of the high degree of sequence identity, dendrotoxins display remarkable differences in their K^+-channel subtype binding specificity. This makes them very useful markers of subtypes of K^+-channels *in vivo* and they are widely used as probes for studying the function of K^+-channels [34].

Phospholipases A_2 with neurotoxic activity affect the K^+-currents at the nerve terminal. However, only in the case of β-bungarotoxin (β-Butx) has direct binding to the K^+-channel been demonstrated. β-Butx binds to the sub-population of α-dendrotoxin-sensitive, voltage-dependent K^+-channels [43, 44].

Scorpion toxins (e.g. chlorotoxin) that bind to voltage-gated Cl^--channels are, like scorpion K^+-channel toxins, polypeptides of 29–41 amino acids, cross linked by three to four disulphides [8, 41].

24.4.3   LIGAND-GATED ION CHANNELS

Typical members of ligand-gated ion channels are the acetylcholine receptors (AChR), which play an important role in the regulation of neurotransmission.

Postsynaptically located nicotinic AChR is the target of α-conotoxin from *Conus* snails, peptides of 13–18 amino acids with two disulphide bonds. Its binding to the alpha subunit of nicotinic AChR inactivates the receptor [27], α-conotoxins selectively block different subtypes of the nicotinic AChR [25, 45].

In the venoms of Elapidae snakes, different α- and κ-neurotoxins (e.g. α- and κ-bungarotoxins, α-cobratoxin, erabutoxin, toxin α) exits which antagonise competitively the binding of acetylcholine to various types of nicotinic AChR. Although α- and κ-neurotoxins are structurally very similar, their native association states as well as their receptor subtype specificity differ [3, 19].

Similar in structure to α- and κ-neurotoxins are the toxins acting on muscarinic AChR present in mamba venoms [46]. Muscarinic toxins antagonise particular subtypes of muscarinic AChR with high selectivity [47].

24.4.4 Ca^{2+}-BINDING PROTEINS

Using a trimeric PLA_2 neurotoxin–taipoxin from *Oxyranus s.scutellatus* venom as affinity ligand, TCBP-49 (taipoxin-associated Ca^{2+}-binding protein of 49 kDa) has been purified from rat brain, and its cDNA cloned and sequenced [48, 49]. Subsequently, an orthologue of TCBP-49, crocalbin, has been characterised in porcine brain using crotoxin, a dimeric PLA_2 neurotoxin from *Crotalus durissus terrificus* [50, 51]. The physiological role of TCBP-49, which is localised in the lumen of the endoplasmic reticulum, is not yet clear. This member of the reticulocalbins, rather than being a Ca^{2+}-storage protein, participates in the Ca^{2+}-dependent regulation of membrane traffic [49]. Recently, one of the neuronal acceptors for ammodytoxin, a snake venom presynaptically neurotoxic PLA_2, was identified as calmodulin [52], a very important and highly conserved EF-hand Ca^{2+}-binding protein which participates in signalling pathways that regulate many physiological processes [reviewed in 53].

24.4.5   PENTRAXINS

In addition to TCBP-49, a protein of 47 kDa from rat brain has been highly enriched on a taipoxin column [48]. cDNA cloning and sequencing revealed 20–30 % identity at the protein level to acute phase proteins of the pentraxin family [54]. Exclusively present in the brain, it was named neuronal pentraxin (NP1). NP1 and the subsequently discovered NP2 are secreted from neurons while neuronal pentraxin receptor (NPR) is an integral membrane pentraxin. It was proposed that NP1, NP2 and NPR constitute a novel neuronal uptake pathway involved in the remodelling of synapses [55].

24.4.6   LECTINS

The muscle or M-type PLA_2 receptor has been isolated from rabbit skeletal muscle, utilising its affinity towards monomeric neurotoxin PLA_2 from *Oxyranus s. scutellatus* venom (OS_2) [56]. Cloning and detailed analysis showed that it is a homologue of the macrophage mannose receptor and represents a new subfamily of the mannose receptor family of Ca^{2+}-dependent (C-type) multi-lectins [57]. The interaction of PLA_2 with the M-type receptor has been related to many of their physiological and pathological activities [58]. The PLA_2-interaction site on the receptor has been assigned to the carbohydrate recognition domain 5 (CRD5) [59]. It was also found that the binding of PLA_2 to the M-type receptor leads to inhibition of catalytic activity [60]. It is not a surprise therefore, that one of the structural classes of the snake venom PLA_2 inhibitors, which play a protective function in sera of various animals, belongs to the C-type lectins [61].

24.4.7   OTHER TARGETS

Snake venom PLA_2 toxins are bound and neutralised, not only by C-type lectins, but also by protein inhibitors which belong structurally to three-fingered proteins and proteins containing leucine-rich repeats [61]. Strongly anticoagulant PLA_2 from *Naja nigricollis* venom binds to coagulation factor Xa and inhibits its prothrombinase activity [62].

24.4.8   TARGETING OF TOXINS TO RECEPTOR SUBTYPES

The narrow receptor target specificity of animal toxins for their receptors enables discrimination among closely related receptor subtypes. Very narrow specificity is important for efficient envenomation. If they cannot discriminate between the ion channels, the toxins will bind to physiologically irrelevant targets reducing the efficiency of their intended role. Many conotoxins may be 'Janus-ligands', with two distinct recognition faces, substantially increasing target specificity [25].

24.4.9   SYNERGISTIC ACTION OF MULTIPLE TOXINS

Most venomous animals possess three or four different major toxins which, by acting co-operatively, very efficiently disable the motor circuitry of the prey. The general physiological strategies of stunning the prey and inducing paralysis by neuromuscular block both require the synergistic action of multiple toxins [25]. Many toxins in animal venoms act synergistically, a phenomenon best understood in venomous *Conus* snails. These possess groups of toxins working together on the same physiological target. Fish-hunting *Conus* snails require for prey capture two groups of multiple toxins ('toxin cabals'), each comprising a set of venom peptides that act synergistically after venom injection [25]. The 'lightning-strike cabal' causes immediate tetanic immobilisation (this requires δ-conotoxin which inhibits inactivation of voltage-gated Na^+ channels, together with κ-conotoxin which inhibits voltage-gated K^+ channels), giving time for the 'motor cabal' (paralysing peptides which potently abolish neuromuscular transmission) of toxins to spread through the body of prey, reach motor axons and neuromuscular junctions, and cause irreversible neuromuscular paralysis [25, 29].

24.5   TOXINS ARE MEMBERS OF LARGE MULTIGENE FAMILIES

Multigene families are widespread in the genomes of eukaryotes and are associated with the evolution of complex adaptations. When functional diversity is advantageous, it can be enhanced by the divergence of multigene family members through the operation of diversifying selection. All animal toxin multigene families contain functionally diversified isoenzymes. For different snake multigene families, such as PLA_2s [9–20], PLA_2 inhibitors [21], serine proteinases [22] and three-fingered toxins [19, 23, 63, 64], a number of functionally diversified isoenzymes have been found by studies at the protein, cDNA and gene levels. In scorpion toxins [65, 8, 26] and in *Conus* toxins [5–7, 29] similar functional diversity has been found.

Snake phospholipases A_2 (PLA_2s) are one of the most intensively studied multigene toxin families and provide a typical example of an animal toxin multigene family. PLA_2 multigene families are expressed specifically in snake venom glands and code for a variety of PLA_2 isoenzymes with widely different pharmacological activities [31, 17, 19] acquired during the course of evolution. A common evolutionary origin of the group II PLA_2 genes in Viperidae snakes is evident from the structure of their genes, since they exhibit extensive conservation of noncoding regions, including all the intron sequences.

Snake group II PLA_2 multigene families [17, 19] have been studied in only a few Viperidae species. The structure and organisation of the complete PLA_2 genes from three *Trimeresurus* species [9, 10, 12], from *Crotalus scutulatus* [13, 14], from *Vipera ammodytes* [15, 16] and from *V. palaestinae* [18] have been determined. They constitute a striking example of extreme conservation of noncoding regions, while the mature protein-coding exons are highly diversified.

5'UTR and signal peptide regions of snake group II PLA_2s, encoded by exons 1 and 2, show very high levels of sequence conservation, while exons 3–5, coding for the mature protein, are highly divergent. The lower percentage of identity at the amino acid level is caused by the prevalence of substitutions at the first and second positions of codons. Comparison of the size and nucleotide sequences of all introns, flanking and untranslated regions, and signal peptides in all known Viperidae PLA_2 genes shows a very high level of conservation, indicating their origin from a single ancestral gene by gene duplication and divergence.

The evolutionary history of snake PLA_2 multigene families may be regarded as the evolution from a single ancestral non-toxic PLA_2 gene to multiple, increasingly diversified PLA_2s (e.g. myotoxins, neurotoxins) through a process of repeated gene duplication. A number of independent gene duplications have occurred during the evolution of PLA_2s in most snake species following speciation, as shown by the number of functionally diversified PLA_2 genes in each snake species. When the same or nearly the same PLA_2s are present in two closely or even distantly related snake species, this is an indication that gene duplication occurred before speciation [Kordiš and Gubenšek, manuscript in preparation].

Highly diversified exons encoding mature protein, extremely conserved signal peptides, introns and other noncoding regions (untranslated and flanking regions) are typical of all the ATMFs. This pattern has been observed, in snake venom multigene families other than the PLA_2s, such as PLA_2 inhibitor genes [21], serine proteinase cDNAs [22], three-fingered toxin genes [19, 23, 63, 64], snake venom metalloproteinase cDNAs [66, 67; Kordiš and Gubenšek, unpublished], and in the *Vipera ammodytes* Kunitz/BPTI proteinase inhibitor multigene family [Župunski, Kordiš and Gubenšek, manuscript in preparation]. In conotoxin cDNAs [5–7] and genes [29], as well as in scorpion toxins [65, 8, 26], the same pattern has been found and is apparently typical for multigene families evolving under diversifying selection.

24.6   ADAPTIVE EVOLUTION OF ANIMAL TOXINS

24.6.1   EVOLUTION OF NEW FUNCTIONS BY GENE DUPLICATION

The duplication of genes and their subsequent functional divergence, leading to the formation of families of evolutionarily related but functionally distinct genes, is a fundamental process of adaptive evolution [68] and one of the most important mechanisms for the evolution of novel gene functions [69]. Gene duplication creates redundancy and allows a gene copy to escape the pressure of negative selection and evolve a new function. Information on the relative roles of natural selection and genetic drift in the functional diversification of genes can be obtained from a comparison of the rates of synonymous and nonsynonymous substitutions in duplicated genes. The patterns of intra- and interspecies variation for the duplicated toxin genes can contribute to understanding of the mechanisms involved in their evolution. A molecular indicator of adaptive evolution is an excess of nonsynonymous over synonymous substitutions, favouring diversity at the amino acid level [70, 71]. Purifying selection is expected to predominate when a duplicated gene has adapted to its new function, allowing the number of synonymous substitutions (dS) to catch up to and eventually exceed the number of nonsynonymous substitutions (dN). In eukaryotes, diversifying selection can still be detected 30–50 million years after gene duplication [71].

Gene duplication followed by functional diversification has been a pivotal mode of adaptive evolution in ATMFs of venomous animals [4]. A number of toxin multigene families contain relatively recently duplicated and functionally diversified genes ideal for studying the process of adaptive diversification of duplicated genes. A substantial amount of direct experimental evidence indicates that members of ATMFs are functionally diversified, producing distinct pharmacological activities [e.g. 26, 27, 3, 19, 17].

Functional diversification of PLA$_2$s in viperid snakes has occurred after gene duplication [17, 19]. Positive Darwinian selection favours amino acid replacements that serve to adapt a duplicate gene to a new function. In the protein-coding exons of PLA$_2$s the dN/dS ratio is higher than one, indicating the action of positive Darwinian evolution [9, 20]. This is compatible with the presence of the diverse pharmacological activities of PLA$_2$s in venom. The nonsynonymous substitutions are spread over the whole mature protein coding region in most viperid PLA$_2$ gene [9–17, 19, 20], the only exception being *V. palaestinae* [18]. Gene duplication of PLA$_2$s has been followed by a period of accelerated accumulation and fixation of amino acid replacements. Replacement rates in Viperidae snake PLA$_2$ genes vary within and between species and dN/dS values are larger than one in most of the species analysed. Accelerated rates of amino acid replacement have been observed to correlate with shifts in toxin function, such as in the evolution of myotoxin to neurotoxin [16].

The excess of nonsynonymous over synonymous substitutions has been observed in most snake venom multigene families studied so far, such as PLA$_2$ inhibitor genes [21], serine proteinase cDNAs [22], three-fingered toxin genes [19, 23], snake venom metalloproteinase cDNAs [Kordiš and Gubenšek, unpublished] and in *Vipera ammodytes* Kunitz/BPTI proteinase inhibitor multigene family [Županski, Kordiš and Gubenšek, manuscript in preparation], as well as in conotoxin cDNAs [5–7] and genes [29]. Conotoxins from different *Conus* species are extremely diversified. Nearly every amino acid in the mature toxin region of conotoxins, excluding the structure-stabilising cysteine residues, has been substituted at least once and most of the amino acid replacements are nonconservative [5–7]. The recent divergence and evolutionary relationships of conotoxins have been recognised from the more conserved prepro regions [5].

The presence of functionally diversified toxin multigene families in the genomes of venomous animals is direct evidence that they have been positively selected. Through insertion/deletion events, nonsense mutations and frameshifts, newly duplicated genes would be eventually converted to pseudogenes.

Without positive selection for maintaining a functional sequence, redundant gene copies would erode over time owing to accumulation of mutations [72]. In some Crotalinae species a few PLA$_2$s pseudogenes have been observed [12, 14, 20].

24.6.2   INTRONS IN TOXIN MULTIGENE FAMILIES ARE HIGHLY CONSERVED

An unusual pattern of molecular evolution has been observed in introns from ATMFs in which the divergence of introns between species is substantially lower than that of exons. Comparison of introns between species typically shows several times higher variation than nonsynonymous sites (dN) in exons [73]. However, in viperid PLA$_2$ genes dN values are several times greater than the corresponding intron values [9, 11, 13–17, 19]. Similar features have been observed in some other ATMFs, such as PLA$_2$ inhibitor genes [21] and the three-fingered toxins [23, 63, 64]. Introns of snake group II PLA$_2$s show unusually low levels of sequence divergence, ranging from 3–7% at the intraspecies level and about 10% at the interspecies level. They may have evolved slowly because of functional constraints [13, 15, 74] or, alternatively, positive selection may have caused rapid divergence of exons, leaving introns relatively unchanged [9, 10]. In the genes of the major histocompatibility complex introns are similarly highly conserved, probably as a result of the homogenising effects of recombination, whereas some exons are highly polymorphic and have evolved by positive selection [75].

24.6.3   VERY HIGH EVOLUTIONARY RATE OF TOXINS

The rate of evolution in ATMFs has been studied only in conotoxins. Highly diversified conotoxins are in fact products of a rapidly duplicating multigene family that has evolved by strong positive selection [5–7]. In conotoxins, the rate of nonsynonymous substitutions among the toxin regions reach 17 nonsynonymous substitutions per site per 10^9 years, that is five times greater than the highest nonsynonymous rate reported by Li [73] for interferon γ in mammals and nearly three times greater than the highest nonsynonymous rates reported for chorion proteins in *Drosophila*. Conotoxins have diversified at an extraordinarily high evolutionary rate, higher than in most proteins. Adaptive evolution for conotoxin size, generating different specificities for particular ion channels, has been proposed on the basis of the considerable length variation among mature conotoxin peptides [5]. In snake venom PLA$_2$ multigene families [Kordiš and Gubenšek, unpublished], similarly high evolutionary rates have been observed and the same pattern is very likely to be found in other venom ATMFs.

24.6.4   ADAPTIVE EVOLUTION DIVERSIFIES DIFFERENT REGIONS OF TOXINS

Adaptive evolution has diversified members of toxin multigene families at different parts of the mature protein. In a few cases, adaptive evolution has operated only on subdomains, such as in the third exon of snake PLA_2 inhibitors [21] and, in *V. palaestinae*, only in the third exon encoding the N-terminal part of PLA_2 [18] as well as in the metalloprotease domain of haemorrhagic metalloproteinases [Kordiš and Gubenšek, unpublished]. In most cases, however, adaptive evolution operates, surprisingly, at the whole mature protein level, as has been observed in most snake PLA_2s [9–17, 19], in snake serine proteinases [22], in three-fingered toxins of Elapidae snakes [19, 23, 63, 64] and in conotoxins [5–7].

24.6.5   WHAT IS THE DRIVING FORCE FOR THE ADAPTIVE EVOLUTION OF TOXINS?

Coevolution of predator and prey may generate evolutionary forces similar to those seen in host-pathogen evolution [71] and provide the means by which ecologically relevant genetic loci may diversify rapidly. The possibility of an 'arms race' between conotoxins and the ion channels and receptors of prey has been proposed [5]. Rapid evolution of conotoxins may lead to substantial variation in venom effectiveness against particular prey types [5–7]. Evolutionary pressures on ATMFs are related to the acquisition of specialised, species-specific pharmacological activities, each toxin family having diverged to interact specifically with physiologically relevant targets [25].

24.6.6   MECHANISMS FOR EVOLVING HYPERVARIABILITY OF ANIMAL TOXINS

Hypervariability is a prominent feature of large multigene families that mediate interactions between organisms, including those coding for animal toxins. Very recently the mechanisms for evolution of hypervariability have been studied in conotoxins [7]. It has been found that mature conotoxins show: (1) an accelerated rate of nucleotide substitutions; (2) a bias for transversions over transitions in nucleotide substitutions; (3) a position-specific conservation of Cys codons within the hypervariable region; and (4) a preponderance of nonsynonymous over synonymous substitutions. The authors proposed that the first three observations argue for a mutator mechanism targeted to mature domains of conotoxin genes, combining a protective activity for Cys codons and a mutagenic polymerase. The error-prone DNA polymerases may act as 'mutases' that are activated under specific conditions or in specific gene regions [76]. The low processivity of DNA polymerase V may fit well with the short intercysteine stretches typical of conotoxins [7]. The high dN/dS ratio is consistent with diversifying selection. Hypervariability of conotoxins and the large number of unique sequences per species might be explained by the combination of a targeted mutator mechanism to generate high variability with the subsequent action of diversifying selection on highly expressed conotoxin variants [7].

24.7 EXTREME STRUCTURAL PLASTICITY OF TOXINS

Three-dimensional structures of proteins are much more conserved than their primary structures, and such structural robustness to residue exchanges allows much scope for variety in evolution. On average only 3–4% of amino acid residues are crucial for maintaining a functional protein structure [77]. In toxins particularly, cysteines forming disulphide bonds have roles in determining and stabilising the three-dimensional structures.

Extreme structural plasticity and the highly specific toxin targeting which has evolved have been found in different animal toxin protein families and protein folds, such as in snake venom PLA_2s [19, 31, 56, 78, 79], in three-fingered toxins of elapid snakes [3, 19, 33, 79], in the Kunitz/BPTI family of protease inhibitors [35, 36, 79, 80], in conotoxins [25, 27, 29, 81] and in scorpion toxins [3, 26, 30].

Snake venom PLA_2s interact with specific PLA_2 receptors [56] via surface residues [78]. The specific binding of PLA_2 to its receptor is realised by a pharmacological site on its surface which is often independent of the catalytic site [31]. The observed amino acid replacements predominate at the molecular surface of snake venom PLA_2s, where they have a significant role in the evolution of new pharmacological effects by altering targeting specificity to various tissues or cells [78, 79].

In all three-fingered toxins the core region containing the four disulphide bridges, part of the β-strands, and the C-terminal loop is very similar. The primary function of the structural core is to maintain the overall structure of these toxins. The diversity of their biological functions is achieved by the exposed loops anchored to this stable core, which can present a variety of arrangements of the functional residues in order to fit different targets [3, 19]. Different members of the three-fingered toxin family act on different targets, but in each case the binding region involves residues from loops I and II. Although the three-fingered toxins share a similar overall fold, their modes of action have little in common [3, 19]. Their different activities reside in the conformational differences in the loop regions, which, and especially their tips, are involved in the functional sites of the three-fingered toxins [3, 19]. Comparisons of several three-dimensional structures has revealed that the orientation of the loops can vary considerably, supporting their role as determinants of biological specificity [33]. The distribution of the charged residues on the accessible surface of the three loops plays a decisive role in the recognition of their specific targets [3, 19]. Three-fingered toxins bind to receptors through a large number of contacts, which produce the high affinities of these toxins for their targets [33]. Interaction sites in three-fingered toxins are spread on different regions of its surface, this pattern being found in all other animal toxins [3].

A large number of Kunitz-type serine proteinase inhibitors are present in snakes. Dendrotoxins specific to mamba snakes have evolved a unique function to block particular subtypes of voltage-dependent potassium channels in neurons by conformational changes [34–36]. Structures of dendrotoxins reveal striking structural homologies with nontoxic Kunitz/BPTI homologues, which have different biochemical activities [80]. Structural homologies of toxic and nontoxic Kunitz/BPTI homologues are direct proof that toxicity can arise by mutations on the loops of the protein scaffold. Such mutations do not affect the overall fold of the dendrotoxins [80]. Since, for the same reason, the mutations do not interfere with protein folding, dendrotoxins can diversify functionally. The dendrotoxins interact strongly with a receptor, through the receptor interaction site, with a diameter of 20–25 Å [80]. Dendrotoxins provide an example of adaptation for diverse but specific biological functions while at the same time maintaining the overall three-dimensional structure of a common ancestor [80].

Among small disulphide crosslinked polypeptides, a number of different folds have been recognised. One of them is the inhibitor cystine knot (ICK) structural motif [81] characterised by a triple-stranded anti-parallel β-sheet stabilised by a cystine knot. The ICK fold is dictated largely by the gaps between the half-cystine residues and their pairing pattern. It is an ideal compact globular scaffold for presenting a variety of functional groups, thereby generating a range of polypeptides with diverse biological targets [81]. Members of the ICK fold have been found in organisms as diverse as spiders, *Conus* snails (conotoxins), plants and fungi. The hypervariability of conotoxins is restricted to the loops between the Cys residues [25, 27, 29]. The variable residues are important for biological activity, since the six Cys residues and one Gly that are conserved are clearly insufficient to specify a functional conotoxin. In this structural family the majority of polypeptides are blockers of the three major voltage-gated ion channels (Na^+, K^+ and Ca^{2+}). Many of the ICK motif structures, especially the ion-channel toxins, display highly basic surfaces. These charged groups play a role in interaction between the toxin and the vestibule and pore of the target ion channel [81]. The ICK motif allows variability in the gap sizes between the half-cystines and sequence variation among the nonbinding residues. The structure–function studies of naturally occurring ICK motif molecules have demonstrated that the residues most important for toxicity are frequently located on only one or two loops [81].

In all scorpion toxins the same α/β 'scorpion fold' with six conserved cysteines involved in disulphide bonds is present. The architecture of the scorpion toxin scaffold, made from an interior core with most side chains on the protein surface, allows effective sequence replacements, and not just insertions, without the loss of structural stability, with the potential for gaining new functions [30].

The tolerance for sequence mutations and the retained stability after multiple substitutions appear as unique properties of all animals toxin scaffolds so far known. The functional variety and structural plasticity of toxin families present in venomous animals indicate that they originate by adaptive evolution. Although the number of examples of adaptive evolution in animal toxins is still limited [5–23] we may expect that the same general mechanism of adaptive evolution will also be found in other prokaryotic and eukaryotic toxins.

ACKNOWLEDGEMENTS

The authors are indebted to Prof. R. Pain for critical reading of the manuscript. This study was supported by programme P0-0501-0106 of the Ministry of Science and Technology of Slovenia.

REFERENCES

[1] Buecherl, W., Buckley, E.E. and Deulofeu, V. (1971) Venomous chilopods or centipedes, p. 169–96 and Spiders, p. 197–277. In *Venomous animals and their venoms*. **Vol. 3**. Academic Press, New York, London.

[2] Kochva, E. (1987) The origin of snakes and evolution of the venom apparatus. *Toxicon* **25**, 65–106.

[3] Ménez, A. (1998) Functional architectures of animal toxins: a clue to drug design? *Toxicon* **36**, 1557–72.

[4] Kordiš, D. and Gubenšek, F. (2000) Adaptive evolution of animal toxin multigene families. *Gene* **261**, 43–52.

[5] Duda, T.F., Jr. and Palumbi, S.R. (1999) Molecular genetics of ecological diversification: duplication and rapid evolution of toxin genes of the venomous gastropod *Conus*. *Proc. Natl. Acad. Sci. USA* **96**, 6820–3.

[6] Duda, T.F., Jr. and Palumbi, S.R. (2000) Evolutionary diversification of multigene families: allelic selection of toxins in predatory cone snails. *Mol. Biol. Evol.* **17**, 1286–93.

[7] Conticello, S.G., Gilad, Y., Avidan, N., Ben-Asher, E., Levy, Z. and Fainzilber, M. (2001) Mechanisms for evolving hypervariability: the case of conopeptides. *Mol. Biol. Evol.* **18**, 120–31.

[8] Froy, O., Sagiv, T., Poreh, M., Urbach, D., Zilberberg, N. and Gurevitz, M. (1999) Dynamic diversification from a putative common ancestor of scorpion toxins affecting sodium, potassium, and chloride channels. *J. Mol. Evol.* **48**, 187–96.

[9] Nakashima, K., Ogawa, T., Oda, N., Hattori, M., Sakaki, Y., Kihara, H. and Ohno, M. (1993) Accelerated evolution of *Trimeresurus flavoviridis* venom gland phopholipase A2 isozymes. *Proc. Natl. Acad. Sci. USA* **90**, 5964–8.

[10] Nakashima, K., Nobuhisa, I., Deshimaru, M., Nakai, M., Ogawa, T., Shimohigashi, Y., Fukumaki, Y., Hattori, M., Sakaki, Y., Hattori, S. and Ohno, M. (1995) Accelerated evolution of the protein-coding regions is universal in crotalinae snake

venom gland phospholipase A2 isozyme genes. *Proc. Natl. Acad. Sci. USA* **92**, 5605–9.

[11] Ogawa, T., Kitajima, M., Nakashima, K., Sakaki, Y. and Ohno, M. (1995) Molecular evolution of group II phospholipases A2. *J. Mol. Evol.* **41**, 867–77.

[12] Nobuhisa, I., Nakashima, K., Deshimaru, M., Ogawa, T., Shimohigashi, Y., Fukumaki, Y., Sakaki, Y., Hattori, S., Kihara, H. and Ohno, M. (1996) Accelerated evolution of *Trimeresurus okinavensis* venom gland phospholipase A2 isozyme-encoding genes. *Gene* **172**, 267–72.

[13] John, T.R., Smith, L.A. and Kaiser, I.I. (1994) Genomic sequences encoding the acidic and basic subunits of Mojave toxin: unusually high sequence identity of noncoding regions. *Gene* **139**, 229–34.

[14] John, T.R., Smith, L.A. and Kaiser, I.I. (1996) A phospholipase A2-like pseudogene retaining the highly conserved introns of Mojave toxin and other snake venom group II PLA$_2$s, but having different exons. *DNA Cell. Biol.* **15**, 661–8.

[15] Kordiš, D. and Gubenšek, F. (1996) Ammodytoxin C gene helps to elucidate the irregular structure of Crotalinae group II phospholipase A2 genes. *Eur. J. Biochem.* **240**, 83–9.

[16] Kordiš, D. and Gubenšek, F. (1997) Bov-B long interspersed repeated DNA (LINE) sequences are present in Vipera ammodytes phospholipase A2 genes and in genomes of Viperidae snakes. *Eur. J. Biochem.* **246**, 772–9.

[17] Gubenšek, F. and Kordiš, D. (1997) Venom phospholipase A2 genes and their molecular evolution. In *Venom phospholipase A$_2$ enzymes: structure, function and mechanism* (R.M. Kini, ed.), Wiley and Sons, Chichester, 73–95.

[18] Kordiš, D., Bdolah, A. and Gubenšek, F. (1998) Positive Darwinian selection in *Vipera palaestinae* phospholipase A2 gene is unexpectedly limited to the third exon. *Biochem. Biophys. Res. Commun.* **251**, 613–19.

[19] Ohno, M., Ménez, R., Ogawa, T., Danse, J.M., Shimohigashi, Y., Frometn, C., Ducancel, F., Zinn-Justin, S., Le Du, M.H., Boulain, J.C., Tamiya, T. and Ménez, A. (1998) Molecular evolution of snake toxins: is the functional diversity of snake toxins associated with a mechanism of accelerated evolution? *Prog. Nucleic. Acid. Res. Mol. Biol.* **59**, 307–64.

[20] Chijiwa, T., Deshimaru, M., Nobuhisha, I., Nakai, M., Ogawa, T., Oda, N., Nakashima, K., Fukumaki, Y., Shimohigashi, Y., Hattori, S. and Ohno, M. (2000) Regional evolution of venom-gland phospholipase A2 isoenzymes of *Trimeresurus flavoviridis* snakes in the southwester islands of Japan. *Biochem. J.* **347**, 491–9.

[21] Nobuhisa, I., Deshimaru, M., Chijiwa, T., Nakashima, K., Ogawa, T., Shimohigashi, Y., Fukumaki, Y., Hattori, S., Kihara, H. and Ohno, M. (1997) Structures of genes encoding phospholipase A2 inhibitors from the serum of *Trimeresurus flavoviridis* snake. *Gene* **191**, 31–7.

[22] Deshimaru, M., Ogawa, T., Nakashima, K., Nobuhisha, I., Chijiwa, T., Shimohigashi, Y., Fukumaki, Y., Niwa, M., Yamashina, I., Hattori, S. and Ohno, M. (1996) Accelerated evolution of crotalinae snake venom gland serine proteases. *FEBS Lett.* **397**, 83–8.

[23] Gong, N., Armugam, A. and Jeyaseelan, K. (2000) Molecular cloning, characterization and evolution of the gene encoding a new group of short-chain alpha-neurotoxins in an Australian elapid, *Pseudonaja textilis*. *FEBS Lett.* **473**, 303–10.

[24] Kardong, K.V. (1996) Snake toxins and venoms: an evolutionary perspectives. *Herpetologica* **52**, 36–46.

[25] Olivera, B.M. (1997) E.E. Just Lecture, 1996. Conus venom peptides, receptor and ion channel targets, and drug design: 50 million years of neuropharmacology. *Mol. Biol. Cell.* **8**, 2101–9.

[26] Possani, L.D., Becerril, B., Delepierre, M. and Tytgat, J. (1999) Scorpion toxins specific for Na$^+$-channels. *Eur. J. Biochem.* **264**, 287–300.

[27] Olivera, B.M., Rivier, J., Clark, C., Ramilo, C.A., Corpuz, G.P., Abogadie, F.C., Mena, E.E., Woodward, S.R., Hillyard, D.R. and Cruz, L.J. (1990) Diversity of Conus neuropeptides. *Science* **249**, 257–63.

[28] Woodward, S.R., Cruz, L.J., Olivera, B.M. and Hillyard, D.R. (1990) Constant and hypervariable regions in conotoxin propeptides. *EMBO J.* **10**, 1015–20.

[29] Olivera, B.M., Walker, C., Cartier, G.E., Hooper, D., Santos, A.D., Schoenfeld, R., Shetty, R., Watkins, M., Bandyopadhyay, P. and Hillyard, D.R. (1999) Speciation of cone snails and interspecific hyperdivergence of their venom peptides. Potential evolutionary significance of introns. *Ann. N.Y. Acad. Sci.* **870**, 223–37.

[30] Vita, C., Roumestand, C., Toma, F. and Ménez, A. (1995) Scorpion toxins as natural scaffolds for protein engineering. *Proc. Natl. Acad. Sci. USA* **92**, 6404–8.

[31] Kini, R.M. (1997) Phospholipase A2 – a complex multifunctional protein puzzle. In *Venom phospholipase A2 enzymes: structure, function and mechanism* (R.M. Kini, ed.), Wiley and Sons, Chichester, 1–28.

[32] Kini, R.M. (1998) Proline brackets and identification of potential functional sites in proteins: toxins to therapeutics. *Toxicon* **36**, 1659–70.

[33] Harel, M., Kleywegy, G.J., Ravelli, R.B., Silman, I. and Sussman, J.L. (1995) Crystal structure of an acetylcholinesterase-fasciculin complex: interaction of a three-fingered toxin from snake venom with its target. *Structure* **3**, 1355–66.

[34] Harvey, A.L. (2001) Twenty years of dendrotoxins. *Toxicon* **39**, 15–26.

[35] Cardle, L. and Dufton, M.J. (1997) Foci of amino acid residue conservation in the 3D structures of the Kunitz BPTI proteinase inhibitors: how do variants from snake venom differ? *Protein Eng.* **10**, 131–6.

[36] Pritchard, L. and Dufton, M.J. (1999) Evolutionary trace analysis of the Kunitz/BPTI family of proteins: functional divergence may have been based on conformational adjustments. *J. Mol. Biol.* **285**, 1589–607.

[37] Daltry, J.C., Wuester, W. and Thorpe, R.S. (1996) Diet and snake venom evolution. *Nature* **379**, 537–40.

[38] Gordon, D., Savarin, P., Gurevitz, M. and Zinn-Justin, S. (1998) Functional anatomy of scorpion toxins affecting sodium channels. *J. Toxicol. Toxin Rev.* **17**, 131–59.

[39] Cestele, S. and Catterall, W.A. (2000) Molecular mechanisms of neurotoxin action on voltage-gated sodium channels. *Biochimie* **82**, 883–92.

[40] Olivera, B.M. and Cruz, L.J. (2001) Conotoxins, in retrospect. *Toxicon* **39**, 7–14.

[41] Possani, L.D., Merino, E., Corona, M., Bolivar, F. and Becerril, B. (2000) Peptides and genes coding for scorpion toxins that affect ion channels. *Biochimie* **82**, 861–8.

[42] Possani, L.D., Selisko, B. and Gurrola, G.B. (1999) Structure and function of scorpion toxins affecting K$^+$-channels. *Persp. Drug. Discov. Design* **15/16**, 15–40.

[43] Black, A.R. and Dolly, J.O. (1986) Two acceptor sub-types for dendrotoxin in chick synaptic membranes distinguishable by beta-bungarotoxin. *Eur. J. Biochem.* **156**, 609–17.

[44] Scott, V.E.S., Parcej, D.N., Keen, J.N., Findlay, J.B.C. and Dolly, J.O. (1990) Alpha-dendrotoxin acceptor from bovine brain is a K$^+$ channel protein. Evidence from the N-terminal sequence of its larger subunit. *J. Biol. Chem.* **265**, 20094–7.

[45] McIntosh, J.M., Santos, A.D., and Olivera, B.M. (1999) Conus peptides targeted to specific nicotinic acetylcholine receptor subtypes. *Annu. Rev. Biochem.* **68**, 59–88.

[46] Segalas, I., Roumestand, C., Zinn-Justin, S., Gilquin, B., Ménez, R., Ménez, A. and Toma, F. (1995) Solution structure of a green mamba toxin that activates muscarinic acetylcholine receptors, as studied by nuclear magneitc resonance and molecular modeling. *Biochemistry* **34**, 1248–60.

[47] Karlsson, E., Jolkkonen, M., Mulugeta, E., Onali, P. and Adem, A. (2000) Snake toxins with high selectivity for subtypes of muscarinic acetylcholine receptors. *Biochimie* **82**, 793–806.

[48] Schlimgen, A.K. Helms, J.A., Vogel, H. and Perin, M.S. (1995) Neuronal pentraxin, a secreted protein with homology to acute phase proteins of the immune system. *Neuron* **14**, 519–26.

[49] Dodds, D., Schlimgen, A.K., Lu, S.-Y. and Perin, M.S. (1995) Novel reticular calcium binding protein is purified on taipoxin columns. *J. Neurochem.* **64**, 2339–44.

[50] Hseu, M.J., Yen, C.-Y., Tseng, C.-C. and Tzeng, M.-C. (1997) Purification and partial amino acid sequence of a novel protein of the reticulocalbin family. *Biochem. Biophys. Res. Commun.* **239**, 18–22.

[51] Hseu, M.J., Yen, C.-Y. and Tzeng, M.-C. (1999) Crocalbin: a new calcium-binding protein that is also a binding protein for crotoxin, a neurotoxic phospholipase A2. *FEBS Lett.* **445**, 440–4.

[52] Šribar, J., Čopič, A., Pariš, A., Sherman, N.E., Gubenšek, F., Fox, J.W. and Križaj, I. (2001) A high affinity acceptor for phospholipase A2 with neurotoxic activity is a calmodulin. *J. Biol. Chem.* **276**, 12493–6.

[53] Chin, D. and Means, A.R. (2000) Calmodulin: a prototypical calcium sensor. *Trends in Cell. Biol.* **10**, 322–8.

[54] Pepys, M.B. and Baltz, M.L.(1983) Acute phase proteins with special reference to C-reactive protein and related proteins (pentaxins) and serum amyloid A protein. *Adv. Immunol.* **34**, 141–212.

[55] Dodds, D.C., Omeis, I.A., Cushman, S.J., Helms, J.A. and Perin, M.S. (1997) Neuronal pentraxin receptor, a novel putative integral membrane pentraxin that interacts with neuronal pentraxin 1 and 2 and taipoxin-associated calcium-binding protein 49. *J. Biol. Chem.* **272**, 21488–94.

[56] Lambeau, G. and Lazdunski, M. (1999) Receptors for a growing family of secreted phospholipases A2. *Trends Pharmacol. Sci.* **20**, 162–70.

[57] Stahl, P.D. and Ezekowitz, R.A.B. (1998) The mannose receptor is a pattern recognition receptor involved in host defense. *Curr. Opin. Immunol.* **10**, 50–5.

[58] Hanasaki, K. and Arita, H. (1999) Biological and pathological functions of phospholipase A(2) receptor. *Arch. Biochem. Biophys.* **372**, 215–23.

[59] Nicolas, J.-P., Lambeau, G. and Lazdunski, M. (1995) Identification of the binding domain for secretory phospholipases A2 on their M-type 180-kDa membrane receptor. *J. Biol. Chem.* **270**, 28869–73.

[60] Ancian, P., Lambeau, G. and Lazdunski, M. (1995) Multifunctional activity of the extracellular domain of the M-type (180-kDa) membrane receptor for secretory phospholipases A2. *Biochemistry* **34**, 13146–51.

[61] Faure, G. (2000) Natural inhibitors of toxic phospholipases A(2). *Biochimie* **82**, 833–40.

[62] Kerns, R.T., Kini, R.M., Stefansson, S. and Evans, H. (1999) Targeting of venom phospholipases: the strongly anticoagulant phospholipase A(2) from *Naja nigricollis* venom binds to coagulation factor Xa to inhibit the prothrombinase complex. *Arch. Biochem. Biophys.* **369**, 107–13.

[63] Lachumanan, R., Armugam, A., Tan, C.H. and Jeyaseelan, K. (1998) Structure and organization of the cardiotoxin genes in *Naja naja sputatrix*. *FEBS Lett.* **433**, 119–24.

[64] Chang, L.S., Huang, H.B. and Lin, S.R. (2000) The multiplicity of cardiotoxins from *Naja naja atra* (Taiwan cobra) venom. *Toxicon* **38**, 1065–76.

[65] Becerril, B., Marangoni, S. and Possani, L.D. (1997) Toxins and genes isolated from scorpions of the genus Tityus. *Toxicon* **35**, 821–35.

[66] Paine, M.J., Moura-da-Silva, A.M., Theakston, R.D. and Crampton, J.M. (1994) Cloning of metalloprotease genes in the carpet viper (*Echis pyramidum leakeyi*). Further members of the metalloprotease/disintegrin gene family. *Eur. J. Biochem.* **224**, 483–8.

[67] Moura-da-Silva, A.M., Theakston, R.D. and Crampton, J.M. (1996) Evolution of disintegrin cystein-rich and mammalian matrix-degrading metalloproteinases: gene

duplication and divergence of a common ancestor rather than convergent evolution. *J. Mol. Evol.* **43**, 263–9.

[68] Hughes, A.L. (1994) The evolution of functionally novel proteins after gene duplication. *Proc. R. Soc. Lond. B Biol. Sci.* **256**, 119–24.

[69] Zhang, J., Rosenberg, H.F. and Nei, M. (1998) Positive Darwinian selection after gene duplication in primate ribonuclease genes. *Proc. Natl. Acad. Sci. USA* **95**, 3708–13.

[70] Hughes, A.L. and Nei, M. (1988) Pattern of nucleotide substitution at major histocompatibility complex class I loci reveals overdominant selection. *Nature* **335**, 167–70.

[71] Hughes, A.L. (2000) *Adaptive evolution of genes and genomes.* Oxford University Press, New York.

[72] Walsh, J.B. (1995) How often do duplicated genes evolve new functions? *Genetics* **139**, 421–8.

[73] Li, W.H. (1997) *Molecular evolution.* Sinauer, Sunderland, Mass.

[74] Forsdyke, D.R. (1995) Conservation of stem-loop potential in introns of snake venom phospholipase A2 genes: an application of FORS-D analysis. *Mol. Biol. Evol.* **12**, 1157–65.

[75] Cereb, N., Hughes, A.L. and Yang, S.Y. (1997) Locus-specific conservation of the HLA class I introns by intra-locus homogenization. *Immunogenetics* **47**, 30–6.

[76] Radman, M. (1999) Enzymes of evolutionary change. *Nature* **401**, 866–9.

[77] Rost, B. (1997) Protein structures sustain evolutionary drift. *Folding and Design* **2**, S19–S24.

[78] Kini, R.M. and Chan, Y.M. (1999) Accelerated evolution and molecular surface of venom phospholipase A2 enzymes. *J. Mol. Evol.* **48**, 125–32.

[79] Alape-Giron, A., Persson, B., Cederlund, E., Flores-Diaz, M., Gutierrez, J.M., Thelestam, M., Bergman, T. and Jörnvall, H. (1999) Elapid venom toxins: multiple recruitments of ancient scaffolds. *Eur. J. Biochem.* **259**, 225–34.

[80] Lancelin, J.M., Foray, M.F., Poncin, M., Hollecker, M. and Marion, D. (1994) Proteinase inhibitor homologues as potassium channel blockers. *Nature Struct. Biol.* **1**, 246–50.

[81] Norton, R.S. and Pallaghy, P.K. (1998) The cystine knot structure of ion channel toxins and related polypeptides. *Toxicon* **36**, 1573–83.

Part V   From Venoms to Treatment

25 The Venomous Function

MAX GOYFFON

Venomous animals are characterised by the presence of a specialised gland or tissue producing a toxic secretion, the venom, connected with an injecting apparatus. This definition is valid for active or 'true' venomous animals. Passive venomous animals such as toads excrete a venom but do not have an injecting tool, and the venom can exert its effects only by the mucous or exceptionally the cutaneous route. These two categories are not clear-cut: venoms are sometimes thrown, some species are or can be venomous and poisonous. Venomous animals appear in almost all phyla, even in almost all classes, but the ratio of venomous species is variable. All the Cnidarians, all the scorpions, almost all the spiders with the exception of a small family (Uloboridæ) are venomous. In contrast, about 20% of the snakes and very few mammals are venomous [1]. In birds, only the genus *Pitohui*, which is endemic in New-Guinea, can be considered venomous.

Venomous animals live principally in intertropical area where a frequent endemism is observed. They constitute a typical element of the fauna of warm deserts. The distribution of families is variable. Concerning snakes, there are no Viperids in Australia, no Crotalids in Africa, but Elapids populate the major part of the intertropical belt of the globe. Almost all the scorpion species dangerous to humans belong to the family Buthidae: everywhere scorpions live, one or more Buthids are encountered, except in Italy, where there are none, which appears as a chorologic break for the Buthidae family. Therefore, everywhere scorpions live, theoretically there may be among them one or more dangerous species. Sometimes, the toxic potential of a species, a genus, or a family decreases in temperate regions, in particular for arachnids. In each phylum, or class, or order of venomous animals, the number of dangerous or medically important species is generally small (Table 25.1).

25.1 VENOMOUS APPARATUS

The venom gland secreting the venom and the injecting tool which constitute the venom apparatus present a high degree of diversity.

Perspectives in Molecular Toxinology
Edited by A Ménez © 2002 John Wiley & Sons, Ltd

Table 25.1 Estimated number of dangerous venomous species for humans.

Phylum or class	Number of species	Number of medically important venomous species
Cnidarians	10 000	60
Mollusks	70 000	20
Spiders	35 000	200
Mites	50 000	< 10
Scorpions	1 500	30
Insects	> 10^6	?(1)
Fish	30 000	500
Snakes	2 500	500
Birds	8 000	3
Mammals	3 500	3–5

(1) The number of Hymenoptera aculeata species and subspecies capable to provoke a severe anaphylactic shock is high. Besides, many Coleoptera, Diptera, Lepidoptera are responsible of accidents sometimes severe. Practically, the genera medically important are relatively limited, but any estimation remains difficult: about 1 000 Hymenoptera aculeata species, according to Meier and White [33]

25.1.1 VENOMOUS GLANDS

They are sometimes unicellular, as in fish producing serous toxic secretions. Most often though they are pluricellular exocrine glands of ectodermic origin. Some amphibians and snakes possess different types of associated or isolated mucous and serous glands. Venoms frequently contain peptide or protein toxins, and venom secretion then depends on a cell secretion cycle. In a few cases, the venom is produced by two glands, the secretions of which are synergistic (Brachinids, Diplopods). The gland can be covered with a sheath of striated muscular fibres, and the animal is able to control the injection of venom: at the time of sexual pairing, the female of the giant African scorpion *Pandinus imperator* is stung by the male without venom injection. The venom is stored in the venomous gland, sometimes in a specific reservoir (Hymenoptera). The volume of the gland is variable, but the venom toxicity or the quantity of venom injected do not depend on the size of the glands. In snakes, Elapids have relatively small venom glands, and 50% of the venom contents are generally injected per bite; Viperids have large venom glands and often inject 10% of the gland volume per bite. The female of the small European black scorpion *Euscorpius carpathicus* has a venom gland smaller than that of the male, and the size of the venom gland appears to be a secondary sexual characteristic in this scorpion species.

25.1.2   INJECTING APPARATUS

This characterises the active venomous animals. Despite a great diversity, these apparatuses can be classified in a limited number of types: stings (scorpions, Hymenoptera aculeata, fish), bites (spiders, centipedes, sea-urchins), urticating apparatus and hairs (Cnidarians, Lepidoptera). The venom injection functions either as an 'injecting pear', or as a syringe with a piston [1]. The former is the most frequent: the muscles surrounding the venom gland compress it and the venom is ejected under pressure, sometimes projected (spitting cobra). The latter is less frequent (Hymenoptera aculeata, Hemiptera). The penetration tool is a sting, simple (scorpions, fish), complex (Hymenoptera aculeata), or claws (Labidognatha spiders, Chilopoda). The opening of the stings, of the venomous fangs, is usually subterminal, which avoids filling and/or obturation when they penetrate the victim. The oral apparatus of the cones (Gasteropoda) is special: it is a harpoon-like radular tooth which is filled with venom when protruded and set in the prey by a pharyngal evaginated proboscis. The tooth acts as a venomous dart implanted in prey (annelid, mollusc, fish) or a human victim [2].

25.2   VENOMS: MODES OF ACTION AND COMPOSITION

Venoms and venomous animals have always fascinated man. Very early in human history they were used as medicines, and it is difficult to date the first scientific studies. It is generally admitted [3] that the pioneers were the Italian scientists Francisco Redi (seventeenth century) and Felix Fontana (eighteenth century). In some cases, the dating is precise. For example, the first experiments on scorpion venoms, in a modern sense, seem to have been carried out by Paul Bert (1865–1885) who wrote: 'I have concluded that [the scorpion venom] is a poison of the nervous system which seems to act in the same time on the excitability of spinal cord which is stimulated and on the peripheral motor nerves which are paralysed' [4]. One century later, Miranda et al. (1960) isolated the first scorpion neurotoxin and initiated the era of extensive research in venomology [5]. Venoms are concentrates of highly active molecules, especially peptides and proteins. Due to their structural and functional diversity, a classification is not easy, some components being multifunctional. A first difficulty is the milking of venomous animals. Very often, the venom is obtained by electric shocks altering the venom gland, which results in a release of molecules not present in the physiological secretion. When the venom apparatus is oral, electric shocks are accompanied by digestive secretions and/or regurgitations. Therefore, it is better to obtain the venom by a bite or a sting of a thin membrane covering a vessel. Many venoms are stable and remain active a very long time if they are stored in a lyophilised form, in darkness. In unfavourable cases (fish venoms), they are labile and difficult to store without any loss of activity.

Classically, the venom components are divided into four categories [6]:
1 neuromuscular toxins
2 toxins that induce cytolysis
3 factors acting on the blood coagulation, at different levels of the coagulation cascade
4 haemorrhagins, that damage the vascular endothelium and are responsible for oedema and capillary haemorrhage.

In practice, two main classes of components can be considered: enzymes and neuro/myotoxins.

Enzymes are constant constituents of venoms. The hyaluronidases, considered as spreading factors helping the tissular diffusion of other components, are present in all venoms. The oral venoms are particularly rich in proteases which exert a predigestive action (spiders, snakes). Protease activities can be specific, on a molecular target, like snake venom proteases that act on coagulation factors (plasminogen activators, thrombin-like enzymes, factor X activators etc...) or not (cytolysis). Rarely, the venoms are poor in enzymes, as in Buthid scorpions whose sting does not provoke any local reaction.

Neurotoxins constitute another important class of venoms components. Many of them are peptides from 30 to 70 amino acid residues acting synaptically, presynaptically or postsynaptically, or on ionic channels of excitable cells (sometimes also of various tissues). The category of phospholipases A2 (PLA2) is complex [7]: they have both enzymatic properties and presynaptic activities. They are monomeric (about 15 kDa) or multimeric (from 2 to 5 chains). Other PLA2s are myotoxic and induce sometimes severe rhabdomyolysis. Whether the effect is pre-or post-synaptic, the neurotoxins provoke a rapid flaccid paralysis which facilitates the capture of a prey.

Many other types of toxin can be found in venoms. The sarafotoxins of the *Atractaspis* snakes are functionnally and structurally similar in structure to endothelins, which are physiological vasoactive mediators. The lethal potential of sarafotoxins is related to their tropism for coronary arteries on which they exert a powerful vasoconstrictor effect. Other vasoactive molecules, bradykinin-potentiating peptides (BPP), which have a similar short amino acid sequence, have been found in the venom of a spider, a scorpion, and a snake. Mambas (*Dendroaspis* sp.) are snakes whose venoms contain a pool of several synergistic molecules: dendrotoxin which facilitates the release of acetylcholine and fasciculin which inhibits acetylcholinesterase.

Venom composition is quite variable according to sex (sometimes), age, and environmental conditions. Chippaux *et al.* [8] have extensively reviewed the variability of snake venoms according to the age/size of the snake, and Daltry *et al.* [9] have described the influence of feeding and accessible preys. The venom of the Australian funnel-web spider *Atrax robustus* is potentially lethal for humans. However, fatal cases have always been due to the males only, the

venom of which contains a particular neurotoxin, robustoxin. The spider
Tegenaria agrestis was described in the USA as responsible for necrotic arachnidism similar to that of *Loxosceles reclusa* [10]. *T. agrestis* was introduced
into North America in the historically recent past from Western Europe, where
such accidents have never been reported, and recent studies of venoms cannot
clearly explain these differences [11]. In France, the Buthid scorpion *Buthus
occitanus* is completely harmless. Rare severe but nonfatal accidents have been
described in Spain. On the other side of the Mediterranean, in Tunisia and
Algeria, fatal cases due to this species are observed every year.

25.3   VENOMS AND MOLECULAR EVOLUTION

One of the most interesting features of venoms is the existence of a molecular
polymorphism in some toxin families, which can lead to phylogenetic considerations [12]. Two examples will be used to illustrate this point: the PLA_2
neurotoxin family and the Buthid scorpions toxins acting on ion channels.

25.3.1 $vPLA_2$s

Among the secreted PLA_2s, ($sPLA_2$), the venom PLA_2s ($vPLA_2$) constitute an
important family, since over 150 $vPLA_2$s have been identified and classified into
three main groups according to their primary structure [13]. Group I comprises
the $vPLA_2$s of Elapidae and Viperidae venoms, group II the $vPLA_2$s of Viperidae venoms and group III *Heloderma*, Hymenoptera, Scorpionidae and Cni-
darian venoms. In a given group, the minimal level of similarity is about 30%;
in a given venom, it can be as high as 99%. The snake venom $vPLA_2$s (groups I
and II) are monomeric or multimeric and composed of two to five chains
generally associated by noncovalent interactions. The β-bungarotoxin of *Bungarus multicinctus* is a dimeric $vPLA_2$ consisting of one chain covalently linked
to a peptide belonging to the Kunitz trypsin inhibitor family, homologous to
the dendrotoxins of the mamba venoms. Whatever the toxin, many isoforms are
frequently found in snake venoms. The effects of $vPLA_2$s are also complex: they
can be neurotoxic (with a presynaptic action), myotoxic, cardiotoxic, cytotoxic,
active on blood coagulation. Therefore, the $vPLA_2$s present an apparent high
diversity in structure and function [14]. However, despite their marked polymorphism, their basic structures are limited in number and it has been shown
that unrelated structures with similar functions possess conserved key functional residues organised in an identical topology and suggesting a convergent
evolution [12]. The $vPLA_2$s comprise subunits the architecture of which is based
on the $sPLA_2$ fold, the Kunitz inhibitor, or three-fingered toxins. Genomic
studies of $vPLA_2$s have shown that the genes coding for the different $vPLA_2$s
found in a same venom could arise from a gene duplication. Besides, exons are

far less conserved than introns [15, 16]. So, it is assumed that the diversity of vPLA$_2$s in snake venoms reflects an accelerated evolution [15–17], and that the venom composition results from a Darwinian selection process [14]. This statement corroborates the current perspectives considering, in Colubroids, the venomous snakes as basic forms and the nonvenomous ones as advanced forms resulting from a secondary loss of the venomous apparatus [18].

25.3.2 TOXINS OF BUTHID SCORPION VENOMS

The venoms of the Buthid scorpions contain numerous peptide toxins active on ionic channels of excitable cells principally, but also sometimes on non-excitable cells [19]. Over 200 toxin sequences are currently known [20], and affect Na$^+$ (about 120), K$^+$ (about 70), Ca^{2+} (2) and Cl$^-$ (5) channels, the Na$^+$ channel toxins being responsible for the signs and symptoms of envenomation in mammals. All these toxins can be divided in two sub-classes: long toxins (Na$^+$, Ca^{2+}) which contain 60–70 amino acid residues, and short toxins (K$^+$, Cl$^-$) with 30–40 residues. The former are stabilised by four disulphide bridges, the latter generally by three, rarely by four. The short toxins with three disulphide bridges have a common structural motif which consists of an antiparallel β-sheet and an α-helix segment stabilised by two disulphide bridges (CSα/β motif) and appear as the core of long toxins. This core was discovered to be present in circulating defensins, which are antimicrobial peptides present in the haemolymph of the scorpions [21] and share a consensus sequence with K$^+$ channel toxins [22]. Usually inducible in arthropods, the defensins are constitutive in scorpions. The most closed-related arthropod defensins are those of Odonata, considered as one of the most ancient Hexapoda orders. Scorpions, which are also considered as very old, one of the most ancient terrestrial arthropods, possess a unique particularity. Their venoms contain toxins active against Eukaryote cells but inactive on microorganisms, whereas their haemolymph contains molecules with the same basic architecture that are active on Prokaryote cell membranes but inactive on Eukaryote cells. In both cases, these toxins act by modifying ionic fluxes of cell membranes. Venom toxins are used for catching prey and thereby nutrition; defensins are efficient protective agents. Therefore, the scorpions were able to diversify efficacious survival systems with remarkable economy by using a molecular polymorphism founded on a basic molecular scaffold [23]. This feature is observed in panchronic animals resisting various aggressive environmental factors, including ionising radiation [24]. Are this particular organic resistance and the panchronic character related?

It is interesting to compare the molecular polymorphism of the two toxin families, the vPLA$_2$s of the snake venoms, and that of the peptide toxins of the scorpions. In snakes, the vPLA$_2$s are considered to be derived from digestive sPLA$_2$s and to have evolved by Darwinian selection to an accumulation of neurotoxic PLA$_2$s in a specific digestive gland, the venom gland. Secondarily, the evolution of snakes has led to species able to capture prey without any venom apparatus. In this view, the accelerated evolution of vPLA$_2$s appears as a marker of the evolutionary potential of the snakes [18]. Scorpions, as panchronic animals, seem to have a weak evolutionary potential. Their toxins appear rather as relics of primitive excitable cell ligands conserved in a highly specialised glands, but they might have disappeared in other organisms concomitantly with the development of a nervous system. Another branch, the defensins which act inefficiently on excitable cells, have persisted freely in arthropods. If such a hypothesis can be proposed, it would be possible to find sequences of scorpion neurotoxins/defensins in common with molecules of primitive organisms (or considered as such) able to act as excitable cell ligands. Such a result was published in 1991, the existence of a structural and functional similarity with the Nef protein of HIV [25], but no other similar results have confirmed this first observation. To conclude, venom toxin polymorphism could be interpreted as a marker of an evolutionary process in snakes, as a relic or 'frozen' state in scorpions, even if one should remember that work on venom scorpions principally concerns the Buthidae family (40% of scorpion species).

25.4 VENOMS AND VITAL FUNCTIONS

Venoms play a primordial role in the nutrition of venomous predators by paralysing and often pre-digesting the prey, but as they are excreted products, they can be also related to the excretion function in a broad sense. Secondary metabolites present in venoms are various and some are chemical messengers. More rarely, venoms interfere with reproduction.

Another important function of venoms is their role in individual protection and/or defence. In this regard, the limits between venoms and defensive or repellent secretions is blurred. Besides, the protective effect of venoms is incomplete: venomous and nonvenomous snakes may be the prey of specialised predators, including nonvenomous snakes. Venomous animals, like others, can therefore employ various protective strategies: disruptive or aposematic colourings, defensive postures [26].

The venomous function, allied with nutrition and defence functions, increases the capacity to survive [27]. It is not surprising that the venomous animals are characteristic of the fauna of warm deserts where prey is rare and not easily accessible.

25.5 VENOMS AND ENVENOMATIONS

Antivenom therapy was discovered over one century ago [28]. It was a direct consequence of the discovery of vaccination and serotherapy of infectious diseases, diphtheria and tetanus, by Behring and Kitasato in 1890. Active and passive immunotherapy, experimentally described by Roux in 1888, were understood early and clearly defined by Ehrlich (1892). As Nuttal (1924) later wrote: 'Rarely in the history of scientific discovery have the results of laboratory researches been followed so rapidly by their practical application, and few indeed are the workers in the domain of application of medical science who have in their lifetime seen comparable benefits accrue to mankind as a direct consequence of their labour'. At the same time, some workers such as Chauveau in France believed microbial toxins and venoms to be very close, and prompted young scientists to apply the discoveries of Behring and Kitasato to snake venoms. Simultaneously, Phisalix and Bertrand, and Calmette, demonstrated in 1894 the antitoxic activity of the blood of animals immunised against the European viper *Vipera aspis* venom (Phisalix and Bertrand) or against the Vietnamese cobra venom, *Naja tripudians (= N. naja* or *N. kaouthia)* (Calmette). Calmette was the first to prepare a commercial antivenin for medical use and is now considered as the promoter of serotherapy [29].

Since these early days, the concept of antivenom therapy and the principles of antivenom preparation have not fundamentally changed, but the purification procedures, the efficacy and the tolerance of antivenoms have greatly improved. Taking into account the protein nature of many venom toxins, envenomations can be classified in three categories according to their responses to serotherapy:

1 *Envenomations due to venom toxins acting on a definite cell targets*: such as ionic channels, synaptic receptors (neurotoxins, myotoxins). Neutralising antibodies are a lure for toxins which are intercepted before reaching the cell target. The envenomation symptoms appear weakly or not at all if the antivenom is administered promptly. This category of envenomations is one of the most successful indications for serotherapy.

2 *Envenomations due to proteolytic enzymes*: haemorrhagic syndromes and skin necrosis, for example, are the consequences of the activity of various proteolytic enzymes, in particular when the venom apparatus is oral. When the targets of the proteases are definite proteins such as fibrinogen, prothrombin, factor X, serotherapy will be efficient as in the previous case. When cell membranes undergo lysis, which may be digestion following the action of one or several proteases, the inflammatory reaction accompanying or following necrosis evolves *per se* and is not sensitive to the neutralising action of antibody antiproteases. It is well known that the local effects of the venoms and the skin damage are not greatly improved by serotherapy, which is, however, regularly active on coagulation disorders. Therefore, this category of envenomations will be treated more or less successfully according to the target of the proteolytic enzymes.

3 *Envenomations and allergic reactions*: some venoms, principally those of Hymenoptera aculeata, can be dangerous because they may induce intense allergic reactions, rather than because of the activity of their neurotoxins or haemolytic toxins. These venoms contain many proteins which are powerful

allergens: sometimes the toxins themselves, more often enzymes (phospholipases, hyaluronidases). The risk of anaphylactic shock is increased by the effects of degranulating peptides (MCD of venom bees, mastoparans of venom wasps). The mechanism of these specific accidents deprives serotherapy of any efficacy.

Antivenins are prepared with a limited choice of venoms, a choice which is based on medical criteria. Even if a paraspecificity of antivenoms is sometimes described, the variability in venom composition constitutes an additional difficulty in their preparation. Besides, many toxins are not proteins, but rather amines, alkaloids, or steroids. However, serotherapy is the only specific therapy, and its design and realisation are still of major importance more one century after its discovery [29].

But there also exists also natural immunity against venoms, especially against the snake venoms, which has long been known in hedgehogs or mongooses, and in snakes against their own venoms [30]. Different mechanisms may explain this phenomenon: a mutation of the gene encoding the target of the venom, making the animal insensitive or less sensitive to it, or the presence of natural circulating antitoxins in resistant animals [31]. Both mechanisms have been described for the Chinese cobra *Naja naja atra*. In the blood of many snakes, protein inhibitors of PLA_2s have been identified. They may share structural and functional similarities with various mammalian proteins. Several natural snake venom inhibitors have been isolated from the marsupials *Didelphis marsupialis* and *Philander opossum*. They are acidic glycoproteins that form a complex with venom toxins in a one/one stoichiometric ratio. One of these inhibitors possesses three Ig-like domains [32]. All natural venom inhibitors are of great interest in studies of the activities of venom toxins and may offer new therapeutic approaches to envenomations. To date, research into natural antivenom immunity has principally concerned snake venoms and more work remains to be done on other venoms.

25.6   CONCLUSION

Toxins are ubiquitous in nature, and are found in the microbial, plant and animal kingdoms. When they diffuse in an ecosystem or appear in a food chain, the consequences can be unforeseeable. In animals, the venoms are produced, stored, used by a particular specialised apparatus of variable complexity and are frequently an important means of survival. They are used in various strategies, actively or passively, and in the latter case the whole organism may be toxic. Their role in defence and protection is therefore major and the venomous function appears as an adaptative character. But venom toxins have become important tools, essentially in biophysiology. Knowledge of ionic channels, of the acetylcholine receptor, of some coagulation mechanisms and of exocytosis has greatly benefited their utilisation. They also constitute basic resistant scaffolds that can be used in the synthesis of new original molecules usable in the engineering of novel ligands for membrane receptors [23]. Venomology is indeed at the dawn of a new age.

ACKNOWLEDGMENTS

I am grateful to Nicolas Vidal, Roland Stockmann and André Ménez for fruitful discussions on biological and evolutionary questions concerning venomous animals and their venoms.

REFERENCES

[1] Goyffon, M. and Heurtault, J. (1995) La fonction venimeuse. 1 vol, Masson, Paris, 283 p.

[2] Le Gall, F., Favreau, P., Richard, G., Benoît, E., Letourneux, Y. and Molgo, J. (1999) Biodiversity of the genus *Conus* (Fleming, 1822): a rich source of bioactive peptides. *Belg. J. Botan.* **129**, 17–42.

[3] Russell, F.E. (1988) Snake venom imunology: historical and practical considerations. *J. Toxicol. Toxin. Rev.* **7**, 1–82.

[4] Bert, P. (1885) Venin du scorpion. *CR. Soc. Biol.* **17**, 136–7.

[5] Miranda, F., Rochat, H. and Lissitzky, S. (1960) Sur la neurotoxine du venin de scorpion. I. Purification à partir du venin de deux espèces de scorpions. *Bull. Soc. Chim. Biol.* **42**, 379–91.

[6] Chippaux, J.P. and Goyffon, M. (1998) Venoms, antivenoms and immunotherapy. *Toxicon* **36**, 823–46.

[7] Faure, G. (1999) Les phospholipases A2 des venins de serpents. *Bull. Soc. Zool. Fr.* **124**, 149–68.

[8] Chippaux, J.P., Williams, V. and White, J. (1991) Snake venom variability: methods of study, results and interpretation. *Toxicon* **29**, 1271–303.

[9] Daltry, J.C., Wuster, W. and Thorpe, R.S. (1996) Diet and snake venom evolution. *Nature* **379**, 537–40.

[10] Vest, D.K. (1987) Necrotic arachnidism in the Northwest United States and its probable relationship to *Tegenaria agrestis* (Walckenaer) spiders. *Toxicon* **25**, 175–84.

[11] Binford, G.J. (2001) An analysis of geographic and intersexual chemical variation in venoms of the spider *Tegenaria agrestis* (Agelenidae) *Toxicon* **39**, 955–68.

[12] Dauplais, M., Lecoq, A., Song, J., Cotton, J., Jamin, N., Gilquin, B., Roumestand, C., Vita, C., Medeiros, C.L.C. de, Rowan, E.G., Harvey, A.L. and Ménez, A. (1997) On the convergent evolution of animal toxins. Conservation of a diad of functional residues in potassium channel-blocking toxins with unrelated structures. *J. Biol. Chem.* **272**, 4302–9.

[13] Valentin, E. and Lambeau, G. (2000) What can venom phospholipases A2 tell us about the functional diversity of mammalian secreted phospholipases A2? *Biochimie* **82**, 815–31.

[14] Thorpe, R.S., Wüster, W. and Malhotra, A. (1997) Venomous snakes. Ecology, evolution and snakebite. 1 vol., Zool. Soc. London, Clarendon Press Oxford, UK, 269 p.

[15] Ogawa, T., Oda, N., Nakashima, K., Sasaki, H., Hattori, M., Sakaki, Y., Kihara, H., Ohno, M. (1992) Unusually high conservation of untranslated sequences in cDNAs for *Trimeresurus flavoviridis* phospholipase A2 isozymes. *Proc. Natl. Acad. Sci. USA* **89**, 8557–61.

[16] Nakashima, K., Nobuhisa, I., Deshimaru, M., Nakai, M., Ogawa, T., Shimohigashi, Y., Fukumaki, Y., Hattori, M., Sakaki, Y. and Hottori, S. (1995) Accelerated evolution in the protein-coding regions is universal in crotalinae snake venom gland phospholipase A2 isozyme genes. *Proc. Natl. Acad. Sci. USA* **92**, 5605–9.

[17] Ohno, M., Ménez, R., Ogawa, T., Danse, J.-M., Shimohigashi, Y., Fromen, C., Ducancel, F., Zinn-Justin, S., Le Du, M.-H., Boulain, J.-C., Tamiya, T. and Ménez, A. (1998) Molecular evolution of snake toxins: Is the functional diversity of snake toxins associated with a mechanism of accelerated evolution. *Progress in Nucleic Acid Resarch and Molecular Biology* **59**, 307–64.

[18] Vidal, N. (2001) Colubroid systematics: evidence for an early apparition of the venomous apparatus followed by extensive evolutionary tinkering. J. Toxicol, in press.

[19] Martin-Eauclaire, M.F., Legros, C., Bougis, P. and Rochat, H. (1999) Les toxines de venins de scorpions. *Ann. Inst. Past/Actual* **10**, 207–22.

[20] Possani, L.D., Merino, E., Corona, M., Bolivar, F. and Becerril, B. (2000) Peptides and genes coding for scorpion toxins that affect ion-channels. *Biochimie* **82**, 861–8.

[21] Ehret-Sabatier, L., Loew, D., Goyffon, M., Fehlbaum, P., Hoffmann, J.A., Van Dorsselaer, A. and Bulet, P. (1996) Characterization of novel cystein-rich antimicrobial peptides from scorpion blood. *J. Biol. Chem.* **271**, 29537–44.

[22] Bontems, F., Roumestand, C., Gilquin, B., Ménez, A. and Toma, F. (1991) Refined structure of charybdotoxin: common motifs in scorpion toxins and insect defensins. *Science* **254**, 1521–3.

[23] Ménez, A., Bontems, F., Roumestand, C., Gilquin, B. and Toma, F. (1992) Structural basis for functional diversity of animal toxins. *Proc. of the Royal Soc. of Edimburgh* **99B**, 83–103.

[24] Goyffon, M. and Roman, V. (2001) Radioresistance of scorpions. In *Scorpion Biology and Research* (P. Brownell and G. Polis, eds), 1 vol. Oxford Univ. Press, New York, USA, 393–405.

[25] Werner, T., Ferroni, S., Saermark, T., Brack-Werner, R., Banati, R.B., Mager, R., Steinaa, L., Kruetzberg, G.W. and Erfle, V. (1991) Nef protein exhibits structural and functional similarity to scorpion peptides interacting with K^+ channels. *AIDS* **5**, 1301–8.

[26] Mebs, D. (1992) Gifttiere: ein Handbuch für Biologen, Toxikologen, Ärzte und Apotheker. 1 vol, Wiss Verb-Ges, Stuttgart, Germany, 272 p.

[27] Mebs, D. (1994) The strategic use of venom and toxins by animals. *Universitas* **36**, 213–222.

[28] Brygoo, E.R. (1985) La découverte de la sérothérapie antivenimeuse en 1894. Phisalix et Bertrand ou Calmette? *Bull. Assoc. Anc. El. Inst. Past.* **4**, 10–22.

[29] Bon, C. and Goyffon, M. (1996) Envenomings and their treatments. I vol. Fond. Marcel Mérieux, Lyon, 343 p.

[30] Phisalix, C. (1897) Venins et animaux dans la série animale. *Rev. Scientif.* **33**, 1–68.

[31] Faure, G. (2000) Natural inhibitors of toxic PLA2. *Biochimie* **82**, 833–40.

[32] Domont, G.B. (2000) Natural immunity mechanisms: inhibition of myotoxin, metalloprotease and PLA2 activities from snake venoms. *Communic. XIIIth World Congress IST*, Paris.

[33] Meier, J. and White, J. (1995) Handbook of clinical toxicology of animal venoms and poisons. 1 vol., C.R.C. Press, Boca Raton, 752 p.

26 Are Inhibitors of Metalloproteinases, Phospholipases A$_2$ and Myotoxins Members of the Innate Immune System?

JONAS PERALES and GILBERTO B. DOMONT[*]

26.1 INTRODUCTION

The absolute values for global incidence of envenomations and their severity remain unknown or misunderstood. In spite of incomplete statistics, fragmentary data show that snake-bites are still a public health problem in many countries, specially if chronic morbidity, like member amputations, deformations and renal insufficiency, are computed. Global estimates of ophidian accidents number five million cases per year, accounting for 50 000 deaths, specially in the rural areas of tropical countries of Asia, South America and Africa [1–3].

Snake venoms are a charming subject of study. Snakes and their venoms are synonymous with treachery and death, history and legends, fear and terror, witchery and science. Particularly charming is the study of the resistance against snake venoms of animals that, unlike humans, are spontaneously protected against the action of toxins provided by these venoms and, much more, whose resistance can be transferred to other experimentation animals. Throughout history, play-writers and novelists could have improved the 'suspense' of their plots had they had a little knowledge of the natural protection against snake venoms. And they could have done that from about two hundred and twenty years ago, beginning in 1781, when Abbé Felice Fontana, the founder of modern toxinology, had his classic text *Traité sur le Venin de la Vipère* translated from Italian to French. This book, remarkable in its clarity of thought, reports accurate experiments on some fundamental snake venom properties, like myotoxic activity and coagulability and, paradoxically, fluidity [4, 5].

[*] Corresponding author

Perspectives in Molecular Toxinology
Edited by A Ménez © 2002 John Wiley & Sons, Ltd

More than a century after the rigorous establishment of the phenomenon of natural resistance against snake venom by Fontana, his celebrated aphorism 'the venom of the viper is not venomous to its own species' became an inspiring broader focus of study. In 1922, Phisalix, in the book *Animaux Venimeux et Venins* [6] reviews experiments to test the resistance of an incredible number of animals – fishes, batrachians, invertebrates, and mammals – to snake venoms, some of which are modern subjects of study, like the hedgehog (*Erinaceus europaeus*) [7], the mongoose (*Herpestes edwardsii*) [8, 9] and the opossum (*Didelphis marsupialis*) [10]. In the last 20 years, the emphasis on collection of data on animals that resist the deleterious action of snake venom changed to the full chemical and biological characterisation of isolated proteins responsible for the protection. The last decade has witnessed the publication of several reviews dealing with inhibitors of snake venom haemorrhagins [11, 12] and of myotoxins and neurotoxins [13, 14].

The main purpose of this chapter is to look back to the last important concepts that were developed in the field of natural inhibitors and to look forward to the perspectives opened up by these discoveries.

26.2 SNAKE VENOM METALLOPROTEINASES (SVMPs)

The term metalloproteinase comprises a wide range of peptidases classified in at least 30 families grouped in five clans according to their metal (normally zinc) chelating consensus sequences. The second clan (MB) has the zinc metal atom bound to three histidine residues: two from the metal-binding consensus sequence (HEXXH) and a third localised six residues after the second histidine [15]. The Snake Venom MetalloProteinases (SVMPs) are classified in the astacin family M12, reprolysin (or adamalysin) subfamily of clan MB. Until 1995, 102 known SVMPs from 32 snake species were described. Many of them are isoenzymes from same snake species or homologous enzymes of different species [16]. These enzymes can be grouped, according to their domain composition into four classes: P-I, which comprises enzymes with the proteinase domain only; P-II, with a proteinase and disintegrin-like domains; P-III, which have proteinase, disintegrin-like and cysteine-rich domains; and P-IV, which besides the three domains found in P-III displays a fourth lectin-like domain. Metalloproteinases are typical components of venomous snakes from the Viperidae family [17]. The Reprolysin subfamily includes also the ADAMs (A Disintegrin-like And Metalloproteinase) also known as MDC (Metalloproteinase, Disintegrin-like, Cysteine-rich), which are, most often, integral multidomain membrane proteins that present around 25% sequence identity when compared to class SVMPs [18]. Although their exact mechanism of action is still unknown, SVMPs induce local haemorrhage, apparently as a primary consequence of the degradation of proteins present in the basal membrane of vessels, which leads to the loss of capillary and venule integrities and, consequently, to blood extravasation [19, 20]. In general, SVMPs can degrade fibrinogen, fibronectin, laminin, collagen IV and nidogen; native molecules of collagen I, II and V are not degraded by SVMPs whereas their gelatines are [21].

26.3 SNAKE VENOM METALLOPROTEINASE INHIBITORS (SVMPIs)

Metalloproteinase inhibitors have high thermal and pH stabilities, seem to be glycoproteins and have an acidic nature with pIs ranging from less than 3.5 to 5.4 [13, 14]. Tables 26.1 and 26.2 summarise the state-of-the-art of the anti-metalloproteinase inhibitors known to exist in/or that have been isolated from the plasma, serum or muscle of snakes and mammals.

26.3.1 CLASSIFICATION

A critical analysis of the data shown in Tables 26.1 and 26.2 indicates that, according to their properties, the best parameters to be used to classify these metalloproteinase inhibitors are their physicochemical and chemical properties, excepting their amino acid sequence, which is used as the second criterion. Through the first, we can classify them into low and high molecular mass; through the second, we can glance at their genic families and, consequently, at the biological implications of their structures.

Twenty-two inhibitors have been characterised to different extents. The low molecular mass group has masses ranging from 40–95 kDa whereas the high molecular mass group ranges from 700–1090 kDa. Independently of the molecular masses, attention must be paid to the experimental presence of oligomers in the native state.

26.3.2 HIGH MOLECULAR MASS CLASS

The only known high molecular mass inhibitor isolated from a snake is NtAH present in the serum of nonvenomous *Natrix tesselata* [22]. In the native state this inhibitor has a molecular mass of 880 kDa made of three different polypeptide chains of 70, 100 and 150 kDa in a still unknown ratio. The other high molecular mass inhibitors are from *Erinaceus europaeus* plasma [23] and muscle [7, 24]. The first showed a molecular mass of 700 kDa by native PAGE. After reduction and SDS-PAGE it furnished two bands of 34 and 39 kDa [23]. No data using only denaturing conditions without reduction were published, which renders difficult the comparison with the erinacin complex. The latter inhibitor, isolated from muscle [24], was shown, after exposure to guanidine hydrochloride, to dissociate into α- and β-subunits oligomers in a 1:2 ratio. The α-subunit

Table 26.1 Data on metalloproteinase inhibitors isolated from snake plasma or serum.

Snake Plasma/serum	Name	MM (kDa)		Complex formation Yes/ Ratio[a]		Amino acid sequence	Super family	pI
		Native	Subunit	Yes	Ratio[a]			
Venomous								
Agkistrodon c. mokassen[1]	—	nd	64–68[b]	nd		nd	nd	4.6
Bothrops asper[2]	BaSAH	66	-	Y	2:1	nd	nd	5.2
Bothrops jararaca[3]	BJ46a	79[c]	46	Y	1:2[d]	cDNA	Cystatin	4.6
Crotalus atrox[4]	—	nd	65–80[b]	Y		nd	nd	nd
Trimeresurus flavoviridis[5]	HSF	nd	70	Y		Total, Edman	Cystatin	4.0
Trimeresurus mucrosquamatus[6]	TMI	nd	47	nd		N-terminal	nd	nd
Vipera palaestinae[7]	—	nd	80	nd		nd	nd	4.7
Non-venomous								
Dinodon semicarinatus[8]	—	59	52	nd		nd	nd	nd
Natrix tesselata[9]	NtAH	880	70, 100, 150[e]	Y	4:1	nd	nd	4.5

a. inhibitor: enzyme (mol:mol); b. from – to; c. dimer; d. inhibitor monomer:enzyme (mol:mol); e. not subunits, but polypeptide chains separated by SDS-PAGE after reduction; nd. not determined. Ref: 1. [29, 30]; 2. [22, 31]; 3. [26, 32]; 4. [33, 34]; 5. [25, 35, 36]; 6. [37]; 7. [38]; 8. [39]; 9. [22]

Table 26.2 Data on metalloproteinase inhibitors isolated from plasma/serum or muscle of mammals.

Animal Plasma/serum	Name	MM (kDa)		Complex formation		Amino acid sequence	Family	pI
		Native	Subunit	Yes	Ratio			
Didelphis albiventris[1]	DA2-II	nd	43-45	nd	nd	N-terminal[a]	nd	nd
Didelphis marsupialis[2]	DM40	95[b]	40	Y	1:1	N-terminal[a]	Ig supergene	<3.5
	DM43	72[b]	43	Y	1:1	Total, Edman	Ig supergene	<3.5
Didelphis virginiana[3]	Oprin	nd	52	Y	nd	Partial, cDNA and Edman[c]	Ig supergene	3.5
Erinaceus europaeus/plasma[4]	βMG	700[d]	34 and 39	nd	nd	nd	nd	nd
Erinaceus europaeus/muscle[5]	Erinacin	1090[d] $\alpha_{10}-2\beta_{10}$	$\alpha = 37.5$[e] $\beta = 34.7$	Y	1:1	α and β, Edman, partial[f]	Ficolin / opsonin P35	nd
Herpestes edwardsii[6]	AHF-1	65	69	Y	nd	Edman, partial[g]	Ig supergene	nd
	AHF-2	65	69	nd	nd	N-terminal[a]		
	AHF-3	65	69	nd	nd			
Lutreolina crassicaudata[7]	nd	nd	48	nd	nd	N-terminal[a]	Ig supergene	nd
Neotoma micropus[8]	nd	nd	54	nd	nd	nd	nd	4.1
Philander opossum[9]	PO41	87	41.3	Y	nd	N-terminal[a]	Ig supergene	<3.5
Sigmodon hispidus[10]	nd	90	nd	nd	nd	nd	nd	5.4
Spermophilus mexicanus[11]	nd	nd	52	nd	nd	nd	nd	4.9

a. Edman; b. dimer; c. ca. 80%; d. MM of the oligomer; e. MM of the monomer after reduction of the 350 kDa subunit; f. α subunit: N-terminal; β subunit: ca. 75%; g. ca. 76%; nd: not determined. Ref.: 1. [40, 41]; 2. [10, 42]; 3. [27]; 4. [23]; 5. [7, 24]; 6. [8, 9, 43]; 7. [44]; 8. [45]; 9. [46]; 10. [47]; 11. [48]

exists as monomers in the erinacin molecule and the β-subunits as decamers cross-linked by disulphide bridges. In summary, erinacin seems to have the most complex structure among all known metalloproteinase inhibitors isolated thus far; its molecular form being α_{10}-$2\beta_{10}$ [7].

26.3.3 LOW MOLECULAR MASS CLASS

The most well-characterised low molecular mass inhibitors are HSF from *Trimeresurus flavoviridis* [25], BJ46a from *Bothrops jararaca* [26], oprin (*Didelphis virginiana*) [27], AHF-1 (*Herpestes edwardsii*) [8, 9] and DM43 (*Didelphis marsupialis*) [10, 28]. HSF (323 residues long) and BJ46a (322 residues) have three and four putative N-glycosylation sites and share 85% sequence identity. Both have an RGD sequence at the N-terminal portion whose role in the inhibition of haemorrhagins is still obscure. Whereas, HSF's sequence was determined by the classical Edman degradation of peptides obtained by enzymatic hydrolyses, BJ46a had its sequence determined from a full-length cDNA. There is no pro-region in the BJ46a structure and no C-terminal processing is required after translation of the coding region into the mature BJ46a subunit [25, 26].

DM43 is a 291 residues long, single-chain glycoprotein, possessing four N-linked putative glycosylation sites and three disulphide bridges [28]. It is a homologue of α1B-glycoprotein (37% identity), a human plasma protein of unknown function and of two other partially sequenced inhibitors, oprin, from *Didelphis virginiana* [27] and AHF1- from the mongoose, *Herpestes edwardsii* [8, 9].

26.3.4 GENIC FAMILY

The second classification criterion indicates that, according to their amino acid sequence, these metalloproteinase inhibitors fall into three main protein families. Thus, HSF and BJ46a belong to the fetuin family of the cystatin superfamily of proteinase inhibitors [25, 26], while DM43, oprin, and AHF-1 fall into the immunoglobulin supergene family [8–10, 27, 28]. Another inhibitor with 75% of its amino acid sequence already determined is erinacin, which was shown to be part of the ficolin/opsonin P35 lectin family [7] (Tables 26.1 and 26.2).

The amino-terminal regions of the α and β-subunits of erinacin have almost identical sequences. Further sequence analysis of the β-subunit revealed high homology with the ficolins- α and -β (TGF-1β-binding proteins from porcine uterus), with P35 (opsonin protein) and with the Hakata antigen. These proteins have been classified as the ficolin/opsonin P35 lectin family. The members of this family share common characteristics like their N-terminal sequence as well as a collagen- (GXY-triplet repeats) and fibrinogen-like domains. Similarly to plasma ficolin and Hakata antigen, erinacin showed a flower bouquet-like structure in electron microscopy. According to this molecular arrangement, the bundle, stem and flower parts of erinacin correspond to the amino-terminal, collagen- and fibrinogen-like domains, respectively [7].

Both HSF and BJ46a have cysteine positions typical of the cysteine pattern observed for the fetuin group, except that HSF has one more cysteine at position 44, substituted for a valine in BJ46a [25, 26]. Although classified as members of the cystatin superfamily, the fetuins, including HSF and BJ46a, are unlikely to inhibit cysteine proteinases since the consensus sequence (QXVXQ) for the cystatin active site has been extensively altered in these molecules. This view was corroborated for HSF, which did not inhibit papain and cathepsin B but did inhibit several metalloproteinases in Habu venom [25].

DM43 possesses three Ig-like domains homologous to those found in the N-terminal region of α1B-glycoprotein, a five-domain protein. More distant relatives are the extra-cellular portions of the natural killer cell inhibitory receptors, KIR2 and KIR3, which have two and three Ig-like domains, respectively. These molecules are membrane-bound receptors which specifically detect the presence of class I HLA molecules on host cells [49]. One of the most striking features of the DM43 amino acid sequence is the presence of a degenerate WSXWS box [50] in each domain. The most authentic of the WSXWS box motifs is observed in domain D2 [28]. Homology modelling based on the crystal structure of a killer cell inhibitory receptor (KIR2DL1, PDB code 1nkr) confirmed that the DM43 conformation may exist as a three-domain (D0 , D1 and D2) protein. Each domain would adopt an Ig fold of the I-type. An acute elbow angle is formed between domains D1 and D2, bringing together six exposed loops on the convex surface formed by them. By analogy with hematopoietic receptors, like the ones for growth hormone [51] and prolactin [52], for which crystal structures of the ligand–receptor complexes are known, these loops are expected to form the binding surface for the ligand. In the case of DM43, this region is predicted to form the metalloproteinase-binding site. Analyses of the molecular surface of the models showed also the presence of exposed hydrophobic clusters on β-strands B and E suggesting that this region corresponds to the homodimeric interface of DM43 in solution [28].

26.3.5 MECHANISMS OF INHIBITION

Erinacin completely inhibited the haemorrhagic activity of *Bothrops jararaca* venom. It formed a one to one complex with jararhagin, a class P-III haemorrhagic SVMP present in this venom, indicating that the inhibitor contains one active binding site for the haemorrhagic proteinase. It is proposed that the fibrinogen-like domain may contribute to the inhibition of the metalloproteinase by recognising the N-acetylglucosamine residue, as reported for P35 and for plasma ficolin. Another suggestion lies in the fact that the collagen-like domain may have an affinity for the haemorrhagic metalloproteinase as a mimic substrate, since SVMPs exhibit activity towards collagen substrates [7].

The mechanism of metalloproteinase inhibition began to be clarified by the work done with BJ46a and DM43. The first is an effective inhibitor of *Bothrops jararaca* whole venom haemorrhagic activity as well as of isolated SVMPs, specifically atrolysin C (P-I) and jararhagin (P-III). BJ46a formed stable non-covalent complexes with these metalloproteinases. The interaction seems to occur with the metalloproteinase domain only, since atrolysin C is a P-I. The stoichiometry of the inhibition is of one inhibitor monomer to two metallopro-teinase molecules. Thus, it was proposed that the BJ46a dimer dissociates during the formation of the enzyme–inhibitor complex and that its monomer has two binding sites for the SVMPs [26].

DM43 completely inhibited the proteolytic activity of jararhagin and bothrolysin (P-I), forming stable noncovalent complexes in a one to one molar ratio. It did not inhibit atrolysin A and poorly inhibited atrolysin C. Also, DM43 did not bind to jararhagin-C, a fragment of jararhagin which lacks the metalloproteinase domain. Again, the inhibition mechanism points to the metalloproteinase domain as the binding site. No proteolysis of either inhibitor or enzyme was observed. Given that the final SVMP:DM43 (subunit) complex has a 1:1 stoichiometry, it is necessary for the DM43 homodimer to dissociate during complex formation. Two generic mechanisms, for which the terms concerted and sequential were coined, have been proposed [28]. In the concerted dissociation model, intermediates with 1:2 and 2:2 stoichiometries (SVMP: inhibitor monomer) are formed along the mechanistic pathway leading to the final dissociation into two 1:1 complexes. In the sequential model, the binding of the first metalloproteinase molecule destabilises the DM43 homodimer, generating a 1:1 complex and a free inhibitor monomer. The latter may either recombine with a second DM43 molecule to form a new homodimer or, alternatively, complex with a second metalloproteinase molecule. In both cases, it was considered reasonable to assume that the final enzyme–inhibitor complex has its dimerisation surface partially blocked or, at least, sterically hidden from the other inhibitor monomer as a consequence of a conformational change, since no stable complex with an oligomeric structure of two inhibitor monomers to one enzyme molecule was detected [28].

26.4 PHOSPHOLIPASES A_2 (PLA$_2$s) AND THEIR RECEPTORS

Although all PLA$_2$ enzymes catalyse the hydrolysis of the same bond, an acyl ester at the *sn*-2 position of glycerophospholipids, they participate in different kinds of important processes like digestion, phospholipid metabolism, signal transduction, membrane remodelling, inflammation, blood coagulation and platelet aggregation. The PLA$_2$s have been classified into several groups on the basis of their primary structure, disulphide bonding pattern, Ca^{2+} requirement and cellular localisation (secreted or cytoplasmatic) [53–55]. However, snake venom PLA_2s belong to two groups only. The PLA_2s of Elapidae snake venoms belong to group I, being equivalent to the pancreatic $sPLA_2$ present in mammals, while PLA_2s from Viperidae belong to group II, which is similar to mammals' nonpancreatic, inflammatory secreted PLA_2 ($sPLA_2$) [55, 56]. Snake venom PLA_2s have been associated with important effects present in the venom as neurotoxic, myotoxic, cardiotoxic, platelet aggregation and oedema-forming activities [57]. These effects may, or may not, be dependent on the catalytic activity. In the case of presynaptic neurotoxins, evidence indicates the existence of a relationship between the PLA_2 catalytic activity and the pharmacological effects of these toxins [58, 59], whereas in the case of myotoxins there is evidence that the myotoxicity is independent of the catalytic activity [60–62]. Many myotoxins with PLA_2 structure, group IIA, have been described in Viperidae venoms; some have a critical substitution at the calcium-binding domain in residue Asp49 (enzymatically active) generating Lys49 myotoxins with extremely low or without catalytic activity [57, 60].

The capacity of the PLA_2s to produce different kinds of effects has been associated not only with its catalytic activity, but also with the presence of specific receptors or protein acceptors of these enzymes in different cells. To date, two main types (M and N) of PLA_2 receptors have been clearly identified. The N-type (neuronal type), more abundant in rat brain [63, 64], and the M-type receptor (muscle type), which was initially detected in rabbit skeletal muscle [65]. The latter receptor has a tandem repeat of eight distinct C-type (Ca^{2+}-dependent lectin) carbohydrate-recognition domains (CRDs) in their extracellular region. It is important to note that CRDs have also been identified in many different proteins, as hepatic lectins, pulmonary surfactant proteins and, interestingly, in some PLA_2 inhibitors isolated from snake blood serum or plasma [59, 66]. It has been shown that the CRDs of the M-type receptors are responsible for $sPLA_2$s binding [59, 67, 68].

26.5 PLA₂ INHIBITORS (PLIS): ANTINEUROTOXIC AND ANTIMYOTOXIC

One of the most important sources of PLIs is snake blood serum. The presence of the inhibitors has been associated with the resistance of the snake to its own venom or to the deleterious effects of their venom PLA_2 components [13, 14].

26.5.1 CLASSIFICATION

Almost all PLIs isolated to date from snake serum or plasma are oligomeric acid glycoproteins, with molecular masses from 75–180 kDa, formed by three to six identical or different subunits (some glycosylated) linked by noncovalent bonds. These PLIs can be classified in three groups based on the homology of their amino acid sequence: PLI α, β and γ. This classification was originally proposed by Ohkura and co-workers [69] for three different PLIs isolated from the blood plasma of the Chinese mamushi, *Agkistrodon blomhoffii siniticus*. The PLIα is homologous to the CRD of C-type lectins (as for example M-type PLA$_2$ receptors and surfactant protein SP-A) [70]. PLIβ has 33% identity to human leucine-rich α$_2$-glycoprotein, a serum protein of unknown function. It has a molecular mass of 160 kDa, is formed by three identical 50 kDa subunits, and has nine leucine-rich repeat (LRR) domains, each with 24 amino acid residues [71]. The PLIγ group is characterised by the presence of two tandem patterns of cysteine residues constituting two internal typical three-fingered shaped motifs of the urokinase-type plasminogen activator receptor (u-PAR), cell surface antigens of the Ly-6 superfamily [72]. Interestingly, it has been proposed that these proteins have an evolutionary relationship with three-fingered neurotoxins and cytotoxins from Elapidae venoms [73]. Recently, some other PLIs have been isolated from Australian elapid blood sera but the reported structural data were insufficient to allow their classification [74]. The members of each group are described in Table 26.3.

26.5.2   SPECIFICITY

It is now clear that the PLI have different specificity levels (Table 26.3). The only PLIβ isolated from *A. b. siniticus* was shown to be a selective inhibitor of the group-II basic PLA$_2$ from Crotalidae venom. It did not, however, inhibit the catalytic activity of other PLA$_2$s from snake venoms, including the neutral and acidic PLA$_2$s of its own venom [69, 71]. The PLIα isolated from *A. b. siniticus* and *T. flavoviridis* [69, 81, 82, 88, 89] have been reported to inhibit with high affinity the group II acidic PLA$_2$s of their own venoms. This is not the case with the PLA$_2$ inhibitor isolated from the plasma of *B. asper* (BaMIP), which inhibited the enzymatic activity of *B. asper* basic myotoxins I and III [83]. Finally, some γ-type PLI isolated from *N. n. kaouthia* [72, 90], *N. ater* [77], *L. semifasciata* [76], *E. quadrivirgata* [85] and *P. reticulatus* [86] sera showed less specificity because they inhibited snake and mammal PLA$_2$s from groups I, II and III. The PLIγ from the serum of *C. d. terrificus* was isolated by two groups, working separately. It was named CNF (<u>C</u>rotalus <u>N</u>eutralizing <u>F</u>actor, standing for the 23.6 kDa subunit) by one group [79] and CICS (<u>C</u>rotoxin <u>I</u>nhibitor from <u>C</u>rotalus <u>S</u>erum, standing for the oligomeric protein) by the other [80]. CICS or CNFn (standing for the oligomer composed of n CNFs) inhibited PLA$_2$s from Viperidae snakes venoms but did not inhibit human nonpancreatic (group II), porcine pancreatic (group I), β-bungarotoxin (group II) or bee venom (group III) PLA$_2$ enzymes [80, 91]. CNFn/CICS inhibited not only crotoxin, the principal neurotoxin from *C. d. terrificus* venom, but also the monomeric

Table 26.3 Data on isolated neurotoxin, myotoxin and PLA$_2$ inhibitors.

ANIMAL	Inhibitors Names	Type	Molecular Mass (kDa)	Activity Inhibited	Specificity
Snake	CSAP	Alb		Neurotoxic	Post-synaptic neurotoxins
Elapidae					
Naja naja atra[1a]			70		
Naja naja kaouthia[2]	NkPLI	γ	90: (25–31×3) He[b]	nd	Broad PLA$_2$ (I-II-III)
Laticauda semifasciata[3]	LsPLI	γ	100: (20–25×3) He	nd	Broad PLA$_2$ (I-II-III)
Notechis ater[4]	NAI	γ	110: (30–25×3) He	Lethal	Broad PLA$_2$ (I-II-III)
Notechis scutatus[5]	NSI	—	110: (22.5–19.8×3) He	Lethal	Broad PLA$_2$ (I-II-III)
Notechis a. serventyi[5]	NAsI	—	(22.5–19.8) He	nd	nd
Oxyuranus scutellatus[5]	OSI	—	(22.7–19.8) He	nd	nd
Oxyuranus microlepidotus[5]	OMI	—	(22.8–19.9) He	nd	nd
Pseudonaja textilis[5]	PTI	—	(22.5–19.8) He	nd	nd
Viperidae					
Agkistrodon b. siniticus[6]	PLIα	α	75: (20×3) Ho	nd	PLA$_2$ II-Acidic
	PLI β	β	160: (50×3) Ho	nd	PLA$_2$ II-Basic
	PLI γ	γ	100: (20–25×3) He	nd	Broad PLA$_2$ (I-II-III)
Crotalus d. terrificus[7]	CICS	γ	130: (23–25×6) He	Neurotoxic	Snake PLA$_2$ II
Trimeresurus flavoviridis[8–9]	PLI-A/PLI-B	α	100: (21–22×4) He	nd	PLA$_2$ II-Acidic
	PLI-I	γ	(27×?)		
Bothrops asper[10]	BaMIP	α	120: (24×5) Ho	Myotoxic	PLA$_2$ II-Basic
Cerrophidion godmani[11]	CgMIP-I	γ	110: (20–25×?) ?	Myotoxic (I)	PLA$_2$ II D49[d]
	CgMIP-II[c]	α	180: (20–25×?) ?	Myotoxic (II)	PLA$_2$ II K49[e]
Vipera palaestinae[12a]	—	—	56	Neurotoxic	nd

continues overleaf

Table 26.3 (continued)

ANIMAL	Inhibitors Names	Type	Molecular Mass (kDa)	Activity Inhibited	Specificity
			Nonvenomous		
Elaphe quadrivirgata[13]	EqPLI	γ	130: (29–30×?) He	nd	Broad PLA$_2$ (I-II-III)
Python reticulatus[14]	PIP	γ	140: (23×6) Ho	Lethal Edematogenic	Broad PLA$_2$ (I-II-III)
			Mammals		
Didelphis marsupialis[15c]	DM64	α$_1$B	64	Myotoxic	PLA$_2$ II K49[d] PLA$_2$ II D49[e]

a – Inhibitory properties were not tested; b – molecular mass native protein: (molecular masses of subunits × number of subunits); c – does not have PLA$_2$ inhibitory properties; d – enzymatically active; e – enzymatically inactive; He = heteropolymer; Ho = homopolymer ; α$_1$B = α$_1$ B-glycoprotein; nd – not determined. Ref.: 1. [75]; 2. [72]; 3. [76]; 4. [77]; 5. [74]; 6. [69]; 7. [78–80]; 8. [81]; 9. [82]; 10. [83]; 11. [61]; 12. [84]; 13. [85]; 14. [86]; 15. [87]

neurotoxins agkistrodotoxin (Agtx) from *Agkistrodon blomhoffii brevicaudus* and ammodytoxin (Atx) from *Vipera ammodytes ammodytes*, the multimeric neurotoxins and their enzymatically active subunits, CbICbII from *Pseudocerastes fieldi*, mojave toxin from *Crotalus scutelatus scutelatus* [91] and four PLA_2 (one basic and three acidic ones) from *Lachesis muta muta* venom [92]. PLIα, PLIβ and CICS/CNFn (PLIγ) are present in Viperidae snakes as well as the PLA_2s that they inhibit, a fact that could indicate that they afford specific protection to this family of snakes against their own venom.

26.5.3 MECHANISM

The formation of a stable soluble complex between PLI and PLA_2 seems to be responsible for the inhibition of the enzymatic activity, independently of the PLI type. The PLIα from *T. flavoviridis* formed a stable complex with a PLA_2 from its own venom in an equimolecular ratio, with or without Ca^{2+} [93]. A few years ago, it was shown that all three CRD-like subunits of the PLIα would be necessary to bind one PLA_2 molecule [88], suggesting that the interaction between the α inhibitors and the M-type receptors with the $sPLA_2$s would be similar. It was shown that the carbohydrate moiety of this PLIα is not necessary for the binding with the PLA_2s. In the case of PLIα from *A. b. siniticus*, the carbohydrate moiety seems to play an important role in the specificity of the PLA_2 inhibition, at least with basic enzymes [94]. The presence of soluble M-type receptors with binding and inhibitory PLA_2 activities has already been reported [55]. It has also been shown that another soluble protein with a CRD domain, the mammalian lung surfactant protein A (SP-A), inhibited a Habu snake venom PLA_2 [95]. These facts suggest that these soluble proteins could behave as physiological $sPLA_2$s inhibitors [55]. In the case of PLIβ, it has been shown that one PLIβ trimer formed a complex with three basic PLA_2 molecules, indicating that one basic PLA_2 molecule would bind stoichiometrically to one subunit of PLIβ. It was also mentioned that the leucine-rich repeat domains present in PLIβ might be responsible for the specific binding to PLA_2, because these LRRs are considered to be motifs involved in protein–protein interaction [71].

The intimate molecular mechanism of inhibition is not completely known. CICS/CNF [79, 80, 91] also forms a complex with crotoxin, the principal heterodimeric neurotoxin from *C. d. terrificus*. CICS interacts only with CB, the catalytic subunit of crotoxin, and not with CA, the nonenzymatic subunit, forming a stable enzymatically inactive complex CICS-CB, where CICS replaces CA. CA alone does not form a complex with CICS. Similar results were obtained with other Viperidae PLA_2 neurotoxins. No complexes between CICS and Elapidae neurotoxins or pancreatic and nonpancreatic mammal PLA_2s could be detected [80, 91]. The interaction mechanism of CICS with crotoxin seems to be similar to that of crotoxin with its membrane protein acceptor [96, 97]. Therefore, it was proposed that CICS acts physiologically as a false crotoxin acceptor to retain the toxin in the snake vascular system, preventing its action on the neuromuscular system [80]. Fortes-Dias *et al.* proposed the existence of a complex between one CNF subunit and CB (1:1 molar ratio) [79]. In contrast, and supported by the molecular mass of the complex determined by several chromatographic and electrophoretic techniques, Perales and colleagues postulated that CICS:CB exists as a complex between one CICS oligomeric molecule and one CB subunit. It was shown that glycosylation of CICS is not necessary for the formation of the CB:CICS complex [80, 98].

Few structural data are available on the PLIγ domains that participate in the interaction with PLA$_2$s. In a binding study with truncated recombinant forms of PLI-I from *T. flavoviridis*, Nobuhisa and co-workers showed that one of the two three-fingered motifs was able to bind to PLA$_2$ isozymes [82]. On the other hand, it was shown that recombinant soluble uPAR, which is structurally similar to PLIγ and contains three intramolecular repeats of the cysteine-rich domain, binds to crotoxin and to its catalytically active subunit CB [59]. These results indicate that the three-fingered motif is important not only for the interaction between PLIγ and PLA$_2$ enzymes, but probably also for the interaction between the receptors and agonists that might have this motif.

26.5.4   INHIBITORS OF SNAKE VENOM PLA$_2$s TOXIC EFFECTS

It has been shown that CICS/CNF, besides inhibiting phospholipase enzymatic activity, also neutralises the lethal effects of *C. d. terrificus* venom, crotoxin and its CB subunit [78–80]. Recently, Faure *et al.* (2000) reported a strong protective effect of CICS against the neurotoxic action of β-neurotoxins (including crotoxin) from some Viperidae venoms on primary and evoked acetylcholine (ACh) release from *Torpedo marmurata* synaptosomes. This inhibition by CICS of the *in vitro* effect of Viperidae β-neurotoxins could correspond to its neutralising effects demonstrated *in vivo* in lethality assays, at least in the case of crotoxin, maybe by blocking the binding to their pharmacologically active synaptosome receptors. Our results are in agreement with the proposition that the catalytic activity of the PLA$_2$ β-neurotoxins is necessary, but not sufficient, for its neurotoxic action. Antilethal and antiedematogenic activities of a PLIγ (PIP) isolated from the nonvenomous snake *Phyton reticulatus* also showed that the recombinant protein conserves a broad specificity for PLA$_2$. The antineurotoxic effects of PIP were not demonstrated [86].

Factors have been noted in the sera of the marsupials *Didelphis marsupialis* and *Philander opossum* that inhibit the release of sarcoplasmic enzymes (CK and LDH) induced by *Bothrops jararacacussu* venom, suggesting neutralisation of its myotoxic effect [99]. More recent work showed that *Pseudechis australis* serum also conferred almost complete protection against myonecrosis induced by its homologous venom [100]. The first well-characterised PLI with antimyotoxic activity was isolated from the blood of *Bothrops asper* (BaMIP) in 1997 [83]. Its N-terminal sequence displays an identity ranging from 68 to 78% when compared with other PLIα, therefore indicating the existence of the CRD structure. In addition to its PLA$_2$ inhibitory activity against *B. asper* basic myotoxins I and III, BaMIP also inhibited the myotoxic and oedema-forming activities *in vivo* and cytolytic activity *in vitro* towards cultured endothelial cells of all four myotoxin isoforms (I-IV) isolated from *B. asper* venom, whether or not they have PLA$_2$ activity. *B. asper* myotoxins I and III are group II PLA$_2$ enzymes with catalytic activity, while myotoxin II is one of the catalytic inactive isoforms. All isoforms are myotoxic *in vivo* and show toxicity against endothelial cells *in vitro*, showing that the enzymatic activity of PLA$_2$ is not needed in these processes. From *Cerrophidion godmani* were isolated, characterised and cloned two myotoxin inhibitors belonging to two different types of PLIs, γ-type CgMIP-I, and α-type CgMIP-II. CgMIP-I specifically neutralised the PLA$_2$, oedema-forming, myotoxic, and cytolytic activities of the enzymically active myotoxin I, whereas CgMIP-II selectively inhibited the toxic properties of enzymically inactive myotoxin II. Such pharmacological selectivity may be explained by critical differences in the structures of myotoxins, inhibitors, or both [61].

Recently, the isolation and partial characterisation of an antimyotoxic factor from *D. marsupialis* serum was reported [62]. It is an acidic glycoprotein with molecular mass of 63.650 amu by MALDI-TOF MS. Sequencing data showed homology with α_1B-glycoprotein and with the SVMP inhibitors DM43 and DM40. DM64 completely inhibited the *in vivo* myotoxic activity of myotoxin-I and myotoxin-II from *B. asper* venom as well as their *in vitro* cytotoxicity. DM64 formed complexes with myotoxins I (D49) and II (K49). Finally, DM64 did not inhibit the PLA$_2$ activity present in myotoxin I nor did it neutralise the anticoagulant effect or the intracerebroventricular lethality associated with this myotoxin. These data indicate that these two pathological effects depend on the enzymatic activity of the toxin. When these results are taken together, it becomes clear that the myotoxicity and the catalytic activity are independent phenomena.

26.6 PERSPECTIVES

Consideration of the perspectives of a research field necessarily reflects some-what idiosyncratic speculations. Thus, we can indulge in conjectural thoughts on the research problems that need to be pursued in this area, on the use of these molecules as research tools and on their trip from the bench to the bedside. The structural richness of these inhibitors can be considered as a source of inspiration in solving research and/or therapeutic problems. Researchers should be aware of the specificity and selectivity of the properties of inhibitors. The combination of these characteristics may be useful in finding experimental conditions to propose the elucidation of still unsolved questions. They should find and use the best of the pest.

The first option, obviously, refers to the most important questions that must be answered and to the problems that deserve to be investigated in the near future. In the case of all natural inhibitors, besides the search for new inhibitory molecules, so that a broader view of their families can be ascertained, a deeper insight into their structures, conformations, and mechanism(s) of action is needed. Within the next few years it will be necessary to know the three-dimensional conformation of inhibitors from different families, their domain organization, their polypeptide fold and the main atoms or groups that determine their specificity, as well as a clear picture of the conformation of enzyme–inhibitor complexes. Certainly, this new knowledge will improve the chemical and/or biological modifications of these natural inhibitors or the synthesis of new ones, with high affinity ligands, that could be used as drugs.

One possible goal for their use is in the therapy of snake envenomation. Metalloproteinase inhibitors have the special property of neutralising local and systemic effects of the SVMPs, a kind of performance attained also by the specific anti-sera prepared against snake venoms still used as the therapeutic weapon against the deleterious action of the venom *in vivo*.

A tempting therapeutic/research tool proposition is to test the SVMPIs against the ADAMs and the PLIs against the PLA$_2$s. The appeal of testing is very strong, since, (a) SVMPs display at least 25% sequence identity with the metalloproteinase domain of the ADAMs [11], and (b) the ADAMs are enzymes that have functional roles in various physiological processes such as proteolysis of the extracellular matrix, extracellular communication and/or intracellular signalling, processing of plasma membrane proteins, proteolysis in the secretory pathway and procytokine conversion [101]. Also, some PLIs could be assayed, for instance, against secretory inflammatory phospholipases or in other physiological PLA$_2$-dependent processes.

Another specific example of the use of these inhibitors as molecular tools derives from the finding that DM64 inhibits the myotoxic activity of *B. asper* myotoxin-I without interfering with its PLA$_2$ activity. This could provide a model system for the mapping of the active myotoxic and enzymatic regions of the toxins [62]. Therapeutically, these data suggest its possible use as an inhibitor of the myotoxic effects of Viperidae venoms without interfering in the activity of endogenous phospholipases A$_2$. In terms of PLA$_2$, the selective action of some PLIs on snake venom enzymes is another important property to be considered. It has been shown experimentally that PLIα and PLIβ specifically inhibit group II snake venom PLA$_2$s. If the venom toxic effect is due to the phospholipase activity, these inhibitors could be assayed for therapeutic use. Examples of inhibitors that neutralise both the toxic and the enzymatic activities are PIP [86] and CICS [80, 91]. These two inhibitors, or regions of their polypeptide chains, could prove useful in the treatment of snake envenomation.

26.7 FINALLY, AN ANSWER TO THE TITLE

Healthy individuals protect themselves against pathogens by means of many different mechanisms: physical barriers, a class of lymphocytes called natural killer (NK) cells, phagocytic cells in the blood and tissues, and various blood-borne molecules like those of the complement system. The goal is to defend individuals against a potentially hostile environment. Defence mechanisms are called natural or innate immunity when they satisfy three requirements: they are present prior to exposure to infectious microbes or other foreign macromolecules; they are not enhanced by such exposures; and they do not discriminate among most foreign substances. Prior to the evolution of vertebrates, host defence against foreign invaders was mediated largely by the mechanisms of natural immunity and, probably, included phagocytic cells and circulating molecules like components of the mammalian complement system. Phagocytes and complement cannot discriminate between distinct antigens and are not specifically enhanced by repeated exposures to the same antigen [102].

Now the aim is to compare this concept to the genetically inherited resistance that certain animals possess against the action of snake venom toxins. So far, this resistance has been noted against metalloproteinases, phospholipases A_2 and myotoxins. It is afforded by two processes. In the first, inhibitors present in the circulating blood or on muscle are responsible for the inhibition of the activity of metalloproteinases, phospholipases A_2 or myotoxins. In the second, toxins are not able to bind to mutated synaptic receptors, a process beyond the scope of this article.

It is our understanding that a conceptual functional framework arises from the data described and that this concept can be translated into natural or innate immunity, that is, the natural resistance to snake venoms expressed by these animals is a component of their innate immune system. All known natural inhibitors can be considered as soluble acceptors of metalloproteinases, phospholipases or myotoxins. This property makes these proteins readily available in the plasma to complex with foreign or self-proteins (with deleterious action against the self) that invade the blood system. As such, they perform the role of molecules that belong to the first line of defence, acting instantaneously against molecular invaders and satisfying the most important condition for a member of the innate immune system.

Analysis of the data allow us to classify these inhibitors into immune system structurally related proteins or into proteins that have no kinship with these kinds of molecules. The first group includes metalloproteinase inhibitors of the immunoglobulin genic family (Table 26.2). Furthermore, DM43, the most well characterised inhibitor of its class, also shows sequence homology with other molecules that belong to the natural immunity system, like natural killer receptors, the KIRs. Additionally, they have Ig-like signatures, as the disulphide bridges that link β-strands B and F and the aromatic residues originating in β-strand C. Conformationally, these conditions are met by the existence of an Ig-like fold [28] and biologically the protection afforded by all serum metalloproteinase inhibitors against foreign toxins indicates that they perform functions of the innate immune system. This group also includes DM64, an inhibitor of myotoxicity related to α1B-glycoprotein [62]. The second group comprises inhibitors that are not structurally related to members of the immune system, but that perform functions of it. This group is represented by erinacin, a high molecular mass SVMPI possessing ficolin/opsonin P35 lectin domains [7] that display structural and functional similarities to collectins. These proteins are involved in the first line of host defence (by opsonisation) against microorganisms possessing certain oligosaccharides at their surface [103]. Again, in terms of sequence and function a metalloproteinase inhibitor of a different protein family is kin to proteins that perform functions of the innate immunity system.

No similarity to proteins of the immune system has been found for the cystatin family of snake venom metalloproteinase inhibitors. The same can be said of the PLA$_2$s and myotoxic inhibitors from snake blood. Their property of being circulating defence proteins, like the complement family of proteins, is the only, albeit very important one that could, at the moment, warrant their position in the innate immune system.

ACKNOWLEDGEMENTS

The authors gratefully acknowledge grants from the Conselho Nacional de Desenvolvimento Científico e Tecnológico (CNPq), Financiadora de Estudos e Projetos (FINEP), Coordenação de Aperfeiçoamento de Pessoal de Nível Superior (CAPES), Fundação de Amparo à Pesquisa do Rio de Janeiro (FAPERJ), Fundação Oswaldo Cruz (Papes-Fiocruz), Universidade Federal do Rio de Janeiro (UFRJ). The authors also thank Dr Richard H Valente and Ana GC Neves-Ferreira for their critical review of the manuscript.

REFERENCES

[1] Warrell, D.A. (1996) *Clinical features of envenoming from snake bites*, in *Envenomings and their treatments*, p. 63–76, Bon, C., and Goyffon, M., ed. Fondation Marcel Mérieux, Lyon.
[2] Chippaux, J.P. (1998) *Bull. WHO* **76**, 515–24.
[3] Chippaux, J.P., and Goyffon, M. (1998) *Toxicon* **36**, 823–46.
[4] Fontana, F. (1781) *Traité sur le venin de la Vipère, sur les poisons Américaines, sur le laurier-cerise et sur quelques autres poisons vegetaux*, Florence, Italy. Gibelin.
[5] Hawgood, B. (1995) *Toxicon* **33**, 591–601.
[6] Phisalix, M. (1922) *Animaux Venimeux et Venins*, p. 744–59. Paris, France. Masson & Cie, Editeurs, Libraires de L'Académie de Médicine.
[7] Omori-Satoh, T., Yamakawa, Y., and Mebs, D. (2000) *Toxicon* **38**, 1561–80.
[8] Qi, Z.-Q., Yonaha, K., Tomihara, Y., and Toyama, S. (1994) *Toxicon* **32**, 1459–69.
[9] Qi, Z.-Q., Yonaha, K., Tomihara, Y., and Toyama, S. (1995) *Toxicon* **33**, 241–5.

[10] Neves-Ferreira, A.G.C., Cardinale, N., Rocha, S.L.G., Perales, J., and Domont, G.B. (2000) *Biochim. Biophys. Acta* **1474**, 309–20.

[11] Fox, J.W., and Bjarnason, J.B. (1998) *Metalloproteinase Inhibitors*, in *Enzymes from Snake Venoms*, p. 599–632, Bailey, G.S., ed. Alaken Inc., Fort Collins, Colorado.

[12] Perez, J.C., and Sánchez, E.E. (1999) *Toxicon* **37**, 703–728.

[13] Domont, G.B., Perales, J., and Moussatché H. (1991) *Toxicon* **29**, 1183–94.

[14] Thwin, M., and Gopalakrishnakone, P. (1998) *Toxicon* **36**, 1471–82.

[15] Rawlings, N.D., and Barrett, A.J. (1995) *Evolutionary families of the metallopeptidases*, in *Methods in Enzymology*, Vol. **248-Part E**, p.369–87, Barrett, A.J., ed. Academic Press, San Diego.

[16] Bjarnason, J., and Fox, J.W. (1995) *Snake venom metalloendopeptidases: Reprolysins*, in *Methods in Enzymology*, Vol. **248-Part E**, p. 345–68, Barret, A.J., ed. Academic Press, New York.

[17] Mebs, D. (1998) *Enzymes in Snake Venoms: An Overview*, in *Enzymes from Snake Venoms*, p. 1–10, Bailey, G.S., ed. Alaken, Inc., Fort Collins, Colorado.

[18] Fox, J.W., and Long, C. (1998) *The ADAMs/MDC family of proteins and their relationships to the snake venom metalloproteinases*, in *Enzymes from Snake Venoms*, p. 151–78, Bailey, G.S., ed. Alaken, Inc., Fort Colins.

[19] Kamiguti, A.S., Hay, C.R.M., Theakston, R.D., and Zuzel, M. (1996) *Toxicon* **34** (6), 627–42.

[20] Markland, F.S. (1998) *Toxicon* **36**, 1749–800.

[21] Baramova, E.N., Shannon, J.D., Bjarnason, J.B., and Fox, J.W. (1989) *Arch. Biochem. Biophys.* **275**, 63–71.

[22] Borkow, G., Gutiérrez, J.M., and Ovadia, M. (1994) *Biochim. Biophys. Acta* **1201**, 482–490.

[23] de Wit, C.A., and Weström, B.R. (1987) *Toxicon* **25**, 1209–1219.

[24] Mebs, D., Omori-Satoh, T., Yamakawa, Y., and Nagaoka, Y. (1996) *Toxicon* **34**, 1313–1316.

[25] Yamakawa, Y., and Omori-Satoh, T. (1992) *J. Biochem.* **112**, 583–589.

[26] Valente, R.H., Dragulev, B., Perales, J., Fox, J.W., and Domont, G.B. (2001) *Eur. J. Biochem.* **268**, 3042–3052.

[27] Catanese, J.J., and Kress, L.F. (1992) *Biochemistry* **31**, 410–418.

[28] Neves-Ferreira, A.G.C., Perales, J., Fox, J.W., Shannon, J.D., Makino, D.L., Garratt, R.C., and Domont, G.B. (2001), *in press*.

[29] Gloyd, H.K. (1933) *Science* **78**, 13–14.

[30] Weinstein, S.A., Lafaye, P.J., and Smith, L.A. (1990) *Toxicon* **28**, 629.

[31] Borkow, G., Gutiérrez, J.M., and Ovadia, M. (1995) *Biochim. Biophys. Acta* **1245**, 232–8.

[32] Tanizaki, M.M., Kawasaki, H., Suzuki, K., and Mandelbaum, F.R. (1991) *Toxicon* **29**, 673–81.

[33] Weissenberg, S., Ovadia, M., Fleminger, G., and Kochva, E. (1991) *Toxicon* **29**, 807–18.

[34] Weissenberg, S., Ovadia, M., and Kochva, E. (1992) *Toxicon* **30**, 591–7.

[35] Omori-Satoh, T., Sadahiro, S., Ohsaka, A., and Murata, R. (1972) *Biochim. Biophys. Acta* **285**, 414–26.

[36] Omori-Satoh, T. (1977) *Biochim. Biophys. Acta* **495**, 93–8.

[37] Huang, K.-F., Chow, L.-P., and Chiou, S.-H. (1999) *Biochem. Biophys. Res. Commun.* **263**, 610–16.

[38] Ovadia, M. (1978) *Toxicon* **16**, 661–72.

[39] Tomihara, Y., Yonaha, K., Nozaki, M., Yamakawa, M., Kawamura, T., and Toyama, S. (1988) *Toxicon* **26**, 420–23.

[40] Farah, M.F.L., One, M., Novello, J.C., Toyama, M.H., Perales, J., Moussatché, H., Domont, G.B., Oliveira, B., and Marangoni, S. (1996) *Toxicon* **34**, 1067–71.

[41] Soares, A.M., Rodrigues, V.M., Borges, M.H., Andrião-Escarso, S.H., Cunha, O.A.B., Homis-Brandeburgo, M.I., and Giglio, J.R. (1997) *Biochem. Mol. Biol. International* **43**, 1091–9.

[42] Perales, J., Muñoz, R., and Moussatché, H. (1986) *An. Acad. brasil. Ciênc.* **58**, 155–62.

[43] Tomihara, Y., Yonaha, K., Nozaki, M., Yamakawa, M., Kawamura, T., and Toyama, S. (1987) *Toxicon* **25**, 685–9.

[44] Perales, J., Moussatché, H., Marangoni, S., Oliveira, B., and Domont, G.B. (1994) *Toxicon* **32**, 1237–49.

[45] Garcia, V.E., and Perez, J.C. (1984) *Toxicon* **22**, 129–38.

[46] Jurgilas, P.B., Neves-Ferreira, A.G.C., Domont, G.B., and Perales, J. (2000) Abstract Book, *XIIIth World Congress of the International Society on Toxinology*, Paris, France: p. 145.

[47] Pichyangkul, S., and Perez, J.C. (1981) *Toxicon* **19**, 205–15.

[48] Martinez, R.R., Pérez, J.C., Sánchez, E.E., and Campos, R. (1999) *Toxicon* **37**, 949–54.

[49] Moretta, A., Bottino, C., Vitale, M., Pende, D., Biassoni, R., Mingari, M.C., and Moretta, L. (1996) *Ann. Rev. Immunol.* **14**, 619–48.

[50] Bazan, J.F. (1990) *Proc. Natl. Acad. Sci. (USA)* **87**, 6934–8.

[51] de Vos, A.M., Ultsch, M., and Kossiakoff, A.A. (1992) *Science* **255**, 306–12.

[52] Somers, W., Ultsch, M., de Vos, A.M., and Kossiakoff, A.A. (1994) *Nature* **372**, 478–81.

[53] Dennis, E.A. (1997) *TIPS* **22**, 1–2.

[54] Kini, R.M. (1997) *Phospholipase A$_2$ a complex multifunctional protein puzzle*, in *Venom phospholipase A$_2$ enzymes: Structure, function and mechanism*, pp. 1–28, Kini, R.M., ed. John Wiley & Sons Ltd.

[55] Lambeau, G., and Lazdunski, M. (1999) *TIPS* **102**, 162–70.

[56] Dennis, E.A. (2000) *Am. J. Respir. Crit. Care Med.* **161**, 532–5.

[57] Arni, R.K., and Ward R.J. (1996) *Toxicon* **34**, 827–41.

[58] Hawgood, B., and Bon, C. (1991) *Reptile venoms and toxins*, in *Handbook of Natural Toxins*, Vol. **5**, p. 3–52, Tu, A.T., ed. Marcel Dekker, New York.

[59] Faure, G. (2000) *Biochimie* **82**, 833–40.

[60] Gutiérrez, J.M., and Lomonte, B. (1995) *Toxicon* **33**, 1405–24.

[61] Lizano, S., Angulo, Y., Lomonte, B., Fox, J.W., Lambeau, G., Lazdunski, M., and Gutiérrez, J.M. (2000) *Biochem. J.* **346**, 631–9.

[62] Rocha, S.L.G., Gutiérrez, J.M., Lomonte, B., Neves-Ferreira, A.G.C., Domont, G.B., and Perales, J. (2000) Abstract Book, *VI Simpósio da Sociedade Brasileira de Toxinologia, São Paulo, Brazil*.

[63] Lambeau, G., Barhanin, J., Schweitz, H., Qar, J., and Lazdunski, M. (1989) *J. Biol. Chem.* **264**, 11503–10.

[64] Lambeau, G., Lazdunski, M., and Barhanin, J. (1991) *Neurochem. Res.* **16**, 651–8.

[65] Lambeau, G., Schmid-Alliana, A., Lazdunski, M., and Barhanin, J. (1990) *J. Biol. Chem.* **265**, 9526–32.

[66] Drikramer, K., and Taylor, M.E. (1993) *Annu. Rev. Cell. Biol.* **9**, 237–64.

[67] Ishizaki, J., Hanasaki, K., Higashino, K., Kishino, J., Kikuchi, N., Ohara, O., and Arita, H. (1994) *J. Biol. Chem.* **269**, 5897–904.

[68] Lambeau, G., Ancian, P., Barhanin, J., and Lazdunski, M. (1994) *J Biol. Chem.* **269**, 1575–8.

[69] Ohkura, N., Okuhara, H., Inoue, S., Ikeda, K., and Hayashi, K. (1997) *Biochem. J.* **325**, 527–31.

[70] Ohkura, N., Inoue, S., Ikeda, K., and Hayashi, K. (1993) *J. Biochem.* **113**, 413–19.

[71] Okumura, K., Ohkura, N., Inoue, S., Ikeda, K., and Hayashi, K. (1998) *J. Biol. Chem.* **273**, 19469–75.
[72] Ohkura, N., Inoue, S., Ikeda, K., and Hayashi, K. (1994a) *Biochem. Biophys. Res. Commun.* **204**, 1212–18.
[73] Fleming, T.J., O'Higuin, C., and Malek, T.R. (1993) *J. Immunol.* **150**, 5379–90.
[74] Hains, P.G., and Broady, K.W. (2000) *Biochem. J.* **346**, 139–46.
[75] Shao, J., Shen, H. and Havsteen, B. (1993) *Biochem J.* **293**, 559–66.
[76] Ohkura, N., Kitahara, Y., Inoue, S., Ikeda, K. and Hayashi, K. (1999) *J. Biochem.* **125**, 375–82.
[77] Hains, P.G., Sung, K.-L., Tseng, A., and Broady, K.W. (2000) *J. Biol. Chem.* **275**, 983–91.
[78] Fortes-Dias, C.L., Fonseca, B.C.B., Kochva, E., and Diniz, C.R. (1991) *Toxicon* **29**, 997–1008.
[79] Fortes-Dias, C.L., Lin, Y., Ewell, J., Diniz, C.R., and Liu, T.-Y. (1994) *J. Biol. Chem.* **269**, 15646–51.
[80] Perales, J., Villela, C.G., Domont, G.B., Choumet, V., Saliou, B., Moussatché, H., Bon, C., and Faure, G. (1995a) *Eur. J. Biochem.* **227**, 19–26.
[81] Inoue, S., Kogaki, H., Ikeda, K., Samejima, Y., and Omori-Satoh, T. (1991) *J. Biol. Chem.* **266**, 1001–7.
[82] Nobuhisa, I., Chiwata, T., Fukumaki, Y., Hattori, S., Shimohigashi, Y., and Ohno, M. (1998) *FEBS Letters* **429**, 385–9.
[83] Lizano, S., Lomonte, B., Fox, J.W., and Gutiérrez, J.M. (1997) *Biochem. J.* **326**, 853–9.
[84] Ovadia, M., Kochva, E., and Moav. B. (1977) *Biochim. Biophys. Acta* **491**, 370–86.
[85] Okumura, K., Masui, K., Inoue, S., Ikeda, K., and Hayashi, K. (1999b) *Biochem. J.* **341**, 165–71.
[86] Thwin, M.M., Gopalakrishnakone, P., Kini, R.M., Armugam, A., and Jeyaseelan, K. (2000) *Biochemistry* **39**, 9604–11.
[87] Rocha, S.L.G., Lomonte, B., Neves-Ferreira, A.G.C., Domont, G.B., Gutiérrez, J.M., and Perales, J. (2000) Abstract Book, *XIIIth World Congress of the International Society on Toxinology, Paris, France*: P147.
[88] Inoue, S., Shimada, A., Ohkura, N., Ikeda, K., Samejima, Y., Omori-Satoh, T., and Hayashi, K. (1997) *Biochem. Mol. Biol. Int.* **41**, 529–37.
[89] Nobuhisa, I., Inamasu, S., Nakai, M., Tatsui, A., Mimori, T., Ogawa, T., Shimohigashi, Y., Fukumaki, Y., Hattori, S., Kihara, H., and Ohno, M. (1997) *Eur. J. Biochem.* **249**, 838–45.
[90] Ohkura, N., Inoue, S., Ikeda, K., and Hayashi, K. (1994b) *Biochem. Biophys. Res. Commun.* **200**, 784–8.
[91] Faure, G., Villela, C., Perales J., and Bon C. (2000) *Eur. J. Biochem* **106**, 4799–808.
[92] Fortes-Dias, C.L., Jannotti, M.L.D., Franco, F.J.L., Magalhães, A., and Diniz, C.R. (1999) *Toxicon* **37**, 1747–59.
[93] Kogaki, H., Inoue, S., Ikeda, K., Samejima, Y., Omori-Satoh, T., and Hamaguchi, K. (1989) *J. Biochem.* **106**, 966–71.
[94] Okumura, K., Inoue, S., Ikeda, K., and Hayashi, K. (1999a) *Biochim. Biophys. Acta* **1441**, 51–60.
[95] Fisher, A.B., Dodia, C., Chander, A., Beers, M.F., and Bates, S.R. (1994) *Biochim. Biophys. Acta* **1211**, 256–62.
[96] Délot, E., and Bon, C. (1992) *J. Neurochem.* **58**, 311–19.
[97] Krizaj, I., Faure, G., Gubensek, F., and Bon, C. (1997) *Biochemistry* **36**, 2779–87.
[98] Perales, J., Villela, C.G., Chermont, S., Domont, G.B., Moussatché, H., Faure, G. and Bon, C. (1995b) Abstract Book, *X Reunião Anual da Federação de Sociedades de Biologia Experimental, São Paulo, Brazil*: P330.
[99] Melo, P.A., and Suarez-Kurtz, G. (1988) *Toxicon* **26**, 87–95.

[100] Ponraj, D., and Gopalakrishnakone, P. (1996) *Toxicon* **34**, 622–3.

[101] Stone, A.L., Kroeger, M., and Sang, Q.X.A. (1999) *J. Protein Chem.* **18**, 447–65.

[102] Abbas, A.K., Lichtman, A.H., and Pober J.S. (1991) *Cellular and Molecular Immunology*, p. 4–6. Philadelphia, USA. WB Saunders, HBJ International Edition.

[103] Matsushita, M., Endo, Y., Taira, S., Sato, Y., Fujita, T., Ichikawa, N., Nakata, M., and Mizuochi, T. (1996) *J. Biol. Chem.* **271**, 2448–54.

27 The Treatment of Snake Bites: Analysis of Requirements and Assessment of Therapeutic Efficacy in Tropical Africa

J.-P. CHIPPAUX

27.1 INTRODUCTION

In addition to the negative image of the snake in most cultures, the high frequency of snake bites and the seriousness of many instances of envenomation go a long way to explaining why this branch of medicine remains a priority in tropical health care establishments.

All epidemiological surveys have confirmed that envenomation is extremely common (Table 27.1: [1]). Although three quarters of all envenomations are due to insects, snake bites are responsible for most mortality (about 80% of all deaths), far more than scorpion stings, for example (less than 15% of deaths) which are basically confined to Central America and the deserts of North Africa and the Middle East (Figure 27.1).

Many clinical studies have illustrated the profound difficulties experienced by physicians called on to treat snake envenomation. It is also worthwhile taking

Table 27.1 The incidence of snake bites in the world.

Region	Population ($\times 10^6$)	Total bites	Total envenomations	Total deaths
Europe	750	25 000	8 000	30
Middle East	160	20 000	15 000	100
USA – Canada	270	45 000	6 500	15
Latin America	400	300 000	150 000	5 000
Africa	800	1 000 000	500 000	20 000
Asia	3 500	4 000 000	2 000 000	100 000
Oceania	20*	10 000	3 000	200
Total	5 900	5 400 000	2 682 500	125 345

* regions inhabited by venomous land snakes

Perspectives in Molecular Toxinology
Edited by A Ménez © 2002 John Wiley & Sons, Ltd

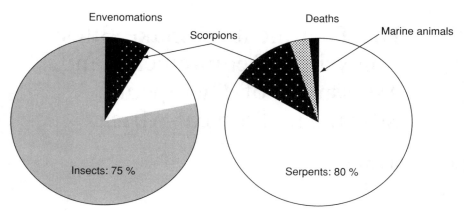

Figure 27.1   Causes of envenomations and deaths in the world

note of the high rate of failures or sub-optimal outcomes in the routine therapeutic modalities used to treat patients who have been bitten by snakes [2].

In Africa, the advent of antivenomous serum in the early 1950s was greeted
by physicians as a source of tremendous hope in the fight to reduce snake bite-
related mortality [3]. Half a century later, we are forced to admit that, despite
huge improvements in the way antivenom is prepared, its place in treatment
remains no more than marginal [2]. Even more striking is the fact that lethality
has not been reduced by more than fifty per cent in peripheral health care
centres – and in many cases by less. The actual number of deaths has little
changed since the 1950s since increases in population density have cancelled out
the effect of the modest therapeutic improvement. It is this surprising and
worrying situation that is analysed in this article.

The number of units of antivenom sold every year is far smaller than the
number of snake bites recorded in health care centres and therefore all the more
smaller than the total number of actual envenomations which occur. If the
average number of subjects bitten by snakes is estimated at 500 000 per annum,
the number of units of antivenom required could be set at two millions.
However, annual sales do not exceed 20 000 units, even in the best years. This
enormous discrepancy could be due to three factors which are not mutually
exclusive:

- **poor recognition of requirements**, due to a lack of either relevant epidemiological data or a failure to understand the real medical problem;
- **insufficient supplies**, due to inadequate production, inefficient distribution,
  or both;
- **poor analysis of the characteristics of consumption**, such as the inability of
  health care structures to treat snake envenomation, low acceptance of
  Western-type medicine, or prejudice against the product on the part of
  either health care providers or the victims and their families.

Thus, three analyses could be undertaken to identify the problems with a view to finding suitable solutions.

Analysis of requirements should be performed first to confirm that the discrepancy reported between the number of cases of envenomation and the number of units of antivenom sold is real and not just the result of the poor distribution of resources, or that it does not reflect some special type of consumption problem.

Analysis of supply should be performed at the level of both producer and prescriber to investigate possible reasons for the discrepancy and derive a fuller understanding of the factors which restrict the availability of antivenom when and where it is needed.

Analysis of consumption should reveal the reasons why a patient might refuse Western-type medicine and, more specifically, the administration of antivenom. For this, it will be necessary to document the circuit followed by bitten subjects seeking treatment and assess the rate of attendance of patients at health care centres – both those who have been bitten by snakes and those with other health problems.

27.2 ANALYSIS OF REQUIREMENTS

It is important to make a distinction between demand and the true requirement because the former can be excessive if it is unjustified or can be difficult to manage if it is spread over a large area. The true requirement is defined by the differential between the actual and the desirable level of health [4]. Assessing demand entails evaluation of the epidemiological picture and requires both indicators and data which are not immediately available but need to be compiled. The desirable level of health will correspond to a compromise between the aspirations of a population which seeks to abolish the risk completely (whose aspirations more or less correspond to the demand) and a realistic set of objectives which take into account the ability to meet the demand.

27.2.1 ACTUAL LEVEL OF HEALTH: INCIDENCE, MORBIDITY AND LETHALITY

Data can be obtained from records consulted in health care centres or can be compiled from notifications of the health care authorities when snake envenomation is dealt with specifically. Such medical data which come from the prescriber are associated with both advantages and disadvantages. Firstly, by definition, the figures are underestimated, both in terms of incidence and lethality: with respect to incidence because all patients who are dealt with by traditional healers are missed by the official statistics – these represent an average of 70% of all subjects bitten by snakes although this figure varies enormously from country to country [1]; with respect to lethality because as soon as the patient's condition becomes critical, the family often (especially in Africa) removes him or her from hospital in order to avoid the costs associated with transport of the cadaver (which are much higher than those for a living person, even a very sick one).

Medical data can be cross-checked against other sources of information but this is fraught with problems. Furthermore, medical statistics are integrated meaning that they represent observations which have been pooled across an entire region and often cover a variety of highly disparate, local conditions. This can have significant consequences, especially if the extent of pooling is great. In Africa, medical information is usually available for administrative units corresponding to large populations (e.g. half a million inhabitants) or large geographical areas (typically 10 000 km^2) which considerably cuts down its usefulness when it comes to analysis, the difficulties associated with the distances involved being compounded by the fact that travel is often complicated. On the other hand, medical data give a good idea of how well the health care system is functioning in the form of a few simple indicators, namely frequency parameters (the overall number of consultations and incidence), treatment parameters (drug consumption and the length of hospital stays), and therapeutic efficacy (mortality).

Such an analysis is often quick, cheap and results in the identification of priority areas. Nevertheless, it is important to recognise inadequacies in the medical information system, inadequacies which can be due to multiple factors. In particular, it is necessary to compare differences in recruitment rate with details on the services offered by each health care centre concerned, and the users' therapeutic choices. For example, a health care centre may be empty because it is isolated, because there are other centres in the vicinity, or because it is underequipped.

A second approach is based on reviewing all the published data relevant to the incidence of snake envenomation as reported by physicians. This method has already been the subject of a number of studies [1] which have revealed both the diversity of the criteria used and the irreproducibility of the results. The specific aims of published investigations go a long way in accounting for these problems. Even epidemiological surveys (which should, in theory, avoid these pitfalls) cannot always take into account bias introduced as a result of ecological, social and economic factors specific to the study area or population. In prospective studies, perfectly defined, constant criteria can be used to assess incidence, morbidity, the clinical features of envenomation and lethality. Serological testing improves the precision and, to an even greater extent, the reliability of the results [5], as long as the method used is sufficiently specific [6].

Therefore, literature reviews have major limitations. For the future, the relevance of prospective studies could be strengthened if all investigators were to use standard indicators (with the same specifications).

A third alternative, proposed some time ago [7], involves collecting the basic information directly from the population as a whole. Investigations based on the interrogation of householders measure the incidence of snake-bites with a fairly high degree of precision. The inhabitants of a village are individually asked about whether they personally have ever been bitten by a snake and also about the number of their friends and relations who they remember having been bitten since they were born; they are also asked about the outcomes of any envenomations they have either experienced or witnessed. After correction for age, the elimination of any repeated data points, and the cross-checking of all replies, it is possible to assess the incidence of snake bites in the village with a satisfactory degree of precision. Of course, morbidity data will tend to be less precise than those obtained using the other two methods, but it would at least remain possible to identify the most important risks in the region concerned.

The pros and cons of the various different methods are summarised in Table 27.2.

The following priority indicators ought to be used:

- **Incidence** is an estimate of the overall number of snake bites occurring in a given population in a given period of time (usually one year). This parameter makes it possible to assess the theoretical requirements with respect to all snake bites. In practice, a snake bite necessitates emergency care. This kind of information can only be obtained by gathering data at the level of the household.

Table 27.2 The advantages and disadvantages of different epidemiological methods.

Method	Indicators	Advantages	Disadvantages
Literature review	Morbidity Hospital lethality Therapeutic practices	Zero cost Immediate results	Limited Skewed
Records (Health care authorities)	Morbidity Hospital lethality Fraction of beds occupied	Low cost Rapid results	Low precision Partial Integrated

Questionnaires
(households)
	Incidence
	Mortality
	Therapeutic habits
	Population at risk
	Circumstances of the
	  bite
		Rapid results
		Moderate cost
		Precise
			Location-
			  specific

Prospective surveys
(Health care centres)
	Morbidity
	Lethality
	Symptoms
	Delay before
	  consultation
	Identification of snake
	Length of hospital stay
		Precise	Slow
			Expensive
			Location-
			  specific

- **Morbidity** describes the number of envenomations (i.e. snake bites followed by the symptoms expected after the dissemination of venom in the body) in a given population in a given period of time. This parameter is useful for planning the nature and the number of courses of treatment which the health care centres will be required to provide. Details on symptoms, outcomes and, more generally, the severity of bites are acquired by means of surveys conducted in medical training seminars, which should be prospective whenever possible.
- **Lethality** (= case fatality rate) is the fraction of bitten subjects who die. This parameter gives a good measure of the quality and suitability of the health care.
- **Mortality** which measures all deaths from snake bite in the entire population over a given period of time gives a measure of the efficacy of the health care system and the extent to which the local population uses this system.

Information on lethality and mortality should be cross-checked against both data collected in the course of household surveys and those collected from the health care authorities.

Other indicators can be assessed if an investigation has specific aims or in order to increase its resolution.

- **The circumstances of the bite**: this information might be helpful for the purposes of prevention.
- **Identification of the guilty snakes** is helpful in programming treatment modalities and in guiding envenomation management (which therapeutic protocols to adopt, planning supplies of the appropriate antivenom and drugs, etc.).
- Better understanding of the **therapeutic options** will affect the overall organisation of the health care system and, possibly, public awareness programs.
- **The lapse of time before consultation**, i.e. the time between the bite and arrival at the health care centre, gives information on the area covered by the centre and therefore its accessibility, and the extent to which the local population uses the health care system.
- **The fraction of hospital beds occupied** by patients who have been bitten by snakes is an important measure of the relative importance of this kind of pathology in a given health care centre.
- **Seasonal variations in snake bite frequency** will determine how orders ought to be scheduled.
- **The length of hospital stays, and the severity of sequels** are useful in assessing the efficacy and relevance of a therapeutic strategy.

These various ways of assessing requirements are obviously complementary (Table 27.3). The first is based on official medical statistics and gives an overall risk evaluation for a given geographical region. The second is based on monitoring the activity of health care establishments and can be used to confirm certain epidemiological findings and keep track of the importance of this type of pathology, the seriousness of envenomations and the efficacy of the medical

Table 27.3 The objectives of epidemiological surveys and the usefulness of various indicators.

Indicator	Method	Information yielded
Incidence	Questionnaires	Population demand
Morbidity	Records	Therapeutic requirements
Mortality	Questionnaires	Relevance of the health care system
Lethality	Records/Prospective surveys	Quality of medical care
Lapse of time between bite and consultation	Prospective surveys	Acceptance of health care system/accessibility
Circuit followed seeking treatment	Questionnaires	Acceptance of health care system/accessibility
Time of hospitalisation	Records/Prospective surveys	Suitability of the therapeutic protocol
Identification of the guilty snake	Questionnaires/Prospective surveys	Requirements
Circumstances of the bite	Questionnaires/Prospective surveys	Prevention
Symptoms	Questionnaires/Prospective surveys	Compilation of a therapeutic protocol
Sequels	Questionnaires/Records/Prospective surveys	Organisation and quality of medical care

response. The final method is based on direct contact with the population concerned and can contribute important complementary information to help with risk prediction, even though the data it produces are often fragmented and highly specific to a given locality. Above all, it gives a fairly accurate idea of the therapeutic habits of rural populations.

It has been possible to define the characteristics of the populations most likely to benefit from immunotherapy on the basis of the results of numerous epidemiological surveys (Table 27.4). It is known that such a population is rural and that agricultural workers are the most at risk. Thus, it is those between 15 and 45 years of age, especially males, who are most commonly bitten. Bites are most likely to occur during the daytime with peaks at the beginning and at the end of the working day. Three quarters of bites are delivered below the knee. Finally, it has been mentioned that snake populations seem to prosper in cultivated fields and the attraction which planted areas have for certain types of venomous snake means that bites are far more common here than in the uncultivated, surrounding bush [8].

Available information suggests that patterns are relatively stable over time but subject to enormous variations in space (Table 27.4). This could mean that ecological conditions and factors related to human behaviour may have a strong influence, and that the epidemiological methods in current use are sufficiently robust to give reliable data, despite the fact that they are often relatively non-rigorous.

Table 27.4 Incidence and seriousness of snake bites in Africa (taken from Chippaux, 1998 and hitherto unpublished data. Reproduced by permission).

Country	Incidence (/100 000 h)	Morbidity (/100 000 h)	Lethality (%)
Benin	216–653[*]	39–315	1.8–4.1
Burkina Faso		40–70	5–11.7
Cameroon		75–200	5–10
Chad		10 (urban)	
Congo	125–430[*]	20	1–6.6
Côte d'Ivoire		130–400	2–28
Ghana		21	
Kenya	150[*]	1.9–67.9	2.6–9.4
Liberia	420	170	
Natal		100	
Niger		10 (urban)	
Nigeria	48–603[*]	100–120	2.1–16
Senegal	15–195	20–150	7
South Africa		34	
Togo		65–200	
Zimbabwe		3.5	1.8–4.8

[*] Household-based surveys

27.2.2 DESIRABLE LEVEL OF HEALTH: A DIFFICULT PARAMETER TO ASSESS

It is easy to think that the desirable level of health corresponds to zero risk. The absence of a realistic attitude on this point is generally accepted and it will be necessary to agree to a minimum consensus position. The focus must be that of health care supply, often judged as inadequate and difficult to improve given social and economic realities in Africa.

27.3 ANALYSIS OF SUPPLY

The supply of antivenom involves two different players, the producer and the prescriber. In contrast to the usual situation in pharmaceutical distribution, producers tend to display excessive caution when it comes to distributing their products. And prescribers are subject to a variety of restrictions – both real and imaginary – which considerably limit their ability to meet the need.

27.3.1 PRODUCERS

The remarkable advantages of the product which is currently available in Africa [9, 10] should not lead to an underestimation of the major problems which condemn immunotherapy to a paradoxically marginal place in the treatment of envenomation. Aside from the poor appreciation of requirements mentioned above and which are perceived by the producer as dispersed over large geographical areas and highly disparate in different localities, these problems considerably inhibit the commercial spirit.

Producers emphasise that antivenom production is very expensive (especially when it comes to highly purified antibody fragments) and that the customers are poor. Difficulties in identifying the needs of the market has led most producers either to abandon this type of product completely or to cut back production in parallel to the clearly declining demand. Thus, out of the 64 producers which were making 159 anti-snake venom products in the 1970s [11], only 22 were active in the year 2000, and a total of only 73 antivenoms were still in production [12]. In sub-Saharan Africa, only one product can be found (not counting those produced in South Africa which are not available elsewhere) whereas thirty years ago there were about twenty on the market. It is true that the spread of polyvalent antivenoms (which are preferred for technical and logistical reasons) replacing the old monovalent products has contributed to the reduction in the overall number of products available but it does not entirely account for the trend. For example, in Cameroon, antivenom sales fell from 10 000 units per annum in the 1970s to just 2000 in the 1990s.

To this it must be added that, although producers understand the actual product in depth, they fail to understand how it is used in the field. Moreover, there are particular problems like the fact that the lack of suitable resources for storing antivenom in tropical countries often makes it difficult to provide them in outlying health care centres. Finally, the difficulties associated with drug distribution in rural Africa represent another problem but this is beyond the scope of this article.

27.3.2 PRESCRIBERS

In contrast to the situation in industrialised countries like France where immunotherapy is often ruled out because of the risk of serious adverse reactions, it is not rejected in Africa. Although prescribers are not usually familiar with the technical improvements made to immunoglobulin-based antivenom, they tend to have great confidence in 'antivenomous serum'. The reasons why antivenom products are not available are more practical:

- High cost – each unit costs the equivalent of one month's income for a family of farmers;
- The product must absolutely be stored in refrigerator;
- The short shelf-life of antivenom.

These reasons are sufficient to prevent the accumulation of stocks of antivenom products. Apart from the under-utilisation of immunotherapy, another

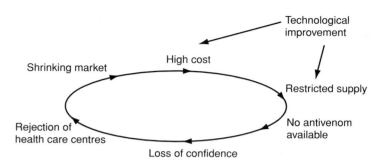

Figure 27.2   The vicious circle underlying the under-availability of antivenom in African dispensaries

factor to be taken into account is the fact that the inevitable delay in commencing treatment due to having to send for a unit (which is in storage far away or in an inaccessible place) prejudices the efficacy of the treatment which can only lead to reduced confidence in the product itself.

In conclusion, antivenom supplies are inadequate and poorly matched to requirements (which are themselves poorly understood), from the point of view of both producer and prescriber. All these factors form a vicious circle (Figure 27.2) which illustrates the complexity of the problem.

27.4   ANALYSIS OF CONSUMPTION

Although as a general rule the medical knowledge of African people is limited, they know of the existence of antivenom and have confidence in its efficacy. They also know how expensive it is, and how difficult to obtain. Therefore, it is less a specific attitude to antivenom (which is widely accepted in its own right) than a general reaction to disease and the health care system which is faced with it. From the point of view of the patient, consumption therefore involves two, interrelated factors: the decision to undergo treatment with its multiple aspects and complexity; and how the health care system is perceived, which varies over time and from place to place.

The analysis has to be extended in terms of how the product is used by the prescriber. In order to insure maximum benefit, antivenom must be administered in the correct way (i.e. in accordance with the Directions).

27.4.1   THE DECISION TO UNDERGO TREATMENT

This decision is influenced by cultural factors, habit, experience and circumstances. It corresponds to a complex choice which can lead to explanations which are very difficult to understand. It has been reported that, in most cases, a traditional healer is consulted first. This common feature is partly due to cultural reflexes because snake bites are usually ascribed to supernatural causes. Some traditional healers have real competence, especially given the low level of provision by the public health care system. Clinical deterioration can stimulate a change in attitude and serious, rapidly progressive envenomation often leads the patient or his or her family to seek help from Western-type practitioners – mostly to treat the complications according to them. On the other hand, the absence of any complications reassures victims as to the sagacity of their choice of traditional medicine, the reputation of which is thereby strengthened.

27.4.2   ACCESS TO CARE

Access is also a crucial factor. The distance to the health care centre and its general accessibility in the broader sense probably plays a role that cannot be ignored – in an emergency, it may be important to go to the nearest facility even if that one is not necessarily the first choice. In Africa, distances are measured in hours rather than in kilometres because a river may be difficult to cross or it may take a long time to go around a hill. Even a direct road to the centre is not a major advantage without a car. Finally, there are other variables such as the centre's reputation, its equipment, the perceived reception and the skill (perceived or real) of its staff in the treatment of envenomation which all need to be considered on a case by case basis. Competition between independent confessional and public dispensaries perfectly illustrates the importance of such factors. In Africa, it is not rare to find an independent confessional dispensary located within a few hundred metres of the public hospital and attracting all the snake bite victims in the district. The way in which patients are received, the availability of drugs (including antivenom) and the staff's reputation all contribute to explaining this preference.

27.4.3   STAFF TRAINING

Analysis of observations made in dispensaries or reported in the literature show that envenomation treatment and the use of antivenom are very poorly understood by the staff of local health care centres. This tends to reduce the efficacy of such therapy and, in the worst cases, makes the use of antivenom positively dangerous. Neither indications, route of administration, dosage nor follow-up are standardised or subject to any norms. This results in failure in cases in which success was possible, and sometimes leads to iatrogenic exacerbation of the patient's condition. One problem is the fact that outdated recommendations designed to mitigate the risk of adverse reactions have not been completely forgotten with the result that treatment is overcautious, insufficiently aggressive and of reduced efficacy. Thus the dose administered may be too low, the wrong route of administration may be used, or too long a delay may occur before administration is commenced. Another problem is that, as a general rule, health care providers are ignorant of the symptoms and pathogenesis of envenomation. This explains why therapeutic indications often lack rigor and also why the rules on patient follow-up and the recommendations on how antivenom ought to be administered are either not known or are incorrectly applied.

27.5 SOLUTIONS

27.5.1 IMPROVED IDENTIFICATION OF REQUIREMENTS

Epidemiological surveys should lead to a good definition of requirements in terms of antivenom.

National surveys based on medical records and reports can be launched straight away with a minimum of expenditure. Such surveys would determine the incidence of envenomation and the associated mortality, i.e. a crude appraisal of the average seriousness of snake bite poisoning. The information given is usually adequate to assess the demand (which does not always cover the real requirement). As already mentioned, most snake bites do not lead to consultation of Western-type health care centres, so the incidence as reported by official health care facilities will tend to be an underestimate.

Hospital-based surveys contribute an indispensable complement concerning treatment modalities and the efficacy of therapeutic strategies. They also give information on the average lapse of time between bite and consultation, the seriousness of envenomation, current therapeutic attitudes in health care centres and the most common symptoms. This should make it possible to draw up standard protocols for both measures to be implemented and therapeutic approaches to be adopted which could be distributed to all levels of the health care facility.

Community-based surveys are indispensable for assessing the true incidence of snake bites which will give an idea of the real requirement. It is this type of survey which measures the difference between the real requirement (the true incidence) and the demand for care (the number of patients seen at health care centres), thus putting the medical information in context and making it possible to fix targets.

27.5.2 IMPROVING SUPPLY

Solving the problems associated with supply would seem to be a more complicated issue.

Improving antivenom

In contrast to the general perception, the important issue is not how to improve the product itself. The use of immunoglobulin fragments certainly enhances both safety and efficacy as well as undeniably reducing the patient's discomfort [13]. The choice of animal in which the immunoglobulins are produced, or whether to opt for Fab or $F(ab')_2$ fragments, remain topics for the experts and their relevance can seem somewhat obscure in the context of deaths due to snake bites. In any case, in Africa, such discussions are beside the point: for one thing, they fail to tackle the real reasons why there is no antivenom in health care centres and moreover, technological advances can only increase the extent of polarisation between industrialised and nonindustrialised countries when it comes to matters of health care. It is even possible to imagine that improvements in antivenom technology could have an adverse effect if increased complexity and cost put these products definitively beyond the reach of public health providers and private individuals in Africa. It ought to be considered that, in technical terms, the product is adequate at this point in time, and that, in any case, the benefits of technological advances in industrialised countries will only reach Africa in the fullness of time.

On the other hand, an effort could be made in terms of how the product is presented: primary and secondary packaging ought to be in keeping with the kind of logic which pertains in African countries and the volume of each unit ought to be matched to the average dose. Packaging ought to be simple, tough and cheap, and the price ought to be adjusted in accordance.

The question of producing locally needs to be raised. Such an initiative would certainly lead to a lower price at the same time as fostering the development of the pharmaceutical industry in countries in which there is a great need for it. It might even attract enough funding to reduce significantly the investment necessary to launch such a venture. The transfer of both technological expertise and skills is perfectly possible. There are two drawbacks to this idea: licensing of the patent and quality control. Nascent industries – especially those located in the unfavourable environment of a developing country – must be subject to particularly stringent monitoring; quality must never be sacrificed for the sake of economy.

Antivenom distribution

The key point seems to be to try to improve the distribution of antivenom, especially in rural areas.

As a first step, an analysis of epidemiological results will give an idea of the state of stocks, in particular the location of storage facilities and their capacities. A number of different criteria should be used:

- The demand or, if possible, the requirements: as a function of the recorded incidence and the severity of envenomation as measured by morbidity and mortality figures;
- The skill of the care-providing team;
- The confidence of the local population in Western-style medicine;
- The infrastructure and equipment at the health care centre.

It will also be necessary to suggest new types of trade practice.

The importance of exchanging expired units should be emphasised. In practice, this is one of the most difficult issues in Africa where people are extremely reluctant to throw away a product which appears to be in perfectly good condition. There thus exists, on the one hand, the risk of having expired antivenom administered and, on the other hand, the risk of reducing product availability by diminishing stocks to avoid waste.

It will also be necessary to monitor the network of agents who introduce extra cost through their profit margins and who have a tendency to restrict stocks to a minimum in order to avoid being left with unsold product. As a result, they have an adverse effect – direct or indirect – on antivenom availability.

Training and improved health care

A key factor is emphasising to health care providers the importance of closely following the Directions when administering antivenom. This is actually a broader question than it appears and includes improving patient care and treatment in general. Comprehensive therapeutic protocols – including indications, dosage, route of administration and patient follow-up modalities – will have to be compiled and distributed in the form of information sheets or instruction booklets. Staff training ought to be provided by both the relevant educational institutions and the producers, profiting at the same time from the credibility of the former and the logistical resources of drug companies. Obviously, the message will need to be coordinated and simple, which will entail collaboration between all the parties concerned.

Strengthening public confidence in the health care system as a whole and in antivenom products in parallel with other types of medicinal product will ultimately lead to improved patient care.

Reducing prices

To react to the problem posed by the under-utilisation of antivenom, due to both its high price and the lack of availability when and where it is needed, it will be necessary to establish a specific funding structure.

Producers could lower the price as much as possible by changing the presentation (thereby reducing cost price) and how distribution is organised (which will, in time, lead to increased sales).

State and local organisations should subsidise the production and distribution of antivenom. Governments ought to be encouraged to adopt measures similar to those applying to rare diseases and develop a policy like that for orphan drugs. More specifically, such a policy should include supporting research, developing investment in such products, promoting their production, helping to set up distribution networks, and contributing to the training of health care providers and thereby improving how antivenom is used.

Businesses, especially agricultural ones, should be encouraged to keep their own stocks of antivenom which would be automatically replaced on expiration.

Health care centres should charge for snake bite treatment on a fixed rate basis rather than charging for the specific antivenom. This initiative might be complicated because, in Africa, the place where patients are treated is not always the same as where the medicinal products are dispensed.

Finally, it ought to be possible to establish a country-based system to adjust the antivenom purchase price so that rich countries help pay for poor countries.

27.6 CONCLUSIONS

Answers to the problem of snake envenomation are not only based on scientific progress and, in fact, practical and operational measures tend to have a greater impact. The observations that antivenom, despite its proven efficacy, is under-utilised, and that morbidity and lethality persist at unacceptable levels, should be stimulating a broader analysis of this problem.

There seem to be two major problems. The first is the lack of available antivenom, the responsibility for which is shared between producer, prescriber and patient. The underlying reason for this is that the product's purchase price is beyond the average patient's purchasing power. The other problem is that the product is often used in an inappropriate way due to the lack of suitable protocols and inadequate training.

A suitable policy covering commercial, financial and educational considerations could improve the situation and significantly lower mortality.

REFERENCES

[1] Chippaux, J.P. (1998) Snake-bites: appraisal of the global situation. *Bull. WHO* **76**, 515–24.
[2] Chippaux, J.P. (1999) L'envenimation ophidienne en Afrique: épidémiologie, clinique et traitement. *Ann IP/actualités* **10**, 161–71.
[3] Giboin, L. (1954) Etude d'ensemble sur l'envenimation ophidienne au Togo pendant les années 1951–1952–1953. *Méd. Trop.* **14**, 542–68.

[4] Salomez, J.L. and Lacoste, O. (1999) Du besoin de santé au besoin de soins. La prise en compte des besoins en planification sanitaire. *Hérodote* **92**, 101–20.

[5] Theakston, R.D.G. (1983) The application of immunoassay techniques, including enzyme-linked immunosorbent assay (ELISA), to snake venom research. *Toxicon* **21**, 341–52.

[6] Ho, M., Warrell, M.J., Warrell, D.A., Bidwell, D. and Voller, A. (1986) A critical reappraisal of the use of enzyme-linked immunosorbent assays in the study of snake bite. *Toxicon* **24**, 211–21.

[7] Chippaux, J.P. (1988) Snakebite epidemiology in Benin (West Africa). *Toxicon* **27**, 37.

[8] Chippaux, J.P. and Bressy, C. (1981) L'endémie ophidienne des plantations de Côte d'Ivoire. *Bull. Soc. Path. Exot.* **74**, 458–67.

[9] Chippaux, J.P., Lang, J., Amadi-Eddine, S., Fagot, P., Rage, V., Le Mener, V. and VAO Investigators (1998) Clinical safety and efficacy of a polyvalent F(ab')$_2$ equine antivenom in 223 African snake envenomations: a field trial in Cameroon. *Trans. R. Soc. Trop. Med. Hyg.* **92**, 657–62.

[10] Chippaux, J.P., Lang, J., Amadi-Eddine, S., Fagot, P. and Le Mener, V. (1999) Treatment of snake envenomation by a new polyvalent antivenom composed of highly purified F(ab')$_2$: results of a clinical trial in Northern Cameroon. *Am. J. Trop. Med. Hyg.* **61**, 1017–18.

[11] Chippaux, J.P. and Goyffon, M. (1983) Producers of antivenomous sera. *Toxicon* **21**, 739–52.

[12] Chippaux, J.P. and Goyffon, M. (2000) Availability of the commercial antivenins. IST-2000.

[13] Chippaux, J.P. and Goyffon, M. (1998) Venoms, antivenoms and immunotherapy. *Toxicon* **36**, 823–46.

Index

Note: Page references followed by 'f' represents a figure and 't' represents a table.